T0142303

Advances in Intelligent Systems and Computing

Volume 1077

The series "Advances in Intelligent Systems and Computing" contains publications on theory, applications, and design methods of Intelligent Systems and Intelligent Computing. Virtually all disciplines such as engineering, natural sciences, computer and information science, ICT, economics, business, e-commerce, environment, healthcare, life science are covered. The list of topics spans all the areas of modern intelligent systems and computing such as: computational intelligence, soft computing including neural networks, fuzzy systems, evolutionary computing and the fusion of these paradigms, social intelligence, ambient intelligence, computational neuroscience, artificial life, virtual worlds and society, cognitive science and systems, Perception and Vision, DNA and immune based systems, self-organizing and adaptive systems, e-Learning and teaching, human-centered and human-centric computing, recommender systems, intelligent control, robotics and mechatronics including human-machine teaming, knowledge-based paradigms, learning paradigms, machine ethics, intelligent data analysis, knowledge management, intelligent agents, intelligent decision making and support, intelligent network security, trust management, interactive entertainment, Web intelligence and multimedia.

The publications within "Advances in Intelligent Systems and Computing" are primarily proceedings of important conferences, symposia and congresses. They cover significant recent developments in the field, both of a foundational and applicable character. An important characteristic feature of the series is the short publication time and world-wide distribution. This permits a rapid and broad dissemination of research results.

**** Indexing: The books of this series are submitted to ISI Proceedings, EI-Compendex, DBLP, SCOPUS, Google Scholar and Springerlink ****

More information about this series at http://www.springer.com/series/11156

Milan Tuba · Shyam Akashe ·
Amit Joshi
Editors

ICT Systems and Sustainability

Proceedings of ICT4SD 2019, Volume 1

Springer

Editors
Milan Tuba
Belgrade, Serbia

Shyam Akashe
ITM University
Gwalior, India

Amit Joshi
Global Knowledge Research Foundation
Ahmedabad, India

ISSN 2194-5357 ISSN 2194-5365 (electronic)
Advances in Intelligent Systems and Computing
ISBN 978-981-15-0935-3 ISBN 978-981-15-0936-0 (eBook)
https://doi.org/10.1007/978-981-15-0936-0

This Springer imprint is published by the registered company Springer Nature Singapore Pte Ltd.
The registered company address is: 152 Beach Road, #21-01/04 Gateway East, Singapore 189721, Singapore

Organization

Organizing Chairs

Mr. Bharat Patel, Chairman, CEDB
Mr. Peter Kent, Chief Executive Officer (CEO), UKEI, UK

Organizing Co-chair

Mr. Nakul Sharedalal, Advisor, GR Foundation, India

Organizing Secretary

Amit Joshi, SITG Ahmedabad and Global Knowledge Research Foundation

Members

Dr. Vijay Singh Rathore, Department of CSE, JECRC, Jaipur, India
Mr. Aman Barot, GR Foundation, India
Dr. Mahipal Singh Deora, BN University, Udaipur, India
Dr. Nisheeth Joshi, Banasthali University, Rajasthan, India
Mr. Nilesh Vaghela, Electromech Corporation, Ahmedabad, India
Mr. Vinod Thummar, SITG, Gujarat, India
Dr. Chirag Thaker, LD College of Engineering, Gujarat, India
Prof. S. Rama Rao, Professor, Goa University, Goa, India
Dr. Nitika Vats Doohan, Indore, India

Dr. Parikshit Mahalle, Sinhgad Group, Pune, India
Dr. Priyanks Sharma, RSU, Ahmedabad
Dr. Nitika Vats Doohan, Indore
Dr. Mukesh Sharma, SFSU, Jaipur
Dr. Manuj Joshi, SGI, Udaipur, India
Dr. Bharat Singh Deora, JRNRV University, Udaipur
Mr. Ricky, P Tunes, Goa University, India
Prof. L. C. Bishnoi, GPC, Kota, India
Dr. Vikarant Bhateja, Lucknow, India
Dr. Satyen Parikh, Dean, Ganpat University, Ahmedabad, India
Dr. Puspendra Singh, JKLU, Jaipur, India
Dr. Aditya Patel, Ahmedabad University, Gujarat, India
Mr. Ajay Choudhary, IIT Roorkee, INDIA

Technical Program Committee

Prof. Milan Tuba, Singidunum University, Serbia
Dr. Durgesh Kumar Mishra, Chairman, Div IV, CSI

Technical Program Committee Co-chairs

Dr. Mahesh Bundele, Dean Research, Poornima University, Jaipur
Dr. Nilanjay Dey, Techno India College of Engineering, Kolkata, India

Advisory Committee

Shri Nitin Kunkolienker, President, MAIT
Prof. S. Rama Rao, Professor, Goa University, Goa
Mr. P. N. Jain, Add. Sec., R&D, Government of Gujarat, India
Prof. J. Andrew Clark, Computer Science University of York, UK
Mr. Vivek Ogra, President, GESIA
Prof. Mustafizur Rahman, Endeavour Research Fellow, Australia
Dr. Kalpdrum Passi, Laurentian University, ON, Canada
Mr. Chandrashekhar Sahasrabudhe, ACM India
Dr. Pawan Lingras, Saint Mary University, Canada
Dr. S. C. Satapathy, Professor in CSE, AP, India
Mr. Niko Phillips, Active Learning, UK, India
Dr. Bhagyesh Soneji, Chairperson, Assocham Western Region
Dr. Dharm Singh, NUST, Windhoek, Namibia

Dr. Vinay Chandna, Principal, JECRC, Jaipur, India
Prof. H. R. Vishwakarma, VIT, Vellore, India
Prof. Siddesh K. Pai, National Institute of Construction Management and Research, Goa, India

Members

Dr. Amit Kaul, UK
Prof. Babita Gupta, College of Business California State University, California, USA
Prof. Ting-Peng Liang, National Chengchi University, Taipei, Taiwan
Prof. Anand Paul, The School of Computer Science and Engineering, South Korea
Prof. Rachid Saadane, Casablanca, Morocco
Prof. Brent Waters, University of Texas, Austin, Texas, USA
Prof. Philip Yang, Price water house Coopers, Beijing, China
Prof. H. R. Vishwakarma, VIT, Vellore, India
Prof. Martin Everett, University of Manchester, England
Dr. Rajeev vaghmare, Principal, SITG, Ahmedabad, India
Prof. Xiu Ying Tian, Instrument Lab, Yangtze Delta Region Institute of Tsinghua University, Jiaxing, China
Prof. Gengshen Zhong, Jinan, Shandong, China
Prof. Mustafizur Rahman, Endeavour Research Fellow, Australia
Prof. Ernest Chulantha Kulasekere, University of Moratuwa, Sri Lanka
Prof. Subhadip Basu, Visiting Scientist, The University of Iowa, Iowa City, USA
Dr. Ashish Rastogi, Higher College of Technology, Muscat, Oman
Prof. Ahmad Al-Khasawneh, The Hashemite University, Jordan
Dr. Basant Tiwari, Professor Ethiopia University, Ethopia
Prof. Jean Michel Bruel, Departement Informatique IUT de Blagnac, Blagnac, France
Dr. Ramesh Thakur, DAVV, Indore, India
Prof. Shuiqing Huang, Department of Information Management, Nanjing Agricultural University, Nanjing, China
Prof. Sami Mnasri, IRIT Laboratory Toulouse, France
Dr. Krishnamachar Prasad, Department of Electrical and Electronic Engineering, Auckland, New Zealand
Prof. Foufou Sebti, Dijon Cedex, France
Dr. Haibo Tian, School of Information Science and Technology, Guangzhou, Guangdong
Mr. Vinod Thummar, SITG, Gujarat, India
Prof. Sunarto Kampus UNY, Yogyakarta, Indonesia
Prof. Feng Jiang, Harbin Institute of Technology, China
Prof. Feng Tian, Virginia Polytechnic Institute and State University, USA
Dr. Savita Gandhi, Gujarat University, Ahmedabad, India

Prof. Abdul Rajak A. R, Department of Electronics and Communication Engineering Birla Institute of Technology and Sciences, Abu Dhabi
Prof. Hoang Pham, Rutgers University, Piscataway, NJ, USA
Prof. Shashidhar Ram Joshi, Institute of Engineering, Pulchowk, Nepal
Prof. Abrar A. Qureshi, Ph.D., University of Virginia's, USA
Dr. Aynur Unal, Stanford University, USA
Dr. Manju Mandot, JRNRVU, Udaipur, India
Dr. Rajveer Shekawat, Manipal University, India
Prof. Ngai-Man Cheung, Assistant Professor, University of Technology and Design, Singapore
Prof. J. Andrew Clark, Computer Science University of York, UK
Prof. Yun-Bae Kim, Sungkyunkwan University, South Korea
Prof. Lorne Olfman, Claremont, California, USA
Prof. Louis M. Rose, Department of Computer Science, University of York
Nedia Smairi, CNAM Laboratory, France
Dr. Vishal Gaur, Govt. Engineering College, Bikaner, India
Dr. Aditya Patel, Ahmedabad University, Gujarat, India
Mr. Ajay Choudhary, IIT Roorkee, India
Prof. Prasun Sinha, Ohio State University Columbus, Columbus, OH, USA
Dr. Devesh Shrivastava, Manipal Uiversity, Jaipur, India
Prof. Xiaoyi Yu, National Laboratory of Pattern Recognition, Institute of Automation, Chinese Academy of Sciences, Beijing, China
Prof. R. S. Rao, New Delhi, India

Preface

The Fourth International Conference on ICT for Sustainable Development (ICT4SD 2019) targets theory, development, applications, experiences and evaluation of interaction sciences with fellow students, researchers and practitioners.

The conference may concern any topic within its scope. Workshops may be related to any topics within the scope of the conference. The conference is devoted to increase the understanding role of technology issues and how engineering has day by day evolved to prepare human-friendly technology. The conference will provide a platform for bringing forth significant research and literature across the field of ICT for Sustainable Development and provide an overview of the technologies awaiting unveiling. This interaction will be the focal point for leading experts to share their insights, provide guidance and address participant's questions and concerns.

The conference was held during 5–6th July 2019, at Hotel Vivanta By Taj, Panaji, Goa, India, and organized by Global Knowledge Research Foundation and supported by The Institution of Engineers, (India), Supporting Partner InterYIT, International Federation for Information Processing, State Chamber Partner Goa Chamber of Commerce & Industry, and National Chamber Partner Knowledge Chamber of Commerce & Industry.

Research submissions in various advanced technology areas were received, and after a rigorous peer review with the help of program committee members and 56 external reviewers for 519 papers from 8 different countries including Algeria, USA, United Arab Emirates, Serbia, Qatar, Mauritius, Egypt, Saudi Arabia, Ethiopia and Oman, 113 were accepted with an acceptance ratio of 0.11.

Technology is the driving force of progress in this era of globalization. Information and Communication Technology (ICT) has become a functional requirement for the socio-economic growth and sustained development of any country. The influence of Information Communication Technology (ICT) in shaping the process of globalization, particularly in productivity, commercial and financial spheres, is widely recognized. The ICT sector is undergoing a revolution that has momentous implications for the current and future social and economic situation of all the countries in the world. ICT plays a pivotal role in empowering people for self-efficacy and how it can facilitate this mission to reach out to grassroots level. Finally, it is concluded that ICT is a significant contributor to the success of the ongoing initiative of Startup India.

In order to recognize and reward the extraordinary performance and achievements by ICT and allied sectors & promote universities, researchers and students through their research work adapting new scientific technologies and innovations, the two-day conference had presentations from the researchers, scientists, academia and students on the research works carried out by them in different sectors.

ICT4SD Summit is a flagship event of GR Foundation. This is the fourth edition. The summit was inaugurated by Dr. Pramod Sawant, Chief Minister of Goa, along with other eminent dignitaries including Shri Manguirsh Pai Raikar, Chairperson, Assocham's National Council for MSME; Shri. Prajyot Mainkar, Chairman, IT Committee of Goa Chamber of Commerce and Industry; Mike Hinchey, President, IFIP and Chair, IEEE, UK and Ireland; Milan Tuba, Vice Rector for International Relations, Singidunum University, Serbia; and Shri Amit Joshi, Director, GR Foundation.

Dr. Pramod Sawant shared his views and aim on creating a world-class IT infrastructure and connectivity for e-governance in the state of Goa. Government is committed to create adequate infrastructure for the industry promotion of e-governance, e-education and streamlining of IT in Goa. He further stated that the infrastructure development and capacity building for promotion of IT is one of the main focus areas.

Shri Manguirsh Pai Raikar, Chairperson, Assocham's National Council for MSME, said that this summit is to provide a common platform and bringing forth significant industries, researches and literatures across the field of ICT for Sustainable Development and provide an overview of the technologies awaiting unveiling along with recognizing and rewarding the extraordinary performance as well achievements by ICT and allied industries, communally.

Shri Prajyot Mainkar, Chairman, IT Committee of Goa Chamber of Commerce and Industry, highlighted and encouraged the entrepreneurship and said that GCCI will continue to support such programmes for larger public benefits with great degree of excellence.

The international dignitaries including Mike Hinchey, President, IFIP and Chair, IEEE, UK and Ireland and Milan Tuba, Vice Rector for International Relations, Singidunum University, Serbia, also highlighted the issues and opportunities in the information processing and education sector.

Belgrade, Serbia — Milan Tuba
Gwalior, India — Shyam Akashe
Ahmedabad, India — Amit Joshi

Contents

About the Editors

Milan Tuba received his B.S. and M.S. in Mathematics, and M.S., M.Ph., and Ph.D. in Computer Science from the University of Belgrade and New York University. He was an Assistant Professor of Electrical Engineering at Cooper Union Graduate School of Engineering, New York. During that time, he was the founder and director of Microprocessor Lab and VLSI Lab, and led scientific projects. Since 1994, he has participated in the scientific projects for the Republic of Serbia's Ministry of Science. Currently, he is the Vice Rector of Singidunum University, Serbia. His research interests include mathematical, queuing theory, and heuristic optimizations applied to computer networks, image processing, and combinatorial problems. He is the author of more than 150 scientific papers and a monograph. He has held ten plenary lectures at conferences at Harvard, Cambridge, Moscow, Paris, and Istanbul. He is also a co-editor, member of the editorial board, scientific committee member, and a reviewer for several international journals and conferences. He is a member of various international and scientific organizations: ACM, IEEE, AMS, SIAM, and FNA.

Shyam Akashe is a Professor at ITM University, Gwalior, Madhya Pradesh, India. He completed his Ph.D. at Thapar University, Punjab, and his M.Tech. in Electronics and Communication Engineering at the Institute of Technology & Management, Gwalior. He has authored 190 publications, including more than 50 papers in SCI-indexed journals. His main research focus is low-power system on chip (SoC) applications in which static random access memories (SRAMs) are omnipresent. He has authored two books entitled Moore's Law Alive: Gate-All-Around (GAA) Next Generation Transistor published by LAMBERT Academic Publishing, Germany, and Low Power High Speed CMOS Multiplexer Design published by Nova Science Publishers, Inc., New York, USA. He has also published over 120 papers in leading national and international refereed journals of repute. Dr. Akashe has participated in numerous national and international conferences and presented over 100 papers.

Amit Joshi is a young entrepreneur and researcher who holds an M.Tech. in Computer Science and Engineering and is currently pursuing research in the areas of Cloud Computing and Cryptography. He has six years of academic and industrial experience at prestigious organizations in Udaipur and Ahmedabad. Currently, he is working as an Assistant Professor in the Department of Information Technology, Sabar Institute in Gujarat. He is an active member of the ACM, CSI, AMIE, IACSIT-Singapore, IDES, ACEEE, NPA, and many other professional societies. He also holds the post of Honorary Secretary of the CSI Udaipur Chapter and Secretary of the ACM Udaipur Chapter. He has presented and published more than 30 papers in national and international journals/conferences of the IEEE and ACM. He has edited three books, on Advances in Open Source Mobile Technologies, ICT for Integrated Rural Development, and ICT for Competitive Strategies. He has also organized more than 15 national and international conferences and workshops, including the International Conference ICTCS 2014 at Udaipur through the ACM-ICPS. In recognition of his contributions, he received the Appreciation Award from the Institution of Engineers (India) in 2014, and an award from the SIG-WNs Computer Society of India in 2012.

Managing Multimedia Big Data: Security and Privacy Perspective

Gayatri Kapil, Zaiba Ishrat, Rajeev Kumar, Alka Agrawal and Raees Ahmad Khan

Abstract Multimedia Big Data (MMBD) is growing day by day because data is created by everyone and for everything from mobile devices, digital camera, digital game, multimedia sensor, video lectures, and social networking sites, etc. It is observed that various organizations and businesses moving toward it and taking out the benefits from it. MMBD has its applications in various fields that containing social networking sites, science and research, business enterprises, surveillance image and video, health care industries, etc. Huge size of data is being transfered from the Internet in each second. Due to increase in MMBD, it is necessary to protect this data from external threats or malicious attacks which can affect privacy, integrity, and genuineness of information systems. Therefore, to protect the integrity, identity, and security of individual's data from these unwanted malicious attacks, an effective security mechanism is much needed. In this paper, the author discussed about the current issues in MMBD and security approaches to deal with these issues and also, focus on multimedia big data security and privacy. It also enumerated the directions to be taken while using MMBD along with security measures.

Keywords Multimedia big data · Security · Privacy · Cryptography and encryption

---

G. Kapil · R. Kumar (✉) · A. Agrawal · R. A. Khan
Department of Information Technology, Babasaheb Bhimrao Ambedkar University (A Central University), Lucknow, India
e-mail: rs0414@gmail.com

G. Kapil
e-mail: gayatri1258@gmail.com

A. Agrawal
e-mail: alka_csjmu@yahoo.co.in

R. A. Khan
e-mail: khanraees@yahoo.com

Z. Ishrat
Department of Electronics and Communication, IIMT College of Engineering, Greater Noida, India
e-mail: hellozaiba@gmail.com

M. Tuba et al. (eds.), *ICT Systems and Sustainability*,
Advances in Intelligent Systems and Computing 1077,
https://doi.org/10.1007/978-981-15-0936-0_1

1 Introduction

In this multimedia environment where the use of Internet and mobile technologies are growing rapidly, the size of Multimedia Big Data (MMBD) is increasing, day by day. Users spend lot of time to communicate with each other for sharing the information through Internet [1], MMBD has massive amount of information in random fashion. Also, it provides huge set of environments for multimedia applications/services such as multimedia retrieval, blessings, etc. [2]. To solve real-world challenges, practitioners of multimedia research focusing on the issues including deploying, handling, excavating, comprehension, and envisaging different types of data in proficient ways. However, different scientists and technological enterprises have various descriptions for this term. Extremely large datasets are also called big data.

Further, according to Apache Hadoop, big data is defined as "datasets that could not be captured, managed, and processed by general systems within an acceptable scope" [3]. It is very critical to represent MMBD model because MMBD is vague, multimodal and miscellaneous and has high level of information's complexity. Further, system cannot comprehend this high-level semantics. Information of multimedia is a type of datasets that is human-centric, assorted, and higher volume than the typical big data. In addition, the typical big data requires less sophisticated algorithms and less computing resources as compared to MMBD. Visibility is appreciated in the media context, content, human-centric, high level, and semantics related to MMBD [1].

In this paper, the authors will indicate issues arising in MMBD and managing through existing security mechanisms. Initially, the authors will discuss the MMBD and focus the security issues of MMBD and then describes the need of security and existing approaches that are used for securing MMBD. After that, we will conclude the paper to achieve better results like exciting research, marketing and better business decisions; organizations need to implement it effectively.

2 Security-Related Issues, Challenges, and Threats in Multimedia Big Data

The first challenge faced by the big data analyzers is looking for ways to decrease the computational time and storage capacity, while keeping the results as accurate as earlier. Parallel computing is practice of utilizing different computing resources at the same time, to provide effective analytics in distributive environment [4–6]. For this purpose, many big data analytics platforms have been developed that include IBM Big Data Analytics, Oracle Big Data Analytics, Microsoft Azure, etc.

In this new era of big data, we have high diversity multimedia data and huge amount of social data that brings new opportunities and challenges. Hence, multimedia big data analytics has attracted a lot of attention in both academic and industrial places in recent years [7, 8]. Because of its critical and valuable insights, it is now considered as an emerging and challenging topic.

2.1 Scalability

With the massive number of users and devices that exists today, there is a need for setting up security policies for these users and devices. Provisioning security policies defines as scalability. One approach is to have a solitary global policy for all the users and another one is to have individual data access policies for individual users. It is not always easy to have a single global policy for all the users if the user base is enormous. Proficiency is very important in these kinds of services along with the promise of security. So, the mechanism for security must be in a position to sustenance high data size and swift data arrival rate. The processing of security should be done on the fly or at a faster rate than the arrival rate of the multimedia data [9].

2.2 Mobility

When it is viewed from the perception of mobility, the benefits comprise availability and easy accessibility. When users access data from different locations, at a different point of time and with the help of different devices, the access privileges may also change because of the change in the locations [1]. For example, take the case of a doctor who tries to access some medical images. He may access them from inside the hospital network or from outside the network like his house. So he/she may be allowed to those medical images when accessed from the local network of the hospital and not while he/she is in off duty. So the spatiotemporal limitations have also been taken in interpretation while setting up the access control and it should be done in a consistent way.

In the case of role management solution, for the user to realize the policy in a better way [4, 6] it should not be made multifarious. The user may not be erudite to use them and may get overwhelmed with it. To obtain the maximum advantage out of it, it should be location aware, which offers real-time and relevant information. Additionally, it should sustenance the requirements related to privacy which includes Supervisory the information, Gathering the information, and Description of Intention and recording. Supervisory refers to the access and use of multimedia data, while gathering defines the data collection in a well-defined manner in accordance with the security policies. Description of intention deals with the questions how and by whom the data is accessed. Recording involves recording the information about those who assemble and use the data.

2.3 Multilevel Protection

There are some security models exist currently which emphasis mainly on file-level protection. But in the case of multimedia data, it is not convenient because usually, multimedia consists of many numbers of elements with various levels of "sensitivity". Surveillance video generally demands a high-level security as compared to some normal multimedia data. So, the mechanism to offer security should be supple and it should sustenance access control on multiple levels which permits the setting of access permissions for different types of multimedia objects in a multimedia stream [10].

2.4 Big Data Maybe Assembled from Assorted End Points

The data in big data arrives from large number of sources including social network users, mobile phone users, and Internet of things. This states that data arriving from different sources must be integrated with careful inspection of data [4, 6].

2.5 Data Accumulation and Distribution

Information gathering and appropriation must be verified inside the settings of a formal and sensible structure. Accessibility and transparency in data i.r.o. current and past use by the users are the important aspects of big data. In certain settings, where such structures are missing or have been arbitrarily created there might be a need of open or walled garden gateways and ombudsman-like jobs for information very still. These framework blends and unexpected mixes call for improved big data structure [7].

2.6 Data Search and Selection Require Privacy and Security Policy

Due to lack of systematic consideration in framework, understanding of needs the data owner should provide secure combinations. A group of educated users, architects, and system defenses are needed. Privacy should be retained for this we require a tool such as Personally Identifiable Information (PII), as there is a channel between data provider and data consumers so we need to implant privacy [10]. Personal identifiable information is a common tool for big data and the feature of big data at said by one of analyst "the ability to derive distinguished perceptions from advanced analytics on data at any scale" [8, 11].

3 Needs of Multimedia Big Data Security

The scope of multimedia big data security is so huge because of the huge amount of multimedia data that is generated every day. Social media and showbiz, surveillance videotapes, consumer multimedia data, etc. are all in compulsion of security measures and policies in place to protect their data from unlawful access. There is a need to integrate many access control policies. Big data is integrated data from different sources. Big data have their own access control policies called "sticky policies". So when they are assimilated with data from other sources, their original policies must be enforced. So there is a need to integrate the policies and resolve the conflicts involved [9].

Authorizations have to be administered automatically, especially to allowance permissions; it is not a practical thing to handle the access controls manually on large data if it needs a fine-grained access control [4]. Therefore, some techniques are needed to allow authorization automatically. Administration of access control policies on multimedia data is diverse. This is a type of access control which is an important one where endorsement is granted or denied-based on the multimedia data content [12]. This type of access control is very significant when dealing with data which needs security such as surveillance camera shots. In the case of large datasets of multimedia type, this can be quite thought provoking because to care the content-based access controls, it involves knowledge of the multimedia contents that require being protected [13].

Prosecutions of policies for access control in big data stores. With some of the hot developments in big data systems, users are permitted to submit capricious jobs with the support of programming languages like Java. This creates many disputes which obstruct the prosecution of a suitable access control mechanism for the users in an effectual way. There are some researches in progress in this field which tried to put on the access control policies to the previously submitted jobs. But still, it desires more work to be done to make it more proficient [10]. Design, develop, and accomplish access control policies automatically. In the case of environments, which are active in nature where there are unremitting changes in the sources, applications and users, designing and developing of access control policies are important in order to ensure unswerving data availability and privacy [13, 14].

4 Different Approaches for Multimedia Big Data Security

4.1 Role Management

In this slant, user-defined policies are formed for a user group. It is a problematic task to grow privacy policies that denote the partialities set by the user in order to accomplish their data. The important facets that have to be measured here include the following questions: who, how, and where. Who defines them? How is it defined?

Where is it going to be getting stowed? For the first question, one answer is to provide power to the user itself so that they govern their data in spreading of the data to other users in a secured and private manner. On the conflicting, an entity can define its policies and the user will be provided with the options either to approve or discard it according to their priorities. For the question how is it defined, one solution is to seizure them in access control systems for an effectual administration of privacy policies within an endeavor. For the third question of where it is stowed, cloud services are a key. Cloud services offer trustworthy storage in an effectual way which cares scalability. It helps in decreasing the cost of the physical infrastructures desired otherwise to save the data [6, 10, 15].

4.2 Location-Based Access Control

Location-based access is given to the users in this tactic. Here, users are obligatory to deliver their current location to the access control system. Subsequently, detailed information of the movement of the user will be seized by the system and analyzed over a period of time. It may cause a lot of issues if there is a fissure in the security of these data [6, 15, 16].

4.3 Encryption

Cryptographic algorithms can be used besides with some permuting techniques on data to evade the unauthorized access of multimedia data.

4.3.1 Video Encryption

Multimedia data can be encrypted by symmetric key cryptography algorithms. AES is one of the fastest algorithms. But when multimedia data is real time, it sustains more computational costs. Video scrambling and selective video encryption are two techniques that are used usually to accomplish video encryption [5].

First of all, we need to consider a multi-stream multimedia system is considered for selective encryption control model. The system consists of clients and stream of data. The media stream of data is irrespective of time t. For the process of mass data this utilizes nodes such as control nodes and sink nodes. The parameters of the nodes in this system are defined as d, c, b, v, e, x, and x, these are vector dimensions. Data stream can be quantified as vector $d = (d1 \ldots dnt)$, which denotes "nt" different data streams in time t. We can also make a copy of data stream by quantifying such can be quantified as vector $c = (c1 \ldots cnt)$ where c is the number of copies of data streams.

4.3.2 Image Encryption

Many data encryption algorithms are available currently. But mainly they are used for text data and may not be applicable for data like images. Also, there are many disadvantages for these algorithms as they may source of major congestion in its process. So, selective image encryption methods are used to avoid congestion. It does not encrypt the entire image, but does for part of the image and thus decreases the computation [17].

4.3.3 Audio Encryption

Secure voice is a name that is employed in cryptography for voice encryption over a variability of communication medium like radio, telephone, etc. The normal encryption algorithms are computationally overpriced in this case too. One of the most important types of audio data is speech when viewed from security perception. Speech needs substantial security unlike other audio types like music or other entertainment types. The selective encryption method can be used in this case too to diminish the computational complexity [5, 17, 18].

4.3.4 Fingerprinting

In this technique, a sole ID will be rooted in each user's copy. This can be later mined to help identify offenders in case of unauthorized leaks. This is a cost-effective method which can put off illegitimate access to data. In order to protect data from illegal access, use and redistribution after an authorized access of a user, sole user information like fingerprints can be rooted to the multimedia data contents into each user's copy to discover the unauthorized users. The rooted information should be indiscernible to the users. It has become a substantial issue in social networks to defend digital multimedia content. Digital fingerprinting can discover out the user who has reorganized the multimedia content illegally. When an illegal copy is detected the fingerprint can be mined from it and using some algorithms, the defector can be easily identified. A watermarking technique will be used to implant the fingerprinting data in the multimedia content [18, 19].

4.4 Overview of Existing Algorithm

From the discussion of Sect. 2, it concludes that there is no static algorithm for different multimedia files in big data. For data stowage security issue in cloud computing, the system is trusting on RSA or DES algorithm. These algorithms are used for encryption of big data prior to stowage, user authentication procedure or structure a secure channel for data transmission. Crossbreed algorithm is the combination of

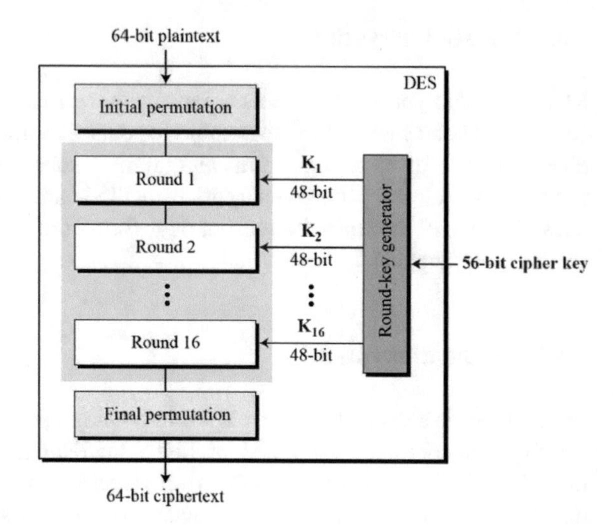

Fig. 1 Steps of data encryption standard

RSA and DES for private key encryption. To encrypt the enormous amount of data hybrid approach is used in which the data is encrypted using symmetric schemes (DES or AES) and the key is transported using asymmetric scheme (RSA).

4.4.1 Data Encryption Standard Algorithm

Data Encryption Standard (DES) uses block cipher. It is used to encrypt data in block length of 64 bits each. Key is 56 bits long, 8 bits are rejected from the key. DES algorithm is based on substitution and permutation and takes place in 16 rounds. In every round, it shifts the key and data bits, permute, XOR and sends it through 8 S-Box. Initial Permutation (IP) is provided with 64 bits plain text. IP divides it into two parts, Left Plain Text (LPT) and Right Plain Text (RPT). These both parts go through 16 rounds and reunite lastly as shown in Fig. 1. The same way in opposite order is used for decryption. 2DES and 3DES are of same procedures excluding that they use two and three different keys, respectively [20].

4.4.2 Blowfish Algorithm

Blowfish algorithm consists of block cipher of 64-nit block. Key length in this algorithm varies from 32 bits to 448 bits. Two main parts of these algorithms are sub key generation and data encryption. Encryption has 16 rounds as used in DES. Each round comprises key depended permutation and data depended substitution [20].

4.4.3 RSA Algorithm

RSA stands Rivest, Shamir & Adleman and was developed in 1977. RSA uses asymmetric keys for cryptography which means it uses different keys for encryption and decryption process [21]. The RSA encrypts the plaintext in the form of blocks and each block has the digital value less than "*n*". Therefore, block size must be less than log2 (*n*) +1. These are the forms of encryption and decryption process if plain text block size is "*M*" and ciphertext block size is "*C*". $\mathbf{C = M^e \bmod n}$ **and** $\mathbf{M = C^d \bmod n}$, Sender has the values of "*e*" and "*n*" and the receiver has the values of "*d*" and "*n*". In this technique, user A has access to public key and user B has access to private key. Both users are located at different stations. Then, two primes *p* and *q* are randomly chosen by each station and kept secret. $n = p * q$ is produced by their product, specifically, the system uses modular arithmetic to send a message to cipher text. The algorithm is based on modular exponentiation. This algorithm is employed in the most widely used Internet electronic communication encryption program [5].

4.4.4 Higher Dimension Chaotic Map

Audio files in big data have composite form and tough task to manage and offer security. For audio files in multimedia, numerous algorithms are used for encryption and decryption. Higher dimension chaotic map uses an algorithm which consists of stream cipher algorithm that will use for encryption of information at bit level thru secret key generator. It uses higher length of keys to improve complexity and enhance strength against brute-force like attacks. Function of this algorithm is quite modest but user can make it composite by increasing its key space. Before the encryption, the analog signal is converted into digital signal [21].

4.4.5 H.264 Video Entropy Coding

For video files in big data, H.264 is the innovative standard video coding, currently, there are three types of video algorithms based on entropy coding selective encryption algorithm, complete encryption algorithm, and encryption.

The best secured is complete encryption but it is very complex, selective algorithms have good features but they will change the performance of entropy and also change the compression properties videos. Entropy coding has good security and compression performance. H.264 is broadly used for big data in video files because of its enactment on Internet. H.264 offers (CAVLC), i.e., Context-Adaptive Variable Length Coding method and (CABAC), i.e., Context-Adaptive Binary Arithmetic Coding. There are following steps that we have to follow for H.264 CAVLC [20].

- All coefficients and trailing ones (Total Coeffs and Trailing Ones) are to be encoded completely.
- Signs of trailing ones are to be encoded.

- Nonzero coefficient levels excluding trailing ones (Levels) are to be encoded.
- Encode total number of zeros preceding last nonzero coefficient (Total Zeros).
- Encode keeps running of zero coefficients before each nonzero coefficient (Run before).

There are some restrictions to described algorithms. It does not have the provision the newer H.265 compression standard. As technologies deviate, the novel standards and formats come into play, the security should also modernize. This algorithm provides balance between security and coding complication. So, it is appropriate for video conferencing, digital rights management, multimedia Storage, and so forth. This algorithm does not show good performance on real-time video which is on big data.

4.4.6 DRM on Identity-Based Encryption

A common tool used to defend cryptographic technique contents is Digital Rights Management (DRM). Identity-based encryption uses secret data receivers solely for simple certificate management. It also makes use of public-key bilinear pairing property which is used for the safety of identity-based encryption and related to elliptical curve discrete log problem [10]. Thru these two characters, identity-based encryption makes the key management simple as well as also decreases the computational cost compared to conventional public-key cryptosystem [11]. DRM which is centered on the identity-based encryption will definitely augment system security such as Eavesdrop Attack, Forgery DRM Module Attack, and Distribution Attack. For Eavesdrop Attack, one private value is fixed inside DRM module. So, the user who steals or copy the neutral DRM format file cannot compute the key value even he copies the encrypted content. If we use some password in key computations, the access control can be implemented [12].

4.4.7 Elliptic Curve Cryptography

In 1985, experts offered an idea for elliptic curves as a novel type of cryptosystem, which concentrates on being an alternative to RSA encryption [13]. Since the system was mainly centered on elliptical curves it was named Elliptic Curve Cryptography (ECC). Cryptography is an electronic technique that is used to guard the valuable data over transmission. To protect our data by using different authentication techniques is prime objective of cryptography. In modern era, ECC has primary application in resource controlled environment as lengthy key size is not feasible in ad hoc wireless networks and mobile networks. The main emphasis while evolving ECC was to have a system that employs as a high-level security public-key cryptography system. For high-level security in a public-key system, two keys are used mostly for encryption and decryption. The first key is known as a public key that is broadly disseminated and second is the personal or private key in which only data is recognized as Global sign. The method of having a public–private key was reflected efficient methods. Low speed and increased bandwidth were the key features of old public-key cryptosystem. ECC provides high-level protection with compressed size [14].

Problems associated with ECC

Elliptic Curve Cryptosystem has a problem in the efficiency related to the private key, private key is required for a secure system which cannot be ignored. Problems associated with ECC was

- When K was static rather than arbitrary, a gathering called failover stream had the capacity to recuperate the private key.
- It was conceivable to recover the private key of a server utilizing Open SSL that utilizes ECC over a binary field by executing a timing attack ("Elliptic Curve Digital Signature Algorithm").
- Another debate reported itself with Elliptic Curve Cryptography the Java class Secure Random had a few organisms in its executions that occasionally produced crashes in the k estimation of the private key $S = k * R$. In any case, this can be avoided using deterministic generation (Elliptic Curve Digital Signature Algorithm) which generates true randomness of K.

For as far back as 20 years, the National Security Agency (NSA) has been encouraging ECC as an increasingly dependable option to RSA; be that as it may, in August 2015, the NSA quit endorsing for the organization of Elliptic Curve Cryptography. This caused guesses that Elliptic Curve Cryptography is not really truly secure [15, 16].

5 Conclusion

The big data development offers a large growth in the field of big data applications but it comes with challenges that have to be overcome also. Further in time to come user requirement of multimedia applications. Multimedia nowadays has an impact on humans, their thinking, and social growth. Economically multimedia also has an impact on human living. This paper emphasizes on survey of various features of multimedia big data, security issues and security approaches that exist nowadays. A lot of research is being approved out in this field of security and some approaches have been identified to secure data from unauthorized access and redistribution such as role-based access control mechanisms and encryption techniques. Therefore, we finally conclude our paper with a view which outcasts the weaknesses of existing protocol ECC and in future this work will be extended for implementation. We hope that paper will provide significant visions into the evolving trends in security and privacy of multimedia big data.

References

1. Wang, Z., Mao, S., Yang, L., Tang, P.: A survey of multimedia big data. China Commun. **15**(1), 155–176 (2018). http://www.eng.auburn.edu/~szm0001/papers/CNCOMM-2017-0044.final.pdf
2. Zhu, W., Cui, P., Wang, Z., Hua, G.: Multimedia big data computing. IEEE MultiMedia **22**(3), 96–106 (2015)
3. Wang, X., Gao, L., Mao, S., Pandey, S.: CSI-based fingerprinting for indoor localization: a deep learning approach. IEEE Trans. Veh. Technol. **66**(1), 763–776 (2017)
4. Zikopoulos, P., Parasuraman, K., Deutsch, T., Giles, J., Corrigan, D., et al.: Harness the Power of Big Data The IBM Big Data Platform. McGraw-Hill Professional, New York, NY (2012)
5. Davidson, J., Liebald, B., Liu, J., Nandy, P., Vleet, T.V., Gargi, U., Gupta, S., He, Y., Lambert, M., Livingston, B., et al.: The YouTube video recommendation system. In: Proceedings of the 4th ACM Conference on Recommender Systems. ACM, pp. 293–296 (2010)
6. Bian, J., Yang, Y., Chua, T.S.: Multimedia summarization for trending topics in microblogs. In: Proceedings of the 22nd ACM International Conference on Information and Knowledge Management. ACM, pp. 1807–1812 (2013)
7. Fleites, F., Wang, H., Chen, S.C.: TV shopping via multi-cue product detection. IEEE Trans. Emerg. Top Comput. **3**, 161–171 (2015)
8. Chang, Y.F., Chen, C.S., Zhou, H.: Smart phone for mobile commerce. Comput. Stand. Interfaces, **31**, 740–747 (2009)
9. Ye, C., Xiong, Z., Ding, Y., Zhang, K.: Secure multimedia big data sharing in social networking using fingerprinting and encryption in the JPEG2000 compressed domain. In: 2014 IEEE 13th International Conference on Trust, Security and Privacy in Computing and Communication (TrustCom) (2014)
10. Tufekci, Z.: Big questions for social media big data: representativeness, validity and other methodological pitfalls. In: Proceedings of the Eighth International Conference on Weblogs and Social Media. Michigan, USA, pp. 505–514 (2014)
11. Smith, J.R.: Riding the multimedia big data wave. In: Proceedings of the 36th International ACM SIGIR Conference on Research and Development in Information Retrieval. ACM, Dublin, Ireland, pp. 1–2 (2013)
12. Toshiwal, R., Dastidar, K.G., Nath, A.: Big data security issues and challenges. Int. J. Innov. Res. Adv. Eng. **2** (2015)
13. Tri Do, T.M., Blom, J., Perez, D.G.: Smartphone usage in the wild: a large-scale analysis of applications and context. In: Proceedings of the 13th International Conference on Multimodal Interfaces, ACM, pp. 313–360 (2011)
14. Chen, M.: A hierarchical security model for multimedia big data. Int. J. Multimed. Data Eng. Manag. **5**, 1–13 (2014)
15. Zhu, W., Cui, P., Wang, Z., Hua, G.: Multimedia big data computing. IEEE Multimed. **22**(3), 96–106 (2015)
16. Tan, W., Yan, B., Li, K., Tian, Q.: Image retargeting for preserving robust local feature: application to mobile visual search. IEEE Trans. Multimed. **18**, 128–137 (2016)
17. Wang, X., Gao, L., Mao, S.: CSI phase fingerprinting for indoor localization with a deep learning approach. IEEE Internet Things J. **3**(6), 1113–1123 (2016)
18. Xiong, H.: Structure-based learning in sampling, representation and analysis for multimedia big data. In: Proceedings of 2015 IEEE International Conference on Multimedia Big Data (BigMM), Beijing, China, pp. 24–27 (2015)
19. Chezian, R.M., Bagyalakshmi, C.: A survey on cloud data security using encryption technique. Int. J. Adv. Res. Comput. Eng. Technol. **1**(5) (2012)
20. Rivest, R.L., Shamir, A., Adleman, L.: A method for obtaining digital signatures and public-key cryptosystems. ACM **21**(2), 120–126 (1978)
21. Gnanajeyaraman, R., Prasadh, K., Ramar: Audio encryption using higher dimensional chaotic map. Int. J. Recent Trends Eng. **1**(2), 103 (2009)

Nature-Inspired Metaheuristics in Cloud: A Review

Preeti Abrol, Savita Guupta and Sukhwinder Singh

Abstract Due to the successful deployment and high performance, metaheuristic review is extensively surveyed in the literature that includes algorithms, their comparisons, and analysis along with its applications. Although, insightful performance analysis of metaheuristic is done by few researchers still it is a "black box". The performance analysis of algorithms is performed. This paper addresses an extensive review of four nature-inspired metaheuristics, namely, ant colony optimization (ACO), artificial bee colony (ABC), particle swarm optimization (PSO), firefly algorithm, and genetic algorithm. It includes introduction to algorithms, its modifications and variants, analysis, comparisons, research gaps, and future work. Highlighting the potential and critical issues are the main objective of intensive research. The metaheuristic review provides insight for future research work.

Keywords Metaheuristic · Particle swarm optimization (PSO) · Ant colony optimization (ACO) · Artificial bee colony (ABC) · Firefly algorithm · Genetic algorithm

1 Introduction

Cloud Computing is a buzz word in the Internet. It delivers elastic execution environment of resources over the Internet with unified resources provisioned for the users on pay per use basis [1]. The researchers are focusing on the performance of the metaheuristics for improvising the quality of service in cloud. Thus, the main target of implementing resource management and resource placement techniques

P. Abrol (✉)
CDAC, Mohali, India
e-mail: preetikpk@gmail.com

S. Guupta · S. Singh
UIET, Chandigarh, India
e-mail: Karanpreet@cdac.in

S. Singh
e-mail: preeti_kpk@yahoo.com

M. Tuba et al. (eds.), *ICT Systems and Sustainability*,
Advances in Intelligent Systems and Computing 1077,
https://doi.org/10.1007/978-981-15-0936-0_2

reasonably and profitably is achieved. For this purpose, cloud service provider needs architecture in order to put everything in a systematic order.

For this reason, the major requirement for the efficient working of cloud is met in middleware architecture. It mainly consists of two major modules, namely Resource Management and Resource Placement which needs attention in regard to the quality parameters. Resource management is a concept in which the tasks are arranged according to the utilization and resources as per their capacity. Initially, these are not arranged as per the QoS requirement.

QoS parameters are the metrics of the tasks to be executed such as execution time, burst time, processing time, utilization, and turnaround time, makespan, and response time provided by the client or by default cloud parameter setting (Fig. 1).

The resource placement allows the mapping of the resources among the given tasks to meet the required QoS. To provide guaranteed QoS, the tasks must be allocated to the resources in such a way that both the cloud owners and cloud users set parameters that are not deviated. This challenge is counted under the global optimization problem.

The basic cloud framework follows a modular pattern. The entire framework is discussed as follows.

The clients send the request of the task set or application along with its QoS parameter configuration setting in automatic or manual form.

The next layer defines the admission control and cloud service provider operations on the request sent, wherein admission control algorithms regulate the traffic of dynamic tasks and CSPs evaluate the QoS parameters of the tasks request.

Resource management module manager (RMM Manager) evaluates the available resources to update its status so as to meet the requirement of the requested tasks. In resource management module (RPM), the evaluated tasks are scheduled and resources are managed as per requirements.

All the steps are accompanied by cloud orchestration at each step, wherein exclusive monitoring of QoS parameters are performed at each layer. This module also includes the QoS negotiation and its enforcement in case of dynamic task request, and accounting maintains the log of the requested tasks, pricing/billing, and risk management in case of failure.

2 Related Work

Related work can be categorized as Middleware Architecture, Resource Management, and Resource Placement. The details are discussed as follows.

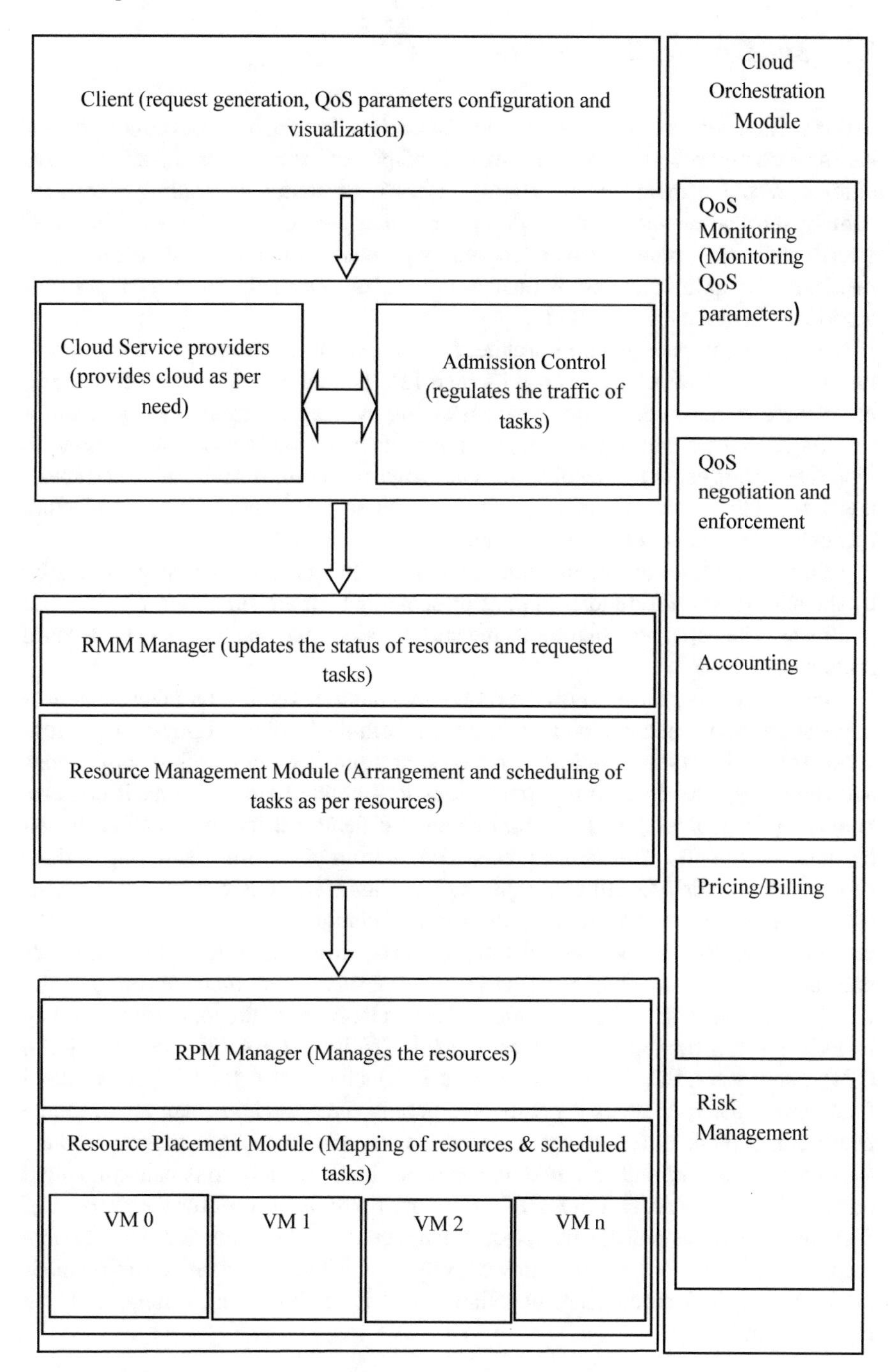

Fig. 1 Cloud orchestrated framework functionality

2.1 Middleware Architecture

In this section, the architecture of the cloud is analyzed on the basis of various policies based on configuring, healing, optimizing, and protecting resources in order to have efficient resource management. Another effective resource scheduling algorithm, namely, self optimization of cloud computing energy-efficient resources (SOCCER) is analyzed which abides the SLA. The author proposed autonomic resource management used for optimizing power consumption and evaluated clustered heterogeneous workloads in cloud environment.

The cloud computing and its architecture for allocating resource by creating virtual machines (VMs) have been presented [2]. The authors also provide insights of resource management policies for sustaining the feature of risk management for fulfilling Service Level Agreement [1no]. An effective cloud service management is identified in this paper that resolves the challenge of resource provisioning and specifies the effective management approach. The authors present conceptual architecture that evidence efficient cloud performance and maintenance.

[7no] The middleware technologies for cloud of things (CoT) have been discussed by the author. Several middleware are presented which are suitable for CoT-based platforms. The paper also highlights the current issues and challenges in CoT-based middleware.

[4no] This paper focused on the cloud architecture, gearing up enterprise-class application to virtualized cloud environment from the traditional physical environment. It highlights the principles, concepts, and practices in creating applications and migrating already existing applications to the cloud environment. It contains variety of technologies that are configured and deployed for the creation of distributed systems [3]. The author presents computing vision for identifying various IT paradigms to deliver utility computing [2]. CloudSim is used for the implementation of Energy-aware Resource Allocation Techniques. The author signifies the role and necessity of middleware [4]. It focuses on various middleware technologies presenting middleware characteristics and features in cloud domain and compared it with various architectural possibilities. The middleware architecture-related survey includes various frameworks such as SHAPE, CHOPPER, EARTH, PURS, ARCS, QRPS, and SOCCER. From this study, it is concluded that for the perfect cloud functioning, the framework must involve mainly three modules, namely, resource management module, resource placement module, and cloud orchestration module. Resource placement and resource management implement various nature-inspired metaheuristics like ACO, PSO, ABC, GA, and firefly algorithm for the purpose of SLA aware resource management, load balancing, task scheduling and task-resource mapping, and for avoiding premature convergence. While Cloud orchestration module performs QoS monitoring, negotiation and its enforcement, billing, and risk management (Fig. 2).

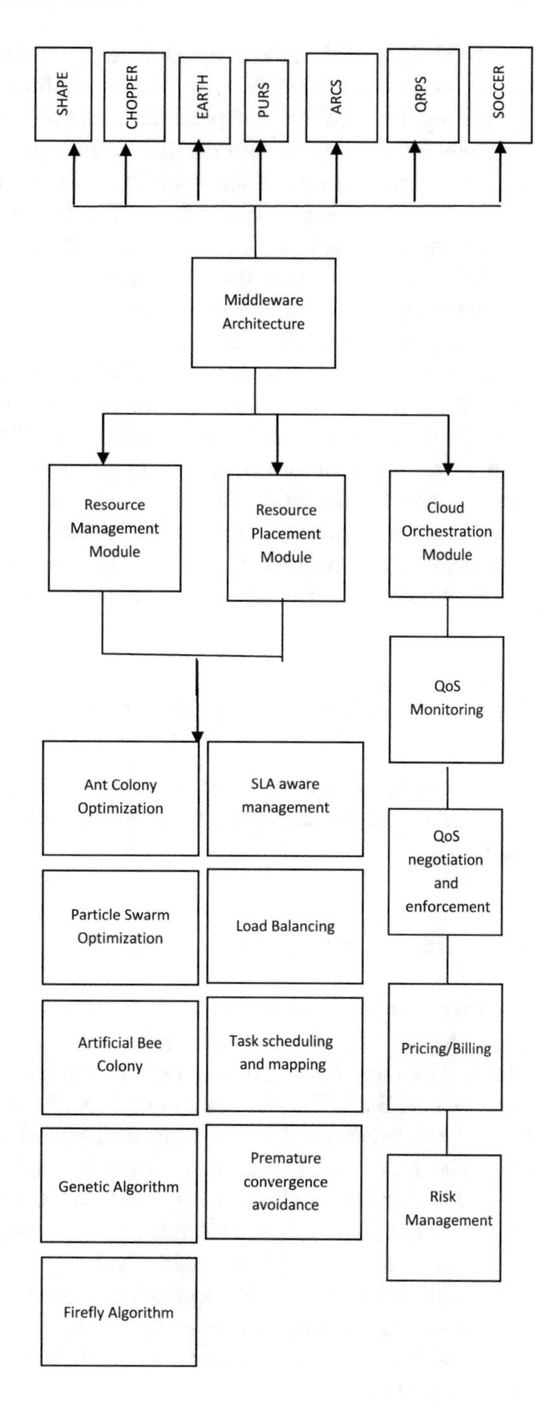

Fig. 2 Overview of cloud middleware frameworks

CHOPPER [100] manages dynamic requests based on user-defined QoS requirements without violating SLA terms. the authors verify concepts, namely, self-optimizing, self-healing, self-protecting, and self-configuring. Resources are automatically managed in CHOPPER as its search and react automatically to faults and maximize resource utilization as well as energy efficiency in order to reduce execution cost, execution time, and SLA violation rate. Authors [5] proposed a technique named partial utility-driven resource scheduling (PURS) for managing SLA for dynamic requests that allow providers in managing VMs in cost-effective and efficient ways. Autonomic resource contention scheduling (ARCS) is discussed in improving resource contention [6].

Another autonomic system called self-healing and self-protection environment (SHAPE) [7] handles and recover various hardware, software, and network faults. Energy-aware autonomic resource scheduling (EARTH) [8] schedules resources automatically and in turn optimizes resource utilization and energy consumption. Hence solves the starvation problem.

QoS-based resource provisioning and scheduling (QRPS) framework for resource management that includes workloads clustering through various workload patterns [9]. Re-clustering of workloads is done by k-means-based clustering algorithm.

2.2 *Nature-Inspired Metaheuristics*

The following sections represent review of ACO based, ABC based, PSO based, GA based, and firefly metaheuristic techniques for resource management and placement (Table 1).

2.2.1 Ant Colony Optimization (ACO)

The ant colony inspired Dorigo in 1992 for the proposal of an algorithm simulating their behavior called the ant colony optimization algorithm (ACO), in his Ph.D. thesis. Whenever an ant moves from its nest to the food or vice versa, it leaves behind the pheromones. Basically, pheromones find shortest distance from food to nest or vice versa. These pheromones have a unique characteristic of evaporation with respect to time. The ants attract to pheromone trails and follow the high pheromones content paths. This is a self-reinforcement process wherein all ants chose on a single optimal solution at one point of time. This methodology is known as convergence. The number of ants acts as agents carrying tasks. Each ant builds the solution of assigning the tasks to the resources. In this way, each ant builds the complete solution of mapping each task to the resources by following the steps to the algorithm (Fig. 3).

The following is the detailed survey of ACO w.r.t resource management and resource placement.

Table 1 Comparison of variants of ACO

Referenced work	Improvement strategy	Performance metrics	Nature of tasks	Purpose
[29]	The convergence of GACO algorithm	Convergence	Independent	Cloud environment
[28]	Genetic algorithm (GA) for ACO initialization	Makespan, DI, utilization	Independent	Cloud environment
[32]	Three constraint conditions	Bandwidth, network latency, execution time	Independent	Cloud environment
[33]	Pheromone rate and convergence time calculated	Pheromone rate	Independent	Cloud environment
[34]	Analysis, design, and implementation of resource scheduling and allocation problems with ACO	Finish time	Independent	Cloud environment
[35]	Scheduling algorithm with efficient resource utilization	Completion time, makespan	Independent	Cloud environment
[97]	ACO resource scheduling metaheuristic converges for global optimal solution	Average execution time	Independent	Cloud environment
[36]	MACO for optimizing resources	Makespan	Independent	Cloud environment
[37]	Solution to the drawbacks of PACO is proposed, such as the parameters selection and the pheromone's update, in SAACO, author uses PSO to make ACO self-adaptive metaheuristic	Makespan and load balance	Independent	Cloud environment

ACO w.r.t Resource Management

This paper [10] improves the shortcoming of ACO algorithm as it falls in local optimal for resource management. Particle optimization merged with ACO calculates solution for improvement in performance. The algorithm also avoids the premature convergence in the local optimal solution.

Further in paper [8], genetic-ACO is proposed for resource management purpose. The choice of the initial pheromone hits convergence rate. Genetic algorithm implemented for the purpose of searching globally and ACO converts it to initial pheromone. Result proves that the proposed algorithm outperforms genetic algorithm and ACO.

Model of ACO algorithm is also considered for the general optimization problem in [11]. The model proves that it can converge to one of the optima if only this

Pseudo code of ACO

Input: Tasks and resources

1. *Initialization Phase*

Set Initial value τij(t)=c, optimal solution=null, iteration t=1 for path between task-resource. [*Initialize food source and ants along with the pheromone.*]

2. Put m ants on first resource randomly. [*Initialize pheromone matrix with a constant (positive value)*]

3. Fitness Evaluation and VM Selection Phase:

For x: =1 to m

Do put the first resource of x th ant.

While tasks choose the food i.e. resource for next task as per eq (1). Evaluate all resources and Insert the selected resource to a temporary list. [*Update the solution given by every ant as per pheromone matrix*]

End Do

4. Pheromone Updating Phase

For x: =1 to m

Do calculate tour length travelled by xth ant as per eq (4).

[*Revise best solution as optimal solution, highest pheromone is better*]

5. For every path, implement the local pheromone as per eq (5).

6. Update global pheromone as per eq (7).

7. Increment t with 1.

8. If (iteration t < $\boldsymbol{t_{max}}$), Pop the temporary lists.

[*Return best Solution*]

Go to step 2

Else display optimal solution.

Fig. 3 Pseudocode of ACO algorithm

optimum is allowed to update the pheromone model. If it is having two optima then it cannot converge to any of the optima.

A step forward is taken by the proof of convergence to optimal control policies in [12]. So the type of convergence analysis implemented is different in ACO and ACL. The convergence analysis mainly relies on the global best update rule. If the best solution is found only then the pheromones get updated.

This paper explored the convergence analysis in ant colony learning (ACL). ACL is represented which combines the control policies of ACO and reinforcement learning [12].

2.3 *Artificial Bee Colony (ABC)*

Foraging behavior of honey bee simulated as ABC. Bee colony contains the following category of bees: employed bees, onlookers and scout bees.

Bees Foraging behavior is specified in brief

- Employed bees find food.
- Employed bees pass information of the food found to onlooker bees.
- Onlooker bees choose the best food.
- Employed bee information chosen converts into scout bee.

Employed bees are sent in search of food sources. One employed bee is allocated for one food. Therefore, employed bees and food are always equal. The employed bee that finds the food source becomes scout bee. Bees communicate the information relevant to the food sources quality with dance called a waggle dance. Onlooker bee chooses the most profitable food. Recruitment of employed bees is done according to rich quality of the food. Basically, characteristics or feedbacks of bees are enlightened:

(i) Positive: With increase in the food, onlooker bees also increase proportionally.
(ii) Negative: Bees stop exploiting low-quality food.
(iii) Fluctuations: New food sources are discovered by random search process carried out by scout bee.
(iv) One-to-many interactions: Bees share information about food with onlookers.

ABC general approach is explained (Fig. 4).

In [13], the accelerated artificial bee colony method specifies two changes, modification rate (MR) and step size (SS) for the promotion of the ability of ABC algorithm

Pseudo code of ABC

Input: Task and VM.

1. initialization of population VM, i =1,2,...n with eq (1)
2. Fitness evaluation.
3. Repeat for each employed bee
4. Compute fitness by greedy procedure with eq (2) and eq (3).
5. Calculate probability values of VM for the solutions.
6. For each onlooker bee, select VM as per probability.
7. Calculate fitness by greedy procedure with eq (3)
8. Replace abandoned solution with random solution produced with eq (4).
9. Save best solution and increment iteration.
10. End

Fig. 4 Pseudocode of ABC algorithm

in order to search locally while focusing on convergence speed. A-ABC performs better with faster convergence when compared with ABC algorithm on various seven benchmark functions.

The convergence analysis [14] of artificial bee colony algorithm is done theoretically. By this, we conclude that artificial bee colony sequence ensures global convergence as algorithm meets two assumptions of the random search algorithm for the global convergence (Table 2).

Reference [15] improves load balancing that utilizes its resources fully and optimizes the scheduling of virtual machine on cloud. Slow convergence due to lacking crossover may lead to limited subspace exploitation ability. Lack of crossover is found in many metaheuristics. Both ACO and ABC are strong in exploration ability, but are weak in exploitation comparatively. Comparison table of ABC is shown in Table 3.

Table 3. Comparison of various ABC-based algorithms

2.4 *Particle Swarm Optimization (PSO)*

Particle Swarm Optimization is swarm intelligent metaheuristic that mimics nature. Kennedy and R. Ederhart developed PSO in 1995. PSO mimics the flocking behavior of birds [16]. In the solution space, the flocking behavior of the birds determines the optimum solution. Particles move toward the local best position (solution) to find pBest as well as tracks global best position, gBest, i.e., the best and the shortest path found by the particle at particular instance. Each particle is initiated with position and velocity in a multidimensional solution space such that it adjusts its velocity and position in the whole population.

Birds find the most optimum (best) path by communicating with each other so that they could reach the food sources. Hence, they learn from the experience of their local best solutions and global best solutions [17]. Reference [18] presents discrete particle swarm optimization algorithm incorporating an opposition-based technique that generates initial population along with load balancing of the processors using greedy algorithm. Makespan, reliability, cost, and flow time are metrics evaluation of the efficiency of DPSO for task scheduling.

Many techniques are invented by researchers for scheduling in cloud computing. A few of them are Metaheuristics, Greedy, Heuristic technique, and Genetic, that are implemented for task scheduling in several parallel and distributed systems. Reference [19] gives a review on the proposals of scheduling. Task scheduling is implemented to maximize utilization of resource hence minimizing the finishing time. DE-GSA [20] minimizes the execution time (makespan). PSO algorithm is represented in Fig. 5. Comparison table of PSO is shown in Table 4.

Table 2 Comparison of variants of ABC

Referenced work	Improvement strategy	Performance metrics	Environment	Task type
[45]	Extensive overview of the original ABC and its variants	Convergence speed	CloudSim	Independent
[46]	HBB-LB method	Makespan, Load balancing, mean wait time, degree of imbalance	CloudSim	Independent
[47]	ABC_FCFS, ABC_LJF, and ABC_SJF in CloudSim	Average makespan	CloudSim	Independent
[14]	Honey bee galvanized load balancing (HBBLB) in CloudSim	Makespan, response time, degree of imbalance, resource utilization	CloudSim	Independent
[48]	Generalized assignment problem (GAP)	Makespan	CloudSim	Independent
[13]	Accelerated-ABC (A-ABC) algorithm	Convergence performance and benchmark function	CloudSim	Independent
[49]	ABC-SA is compared with ABC	Convergence	CloudSim	Independent
[50]	BCO including the strategies for BCO parallelization	Convergence	CloudSim	Independent
[51]	gbest-guided ABC algorithm with benchmark function	Convergence	CloudSim	Independent
[52]	Chaotic reverse learning	Convergence and benchmark functions	CloudSim	Independent
[53]	Self-adaptive ABC	Convergence	CloudSim	Independent

2.5 Genetic Algorithm (GA)

GA described by John Holland in 1960s. Genetic Algorithm targets on NP-hard problems. GA is a nature-inspired algorithm that generates the population which includes individual solution called chromosomes. This population is evolved with generations such that chromosome evaluation is done by fitness evaluation. For developing next

Pseudo code of PSO

1: Set dimension = task set where t ϵ T

2: Initialize position and velocity. [Initialize the population generation.]

3: For individual particle, compute fitness

Cost (M) = max (M_j) ∀j ∈ P (1)

Where $C_{total}(M_j) = C_{exe}(M_j) + C_{tx}(M_j)$

Such that $C_{exe}(M_j)$ specifies total cost of task list and $C_{tx}(M_j)$is access cost.

4: If fitness of pbest is greater, then current fitness = new pbest.

5: Select best particle =gbest.

6: calculate velocity and positions of each particle

$$v_i(k+1) = w * v_i(k) + c_1 * r_1 * (pbest(k) - x_i(k) + c_2 * r_2 * gbest(k) - x_i(k)) \quad (2)$$

$$x_i(\mathrm{k}+1) = x_i(\mathrm{k}) + v_i(\mathrm{k}+1)$$

Where, v_i (k) and x_i (k) are velocity, $r_1$and r_2 are random number, c_1 and c_2 are cognition and w= inertia weight.

7: else, repeat step 3

Fig. 5 PSO algorithm

generation, new chromosomes are developed by merging two chromosomes from present generation with the help of mutation (Fig. 6).

Genetic algorithm is implemented to optimization problems such as graph coloring, pattern recognition, traveling salesman, and other problems. An optimized genetic algorithm is proposed that schedules the independent tasks in [21]. Authors [22] propose genetic algorithm. The implementation of the proposed algorithm proves that it outperforms the standard classic genetic algorithm.

A comparison of various task scheduling algorithms is presented in [23]. Reference [24] enlightens an elaborative idea of genetic algorithm along with its variants developed for task scheduling and GA scheduler is also proposed that generates population using enlarged Max Min which, in turn, reduces the makespan and balance load of resources. Comparison table of GA is shown in Table 5.

Table 4 Comparison of variants of PSO

Referenced work	Improvement strategy	Optimization criteria	Encoding scheme	Initial population generation	Nature of tasks
[32]	An amalgamation of the PSO algorithm and the Cuckoo search (CS) algorithm; called PSOCS	Makespan, utilization, execution time	$1 * n$ vector representation	Randomly	Independent
[17]	Hybrid particle swarm optimization scheduling heuristic	Makespan, resource utilization	$1 * n$ vector representation	Randomly	Independent
[38]	Improved accelerate particle swarm algorithm (IAPSO)	Makespan	Matrix representation	Random	Independent
[39]	Particle swarm optimization based on service and cost (PSO-SC)	Makespan, cost	Matrix representation	Workflow	Independent
[16]	Hybrid discrete particle swarm optimization (HDPSO)	Makespan	Matrix representation	Workflow	Independent
[40]	Load balancing mutation particle swarm optimization (LBMPSO)	Execution time, cost, response time	Matrix representation	Workflow	Independent
[41]	Prune PSO	Total finish time	Matrix representation	Workflow	Independent

(continued)

Table 4 (continued)

Referenced work	Improvement strategy	Optimization criteria	Encoding scheme	Initial population generation	Nature of tasks
[42]	Parallel particle swarm optimization (PPSO)	Execution time, convergence	Matrix representation	Randomly	Independent
[43]	Comparison of PSO and BRS	Makespan, cost, execution time, convergence	Matrix representation	Workflow	Independent
[18]	Discrete particle swarm optimization algorithm	Makespan, flow time, and reliability cost	Matrix representation	Random	Independent
[31]	Algorithm based on PSO and ACO	Execution time	Matrix representation	Random	Independent
[44]	Hybrid of Particle swarm optimization, tabu search, and simulated annealing	Makespan	Matrix representation	Random	Independent

2.6 *Firefly Algorithm*

The bioluminescence leads to the flashing light of fireflies. Researchers highlighted the reason and significance of firefly's flashing light and mating phase [25]. With this flashing light, fireflies attract their mating partner. The firefly attracts each other. It attracts both males and females to each other. Intensity I decreases as distance r increases as per the inverse square law [25]. Air acts as light absorbent such that light becomes dim as distance increases. Firefly algorithm mimics the flashing feature of fireflies. Mainly it follows three idealized rules (Fig. 7):

- The sttractive characteristic of the firefly is equal to the brightness.
- The brightness of firefly gets affected by objective directly [25].

For global optimization problem, brightness is directly proportional to objective. The brightness describes fitness function in other nature-inspired algorithms.

Pseudo code of GA

1. **Start:** In this step, random population of n chromosomes is produced. In the initial phase, VM are not present in the system. Such that VM=0.

2. **Fitness:** Calculate fitness f(x) for chromosome in population. As the number of VMs is added, the algorithm computes its capacity as per equation (1).

3. **New population:** develops the new population till the creation of new generation is done to calculate the optimal mapping result.

3.1. **Selection:** selects parent from the population as per fitness as per equation (2).

3.2. **Crossover:** with the help of the crossover probability, develop the new chromosome

3.3. **Mutation:** Mutate new child with equation (3).

3.4. **Accepting:** new Chromosome has become member.

4. **Replace:** use new generation.

5. goto step 2.

Fig. 6 Pseudocode of genetic algorithm

Fig. 7 Pseudocode of firefly algorithm

Pseudo code of Firefly algorithm
1. Initialize firefly population
2. Initialize absorption coefficient γ of light
3. While (t<Maximum Generations)
4. For i=1: n (number of fireflies)
5. For j=1: I, compute Light intensity I_i at x_i, f(x_i)
6. If (I_i > I_j) , Move firefly I towards j in all d dimensions
7. Else Move firefly randomly.
8. Determine solutions and update intensity
9. Sort fireflies as per light intensity and find the current best

3 Observations

From survey of the above-discussed metaheuristics, the following observations have been made during study.

3.1 Improvement in Optimal Solution with Population

Optimal solutions algorithms like ABC and PSO are enhanced with population generation through different ways. Java-based GA [26] and Modified Genetic Algorithm created initial population with Random, Permutation-based representation to improve makespan and Cost metrics. [27] created random initial particles tournament selection genetic algorithm (TS-GA), whereas [26] used Java-based GA used random, permutation-based representation for population initialization. If the initial population is further enhanced, i.e., their initial values or characteristics such as the processing time and burst time are controlled, significant change in performance rate can be attained in comparison to those normal initial population generated.

3.2 Combination of Metaheuristics for Quality Improvisation

The quality of algorithm can be improvised by mixing other swarm-based algorithms, hence overcoming the drawbacks of the metaheuristics by the strength of other metaheuristics. With this method, the drawback of one algorithm can be covered up by the strength of another algorithm. The genetic algorithm combined with ACO [28, 29] to make GACO that avoids premature convergence leading to local optimal solution. Hence, upgrading the metric leads to efficient resource scheduling. The combination of genetic algorithm with fuzzy theory proposed new technique that improves performance [30]. Algorithm based on PSO and ACO got the advantages of both the algorithms [31]. The new hybrid algorithm can further improve the solution of population-based metaheuristic algorithms.

3.3 Modification in Control Parameter for Quality Improvisation

Researchers have emphasized on the modification of the control parameters of the metaheuristics. For example in ACO and its variants, various control parameters are used in evaporation rate of pheromone. The control parameters are responsible for selection of path by the ants, hence affects the shortest path selection process in ACO.

3.4 Avoiding Premature Convergence

From the research, it is observed that the premature convergence is found in almost all the Meta heuristics that decay the performance of the same. Some conceptualization of control parameter should be made in order to avoid or delay this issue of premature convergence so as to meet the global optimal solution. For example in PSO, the parameters pbest and gbest are responsible for converging the individuals to a single local optimal solution. Hence, exploration decreases and exploitation of the solution increases leading to the individual converging to a single solution and overloading it.

3.5 Providing Quality of Service

In cloud, meeting SLA and QoS guarantee is a challenging job. Due to the dynamic nature of cloud, limitations of cloud worsen the performance. For the desired standard of QoS, Fog computing can fill these gaps. Fog computing is far better than cloud computing in terms of hard real-time, mobility, and bandwidth.

4 Conclusion

Extensive review of the metaheuristic techniques implemented in scheduling in cloud is enlightened. Review done is in the direction of improving the convergence speed and makespan of the metaheuristic. The prevailing issues in metaheuristic techniques are also highlighted. Moreover, various resource management and placement algorithms are focused on solving global optimization problem.

Comparative analysis of metaheuristics compares the technique used in improving the performance, optimization criteria, nature of tasks, and the environment. The challenge is to minimize the makespan without degrading performance and violating SLA constraints.

Acknowledgements The author thank Mr. Sukhwinder Singh for his support, help, and guidance. We extend our gratitude toward for sparing his valuable time in helping us.

References

1. Kalra, M., Singh, S.: A review of metaheuristic scheduling techniques in cloud computing. Egypt. Inform. J. **16**, 275–295 (2015) http://dx.doi.org/10.1016/j.eij.2015.07.001
2. Lin, W., Wu, W., Wang, J.Z.: A heuristic task scheduling algorithm for heterogeneous virtual clusters. Hindawi Publ. Corp. Sci. Program. **2016**, Article ID 7040276 (2016). http://dx.doi.org/10.1155/2016/7040276
3. Tawfeek, M., El-Sisi, A., Keshk, A., Torkey, F.: Cloud task scheduling based on ant colony optimization. Int. Arab J. Inf. Technol. **12**(2), 129–137 (2015)
4. Artificial Bee Colony Optimized Scheduling Framework based on resource service availability in Cloud Manufacturing. In: 2014 International Conference on Service Sciences. 2165-3828/14 $31.00 © 2014 IEEE https://doi.org/10.1109/icss.2014.16

5. Ju-Hua, W.: Research of resource allocation in cloud computing based on improved dual bee colony algorithm. Int. J. Grid Distrib. Comput. **8**(5), 117–126 (2015). http://dx.doi.org/10.14257/ijgdc.2015.8.5.11 ISSN: 2005-4262 IJGDC Copyright © 2015 SERSC, ISSN: 2005-4262 IJGDC Copyright © 2015 SERSC
6. Seddigh, M., Taheri, H., Sharifian, S.: Dynamic prediction scheduling for virtual machine placement via ant colony optimization, SPIS2015, 16–17 Dec 2015, Amirkabir University of Technology, Tehran, IRAN, 978-1-5090-0139-2/15/$31.00 ©2015 IEEE, pp. 104–109
7. Salah Farrag, A.A., Mahmoud, S.A., EI Sayed, M., EI-Horbaty: intelligent cloud algorithms for load balancing problems: a survey. In: 2015 iEEE Seventh International Conference on Intelligent Computing and Information Systems (ICiCIS 'J 5), pp. 210–217 (2015)
8. Liu, C.-Y., Zou, C.-M., Wu, P.: A task scheduling algorithm based on genetic algorithm and ant colony optimization in cloud computing. In: 2014 13th International Symposium on Distributed Computing and Applications to Business, Engineering and Science, 978-1-4799-4169-8/14 $31.00 © 2014 IEEE https://doi.org/10.1109/dcabes.2014.18
9. Bolaji, A.L., Khader, A.T., Al-betar, M.A., Awadallah, M.A.: Artificial bee colony algorithm, its variants and applications: a survey. J. Theor. Appl. Inf. Technol. **47**(2), 234–259, 20 Jan 2013. © 2005 – 2013 JATIT & LLS., ISSN: 1992-8645 www.jatit.org E-ISSN: 1817-3195 434
10. Wen, X., Huang, M., Shi, J.: Study on resources scheduling based on ACO algorithm and PSO algorithm in cloud computing. In: 2012 11th International Symposium on Distributed Computing and Applications to Business, Engineering & Science, pp. 219–223, 978-0-7695-4818-0/12 $26.00 © 2012 IEEE https://doi.org/10.1109/dcabes.2012.63
11. Yu, X., Zhang, T.: Convergence and runtime of ant colony optimization model. Inf. Technol. J. **8**(3), 354–359, ISSN1812-5638 (2009)
12. van Ast, J., Babuška, R., De Schutter, B.: Convergence Analysis of Ant Colony Learning. Technical report 11-012
13. Ozkis, A., Babalik, A.: Accelerated ABC (A-ABC) algorithm for continuous optimization problems. Lect. Notes Softw. Eng. **1**(3), 262–266, August 2013, https://doi.org/10.7763/lnse.2013.v1.57
14. Jyothi, D., Anoop, S., Jyothi, D., et al.: Bio-inspired scheduling of high performance computing applications in cloud: a review. (IJCSIT) Int. J. Comput. Sci. Inf. Technol. **6**(1), 485–487 (2015), ISSN- 0975-9646
15. Rathore, M., Rai, S., Saluja, N., Rathore, M., et al.: Load balancing of virtual machine using honey bee galvanizing algorithm in cloud. (IJCSIT) Int. J. Comput. Sci. Inf. Technol. **6**(4), 4128–4132 (2015), ISSN- 0975-9646
16. Devi, P., Kalra, M.: Workflow scheduling using hybrid discrete particle swarm optimization (HDPSO) in cloud computing environment. Int. J. Innov. Res. Comput. Commun. Eng. **3**(12), 12301–12307 (An ISO 3297: 2007 Certified Organization) December 2015 Copyright to IJIRCCE https://doi.org/10.15680/ijircce.2015.031205912301, ISSN(Online): 2320-9801 ISSN (Print): 2320-9798, https://doi.org/10.15680/ijircce.2015.0312059
17. Gomathi, B., Krishnasamy, K.: Task scheduling algorithm based on hybrid particle swarm optimization in cloud computing environment. J. Theor. Appl. Inf. Technol. **55**(1), 33–38, 10 Sept 2013, © 2005 – 2013 JATIT & LLS. All rights reserved. ISSN: 1992-8645 www.jatit.org E-ISSN: 1817-3195
18. Sarathambekai, S., Umamaheswari, K.: Intelligent discrete particle swarm optimization for multiprocessor task scheduling problem. J. Algorithms Comput. Technol., 1–10. https://doi.org/10.1177/1748301816665521
19. Thaman, J., Singh, M.: Current perspective in task scheduling techniques in cloud computing: a review. Int. J. Found. Comput. Sci. Technol. (IJFCST) **6**(1), 65–85, (2016). https://doi.org/10.5121/ijfcst.2016.6106
20. Sharma, A., Tyagi, S.: Differential evolution—GSA based optimal task scheduling in cloud computing. IJESRT Int. J. Eng. Sci. Res. Technol., 1447–1451. IC™ Value: 3.00 Impact Factor: 4.116 http://www.ijesrt.com © International Journal of Engineering Sciences & Research Technology [1447]

21. Zhao, C., Zhang, S., Liu, Q.: Independent Tasks Scheduling Based on Genetic Algorithm in Cloud Computing, 978-1-4244-3693-4/09/$25.00 ©2009 IEEE
22. Kumar, P., Verma, A.: Independent task scheduling in cloud computing by improved genetic algorithm. Int. J. Adv. Res. Comput. Sci. Softw. Eng. IJARCSSE **2**(5), 111–114 May 2012 ISSN: 2277 128X (2012)
23. .
24. Singh, M., Marken, R.: A survey on task scheduling optimization in cloud computing. Int. J. Adv. Res. Comput. Sci. Softw. Eng. **6**(5), 850–855 (2016) ISSN: 2277 128X
25. Arora, S., Singh, S.: The firefly optimization algorithm: convergence analysis and parameter selection. Int. J. Comput. Appl. (0975 – 8887) **69**(3), 48–52 (2013)
26. Kaur, S., Verma, A.: An efficient approach to genetic algorithm for task scheduling in cloud computing environment. I.J. Inf. Technol. Comput. Sci. **10**, 74–79 Published Online September 2012 in MECS (http://www.mecs-press.org/) https://doi.org/10.5815/ijitcs.2012.10.09 Copyright © 2012 MECS
27. Hamad, S.A., Omara, F.A.: Genetic-based task scheduling algorithm in cloud computing environment. (IJACSA) Int. J. Adv. Comput. Sci. Appl. **7**(4), 550–556 (2016)
28. Xianfeng, Y., HongTao, L.: Load balancing of virtual machines in cloud computing environment using improved ant colony algorithm. Int. J. Grid Distrib. Comput. **8**(6), 19–30 (2015). http://dx.doi.org/10.14257/ijgdc.2015.8.6.03, ISSN: 2005-4262 IJGDC Copyright © 2015 SERSC
29. Zhang, D.: Convergence analysis for generalized ant colony optimization algorithm. In: Proceedings of the 11th Joint Conference on Information Sciences, pp. 1–6. Atlantis Press (2008)
30. .
31. Wang, C., Chen, K.: Research on the task scheduling algorithm optimization based on hybrid PSO and ACO in cloud computing. Comput. Model. New Technol. **17**(5A), 12–16 (2013)
32. Al-maamari, A., Omara, F.A.: Task scheduling using hybrid algorithm in cloud computing environments. IOSR J. Comput. Eng. (IOSR-JCE) **17**(3), 96–106. e-ISSN: 2278-0661, p-ISSN: 2278-8727, Ver. VI (May–Jun. 2015), www.iosrjournals.org https://doi.org/10.9790/0661-173696106 www.iosrjournals.org
33. Huang, H., Wu, C.-G., Hao, Z.-F.: A pheromone-rate-based analysis on the convergence time of ACO algorithm. IEEE Trans. Syst. Man Cybern. Part B Cybern. **39**(4), 910–924, August 2009, 1083-4419/$25.00 © 2009 IEEE
34. Zhu, L., Li, Q., He, L.: Study on cloud computing resource scheduling strategy based on the ant colony optimization algorithm. IJCSI Int. J. Comput. Sci. Issues **9**(5, 2), 54–58. September 2012 ISSN (Online): 1694-0814 www.IJCSI.org
35. Maruthanayagam, D., Arun Prakasam, T.: Job scheduling in cloud computing using ant colony optimization. Int. J. Adv. Res. Comput. Eng. Technol. (IJARCET) **3**(2), 540–547, February 2014 540 ISSN: 2278 – 1323 All Rights Reserved © 2014 IJARCET ISSN: ISSN 2278 – 1323 All Rights Reserved © 2014
36. Brintha, N.C., Winowlin Jappes, J.T., Benedict, S.: A modified ant colony based optimization for managing cloud resources in manufacturing sector. ISSN 978-1-4673-6615-1/16/$31.00 © 2016 IEEE
37. Sun, W., Ji, Z., Sun, J., Zhang, N., Hu, Y.: SAACO: a self adaptive ant colony optimization in cloud computing. In: 2015 IEEE Fifth International Conference on Big Data and Cloud Computing, CFP1552Z-CDR/15 $31.00 © 2015 IEEE, pp. 148–154 (2015). https://doi.org/10.1109/bdcloud.2015.53
38. Li, Z., Wang, C., Lv, H., Xu, T.: Application of PSO algorithm based on improved accelerating convergence in task scheduling of cloud computing environment. Int. J. Grid Distrib. Comput. **9**(9), 269–280 (2016). http://dx.doi.org/10.14257/ijgdc.2016.9.9.23 ISSN: 2005-4262 IJGDC Copyright © 2016 SERSC ISSN: 2005-4262 IJGDC Copyright © 2016 SERSC
39. Xue, S., Shi, W., Xu, X.: A heuristic scheduling algorithm based on PSO in the cloud computing environment. Int. J. u- and e- Serv. Sci. Technol. **9**(1), 349–362 (2016). http://dx.doi.org/10.14257/ijunesst.2016.9.1.36 ISSN: 2005-4246 IJUNESST Copyright © 2016 SERSC, ISSN: 2005-4246 IJUNESST Copyright © 2016 SERSC

40. Jaglan, P., Diwakar, C.: Partical swarm optimization of task scheduling in cloud computing. IJESRT Int. J. Eng. Sci. Res. Technol, 833–840
41. HaghNazar, R., Rahmani, A.M.: Prune PSO: A new task scheduling algorithm in multiprocessors systems. In: International Conference on Networking and Information Technology, 978-1-4244-7578-0/$26.00 © 2010 IEEE, pp. 161–165
42. Singh, S.K., Kumar, R.: Scheduling in multiprocessor systems using parallel PSO. In: International Conference on Computing, Communication and Automation (ICCCA2015) ISBN: 978-1-4799-8890-7/15/$31.00 ©2015 IEEE
43. .
44. Zhang, X.-F., Koshimura, M., Fujita, H., Hasegawa, R.: Hybrid particle swarm optimization and convergence analysis for scheduling problems. In: GECCO'12 Companion, pp. 307–314, 7–11 July 2012, Philadelphia, PA, USA. 2012 ACM 978-1-4503-1178-6/12/07 …$10.00
45. Angel Preethima, R., Johnson, M.: Survey on optimization techniques for task scheduling in cloud environment. Int. J. Adv. Res. Comput. Sci. Softw. Eng. Res. Pap. **3**(12), 413–416, December 2013 ISSN: 2277 128X. Available online at: www.ijarcsse.com, © 2013, IJARCSSE All Rights Reserved
46. Hesabian, N., Javadi, H.H.S.: Optimal scheduling in cloud computing environment using the bee algorithm. Int. J. Comput. Netw. Commun. Secur. **3**(6), 253–259, June 2015. Available online at: www.ijcncs.org E-ISSN 2308-9830 (Online)/ISSN 2410-0595 (Print)
47. Kruekaew, B., Kimpan, W.: Virtual machine scheduling management on cloud computing using artificial bee colony. In: Proceedings of the International MultiConference of Engineers and Computer Scientists 2014, vol. I, IMECS 2014, 12–14 Mar 2014, Hong Kong, ISBN: 978-988-19252-5-1 ISSN: 2078-0958 (Print); ISSN: 2078-0966 (Online) IMECS 2014
48. Baykasoğlu, A., Özbakır, L., Tapkan, P.: Artificial bee colony algorithm and its application to generalized assignment problem. In: Chan, F.T.S., Tiwari, M.K. (eds.) Swarm Intelligence: Focus on Ant and Particle Swarm Optimization, pp. 113–144, 532, ISBN 978-3-902613-09-7 (2007)
49. Yurtkuran, A., Emel, E.: An enhanced artificial bee colony algorithm with solution acceptance rule and probabilistic multisearch. Comput. Intell. Neurosci. (2015)
50. Davidović, T., Teodorović, D.: Bee colony optimization part I: the algorithm overview. Yugoslav J. Oper. Res. **25**(1), 33–56 (2015). https://doi.org/10.2298/yjor131011017d Invited survey
51. Zhu, G., Kwong, S.: Gbest-guided artificial bee colony algorithm for numerical function optimization. Appl. Math. Comput. ISSN 0096-3003/$—see front matter_2010 Elsevier Inc. All rights reserved. https://doi.org/10.1016/j.amc.2010.08.049
52. Liu, W.: A multistrategy optimization improved artificial bee colony algorithm. Sci. World J. 2014: 129483. Published online 2014 Apr https://doi.org/10.1155/2014/129483 PMCID: PMC3997130
53. Bacanin, N., Stanarevic, N.: Guided artificial bee colony algorithm Milan TUBA. In: Proceedings of the European Computing Conference, pp. 398–404, ISBN: 978-960-474-297-4
54. Zhaofeng, Y., Aiwan, F.: Application of ant colony algorithm in cloud resource scheduling based on three constraint conditions. Adv. Sci. Technol. Lett. **123**(CST 2016), 215–219 (2016). http://dx.doi.org/10.14257/astl.2016.123.40, ISSN: 2287-1233 ASTL Copyright © 2016 SERSC
55. Lin, J., Zhong, Y., Lin, X., Lin, H., Zeng, Q.: Hybrid Ant Colony Algorithm Clonal Selection in the Application of the Cloud's Resource Scheduling. arXiv:1411.2528v1 [cs.DC] 10 Nov 2014
56. Jemina Priyadarsini, R., Arockiam, L.: A Framework to Optimize Task Scheduling in Cloud Environment. ISSN 0975-9646
57. Selvaraj, S., Jaquline, J., Selvaraj, S., et al.: Ant colony optimization algorithm for scheduling cloud tasks. Int. J. Comput. Technol. Appl. **7**(3), 491–494 IJCTA| May-June 2016 Available online@www.ijcta.com 492, iSSN:2229-6093, May-June 2016
58. Mod, P., Bhatt, M.: ACO based dynamic resource scheduling for improving cloud performance. Int. J. Sci. Eng. Technol. Res. (IJSETR) **3**(11), 3012–3018, November 2014, ISSN: 2278 – 7798 All Rights Reserved © 2014 IJSETR

59. Ma, H., Zhang, M.: An improved genetic-based approach to task scheduling in inter- cloud environment. Int. J. Comput. Sci. Softw. Eng. (IJCSSE) **5**(3), 28–35. March 2016 ISSN (Online): 2409-4285 www.IJCSSE.org
60. Abdullahi, M., Ngadi, M.A.: Hybrid symbiotic organisms search optimization algorithm for scheduling of tasks on cloud computing environment, 1–29. https://doi.org/10.1371/journal.pone.0158229 June 27, 2016
61. Zhang, D.: Convergence analysis for generalized ant colony optimization algorithm. In: Proceedings of the 11th Joint Conference on Information Sciences, pp. 1–6 (2008) Published by Atlantis Press © the authors
62. Kaushal, J.: Advancements and applications of ant colony optimization: a critical review. Int. J. Sci. Eng. Res. **3**(6), 1–5. June-2012 1 ISSN 2229-5518 IJSER © 2012 http://www.ijser.org
63. Blum, C.: Ant colony optimization: introduction and recent trends, pp. 353–373. ISSN 1571-0645/$ – see front matter 2005 Elsevier B.V. All rights reserved. https://doi.org/10.1016/j.plrev.2005.10.001
64. George, S.: Truthful workflow scheduling in cloud computing using hybrid PSO-ACO. In: 2015 International Conference on Developments of E-Systems Engineering, 978-1-5090-1861-1/15 $31.00 © 2015 IEEE https://doi.org/10.1109/dese.2015.62
65. Al Buhussain, A., Robson, E., De Grande, Boukerche, A.: Performance analysis of bio-inspired scheduling algorithms for cloud environments. In: 2016 IEEE International Parallel and Distributed Processing Symposium Workshops/16 $31.00 © 2016 IEEE https://doi.org/10.1109/ipdpsw.2016.186pp. 776–786. 978-1-5090-3682-0/16 $31.00 © 2016 IEEE
66. Lin, R., Li, Q.: Task scheduling algorithm based on pre-allocation strategy in cloud computing. In: 2016 IEEE International Conference on Cloud Computing and Big Data Analysis, 978-1-5090-2594-7116/$31.00 ©2016 IEEE
67. Ku, H.-H., Huang, S.-Y.: Digital Convergence Services for Situation-aware POI Touring, pp. 108–116. 978-1-4799-2652-7/14 $31.00 © 2014 IEEE https://doi.org/10.1109/waina.2014.27
68. Sun, W., Zhang, N., Wang, H., Yin, W., Qiu, T.: PACO: a period ACO-based scheduling algorithm in cloud computing. In: 2013 International Conference on Cloud Computing and Big Data, 978-1-4799-2829-3/13 $26.00 © 2013 IEEE https://doi.org/10.1109/cloudcom-asia.2013.85
69. Tawfeek, M., El-Sisi, A., Keshk, A., Torkey, F.: Cloud task scheduling based on ant colony optimization. Int. Arab J. Inf. Technol. **12**(2), 129–138 (2015)
70. Duan, P., Ai, Y.: Research on an improved ant colony optimization algorithm and its application. Int. J. Hybrid Inf. Technol. **9**(4), 223–234 (2016). http://dx.doi.org/10.14257/ijhit.2016.9.4.20 ISSN: 1738-9968 IJHIT Copyright © 2016 SERSC, ISSN: 1738-9968 IJHIT Copyright © 2016 SERSC
71. Kaur, N., Kumar, A.: Dependent task scheduling with artificial bee colony optimization. J. Innov. Comput. Sci. Eng. **6**(1), 46–51 July–Dec 2016@ ISSN 2278-0947
72. Benali, A., El Asri, B., Kriouile, H.: A pareto-based artificial bee colony and product line for optimizing scheduling of VM on cloud computing. 978-1-4673-8149-9/15/$31.00 ©2015 IEEE
73. Karaboga, D., Akay, B.: A comparative study of artificial bee colony algorithm. Appl. Math. Comput., 108–132, 0096-3003/$—see front matter_2009 Elsevier Inc. All rights reserved. https://doi.org/10.1016/j.amc.2009.03.090
74. Karaboga, D., Gorkemli, B., Ozturk, C., Karaboga, N.: A Comprehensive Survey: Artificial Bee Colony (ABC) Algorithm and Applications. Springer Science + Business Media B.V. (2012). https://doi.org/10.1007/s10462-012-9328-0
75. Gunasekaran, S., Sonialpriya, S.: Licensed under creative commons attribution CC BY comparison of advanced optimization algorithm for task scheduling in cloud computing. Int. J. Sci. Res. (IJSR) **4**(3), 1572–1577, ISSN (Online): 2319-7064 Index Copernicus Value (2013): 6.14 | Impact Factor (2013): 4.438 www.ijsr.net

76. Gupta, T., Kumar, D.: Optimization of clustering problem using population based artificial bee colony algorithm: a review. Int. J. Adv. Res. Comput. Sci. Softw. Eng. Res. Pap. **4**(4), 491–502, April 2014 ISSN: 2277 128X. Available online at: www.ijarcsse.com, IJARCSSE All Rights Reserved
77. Singh, A.: A Survey on cloud computing and various scheduling algorithms. Volume 4, Issue 2, February 2016 Int. J. Adv. Res. Comput. Sci. Manag. Stud. Res. Artic. Surv. Pap. Case Study **4**(2), 209–212. Available online at: www.ijarcsms.com ISSN: 2321-7782 (Online) 2016, IJARCSMS
78. Ahluwalia, A., Sharma, V.: Differential evolution based optimal task scheduling in cloud computing. Int. J. Adv. Res. Comput. Sci. Softw. Eng. IJARCSSE **6**(6), 340–347 (2016) ISSN: 2277 128X
79. Durga Lakshmi, R., Srinivasu, N.: A dynamic approach to task scheduling in cloud computing using genetic algorithm. J. Theor. Appl. Inf. Technol. **85**(2), 124–135, 20 Mar 2016. © 2005 – 2016 JATIT & LLS. All rights reserved. ISSN: 1992-8645 www.jatit.org E-ISSN: 1817-3195
80. Kaleeswaran, A., Ramasamy, V., Vivekanandan, P.: Dynamic scheduling of data using genetic algorithm in cloud computing. Int. J. Adv. Eng. Technol. **5**(2), 327–334 (2013). ©IJAET ISSN: 2231-1963
81. Javanmardi, S., Shojafar, M., Amendola, D., Cordeschi, N., Liu, H., Abraham, A.: Hybrid Job scheduling Algorithm for Cloud computing Environment, adfa, p. 1. © Springer, Berlin (2014)
82. Wang, T., Liu, Z., Chen, Y., Xu, Y., Dai, X.: Load balancing task scheduling based on genetic algorithm in cloud computing. In: 2014 IEEE 12th International Conference on Dependable, Autonomic and Secure Computing, pp. 146–152, 978-1-4799-5079-9/14 $31.00 © 2014 IEEE https://doi.org/10.1109/dasc.2014.35
83. Wang, B., Li, J.: Load balancing task scheduling based on multi-population genetic algorithm in cloud computing. In: Proceedings of the 35th Chinese Control Conference, 27–29 July 2016, Chengdu, China, pp. 5261–5266
84. Varghese, B.M., Joshua Samuel Raj, R.: A survey on variants of genetic algorithm for scheduling workflow of tasks. In: 2016 Second International Conference on Science Technology Engineering and Management (ICONSTEM), pp. 489–492, 978-1-5090-1706-5/16/$31.00 ©2016 IEEE
85. Savitha, P., Geetha Reddy, J.: A review work on task scheduling in cloud computing using genetic algorithm. Int. J. Sci. Technol. Res. **2**(8), 241–245, August 2013, ISSN 2277-8616 241 IJSTR©2013 www.ijstr.org
86. Kaur, R., Kinger, S.: Enhanced genetic algorithm based task scheduling in cloud computing. Int. J. Comput. Appl. (0975 – 8887) **101**(14) (2014)

An Analytical Approach to Document Clustering Techniques

Vikas Choubey and Sanjay Kumar Dubey

Abstract Clustering is a technique that group data together based on their similarity and apart based on their dissimilarity. When this technique is applied to documents and the terms within these documents retrieval of similar documents become easy and efficient. Document clustering is being researched and utilized for many years but is yet far from being optimal. To study and analyze different document clustering algorithm, a theoretical literature review and analysis was performed and the results are presented in this paper. This paper comprises of theoretical review of papers. 95 papers were identified and out of these 30 were selected. Various techniques or algorithms and modifications to previous algorithms proposed for document clustering by various researchers are compiled and presented with the intent that it will aid the researchers in finding out the current and future scope of research in information retrieval systems and document clustering technologies.

Keywords Document clustering · Information retrieval system · Clustering · K-Means · Precision · Algorithm

1 Introduction

The main objective is to analyze the text document clustering techniques used in information retrieval systems. The clustering technique is crucial and dictates the performance and overhead required for retrieval. Performance metrics used in information retrieval systems are called recall and precision. Better clustering means better recall, better precision, and lesser overhead. Users enter search queries to retrieve the required documents. These queries are then interpreted by the information retrieval systems and based on this interpretation, documentsthat correlate with the search queries, to satisfy the users' information needs, are retrieved from the databases and

V. Choubey (✉) · S. K. Dubey
Department of Computer Science & Engineering, Amity University, Noida, Uttar Pradesh, India
e-mail: vikas.cby@gmail.com

S. K. Dubey
e-mail: sanjukundan@gmail.com

M. Tuba et al. (eds.), *ICT Systems and Sustainability*,
Advances in Intelligent Systems and Computing 1077,
https://doi.org/10.1007/978-981-15-0936-0_3

presented to users for further relevance inspections. This mapping of the user-entered search queries to documents present in the databases can be refined if the clustering of the documents is done effectively. This paper reviews document clustering techniques used in information retrieval systems. Mostly, documents have moderate length, address a solitary topic, and utilize unexceptional language. These documents have been the focal point of the information retrieval systems and data mining network for quite a long time. Therefore, the vast majority of the document clustering algorithms to date relate to clustering these standard documents. The objective of clustering was to aid the locating of data. Clustering of words started with age of thesauri. The objective of clustering is to give a collection of comparable items into a "class" under a general title. The term "class" is every now and then utilized as an equivalent word for the term "Cluster". Clustering is the division of information into gatherings of comparative items. Each gathering, called bunch comprises items that are similar among themselves and unlike the objects of alternate gatherings. At the end of the day, the objective of a decent record clustering plan is to limit the intra-cluster distance between documents, while maximizing the inter-cluster distances. A distance measure lies at the core of the document clustering. Clustering is the most widely recognized type of unsupervised learning and this is the real distinction among clustering and characterization. No supervision implies that there is no human master who allocated documents to classes. In clustering it is the appropriation and cosmetics of the information that will decide group enrollment. Clustering once in a while incorrectly alluded to as programmed grouping, notwithstanding, this is off base since the clusters found are not known before the handling though if there should arise an occurrence of characterization the classes are pre-characterized. In clustering, it is conveyance and the idea of information that will decide the group assignment, contrary to the order where the classifier takes in the relationship among items and classes from a training set.

The process of clustering documents follows the following steps:

a. Define the area for the clustering process. In the event that thesaurus is being made, this equates to deciding the extent of the thesaurus, for example, "processor terms". In the event that document clustering is performed, it is the assurance of the arrangement of items to be clustered. Characterizing space for the clustering recognizes those items to be utilized in the grouping procedure and decrease the potential for incorrect information that could initiate mistakes in the clustering procedure.
b. Once the area is resolved, decide the ascribes of the objects to be clustered. In the event that a thesaurus is being created, decide the explicit words in the items to be utilized in the grouping procedure. Essentially, if reports are being clustered, the clustering procedure may concentrate on explicit zones inside the items.
c. Determine the quality of the relationships between the traits whose co-event in objects proposes those items ought to be in a similar class. For thesauri, this is figuring out which words are equivalent words and the quality of their relationships. For documents it might characterize a closeness work dependent on words co-events that decide the comparability between two items.

d. At this point, the absolute arrangement of items and the quality of the relationship between the items have been resolved. The last step is applying an algorithm to decide the class(es) to which every item will be assigned.

2 Methodology

2.1 Research Questions

The aim of this review is to find out the various document clustering techniques, analyze their properties, and find the challenges and issues in the document clustering process as a whole. Research papers from IEEE Explore, ACM digital libraries Springer and Elsevier were considered for this study. Various keywords related to text document clustering algorithm, clustering techniques, and information retrieval system were used for retrieving related papers.

For this study the research questions mentioned below were framed:

RQ1. What are the categories of clustering techniques and what are the various document clustering techniques present? [Literature Review]
RQ2. What challenges and issues do engineers face during the entire process of document clustering? [Challenges and Issues section]

These questions allow identifying research gaps in the related area.

3 Literature Review

3.1 Categories of Clustering Techniques in Information Retrieval Systems

(a) Agglomerative clustering—is the bottom-up approach of clustering where clusters are paired up to form a parent cluster and this process continues until there is a single parent cluster which is a combination of all child clusters.
(b) Divisive clustering—is the top-down approach of clustering and completely opposite to agglomerative clustering. All data starts in one cluster and is divided into sub-clusters recursively.
(c) Monothetic clustering—to calculate the distance between observations every feature is considered sequentially, i.e., one after the other.
(d) Polythetic clustering—to calculate the distance between observations all the features are considered simultaneously, i.e., all at once.
(e) Hard clustering—each pattern is allocated completely to a single cluster.

(f) Fuzzy clustering—degree of membership is used to cluster patterns into multiple clusters.
(g) Deterministic clustering—features of a pattern lead to the same cluster every time.
(h) Stochastic clustering— the same features of a pattern lead to different cluster every time.
(i) Incremental clustering—when the algorithm is not optimized temporally and/or spatially it is used in an incremental fashion.
(j) Non-incremental clustering—when the algorithm is designed optimally such that it can perform clustering in one iteration.
(k) Single-link clustering—joins two patterns containing the two closest documents that are not yet in the same cluster.
(l) Complete-link clustering—joins two patterns with the minimum most-distant pair of documents. In this way, clusters are kept small and compact since all documents within a cluster have a maximum distance to each other.
(m) Group Average clustering—joins two patterns with the minimum average document distance [1].
(n) Partitional clustering—certain greedy heuristics are used to divide data into several subsets, this results in high-quality cluster with appropriate data.

Anbarasi et al. proposed K-Means which has precise recall and precision values and is fit for large data [2]. Sedding proposed Bisecting K-Means, high-quality purity and precision, entropy has brought down value, subsequently better grouping and its computational cost is low [3]. Sarkar used PSO + K-Means which proved to be better than K-Means and PSO individually, lower intra-cluster and high inter-cluster and supports parallel computation [4]. Aktar et al. proposed Genetic algorithm + K-Means, it is 12.5% better than GA, 50% better than K-Means, it can be used to solve multidimensional problems [5]. Meena et al. proposed Genetic algorithm + DDE clustering which uses fewer iterations and produces better fitness values. It solves multidimensional problems also [6]. Trappey et al. proposed Fuzzy + Ontological semantic clustering which has 25% better precision and 11% better recall, it also supports cluster overlapping [7]. Thilagavathi et al. proposed K-Means + Semantic clustering and produced 20% better recall and 10% better precision. It is simple and efficient [8]. Baghel et al. proposed Frequent Concept-Based clustering, but it only works for small data. It is computationally tough but easy to understand [9]. How Jing et al. used Naïve Bayes which produced sufficient precision and recall. It uses Bayes theorem of probability [10]. Reddy used Cliques and produced better precision and recall, but redundant intermediate clusters and thus requires a lot of memory, used Connected Components which is Single-link and produced huge clusters, therefore, has reduced precision, used Stars algorithm which had the drawback of producing different clusters with same data depending on seed, therefore, it requires seed selection optimization, Reddy et al. also used Strings algorithm which had the same drawbacks as the Stars algorithm [11]. Abualigah et al. used objective functions + hybrid Krill herd algorithm for clustering, 2 objective functions used for clustering, initial solution derived from K-Means, it had accuracy, precision, recall, F-measure, and convergence behavior better than other clustering algorithms [12]. Lydia et al. proposed KNMF + key phrase extraction, it was inspired by recently updated NMF, Key Phrase Extraction Algorithm + Iterated Lovin's stemming provides 5% more reduction, thus 60% document terms get minimized to roots [13]. Altameem et al. proposed Improved monkey search + hybrid fuzzy set, with its feature selection method improved features reduction and better selection, therefore, improved overall efficiency [14]. Dalal et al. proposed PSO + Semantic Graph algorithm and facilitated automatic text summarization, extracts subject-verb-object triples, then used in semantic graphs. PSO trained classifier used to obtain text summary [15]. Ahmad et al. used Word Movers Distance, it is a Pipeline architecture, vector representation of words-based algorithm and it performs best with $F1$-score = 90% [16]. Lakshmi et al. proposed DIC DOC K-means which had dissimilarity-based initial centroid selection and it performs better than K-means, K-means and WAI + K-means [17]. Megarchioti et al. used Big K Clustering + Hadoop Map Reduce technique which is a variation of K-means + Hadoop for large documents and performs better than K-means [18]. Ibraheem, et al. proposed Memetic Algorithm, Memetic scheme of hybrid feature selection applied on K-means and Spherical K-means, also combines Genetic Algorithm-based wrapper FS + Relief-F filter which improves result of K-means and Spherical K-means [19]. Zhu et al. proposed CARDBK, it solves the initial centroid selection of K-means by sorting centroids by their distances. It is a stable, linearly scalable and fast algorithm [20]. Mohammad, et al. used Krill Herd which is Novel Swarm-based optimization and has better performance than other optimization algorithms [21]. Akter et al. used GA, introducing variation in the cluster centroids during crossover operations, also partition the population and run the genetic algorithm on each of them. It frees the population from being trapped in local minima, also partitions the population and runs the genetic algorithm on each of them [22]. Chen proposed FSTVM + K-Means algorithm in which FSTVM is used to represent the documents, automatic dimension reduction, reconfigured similarity measuring formula. It has high-quality initial centroids and clustering results are steadier [23]. Al-Jadir proposed ACMDHS in which differential evolution harmonic search is used, crossover and mutation probabilities are dynamically tuned and to enhance local search capabilities memetic optimization. This technique is almost at par with its nonadaptive counterpart while both of the above are superior to K-Means, DHS, HS, MDHS, CMDHS, CGABC, DEMC techniques [24]. Seshadri et al. used Singular Value Decomposition-based algorithm. It is a parallel generalization of single value decomposition-based technique and used for hierarchical clustering thereby resulting in levels of clusters instead of flat sets. The automatic number of levels and number of clusters per level decision performs faster than its sequential counterpart, better rand index, f-measure, recall, and precision, can handle large data sets as well [25]. Saini et al. proposed SOMODE. SOM is used to develop new genetic operators, Simultaneous optimization of cluster validity indices namely, Pakhira-Bandyopadhyay-Maulik index and Silhouette index. It performs better than VAMOSA, MOCK, NSGA-II-Clust, SOGA-based clustering technique, single-linkage clustering, and K-Means as measures by statistical test [26]. Rani et al. proposed Context well informed clustering. Clustering based on query analysis and regular updating of new information.

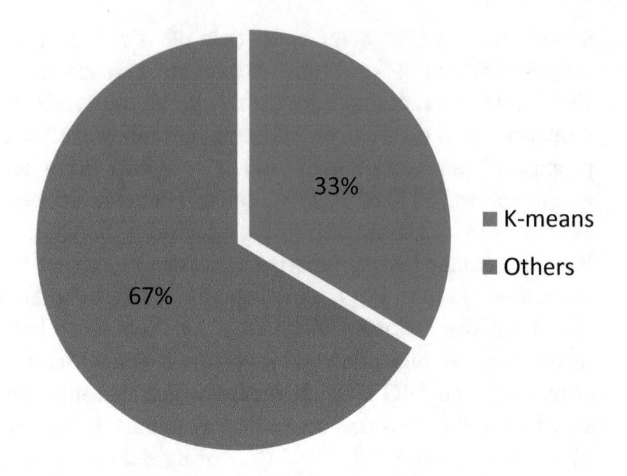

Fig. 1 33% clustering algorithms depend on K-means clustering technique

Applicable to distributed environment, enhanced paradigm for query and context-based clustering [27]. Handa et al. used Ranking method to rank document based on frequent queries an efficient ranking method is used which reduced communication overheads and improved efficiency of 82% in cloud-based system [1]. Gonzàlez et al. proposed Graph-based and probability-based clustering. Non-weighted transformed into a weighted technique, probability-based clustering, and graph-based techniques. Weighted graph-based clustering technique outperformed others [28] (Fig. 1).

4 Challenges and Issues

i. **Context**: The same words possess different meanings in different contexts. Therefore, semantic relations need to be taken into account.
ii. **Dimensionality**: A word that may carry multiple meanings is mapped onto feature space it causes an issue of high Dimensionality.
iii. **Synonyms and Homographs**: when multiple words carry the same meaning and when the same word carries different meaning a need for intensive preprocessing occurs.
iv. **Feature Selection**: selecting features to be considered in similarity measurement.
v. **Similarity Measure**: techniques to be considered for measuring the similarity of documents.
vi. **Algorithm**: algorithm selection which will utilize the similarity measure to create clusters and include documents in those clusters.
vii. **Time and Space**: algorithm used for clustering needs to be efficient in its consumption of space and time so that it is feasible in terms of memory and CPU resources
viii. **Evaluation**: evaluating the clustering method and clusters formed.

5 Conclusion and Future Scope

Upon this literature review and theoretical analysis, it can be concluded that most of the research on document clustering has been carried out using the K-Means technique of clustering. Slight modifications and optimizations have proved to improve the efficiency and accuracy of most algorithms. Feature selection techniques, similarity calculation functions or formulae, initial cluster head or center or centroid determination play an important role in structuring the final result, i.e., the number of clusters and the constituents of those clusters. The choice of the clustering algorithm, optimization techniques, similarity calculation technique, and centroid needs to be carried out only with proper knowledge of the infrastructure to be used for and application of the clusters.

References

1. Handa, R., Rama Krishna, C., Aggarwal, N.: Document clustering for efficient and secure information retrieval from cloud. Concurr. Comput. Pract. Exp. e5127
2. Anbarasi, M.S., et al.: Ontology oriented concept-based clustering. IJRET Int. J. Res. Eng. Technol. **3**(2) (2014)
3. Sedding, J., Kazakov, D.: WordNet-based text document clustering. In: Proceedings of the 3rd Workshop on Robust Methods in Analysis of Natural Language Data. Association for Computational Linguistics (2004)
4. Sarkar, S., Roy, A., Purkayastha, B.S.: A comparative analysis of particle swarm optimization and K-means algorithm for text clustering using Nepali Wordnet. Int. J. Nat. Lang. Comput. (IJNLC) **3**(3) (2014)
5. Akter, R., Chung, Y.: An evolutionary approach for document clustering. IERI Procedia **4**, 370–375 (2013)
6. Meena, K.Y., Singh, P.: Text documents clustering using genetic algorithm and discrete differential evolution. Int. J. Comput. Appl. **43**(1), 0975–8887 (2012)
7. Trappey, A.J.C., et al.: A fuzzy ontological knowledge document clustering methodology. IEEE Trans. Syst. Man Cybern. Part B (Cybern.) **39**(3), 806–814 (2009)
8. Thilagavathi, G., Anitha, J.: Document clustering in forensic investigation by hybrid approach. Int. J. Comput. Appl. **91**(3) (2014)
9. Baghel, R., Dhir, R.: A frequent concepts-based document clustering algorithm. Int. J. Comput. Appl. **4**(5), 6–12 (2010)
10. Jing, H., et al.: Semantic naïve Bayes classifier for document classification. In: Proceedings of the Sixth International Joint Conference on Natural Language Processing (2013)
11. Aggarwal, C.C., Reddy, C.K. (eds.): Data Clustering: Algorithms and Applications. CRC Press, New York (2013)
12. Abualigah, L.M., Khader, A.T., Hanandeh, E.S.: A combination of objective functions and hybrid Krill herd algorithm for text document clustering analysis. Eng. Appl. Artif. Intell. **73**, 111–125 (2018)
13. Lydia, E.L., et al.: Charismatic document clustering through novel K-Means non-negative matrix factorization (KNMF) algorithm using key phrase extraction. Int. J. Parallel Program. 1–19 (2018)
14. Altameem, T., Amoon, M.: Hybrid tolerance rough fuzzy set with improved monkey search algorithm-based document clustering. J. Ambient Intell. Humanized Comput. 1–11 (2018)
15. Dalal, V., Malik, L.: Data Clustering Approach for Automatic Text Summarization of Hindi Documents using Particle Swarm Optimization and Semantic Graph

16. Ahmad, A., Amin, M.R., Chowdhury, F.: Bengali document clustering using word movers distance. In: 2018 International Conference on Bangla Speech and Language Processing (ICBSLP). IEEE (2018)
17. Lakshmi, R., Baskar, S.: DIC-DOC-K-means: dissimilarity-based Initial Centroid selection for DOCument clustering using K-means for improving the effectiveness of text document clustering. J. Inf. Sci. 0165551518816302 (2018)
18. Megarchioti, S., Mamalis, B.: The BigKClustering approach for document clustering using Hadoop MapReduce. In: Proceedings of the 22nd Pan-Hellenic Conference on Informatics. ACM (2018)
19. Al-Jadir, I., et al.: Enhancing digital forensic analysis using memetic algorithm feature selection method for document clustering. In: 2018 IEEE International Conference on Systems, Man, and Cybernetics (SMC). IEEE (2018)
20. Zhu, Y., Zhang, M., Shi, F.: Application of algorithm CARDBK in document clustering. Wuhan Univ. J. Nat. Sci. **23**(6), 514–524 (2018)
21. Abualigah, L.M., et al.: A krill herd algorithm for efficient text documents clustering. In: 2016 IEEE Symposium on Computer Applications & Industrial Electronics (ISCAIE). IEEE (2016)
22. Akter, R., Chung, Y.: An improved genetic algorithm for document clustering on the cloud. Int. J. Cloud Appl. Comput. (IJCAC) **8**(4), 20–28 (2018)
23. Chen, Y., Sun, P.: An optimized K-Means algorithm based on FSTVM. In: 2018 International Conference on Virtual Reality and Intelligent Systems (ICVRIS). IEEE (2018)
24. Al-Jadir, I., et al.: Adaptive crossover memetic differential harmony search for optimizing document clustering. In: International Conference on Neural Information Processing. Springer, Cham (2018)
25. Seshadri, K., Viswanathan Iyer, K.: Design and evaluation of a parallel document clustering algorithm based on hierarchical latent semantic analysis. Concurr. Comput. Pract. Exp. e5094
26. Saini, N., Saha, S., Bhattacharyya, P.: Automatic scientific document clustering using self-organized multi-objective differential evolution. Cogn. Comput. 1–23 (2018)
27. Rani, M.S., Babu, G.C.: Efficient query clustering technique and context well-informed document clustering. In: Soft Computing and Signal Processing, pp. 261–271. Springer, Singapore (2019)
28. Gonzàlez, E., Turmo, J.: Unsupervised document clustering by weighted combination. LSI Research Report LSI-06-17-R, Departament de Llenguatges i Sistemes Informàtics, Barcelona (2006)
29. Gupta, A., Gautam, J., Kumar, A.: A survey on methodologies used for semantic document clustering. In: 2017 International Conference on Energy, Communication, Data Analytics and Soft Computing (ICECDS). IEEE (2017)
30. Jain, A.K., Narasimha Murty, M., Flynn, P.J.: Data clustering: a review. ACM Comput. Surv. (CSUR) **31**(3), 264–323 (1999)

Sentiment Analysis of Amazon Mobile Reviews

Meenakshi, Arkav Banerjee, Nishi Intwala and Vidya Sawant

Abstract Sentiment analysis is used to derive the emotion/opinion that is being conveyed in a text. This helps in determining whether the author's intent is positive or negative. Its applications are vast and help in analyzing product reviews, popularity of a brand, and in determining people's opinions on any subject. Due to the complexities involved in the human language such as subjectivity, metaphors, sarcasm, and multiple sentiments, it becomes difficult to categorize our opinions, computationally. The goal of this project is to conduct sentiment analysis on Amazon product reviews using various natural language processing (NLP) techniques and classification algorithms. The dataset consists of 400,000 reviews of unlocked mobile phones sold on Amazon.com. We will achieve the result by preprocessing the reviews and converting them to clean reviews, after which using word embedding, the word reviews were converted into numerical representations. Then, we finally fit the numerical representations of reviews to the Naïve Bayes, logistic regression, and random forest algorithm. The results and accuracy of all these classifiers are compared in this paper. This will be helpful for a brand/company to understand the general opinion toward their product which in turn will help them in evaluating the improvements required.

Keywords Bag of words · Logistic regression · Natural language processing · Naïve Bayes · Random forest · Sentiment analysis

Meenakshi · A. Banerjee (✉) · N. Intwala · V. Sawant
Mukesh Patel School of Technology Management and Engineering (Mumbai Campus), SVKM's NMIMS University, Mumbai 400056, Maharashtra, India
e-mail: arkav1897@gmail.com

Meenakshi
e-mail: minakshisingh116@gmail.com

N. Intwala
e-mail: nishiintwala1@gmail.com

M. Tuba et al. (eds.), *ICT Systems and Sustainability*,
Advances in Intelligent Systems and Computing 1077,
https://doi.org/10.1007/978-981-15-0936-0_4

1 Introduction

Sentiment analysis is contextual mining of text which helps in understanding the general opinion toward a particular product, by monitoring online data and analyzing the reviews. Another name for this is opinion mining and it is a field within natural language processing. Sentiment analysis is usually used to gauge the common feeling or emotion toward any cause or product so that the organizations can understand whether the consumer's opinion is positive or negative. This can help them in improving their defects and change their strategies to get optimum results. By understanding why a particular product got positive reviews and what makes that product unique, any business or manufacturers can take advantage of this fact and create more products with similar features. Similarly, understanding the deficiencies and the exact problems that the user has with the product will help them in reducing such faults. However, it can be tricky to explain to a machine the modulation, cultural variations, slang, and misspellings that are often found in online reviews. For example, it can be difficult for a machine to understand sarcasm or new abbreviations used by the users. Our approach is an automatic approach which makes use of machine learning. In this, the first step is training, in which the model associates a particular input to a corresponding output. Using a feature extractor, it creates a feature vector. Extraction of the vector, which is converting the words into a numerical representation, is done by bag of words. After this, it is fed into the classification algorithm. In our project, we aim to classify reviews on Amazon.com into positive and negative reviews (Fig. 1).

2 Related Work

In our project, we have referred to a paper by Ankit and Nabizath Saleena in which they proposed the methodology for performing sentiment analysis on Twitter data. In this paper, the proposed methodology has four modules. The first step is data preprocessing to decrease the size of the feature set, after which for feature representation, they have used bag of words to convert the tweets into their numerical representations, which are called the feature vectors. After this, they have explained their ensemble classifier in which they have proposed to aggregate multiple base classifiers into one ensemble classifier which helps in enhancing the accuracy of the base classifiers. From the results, it is evident that ensemble classifier works better than the stand-alone classifiers. They have implemented the project using *Python, Scikit-learn* library for feature representation, classification, similarity measures, and evaluation purposes. They have also used *Natural Language Toolkit (NLTK)*. Another paper referred by us is by Mrs. Smita Bhanap and Dr. Seema Kawthekar in which they have classified tweets into negative, positive, and neutral tweets using Naïve Bayes classifier with bag of words. Another paper in which various techniques for feature extraction used for text classification has been reviewed by Resham N. Waykole,

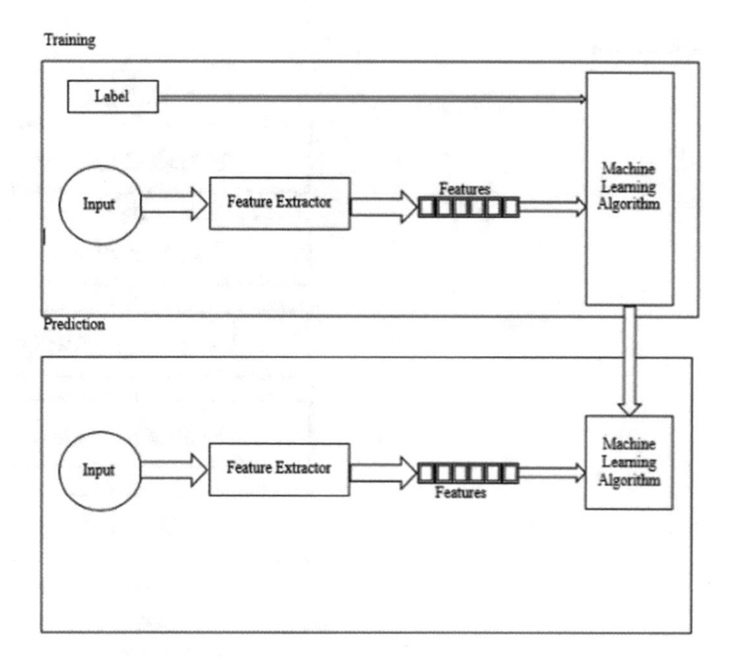

Fig. 1 Basic framework of the system

Anuradha D. Thakare. The feature extraction methods that they have implemented include bag of words, tf-idf, and word2vec which have been compared for text analysis. The extracted features are evaluated against logistic regression and random forest classifier. Their results showed that word2vec is a better method for feature extraction along with random forest classifier for text classification.

3 Proposed Method

3.1 Overview of the Procedure

See Fig. 2.

3.2 Dataset Collection

Data collection is the main aspect of machine learning that makes training of the algorithm possible. Usually, data gathering and processing consume most of the time involved in the whole machine learning process. For our project, we have acquired a

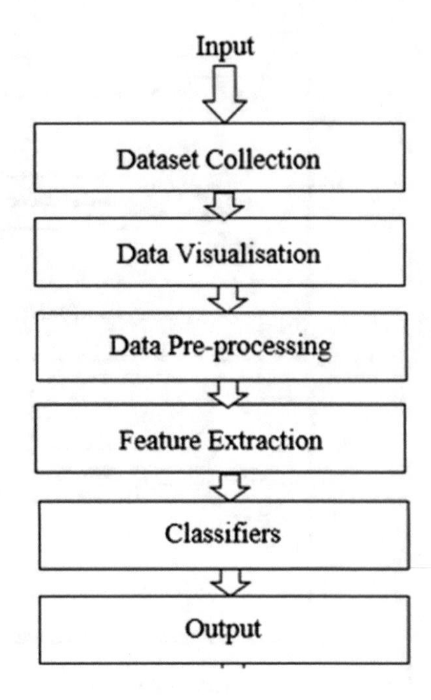

Fig. 2 Overview of the model

dataset consisting of Amazon mobile reviews containing 400 thousand reviews. The dataset is public and available on Kaggle. The dataset has the following fields:

1. Product Title,
2. Brand,
3. Price,
4. Rating (1–5),
5. Review text, and
6. Number of people who found the review helpful.

3.3 Data Visualization

Data visualization is the graphical representation of information and data. Visual aids like charts, graphs, and maps help us visualize data by assisting us in understanding trends, outliers, and patterns in data. We have used Jupyter Notebook in Python in order to visualize our data.

Figure 3 shows that reviews with rating of 5 dominate the distribution. Figure 4 shows top 20 brands with the largest numbers of reviews. The top 3 brands, Samsung, Blu, and Apple, dominate the total number of reviews. Figure 5 shows top 50 products with the largest number of reviews. Figure 6 shows the distribution of review length (number of characters). The number of reviews falls exponentially with the increase

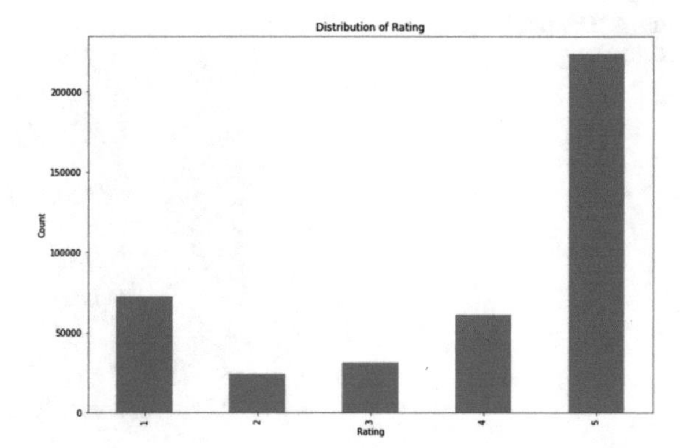

Fig. 3 Distribution of rating

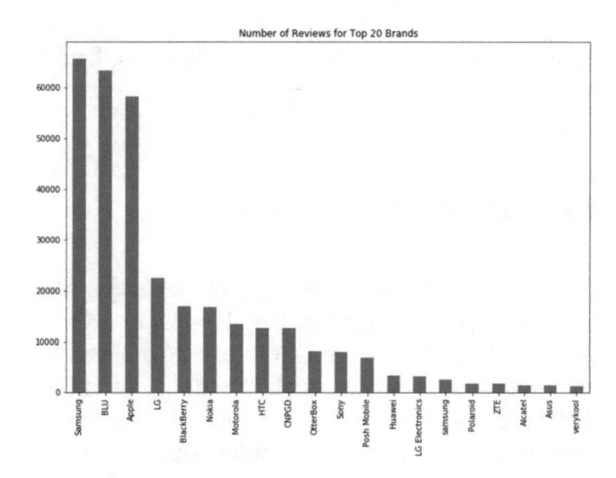

Fig. 4 Number of reviews for top 20 reviews

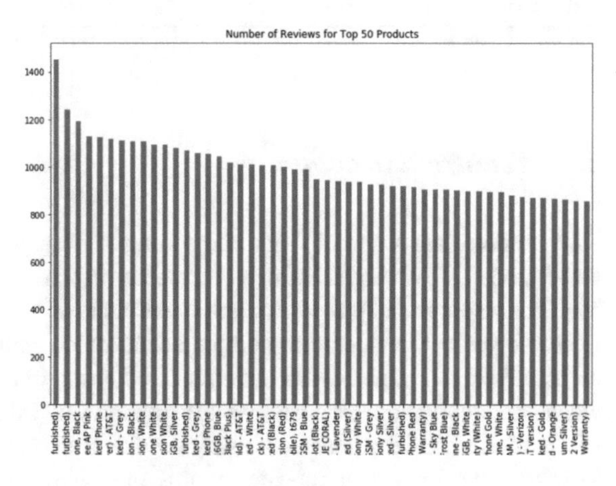

Fig. 5 Number of reviews for top 50 products

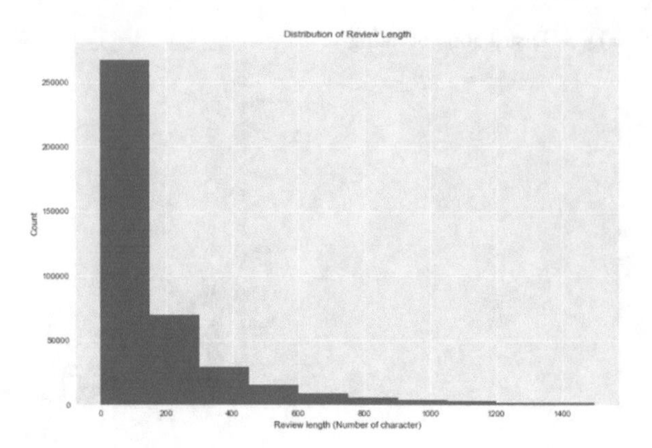

Fig. 6 Distribution of review length

of review length. In some of the models described later, we have truncated certain reviews that are longer in length.

3.4 Dataset Preprocessing

Data preprocessing can be thought as performing a function on raw data to prepare it for another processing procedure. It transforms the data into a simpler and effective format and also can be used for further processing. In our case, we have used various preprocessing techniques such as removal of HTML tags, non-characters such as digits and symbols, stop words such as "the" and "and", and converted words to lower case. We have used stemming to convert certain words into their root words in order to maintain uniformity. For example, with stemming, a review which contained the word "watching" was converted to its root word "watch".

3.5 Feature Extraction

One common approach of word embedding is frequency-based embedding such as BoW model. BoW model learns a vocabulary list from a given corpus and represents each document based on some counting methods of words. We initially implemented CountVectorizer in sklearn to compute occurrence counting of words which implements tokenization as well as occurrence counting in a single class. However, some words frequently appeared but had little meaningful information about the sentiment of a particular review. Instead of using occurrence counting, we used tf-idf transform to reduce the impact of repeated words in a given corpus. We used TfidfVectorizer through sklearn library which implements both tokenization and tf-idf weighted counting in a single class. Another approach of word embedding is prediction-based embedding such as word2vec embedding. Word2vec consists of Continuous Bag of Words (CBoW) and skip-gram model. They consist of shallow neural networks which can learn weights for word vector representations. The word2vec model parses review text to sentences (word2vec model takes a list of sentences as inputs) and creates a vocabulary list. It then transforms each review into numerical representation by computing average feature vectors of words.

3.6 Classifiers

Once we have numerical representations of the text data, we are ready to fit the feature vectors into the following supervised learning algorithms:

1. Multinomial Naïve Bayes—Naive Bayes is based on applying Bayes theorem with a strong assumption that a feature will always be independent of other features. This is done in order to predict the category of a given sample. Naïve Bayes calculates the probability of each category using Bayes theorem. The category having the highest probability becomes the output. In our case, we have used a combination of occurrence counting for feature extraction and multinomial Naïve Bayes for classification.
2. Logistic Regression—Logistic regression is a popular machine learning algorithm borrowed from the field of statistics which can be used for binary classification problems. In our case, we have used a combination of TfidfVectorizer for feature extraction and logistic regression for classification.
3. Random Forest—This is an ensemble learning method which can be used for classification. It operates by building up many decision trees at the time of training and it will output the class that is the mode of the classes (classification) of each tree. In our case, we have used a combination of word2vec embedding for feature extraction and random forest for classification.

4 Results and Discussion

Among all the reviews, we have used a total of 27,799 random reviews for training and 3089 random reviews for testing. We have used three different parameters (precision, recall, and $F1$ score) for checking the accuracy. The results are as follows (Table 1).

As we can see, using logistic regression along with TfidfVectorizer gives us the highest accuracy in our analysis.

The confusion matrices for the three combinations have been shown in Fig. 7. As we can see, logistic regression gives us the highest number of true positives and true negatives in the confusion matrix and lesser number of false positives and false negatives, thereby giving us a higher accuracy (Figs. 8 and 9).

We have generated a word cloud for a particular brand to get an intuition of words that appear in different sentiments. For example, reviews with positive sentiments

Table 1 Results

Techniques	Accuracy	Positive class			Negative class		
		Pre	Rec	$F1$	Pre	Rec	$F1$
Random forest	0.92	0.94	0.96	0.95	0.87	0.82	0.84
Naïve Bayes	0.91	0.93	0.96	0.95	0.87	0.80	0.83
Logistic regression	0.93	0.94	0.96	0.95	0.89	0.83	0.86

```
Confusion Matrix : Confusion Matrix : Confusion Matrix :
[[ 622  156]        [[ 636  142]        [[ 648  130]
 [  96 2215]]        [  97 2214]]        [  83 2228]]
```

Fig. 7 Confusion matrix

Fig. 8 Word cloud for positive sentiments

Fig. 9 Word cloud for negative sentiments

on the brand "Apple" contain words such as "good", "great", and "perfect", while the reviews with negative sentiments contain words such as "repair", "disappointed", and "bad".

5 Conclusion and Future Scope

We have compared various classifiers for the given dataset of mobile reviews from Amazon such as logistic regression, Naïve Bayes, and random forest. We have used different feature extraction techniques with each classifier to check accuracy for binary sentiment analysis. As we can see, logistic regression along with a combination of TfidfVectorizer gives us the highest accuracy. In order to improve the accuracy, we could have used more number of samples in our training set. We could have also classified reviews into higher number of classes instead of binary classification. Even though our work is in the domain of mobile reviews, it can be extended to the analysis of data on other review sites and social network platforms as well.

References

1. Waykole, R.N., Thakare, A.D.: A Review of feature extraction methods for text classification. Int. J. Adv. Eng. Res. Dev. **5**(04) e-ISSN (O): 2348–4470, p-ISSN (P): 2348-6406 (2018)
2. Poecze, F., Ebsterb, C., Strauss, C.: Social media metrics and sentiment analysis to evaluate the effectiveness of social media posts. In: The 9th International Conference on Ambient Systems, Networks, and Technologies, ANT 2018, Procedia Computer Science, vol. 130, pp. 660–666 (2018)
3. Ankit, Saleena, N.: An ensemble classification system for twitter sentiment analysis. In: International Conference on Computational Intelligence and Data Science, ICCIDS 2018, Procedia Computer Science, vol. 132, pp. 937–946 (2018)
4. Chong, W.Y., Selvaretnam, B., Soon, L.-K.: Natural language processing for sentiment analysis: an exploratory analysis on tweets. In: 4th International Conference on Artificial Intelligence with Applications in Engineering and Technology (IEEE), vol. 43 (2014)
5. Devi, D.N., Kumar, C.K., Prasad, S.: A feature-based approach for sentiment analysis by using support vector machine. In: 2016 IEEE 6th International Conference on Advanced Computing (IACC), pp. 3–8 (2016)
6. Alessia, D., Ferri, F., Grifoni, P., Guzzo, T.: Approaches, tools and applications for sentiment analysis implementation. Int. J. Comput. Appl. (0975–8887) **125**(3) (2015)
7. Godsay, M.: The process of sentiment analysis: a study. Int. J. Comput. Appl. (0975–8887) **126**(7) (2015)
8. Medhat, W., Hassan, A., Korashy, H.: Sentiment analysis algorithms and applications: a survey. Ain Shams Eng. J. **5**, 1093–1113 (2014)
9. Bhanap, S., Kawthekar, S.: Twitter sentiment polarity classification & feature extraction. IOSR J. Comput. Eng. (IOSR-JCE), e-ISSN: 2278-0661, p-ISSN: 2278-8727, pp. 01–03
10. Taboada, M., Brooke, J., Tofiloski, M., Voll, K., Stede, M.: Lexicon-based methods for sentiment analysis. Comput. Linguist. **37**(2), 267–307 (2011)
11. Pang, B., Lee, L.: Opinion mining and sentiment analysis. Found. Trends Inf. Retr. **2**(1–2), 1–135 (2008)

12. Wilson, T., Wiebe, J., Hoffmann, P.: Recognizing contextual polarity in phrase-level sentiment analysis. In: Proceedings of the Conference on Human Language Technology and Empirical Methods in Natural Language Processing, pp. 347–354 (2005)
13. Pang, B., Lee, L., Vaithyanathan, S.: Thumbs up?: sentiment classification using machine learning techniques. In: Proceedings of the ACL-02 Conference on Empirical Methods in Natural Language Processing, vol. 10, pp. 79–86 (2002)

Domain-Based Ranking of Software Test—Effort Estimation Techniques for Academic Projects

Jatinderkumar R. Saini and Vikas S. Chomal

Abstract The significant segments of software project development are requirement engineering, designing, coding, testing, deployment, and maintenance. These phases are implemented irrespective of type of software methodology such as traditional or agile followed during the establishment of software as well as effectually documented. Also this practice is executed in all technical environment, i.e., software companies as well as academic courses were software projects are developed. Software project development is having utmost prominence and credit in educational prospectus for computer science and engineering. Apart from fundamental stages such as requirement engineering, designing, coding, testing, deployment, and maintenance it is found that IT industries also strictly and mandatory focuses on issues such as software risk management, software scheduling, and tracking as well as effort estimation and distribution related to time, budget, and testing. Also these IT industries consider and execute various testing techniques during software project development. For this proposed research framework 122 large software projects documentation were taken into consideration. For simplicity and better comprehensive research domain-based classification was used to classify these software projects into four main heads termed as—(a) Desktop Application, (b) Web Desktop, (c) Mobile Application, and (d) Portal. Similarly, the test effort estimation such as (a) Delphi technique, (b) Analogy-Based Estimation, (c) Software Size-Based Estimation, (d) Test Case Enumeration-Based Estimation, (e) Task or Activity-Based Estimation, and (f) Use Case Test Points are considered for this research purpose. We found that these testing techniques were not considered by students while developing academic software projects as well as no due weightage is given to these testing techniques and test effort by academic courses of computer science and engineering. The main objective behind ranking these techniques was to indicate that which technique is more suitable for software projects carried out by postgraduate courses in computer

J. R. Saini
Symbiosis Institute of Computer Studies and Research, Pune, Maharashtra, India
e-mail: saini.expert@gmail.com

V. S. Chomal (✉)
TMES Institute of Computer Studies, Mandvi, Surat, Gujarat, India
e-mail: vikschomal80@gmail.com

M. Tuba et al. (eds.), *ICT Systems and Sustainability*,
Advances in Intelligent Systems and Computing 1077,
https://doi.org/10.1007/978-981-15-0936-0_5

science. To achieve this purpose a survey is conducted as a part of research by considering 31 Computer Science Academicians to rank these software test effort estimation techniques. The result of experimentation shows that highest rank is allotted to Use Case Points-Based Estimation whereas second highest rank is assigned to Task-Based Estimation while Test Case-Based Estimation gained third maximum rank.

Keywords Academic courses · Estimation · Software Development Life Cycle (SDLC) · Software Engineering (SE) · Software project · Testing techniques

1 Introduction

Boehm states that Software Engineering (SE) provides various engineering principles that are to be kept in mind as well as followed so that an economical, reliable, and quality software is developed [1]. Bareisa et al. defined that software engineering not only involves around real-world entity but it is also developed, operated, and maintained by real-world entity [2]. Apart from consideration of technical tasks, SE also teaches decision-making skills such as definition, organization, and planning of multiple interrelated activities as well as collective expertise, i.e., conversation of information and synchronization of people working to fulfill a common goal [3]. SE is one of the principal subjects imparted in all computer science and engineering courses including undergraduate as well as postgraduate. It is compulsory for students of computer science and engineering courses to undertake full time or partial software project development training in IT industries and time duration for this training is 4–6 months. Through this software project development students provide solution to the real-world problems prevailing in domain such as scientific, principle based, corporate and commercial, knowledge and intelligence based as well as decision-making. During this project development, students need to execute all major stages of SDLC and at the same time document the artifacts related to the implementation in software project documentation. Software documentation is a crucial article of software project which provides evidence of software project development. Sommerville [4] states that software project documentation is—(a) a communication medium between members of the development, (b) evidence storehouse to be used by maintenance engineers, (c) provides information to organization that helps them to plan, budget and schedule the software development process, and (d) a tool which can be utilized to guide and train the users of the system so that they can easily use and govern the system. For this research software project, documentation prepared by final year students of Master's Degree in computer science were considered as a source of study and data collection. The main objective of investigation of these software project documentation was to study and analyze testing efforts and techniques adopted and carried out during their software project development.

The foremost observation found in these academic software project documentation was that students exercised all fundamental steps of SDLC during their project development. But the important finding that motivated us for this research was that students do not focus on testing techniques and effort. As academicians, the motivation behind the present research is, since we found, that testing techniques and efforts were not considered by students for software project development. Also incorrect and incomplete test cases were documented by students in their software project documentation without the use of any formal testing techniques and mechanism. Further, after accomplishment of these academic software projects majority of students start their professional career within small, medium, and multinational IT companies. As per contemporary scenario, IT companies prefer those intellectual having basic knowledge in every sphere of software development concerned with designing, coding, and testing. In this perception through this paper, we studied, focused, and analyzed on testing effort and techniques that academic domain of computer science and engineering must include in their curriculum that deals with software project for producing compatible as well as intellectual IT professionals. The preferred test effort estimation included for this research are (a) Delphi technique, (b) Analogy-Based Estimation, (c) Software Size-Based Estimation, (d) Test Case Enumeration-Based Estimation, (e) Task or Activity-Based Estimation, and (f) Use Case Test Points. The paper is structured as described here: Sect. 2 presents literature review related to test effort and testing techniques followed by well-organized and systematic research methodology in Sect. 3. Section 4 represents Experiments and Results and lastly Conclusion.

2 Literature Review

Literature review related to research is presented in Table 1 which includes details such as research scholar citation, objective of research study, and domain area for which related experimentation is conducted. The domain area is basically classified into two heads—(a) Industrial and (b) Academic.

3 Methodology

The research was commenced with foremost inspiration by identifying the following Research Questions (RQ) below is to justify the objective of this research to find the significant indication of testing effort estimation and techniques practices going on in academic software project development:

RQ1: What are the current practices in testing effort and test estimation?
Motivation RQ1: Finding about overall scenario of test effort estimation as well as testing techniques followed by students' academic projects

RQ2: How we as an academician can improve the current trends in testing techniques, effort, and test estimation?

Table 1 I. Summarization of the literature review

Sr. No.	Author(s)	Research focus	Domain area
1	Abran et al. [5]	The research work provides an approach to estimate the effort needed in test estimation which includes verification and validation of software projects. Functional requirements were considered for this estimation	Industrial
2	Bertolino [6]	The author focuses on importance of testing in software project development. Also, a detail explanation on testing, test effort estimation is presented in research work	Industrial
3	Srivastava [7]	Author states that test effort estimation and use of appropriate testing techniques hold an important place in software project development. For execution of research, the author makes use of fuzzy logic attempt to estimate reliable software testing effort	Industrial
4	Karunakaran and Sreenath [8]	Presented a comparative review on various effort estimation techniques	General
5	Butt et al. [9]	The author(s) proposed a software testing framework producing quality software products developed under agile methodology	General
6	Jayakumar and Abran [10]	The research proposal presents a detail explanation of software test estimation techniques. Also, difficulties and challenges that are faced in software test estimation are highlighted	General
7	Jayakumar and Abran [11]	The research work discovered software testing from the viewpoint of estimation of efforts for functional testing	Industrial

(continued)

Table 1 (continued)

Sr. No.	Author(s)	Research focus	Domain area
8	Kafle [12]	Studied and reviewed software test effort and techniques required in software project development. The research was done in order to address software test effort estimation. The research was conducted considering case study for which empirical evidence from five software development companies of Nepal was considered	Industrial
9	Jamil et al. [13]	Software testing is an unavoidable activity of SDLC. Their research is done with the aim to discuss different testing techniques for the superior quality assurance purposes	General
11	Chemuturi [14]	The study focuses on explanation, analysis, and detail comparison of various test effort estimations such as (a) Delphi technique, (b) analogy-based estimation, (c) software size-based estimation, (d) test case enumeration-based estimation, (e) task or activity-based estimation, and (f) testing size-based estimation	General
12	Sabev and Grigorova [15]	The research study describes an approach through which prioritization of test cases can be done and test effort can be calculated and executed using manual and automate software testing tools	Industrial
13	Hermansky [16]	The author states the importance of (a) Experience-Based and (b) Model-Based test effort estimation. Also experiment is conducted and result is presented using both techniques on software project developed in companies	Industrial
14	Chaudhary and Yadav [17]	They proposed an approach which can be used to estimate size and effort required in testing project. Test case was utilized for experimentation. Also they consider major factors related to test estimation for successful estimation	Industrial

(continued)

Table 1 (continued)

Sr. No.	Author(s)	Research focus	Domain area
15	Chaudhary and Yadav [18]	In research, the author(s) make use of two test effort estimations known as parametric estimating and test case point estimations for finding the optimum way for effort estimation using absolute error and relative error calculation techniques	Industrial
16	Chauhan and Singh [19]	In this paper, the author(s) presented different testing techniques. Various research related to software test effort estimated are summarized	General
17	Saravana Kumar [20]	In research work focuses on analysis of various software test estimation techniques. Also, different challenges that are faced during their implementation is also specified. A detailed analysis of strength and weakness of test effort estimation is also presented	Industrial
18	Sangeetha and Dalal [21]	They studied, analyzed, and presented their research on estimation related to cost, budget, effort, and test. As per author(s) suggestions and conclusion these estimation play a vital role in software project development as well as helpful in making a successful and quality software	Industrial
19	Nageswaran [22]	The study focuses on estimating test effort by executing Use Case Points (UCP) approach	Industrial
20	Nguyen [23]	The researcher makes use of Test Case Point Analysis to estimate the size and effort of software testing. Major attributes such as checkpoints, preconditions as well as test data and type of test applied were considered during experiment	Industrial
21	Afzal [24]	The main objective of research was to examine the metric support for software test planning and test design processes	Industrial

Motivation RQ2: Finding about measures suggested for improving current procedure of testing techniques, efforts followed by students in academic software project development in testing.

Software project documentation prepared by final year students of Master Degree level course were considered as primary source of data collection for present research work. These documentation consists of all technical artifacts carried out during software project development. The documentations were collected from college library. The time period of these software project developments was 6 months. We inspected and assessed 122 large software project documentation of software's developed during a period of academic years from 2014–2015 to 2017–2018. The main objective behind analysis of these documentation was to learn about implementation of software test effort and techniques executed by students in their software project development. For inclusive and exhaustive research, linear sequential phase was followed. These phases are mentioned below as well graphically represented in Fig. 1.

1. Exploring Academic Software Project Documentation.
2. Listing Software Project Definition.
3. Classification of listed Software Project Definition.
4. Exploring various existing Software Test Effort Techniques for Estimation.
5. Survey to find the best Software Test Effort Techniques.
6. Findings and Analysis of Results.

Initially, the first step was to explore these 122 software project documentation. Thereafter listing of software project definition was done and presented in Table 2.

After listing these software projects, the next step was to identify and extract the basic details of software project development. As each project under examination is entirely different from other projects, this exploration was repetitively performed for each of the 122 project documentation. The basic background details of software project are summarized in Table 3.

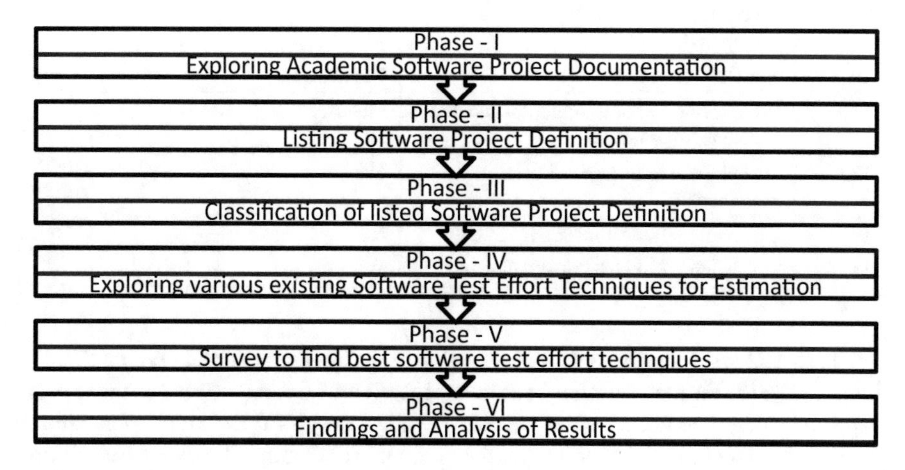

Fig. 1 Diagrammatic representation of research methodology

Table 2 List of software project definition

Sr. No.	Project definition	Project development year
1	Organization MIS	2014–15
2	Production monitoring system	2014–15
3	Health care android applications	2015–16
4	Tours and tourism tracking system	2015–16
5	AdRelease	2016–17
6	CIP—company information portal	2016–17
7	…	…
8	…	…
121	Work flow management system	2017–18
122	APMC system	2017–18

Table 3 Academic software project background

Sr. No.	Details studied	Findings
1	Project development year	2014–15 to 2017–18
2	Size of development team	1–3 members
3	Type(s) of software developed	Desktop application Web application Mobile application Portal
4	Front end used	.Net, ASP.Net, PHP, Spring Framework
5	Back end used	Oracle, MySQL Server,
6	Process model used	Waterfall, Prototype
7	Project plan prepared or not	No
8	Do estimate of project is prepared or not?	No
9	If yes, how and which type of estimation is prepared	–
10	Which level of testing done?	Unit Testing
11	What method(s) of testing implemented (Manual/Automated)?	Manual
12	Do evidence of testing artifacts presented?	Yes
13	Do testing effort estimated, if yes how?	No
14	Do testing effort techniques followed?	No

These 122 software projects were developed during the academic period 2014–15 to 2017–18. Further, domain-based classification of these software projects was done and categorized into four broad categories, namely—(a) Desktop Application, (b) Web Application, (c) Mobile Application, and (d) Portal. The extensively used platforms for developing these software projects are .Net, ASP.Net, PHP, Spring

Framework as well as Oracle, MySQL Server were common back end found. Waterfall and Prototype process models are taken into consideration while implementation of these software projects. The other observation found through study of these project documentations was that no corresponding project plan was prepared before commencement of these projects. The project plan consists of very crucial information such as software project phases, activities as well as milestones required to be executed during software project establishment. Similarly, no formal mechanism was found for estimation of various phases to be followed during SDLC. Our study basically deals with software test effort estimation and techniques, hence to extract information regarding these attributes were of utmost interest. It was revealed that result of unit testing using Test cases was highlighted. The unit testing was manually executed using these test cases and use of any automated software testing tools was not found nor testing efforts as well as techniques were employed.

During the literature review, the following Test Effort Estimation Techniques were found and are also most suitable for our research domain. They are (a) Delphi Technique, (b) Analogy-Based Estimation, (c) Software Size-Based Estimation, (d) Test Case Enumeration-Based Estimation, (e) Task or Activity-Based Estimation, and (f) Use Case Test Points. Brief interpretation of these estimation techniques is as follows:

(a) Delphi Technique: In this estimation, method experts play an important role. They are required to decide individual estimates for an individual set of requirements based on their own prior experience. During this process several iterations take place so that experts involved acquire knowledge and reasoning from other experts, and revise their estimates in subsequent iterations. At last concluding estimate is chosen from the narrowed range of values estimated by experts in the last iteration.
(b) Analogy-Based Estimation: In this technique, comparison is done between components of software under test with standard components. The comparison is carried out considering historical data. The overall estimate of all the components of the software to be tested is further adjusted based on project-specific factors and the management effort required, such as planning and review.
(c) Software Size-Based Estimation: In this mechanism, size is taken into consideration and a regression model is built taking historical data into consideration. Size is considered as one of the significant input parameters. Total Lines of Code (LOC) corresponds to actual Software Size. Further, various tools and techniques can be used to convert size into effort estimation.
(d) Test Case Estimation: In this approach, all test cases are listed out first. After listing all the test cases the next step is to estimate testing effort required for each test case. For this, these test cases are categorized into three heads, namely—(a) Best Case, (b) Normal Case, and (c) Worst Case for estimating effort required for each individual test case and lastly expected effort for each case is computed.
(e) Task or Activity-Based Estimation: In this procedure, project is viewed considering tasks or phases to be performed in executing the project. A software project is implemented and executed in stages. Phases in a testing project includes—(1)

Project Initiation, (2) Project Planning, (3) Test Planning, (4) Test Case Design, (5) Setup Test Environment, (6) Conduct Testing, (7) Integration Testing, (8) System Testing, (9) Log and report test results, (10) Prepare Test Report. The steps required in task-based effort estimation are—(a) Each task is assigned an predefined durations, (b) For assigning time duration make use of three-time estimates such as—Best Case, Normal Case, and Worst Case, (c) Expected Time is computed using formula = [Best Case + Worst Case + Normal Case)]/6, where 6 indicates total number of steps followed in this algorithm, (d) Make necessary adjustments for project complexity, familiarity with the platform used for developing project, expertise of developer team. and use of various tools and techniques, (e) Next step is to sum up the total effort estimate of the project, (f) Make use of Delphi Technique to validate the estimate if necessary.

(f) Use Case Test Points: In this technique Use Case is taken into consideration for estimation for software. Use Case Test Points are calculated as the sum of the actors multiplied by each actor's weight from an actors' weight table and the total number of use cases multiplied by a weight factor, which depends on the number of transactions or scenarios for each use case. Weights as signed to each of the technical and environmental factors are used to convert unadjusted use case points to adjust use case points.

4 Experiments and Results

In the present work, we considered academic projects of 4 years from 2014–2015 to 2017–2018. The preliminary information regarding this academic projects is presented in tabular format in Table 4 as well graphically represented in Fig. 2.

From Table 4 we observed that 122 large software projects were considered for research study. Further, it was found that 28 software projects developed during the academic year 2014–2015, 31 software projects during the year 2015–2016, similarity 25 during the year 2016–2017 whereas 38 software projects developed during the academic year 2017–18. Further, these 122 large software projects were basically categorized into four broad heads as presented in Table 3, namely—(a) Desktop Application—applications that execute stand alone in a desktop or laptop

Table 4 Year-wise academic software project development

Sr. No.	Year of project development	Total project developed
1	2014–15	28
2	2015–16	31
3	2016–17	25
4	2017–18	38
Total		122

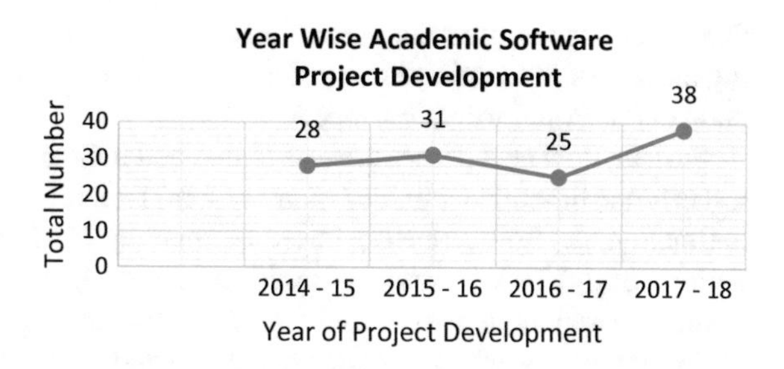

Fig. 2 Year-wise academic software project development

computers and do not require Internet connectivity to utilize the services, (b) Web Application—applications that require Internet connectivity as well as operates on web browser and utilize web technology to execute the functionality, (c) Mobile Application—applications popularly referred as mobile apps and designed to execute of smartphones and (d) Portal—organizational or enterprise-specific applications. Considering these four heads next task was to compute total number of software projects developed under these four heads. The derived information is presented in Table 5 as well as demonstrated graphically in Fig. 3.

Based on the values presented in Table 5 and corresponding Fig. 3, it is found that Web Application (65) is the domain area in which most referred by students for software project development while Mobile Application (27) was found to have achieved

Table 5 Domain-based classification summation of software project developed

Sr. No.	Year of project development	Desktop application	Web application	Mobile application	Portal	Total project developed
1	2014–15	3	19	2	4	28
2	2015–16	9	15	7	0	31
3	2016–17	4	11	8	2	25
4	2017–18	5	20	10	3	38
Total		21	65	27	9	122

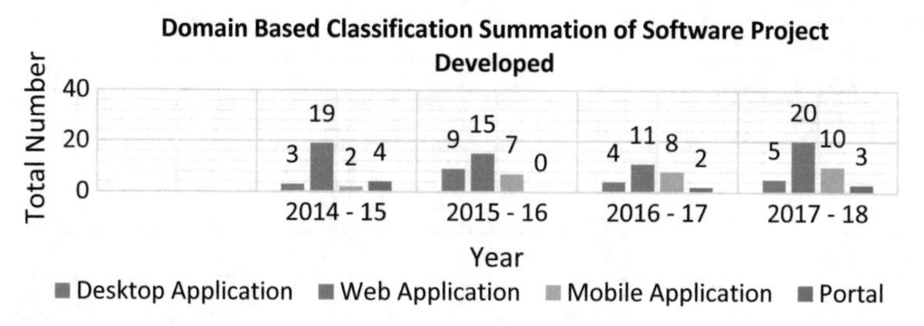

Fig. 3 Domain-based classification summation of software project developed

the second highest preference. Similarly, 21 Desktop Application was developed during these four academic years starting from 2014–15 to 2017–18 while the lowest 9 numbers of Portal were developed.

Further, now next procedure was to find and indicate that out of these six software test effort estimation technique, namely—(a) Delphi Technique, (b) Analogy-Based Estimation, (c) Software Size-Based Estimation, (d) Test Case Enumeration-Based Estimation, (e) Task or Activity-Based Estimation, and (f) Use Case Test Points which estimation technique is more suitable for software projects carried out in academic environment. For which survey was conducted with 31 academicians teaching in graduate and undergraduate colleges of the Faculty of Computers. All academicians fulfilled the criteria of holding at least the Master's Degree with First Class and an experience of at least 10 years in the academic domain. In survey, these academicians were provided with basic details such as four broad categories under which software projects are classified, namely—(a) Desktop Application, (b) Web Application, (c) Mobile Application, and (d) Portal and six software test effort estimation technique, namely—(a) Delphi Technique, (b) Analogy-Based Estimation, (c) Software Size-Based Estimation, (d) Test Case Enumeration-Based Estimation, (e) Task or Activity-Based Estimation, and (f) Use Case Test Points. Further, academicians were asked to rank these test effort estimation considering their relevance in academic software project development. For which scale on basis of 1–6 were to be assigned, where 1 indicates the most appropriate and 6 indicates the least appropriate. The result of the survey is presented in Table 6. The Column "CSA" in Table 6 specifies for the score provided by nth Computer Science Academicians with ranging from 1 to 31.

The next step was to find out maximum rank counts of 1s which is presented in Table 7 and same is demonstrated graphically in Fig. 4.

Table 6 Ranking by respondents

Sr. No.	CSA	Software test effort estimation					
		Delphi technique	Analogy based	Software size based	Test case based	Task based	Use case test points
1	CSA1	6	4	5	3	2	1
2	CSA2	5	6	4	3	2	1
3	CSA3	6	5	4	3	1	2
4	CSA4	6	4	5	3	2	1
5	CSA5	5	6	3	4	1	2
...	...	...	...	...	...	...	...
...	...	...	...	...	...	...	...
29	CSA29	5	6	3	4	1	1
30	CSA30	6	5	4	3	2	1
31	CSA31	5	6	4	3	1	1

Table 7 Statistical measures of count of (1s) by respondents

Sr. No.	CSA	Software test effort estimation					
		Delphi technique	Analogy based	Software size based	Test case based	Task based	Use case test points
1	CSA1	6	4	5	3	2	1
2	CSA2	5	6	4	3	2	1
3	CSA3	6	5	4	3	1	2
4	CSA4	6	4	5	3	2	1
5	CSA5	5	6	3	4	1	2
…	…	…	…	…	…	…	…
29	CSA29	5	6	3	4	1	1
30	CSA30	6	5	4	3	2	1
31	CSA31	5	6	4	3	1	1
Max count of (1's)		0	0	0	0	11	20

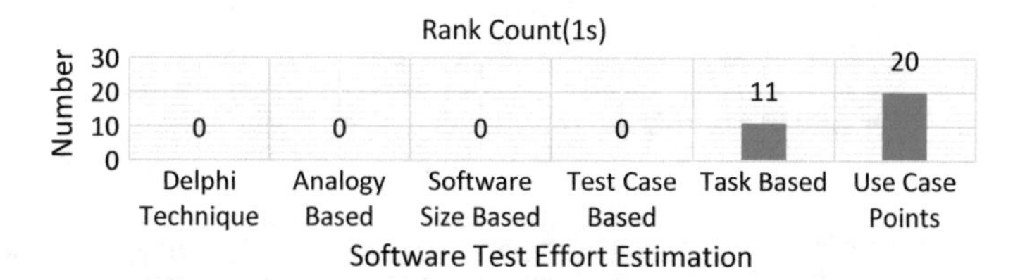

Fig. 4 Count of (1s) ranked by respondents

Based on the result presented Table 7 and Fig. 4 it has been found software test effort estimation technique Use Case Points (20) ranked 1 with maximum number of count (1's) while Task-Based Estimation (11) was found to have achieved the second highest number of count (1's) whereas other software test effort estimation were having zero rank count of (1s). Similarly, rank count for maximum 2s, 3s up to 6s for each individual parameter was carried out and result is presented in Table 8 and graphically in Fig. 5.

From Tables 7 and 8 it can be wrapped up that software test effort estimation technique Use Case Points ranked 1 with maximum number of count (1's) while Task-Based Estimation was found to have achieved the second highest number of count (2's). Similarly, Test Case-Based Estimation was found to have maximum number of count (3's). Whereas Software Size-Based Estimation gained maximum count of (4s) followed by Delphi Technique having maximum count of (5's) and Analogy Based was having maximum count of (6's). At last, top three software test

Table 8 Summation of rank count from rank 1 to rank 10

Rank	Delphi technique	Analogy based	Software size based	Test case based	Task based	Use case test points
1	0	0	0	0	11	20
2	0	0	0	0	20	11
3	0	0	7	24	0	0
4	1	6	17	7	0	0
5	16	11	4	0	0	0
6	14	14	3	0	0	0

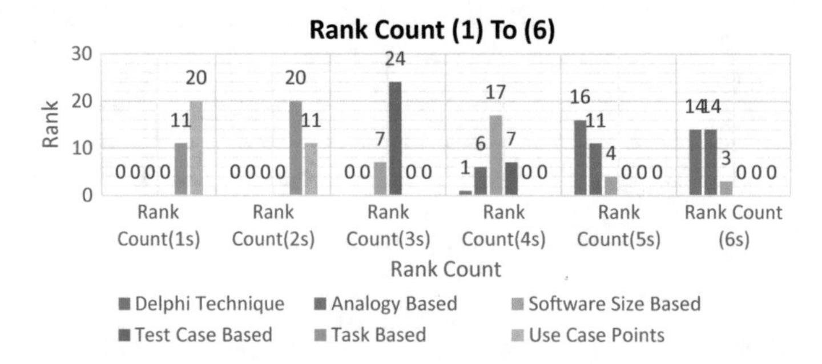

Fig. 5 Count of rank (1s) to rank (6s)

Table 9 Top three ranked software test effort estimation techniques

Sr. No.	Rank	Software test effort estimation techniques
1	1	Use case points-based estimation
2	2	Task-based estimation
3	3	Test case-based estimation

effort estimation techniques based on rank count (1's), (2'), and (3's) are presented into Table 9.

5 Conclusion and Future Work

In the present study, we considered 122 large software project documentation which were used to explore how students handle as well as execute software testing. Further, the main intention was also to find out whether formal mechanism of software test effort estimation and techniques are included in academic courses of computer science and engineering. The elementary objective of analysis these software test effort estimation was to verify whether appropriate and accurate amount of effort in the area of testing is devoted by students in various phases of software development.

These software projects were developed during four consecutive years from 2014–15 to 2017–18. These projects were further classified into four broad categories by using Domain-Based Classification and they are mainly—(1) Desktop Application, (2) Web Application, (3) Mobile Application, and (4) Portal. This classification was done to observe domain preference of students and it was observed that Web Application is having utmost preference whereas Mobile Application having second maximum preference followed by Desktop Application and lastly Portal. The software test effort estimation techniques included in proposed framework are—(a) Delphi technique, (b) Analogy-Based Estimation, (c) Software Size-Based Estimation, (d) Test Case Enumeration-Based Estimation, (e) Task or Activity-Based Estimation, and (f) Use Case Test Points. Our main objective is to find most appropriate software test effort estimation for academic context. For which survey was conducted with 31 academicians teaching in graduate and undergraduate colleges of the Faculty of Computer Science. All academicians fulfilled the criteria of holding at least the Master's Degree with First Class and an experience of at least 10 years in the academic domain. In survey, these academicians were provided with basic details such as four broad categories under which software projects are classified, namely—(a) Desktop Application, (b) Web Application, (c) Mobile Application, and (d) Portal and six software test effort estimation techniques, namely—(a) Delphi Technique, (b) Analogy-Based Estimation, (c) Software Size-Based Estimation, (d) Test Case Enumeration-Based Estimation, (e) Task or Activity-Based Estimation, and (f) Use Case Test Points. Rank is assigned to these software test effort estimations on scale of 1–6 where 1 significances most suitable and 6 impacts least suitable.

The result shows that maximum rank was assigned to Use Case Test Points, the second highest rank was assigned to Task-Based Estimation whereas Test Case-Based achieved third maximum rank. Similarly, Analogy-Based Estimation was given thelowest importance whereas Delphi Technique was assigned second minimum rank. Further, rank assigned to these software test effort estimation techniques is applicable in context to these 122 large software projects and on the basis of rank assigned by 31 Computer Science Academicians. During examination of these software projects documentation, it is observed that good number of students genuinely put efforts as well as their expertise in developing academic software projects. Hence, we believe that software development-oriented courses should focus on these software test effort estimation techniques in order to impart students a formal mechanism as well fruitful practical knowledge in software testing directions which will be helpful in modeling students better IT professional career as well as producing healthy IT intellectual by academic environment. In future, we will execute and implement these top three ranked software test effort estimation techniques, namely—(a) Use Case Test Points, (b) Task-Based Estimation, and (c) Test-Based Estimation on large software projects documentation in order to provide better ways of executing as well as presenting software testing, test data, and test cases in academic software projects documentation.

References

1. Boehm, B., Basili, V., Rombach, H., Zelkowitz, M.: Foundations of Empirical Software Engineering: The Legacy of Victor R. Basili. Springer, New York (2005)
2. Bareisa, E., Karciauskas, E., Limanauskienė, V., Marcinkevicius, R., Motiejūnas, K.: Software engineering process and its improvement in the academy. In: Information Technology and Control, ISSN 1392 – 124, vol. 34, no. 1 (2005)
3. Haapio, T.: Improving Effort Management in Software Development Projects. Publications of the University of Eastern Finland Dissertations in Forestry and Natural Sciences (2011)
4. Sommerville, I.: Software Documentation, no. 3. Lancaster University, UK (2000)
5. Abran, A., Garbajosa, J., Cheikhi, L.: Estimating the test volume and effort for testing and verification & validation. In: International Workshop on Software Measurement—IWSM-Mensura Conference, 5–9 Nov, pp. 216–234. UIBUniversitat de les Illes Baleares, Spain (2007)
6. Bertolino, A.: Software Testing Research: Achievements, Challenges, Dreams, Future of Software Engineering (FOSE'07), 0-7695-2829-5/07 (2007)
7. Srivastava, P.R.: Estimation of software testing effort: an intelligent approach. In: 20th International Symposium on Software Reliability Engineering (ISSRE) (2009)
8. Karunakaran, E., Sreenath, N.: Survey on software effort estimation technique—a review. Int. J. Sci. Eng. Res. **6**(12), ISSN 2229-5518 IJSER (2015)
9. Butt, F.L., Bhatti, S.N., Sarwar, S.: Optimized order of software testing techniques in agile process—a systematic approach. Int. J. Adv. Comput. Sci. Appl. **8**(1) (2017)
10. Jayakumar, K.R., Abran, A.: A survey of software test estimation techniques. J. Softw. Eng. Appl. **6**, 47–52 (2013). https://doi.org/10.4236/jsea.2013.610A006
11. Jayakumar, K.R., Abran. A.: Estimation models for software functional test effort. J. Softw. Eng. Appl. **10**, 338–353. ISSN Online: 1945-3124, ISSN Print: 1945-3116 (2017). https://doi.org/10.4236/jsea.2017.104020
12. Kafle, L.P.: An empirical study on software test effort estimation. Int. J. Soft Comput. Artif. Intell. **2**(2). ISSN: 2321-404X (2014)
13. Jamil, M.A., Arif, M., Abubakar, N.S.A., Ahmad, A.: Software testing techniques: a literature review. In: 6th International Conference on Information and Communication Technology, IEEE 177 (2016). https://doi.org/10.1109/ict4m.2016.40
14. Chemuturi, M.: Test Effort Estimation. http://chemuturi.com/Test%20Effort%20Estimation.pdf (2012)
15. Sabev, P., Grigorova, K.: Manual to automated testing: an effort-based approach for determining the priority of software test automation. World Acad. Sci. Eng. Technol. Int. J. Comput. Syst. Eng. **9**(12) (2015)
16. Hermansky, P.: Analysis of current practice in estimating software testing efforts. Master Thesis Submitted To—Software Technology and Management, Faculty of Electrical Engineering Department of Economics, Management and Humanities (2016)
17. Chaudhary, P., Yadav, C.S.: An approach for calculating the effort needed on testing projects. Int. J. Adv. Res. Comput. Eng. Technol. **1**(1) (2012)
18. Chaudhary, P., Yadav, C.S.: Optimizing test effort estimation-a comparative analysis. Int. J. Sci. Res. Eng. Technol. (IJSRET) **1**(2), 018–020 May 2012 www.ijsret.org ISSN 2278 – 0882 IJSRET (2012)
19. Chauhan, R.K., Singh, I.: Latest research and development on software testing techniques and tools. Int. J. Curr. Eng. Technol. E-ISSN 2277 – 4106, P-ISSN 2347 – 5161 (2014)
20. Saravana Kumar, S.: A survey of software test estimation techniques. Int. J. Emerg. Trends Sci. Technol. IJETST **03**(08), 32–39, ISSN 2348-9480
21. Sangeetha, K., Dalal, P.: A review paper on software effort estimation methods. Int. J. Sci. Eng. Appl. Sci. (IJSEAS) **1**(3), ISSN: 2395-3470 (2015)
22. Nageswaran, S.: Test Effort Estimation Using Use Case Points, Quality Week 2001, San Francisco, California, USA, June 2001 Suresh Nageswaran 1 Copyright(c) 2001, Cognizant Technology Solutions (2001)

23. Nguen, V.: Test case point analysis. White Paper—QASymphony. http://www.qasymphony.com/media/2012/Test-Case-Point-Analysis1.pdf (2012)
24. Afzal, W.: Metrics in software test planning and test design processes. Master Thesis Software Engineering Thesis no: MSE-2007:02, Submitted To—School of Engineering, Blekinge Institute of Technology, Sweden (2007)

Life Cycle Costing Model for Equipment for Indian Naval Shipbuilding Programme

Alok Bhagwat and Pradnya C. Chitrao

Abstract Indian Navy has embarked upon an ambitious fleet expansion and modernization programme to be a 'modern and multi-dimensional Navy'. Further, India has leapfrogged into the league of select few nations capable of building Aircraft Carrier and Nuclear Submarine. Indian Navy has plans to induct 200 new ships by year 2030. A majority of these platforms will see indigenous design. While the warships and submarines have indigenous design, a large volume of equipment and systems going in the platforms are imported. The Indian industry is gearing up for indigenous development of these equipment and is competing with foreign equipment manufacturers for winning the contracts. Whereas, the International navies e.g. USA, Canada, Australia, etc., evaluate the competing suppliers on the basis of 'Total Ownership Cost' or 'Life Cycle Cost (LCC)' of an equipment, in Indian Navy, the contracts are decided on the equipment acquisition cost alone. This is an erroneous process as the 'Operation & Maintenance (O&M) costs' of an equipment is significant given the equipment life cycle of 25–30 years. The O&M costs are much lower for an indigenously sourced equipment as compared to an imported equipment. Therefore, an indigenous supplier is at a disadvantage on one hand and the country pays up higher amounts in the operation cycle of the ships to foreign equipment suppliers on the other hand. The research study aims to develop a LCC model. The authors have undertaken a study towards this. The study has been undertaken by conducting unstructured interviews with senior personnel at Mumbai associated with naval shipbuilding and operations. The findings of this study will be carried forward in a study of larger dimension in near future. This paper aims to outline the study and its findings towards estimation of Life Cycle Costing process for Naval equipment.

A. Bhagwat (✉)
Symbiosis International (Deemed) University, Lavle, Pune 412115, India
e-mail: bhagwat.alok@gmail.com

P. C. Chitrao
Symbiosis Institute of Management Studies (SIMS), Constituent of Symbiosis International (Deemed) University, Pune 412115, India
e-mail: pradnyac@sims.edu

M. Tuba et al. (eds.), *ICT Systems and Sustainability*,
Advances in Intelligent Systems and Computing 1077,
https://doi.org/10.1007/978-981-15-0936-0_6

Keywords Make in India · Indian naval shipbuilding programme · Indigenous submarine construction · Indian naval indigenization plan · Life Cycle Cost

1 Introduction

Indian Navy has embarked upon an ambitious fleet expansion and modernization programme. This includes building new platforms viz. Aircraft Carrier, Destroyer, Frigate, Corvette, Patrol Vessel, Nuclear Submarine and Conventional Submarine to name a few. Indian Navy expects to induct close to 200 warships by year 2030 [5]. It may be mentioned here that, underlying the need for a 'modern and multi-dimensional Navy', Prime Minister Modi has stressed that India would continue to actively pursue and promote its geopolitical, strategic and economic interests on the seas, in particular the Indian Ocean [6].

It is no secret that India spends significant sums in importing equipment for warship building programmes. Additionally, a large chunk of budget is spent in importing critical technology equipment while undertaking modernizations and mid-life upgrades of warships. It has been realized long ago that India must pay greater attention to the modernization and expansion of its naval forces and their ability to meet a wide range of threats. A large part of this maritime modernization policy has another domestic political aspect to it: the concept of the 'Indianization of the Navy'.

Therefore, it is no surprise that the theme of indigenization is central to warship building plans of the Indian Navy which had embarked upon Indigenous shipbuilding journey long ago. The Indigenization theme of Indian Navy got a strong boost by the 'Make in India' initiative. Indian Navy came out with Indian Naval Indigenization Plan (INIP) 2015–2030. The INIP clearly brings out the areas in which indigenous development is desired [4].

While there is a stress and emphasis in developing Naval equipment indigenously, the policy framework has some lacunae. The equipment contracts are awarded solely on the basis of the Acquisition cost of an equipment. The Operation & Maintenance (O&M) costs are not considered while deciding the lowest (L1) bidder. This is an erroneous reflection of the competitiveness of the bidders. Most modern countries viz. USA, Australia, France and Canada evaluate the Defence contracts on the basis of 'Total Ownership Cost' or 'Life Cycle Cost'. Indian Navy considers equipment procurement on the basis of acquisition cost as there is no clear model for estimation of Life Cycle Cost of an equipment. This puts an indigenous supplier at a disadvantage as the O&M costs for an indigenous equipment is much lower than an imported equipment given the life cycle time of 25–30 years. Further the country spends exorbitant sums in maintenance of an imported equipment.

Indian warship building programme has financial outlay to the tune of Rs. 90,000 Cr. Indian Navy is in the process of placing orders for 22 ships Indian shipyards at a total cost of Rs. 15,000 Cr. Indigenous submarine construction programme P75 (I) is on the anvil at a cost of Rs. 75,000 Cr. These ships and submarines have significant potential for indigenization of equipment listed in naval indigenization plan.

DPP 2016 stipulates estimation of Life Cycle Cost (LCC) for an equipment. LCC includes acquisition and exploitation costs. In absence of a clear model, LCC is not considered while deciding L1 Bidder of the equipment.

The research study aims to develop a model for estimation of Life Cycle Cost of a naval equipment. The authors have undertaken a study by conducting unstructured interviews with senior personnel associated with naval shipbuilding and operations to arrive at a Life Cycle Costing model.

Life Cycle Cost includes the Acquisition cost and exploitation costs. Exploitation costs are also termed as the Sustenance Costs or Operation & Maintenance (O&M) Costs. The Life Cycle Cost of an equipment is not clearly defined in Indian Naval procurement documents. At an estimate, the acquisition cost is about 20–40% of the LCC and the sustaining cost would be about 60–80% of the LCC [1].

2 Literature Review

Barringer [2] argued that the LCC economic model provides better assessment of long-term cost-effectiveness of projects that can be obtained with only first cost decisions. He further explained that business must summarize LCC results in net present value (NPV) format considering depreciation, taxes and time value of money. He went on to justify that LCC helps change provincial perspectives for business issues with emphasis on enhancing economic competitiveness by working for the lowest long term cost of ownership which is not an easy answer to obtain. He further listed down Life Cycle Costing process by breaking down into 11 steps and their interdependence. He then applied the steps to analyse the three alternatives of a pump operation to elaborate the application of steps.

Langdon [7] carried out a literature review of Life Cycle Costing and life cycle assessment. It mainly focused on civil construction projects in European Union and aimed at to assess the relevance and impact of Life Cycle Costing towards sustainable construction. It covers various aspects of Life Cycle Costing in details and is expected to contribute towards formulation of life cycle model in the proposed research study.

Prabhakar and Sandborn [9] argue that long life cycle electronic systems typically utilize commercial-off-the-shelf (COTS) parts which subject them to issues viz. high frequency of obsolescence, reliability concerns and supply chain disruptions. They further argue that initial selection decision is often driven by initial costs with little or no insight in the total cost of ownership (TCO). They have suggested a part TCO model and its use in Lifetime Buy and decision Reuse case studies.

Arshad [1] defined Life Cycle Cost of a ship as the total cost of ownership of a ship and its equipment. He elaborated that the LCC of a ship includes its conception, acquisition, operation. Maintenance, upgrades and decommissioning. Accordingly, LCC is normally calculated by summing up the cost estimates from inception to disposal or in other words cost estimates from cradle to grave. He went on to argue that while LCC analysis is carried out for commercial ships to estimate the net profit and return of investment, in the case of Naval Shipbuilding, it is mainly used to choose the most cost-effective option from all the options available.

Kanwal [5] found out Indian Navy's ambitious plan of expanding by 150 ships in next ten to fifteen years. Accordingly, Indian Navy has about 50 ships on order and about 100 more in acquisition pipeline.

Behal [3] found out Indian Naval warship building programmes till year 2027 and highlighted the financial outlay planned for the same. She has highlighted that the order book of Govt. defence shipyards stands at Rs. 86,660 Crore and that of Private shipyards at Rs. 6300 Crore. She brings out a 10% YoY increase in Naval Capital Budget, expected to take it to Rs. 70,000 Cr by Year 2026–27. Highlighting the emphasis laid by Defence Ministry on indigenization, she points to the Indian Naval Indigenization Plan (INIP) 2015–2030, promulgated by Ministry of Defence in year 2015. She explains that the Indigenization Plan particularly focuses on opening up opportunities for private sector and inviting the Indian industry to produce locally, what the Navy has been importing.

DoI [4] lists down the areas of indigenization where Indian Navy has desired import substitution by encouraging Indian industry to establish in-country manufacturing set-up.

Kennedy and Pant [6] analysed the Maritime Security Strategy Document and found out the desire on the part of Indian Navy to modernize and increase its warfighting capabilities across all aspects of naval capability. They found out that the document makes a powerful and thoughtful case for why India must pay greater attention to modernization and expansion of its naval forces. A large part of the maritime modernization policy has another domestic political aspect to it: the concept of the 'Indianization of the Navy'. The authors also found out that the theme of Indigenization runs throughout the statement.

MoD [8] laid down rules and procedures for defence procurement.

3 The Research Gap

The brief literature review clearly brings out the modernization andfleet expansion plans of the Indian Navy and the current shipbuilding orders under execution in the defence andprivate shipyards. However, there is a large gap as regards indigenous manufacture of equipment of critical technology in naval shipbuilding. Indian industry can benefit by associating with indigenous development of the critical technologies that are still being imported for the Naval Warship building.

It is pertinent to mention here that the technologies involved in the imported equipment for Warship building are mature in the civil industry in the country. To cite an example, the Indian Naval submarines use electric propulsion equipment comprising of a high power electric motor and associated drive. The civil industry in India uses a large number of very high power motors and associated drives. In fact, Indian manufacturers viz. BHEL, Bharat Bijlee, IEC, to site a few, regularly manufacture motors of higher power than those used in Naval submarines. But the complete propulsion package of a submarine (motor and drive) is imported. There is, thus, a strong case in laying down a robust process to establish successful, profitable and sustainable manufacturing set-up in country for these technologies. It will involve ruggedization of the industrial equipment to suit marine environment onboard warship. It will involve a capital outlay in setting up test and repair facility dedicated to warship equipment. But the development cost can be offset by lower operation & maintenance costs incurred by an indigenous OEM as compared to a foreign OEM. Therefore, there is a need to evaluate competing suppliers on the basis of Life Cycle Cost, and not the acquisition cost alone. The Life Cycle Costing finds a mention in DPP-2016, mandating the OEM to provide the estimated Life Cycle Cost of the product and the basis thereof. The procurement agencies of the Navy, therefore, should consider the Life Cycle Cost for award of a contract. But this practice is not established as yet as there is a gap in estimation of the Life Cycle Cost. Therefore, till date a majority of contracts are awarded on the basis of acquisition cost. The supplier is liable for equipment performance only up to warranty period. As per the DPP-2016, the following factors are to be considered for arriving at the Life Cycle Cost.

- Operational hours per year
- MTBF (Mean Time Between Failures)
- Requirement of maintenance spares
- Mandatory replacements during preventive maintenance schedules, etc.

However, as mentioned above, the concept of Life Cycle Costing is never applied while deciding the award of contracts where an Indian supplier is competing with a foreign OEM. Thus, there is a need to study and formulate a Life Cycle Costing model for select technologies in Indian Navy's Warship building programme. The proposed study is aiming to bridge this gap by developing a model for estimation of Life Cycle Cost.

4 The Research Problem

Indian Navy has identified technologies for indigenous development in its ships and submarine construction programme. A number of Indian manufacturers is competing with foreign OEMs for getting orders for these equipment. The DPP-2016 mandates that Life Cycle Costs should be considered for deciding the L1 bidder. However, in the absence of a clear model for establishing Life Cycle Cost, the L1 bidder is decided on the basis of acquisition cost alone. This puts the Indian supplier at a great disadvantage. Therefore, there is a need to establish Life Cycle Costing model for technologies being imported by Indian Navy. The following technology has been identified for the study:

- Integrated Platform Management System

The IPMS is being sourced into Indian Navy on competitive bidding. However, the decision is taken based on the Acquisition Cost alone. The Operation & Maintenance Costs are not taken into consideration while deciding the lowest (L1) bidder. Therefore, there are lacunae in arriving at the competitive cost.

5 Research Objective

The study, therefore, seeks to estimate the Life Cycle Cost of IPMS in Warship building programme of Indian Navy.

6 Scope and Limitations

The study is limited to IPMS technology. This technology is explained in brief to bring clarity on its scope and significance for Indian Navy.

6.1 Integrated Platform Management System (IPMS)

A warship has a plethora of equipment and systems to operate and carry out mission objectives. With the advent of technology in networking and data acquisition, Indian Navy inducted a central control system in its warships to provide centralized monitoring and control of various machinery, equipment and systems. The features like embedded training simulator, fire detection and fire fighting, flood detection and control, stability management and dynamic analysis of main machinery provided a potent tool in the hands of the operator and maintainer to be able to effectively utilize the ship machinery like never before. The IPMS was first inducted into Indian Navy in year 2000 by import. Subsequently, all the ships of Indian Navy are now equipped with IPMS. Even the existing ships are being retrofitted and modernized with IPMS during refit. The IPMS for first two projects was imported. But in subsequent projects, Indigenous firms also developed this system in-house and competed with foreign firms. Today the equipment is being sourced by Indian Navy from indigenous as well as foreign firms on competitive basis.

The IPMS is being supplied by following firms:

- L3 Mapps, a Canadian firm with Indian set-up
- Larsen & Toubro (L&T) with total in-house capability
- Tata Group with a Russian firm as technology partner
- BHEL with an Italian firm as technology partner

The estimation of Life Cycle Costs is limited to stipulations in line with DPP-2016

- Operation hours per year
- Mean Time Between Failure (MTBF)
- Requirement of Maintenance Spares
- Mandatory replacement during preventive maintenance

The other Factors are

- Equipment life 25–30 years
- Obsolescence management—two to three modernization cycles during exploitation phase.

7 Research Methodology

7.1 Primary Data

- Select a high technology equipment that is being sourced in Indian Navy from indigenous as well as foreign sources. The researcher has selected IPMS as the study area.
- Analyse Life Cycle Cost elements involved in above technologies in context of Indian Navy's warship building programmes and suggest a method to estimate Life Cycle Cost.
- Unstructured Interviews with senior top management officials of Indian Navy, Defence Shipyards and Indian Industry in Mumbai for following:
  - Validate of competitive procurement process for induction of IPMS in warship building programme
  - Validation of Life Cycle Costing model prepared subsequently
  - Validation of Life Cycle Costs estimates using model above for an import option and indigenous option
- Compare the cost to arrive at the competitive option.

A questionnaire was prepared for unstructured interview with senior officials.

7.2 Secondary Data

- Study of Defence Procurement Procedure-2016 document and Indian Naval Indigenization Plan document for relevance of 'Make in India' concept with stipulations for Life Cycle Costing in focus

- Study Life Cycle Costing models applicable for foreign advanced Navies (USA, Canada, France, etc.) and establish a model suitable to Indian Navy for the selected technologies
- Literature Review for identifying commercial potential of the selected technology area.

8 Findings

8.1 Primary Findings

The research study was carried out by conducting unstructured interviews with top management persons from agencies associated with Indian Naval shipbuilding programme, including a prominent industry. The findings are appended in succeeding paragraphs.

Selection of Suppliers. For critical technology equipment, Indian Navy has a stringent selection process for shortlisting prospective suppliers. Foreign firms must have a proven track record of supplying similar equipment to a modern Navy and demonstrate intent of achieving certain indigenization levels. An Indian form must have partnership with a global technology leader with proven track record. For IPMS, Indian Navy has shortlisted four firms, one foreign firm with an Indian set-up and three Indian firms.

Pre-Bid Process. Before issuing a tender, the Indian Navy and shipyard draw out the technical specifications. These specifications are forwarded to the shortlisted suppliers. The comments from suppliers are sought and a meeting is held to discuss various issues threadbare. Thus, all the suppliers are brought to a level playing field and the specifications are revised in line with the discussions. This is done to ensure that the tendering process is smooth and a level playing field is ensured to all suppliers.

Tendering Process. The tenders are submitted in sealed bids and the technical evaluation commences to validate the offers of the firms and their compliance with the tender terms. Once the technical evaluation is completed, the selected firms are invited for opening the commercial bids and the winner (L1 Bidder) is decided based on the lowest quote.

The suppliers are required to indicate the spares list, MTBF of the equipment, the maintenance schedules and the rate of technicians for carrying out defect rectification and maintenance.

Elements of Life Cycle Costing Model. The suppliers' responsibility is limited to delivering the equipment, commissioning it onboard and providing warranty support for 5 years. After that period, the Indian Navy has to pay for the maintenance of the equipment. The equipment life is stipulated to be 25–30 years. But there is an obligation on the supplier to support the equipment for the entire life of the equipment on commercial terms.

Therefore, it is clear that the bid selection process is erroneous as the supplier is commercial liable only for about 15% (five years out of 30 years) of the operational life of the equipment. For balance 85% of the operational life, the supplier is free to quote at its will. The following elements largely determine the Life Cycle Cost of the equipment.

- **MTBF**

The Mean Time Between Failures is an important measure of the future costs that would be incurred in keeping the equipment operational. The equipment comprises of assemblies, sub-assemblies, Kits and modules. These modules have a failure rate that is specified by the manufacturer. Based on the MTBF of each module and the number of hours of operation, an estimate of modules required to be replaced in the life of the equipment can be estimated.

- **Scheduled Maintenance**

The equipment has to undergo scheduled maintenance by the suppliers' technicians regularly. The extent of maintenance is specified and an estimate of the cost that would be incurred towards this activity can be estimated. This cost element is dependent upon the material required for the maintenance and deputation cost of the technicians.

- **Breakdown Maintenance**

This is a difficult area as the prediction of equipment failure is difficult. But this also form an important element of Life Cycle Cost.

- **Obsolescence Management**

During the life of the equipment of 25 years, it is expected to undergo two cycles of obsolescence necessitating major overhaul of the equipment. Since the cost of the original equipment is known, an estimate of these overhauls can be made.

- **Support System on Shore**

A complex equipment requires test, repair and training facility in naval establishments. This also forms an important aspect of the Life Cycle Cost.

As an example, it is seen that the technicians' charges for the foreign firm if around USD 2500 per day as against Rs. 50,000 for an Indian firm. Therefore, in the event of the order going to the foreign firm on Acquisition cost basis, Indian Navy has to pay three-to-four times in Foreign Exchange (FE) for inviting the technicians of the supplier for 25 years in the Life Cycle of the equipment. If this cost is considered in the overall cost, the foreign firm would not remain L1.

The research study has brought out the elements of the Life Cycle Costs to be considered in bid evaluation process and the technicians' deputation cost is a major factor. The Main study will bring out the exact elements of the Life Cycle Cost and suggest a roadmap for indigenous development of critical technologies for Naval shipbuilding.

8.2 Secondary Findings

These are as given below.

Problems with Imports

- Inadequate support post warranty period
- Exorbitant cost of deputation of service engineer
- High downtime of equipment
- Huge foreign exchange outgo

Boost to Indigenization

- Indian industry has been supporting indigenization drive of Indian Navy
- Indian Navy has promulgated equipment list form indigenous development
- Life Cycle Costs (LCC) must be the basis of deciding 'L1' bidder.

9 Recommendation

Indian Industry should undertake urgent capability building in areas of technologies for supporting Indian Navy's indigenous warship building programme. Indian Navy should evaluate bidders on the basis of Life Cycle Cost of an equipment taking into account O&M (operation and maintenance) costs also. This practice is being followed in modern Navies globally. Life Cycle Cost is the true indication of competitiveness of a supplier. Life Cycle Cost is expected to be significantly lower for an indigenous supplier as compared to a foreign supplier.

10 Benefits of the Study

Inclusion of Life Cycle Costing in the tender evaluation is bound to benefit Indian industry as well as Indian Navy. The operation & maintenance cost of the equipment is expected to go down significantly, thus saving the nation precious foreign exchange. The equipment availability for the Indian Navy is expected to improve with Indian technicians carrying out maintenance activities. The Indian industry is expected to gain financially as a bigger share of wallet of warship building outlay will be ploughed back in the country. The findings of this study will be applied in estimation of the Life Cycle Costing model. After validation of this model by experts from Navy, shipyards and industry, this will be applied for a bigger study to suggest a road map for indigenization of a critical technology that is widely used on naval ships and submarines.

References

1. Arshad, T.K.: Ship's Life Cycle Cost. Published in https://tkays.com/2013/01/31/ships-life-cycle-costing/ (2013)
2. Barrington, H.P.: A Life Cycle Cost Summary. Published in International Conference of Maintenance Societies (ICOMS—2003). (http://www.barringer1.com/pdf/LifeCycleCostSummary.pdf) (2003)
3. Behal, R.: Naval Shipbuilding Financials. Published in http://www.defproac.com/?p=2700 (2015)
4. DoI (Directorate of Indigenization) IHQ MoD (N): Indian Naval Indigenization Plan (INIP) 2015–2030, pp. 6.7.8 (2015)
5. Kanwal, G.: India's Military Modernization: Plans and Strategic Underpinnings, p. 6. The National Bureau of Asian Research for the Senate India Caucus (2015)
6. Kennedy, G., Pant, H.V.: Indian Navy's Maritime Security Strategy—An Assessment, pp. 1, 3, 4, 5. National Maritime Foundation (2016)
7. Langdon, D.: Literature Review of Life cycle Costing (LCC) and Life Cycle Assessment (LCA). (Published in http://www.tmb.org.tr/arastirma_yayinlar/LCC_Literature_Review_Report.pdf) (2006)
8. MoD (Ministry of Defence): Defence Procurement Procedure—2016 (2016)
9. Prabhakar, V.J., Sandborn, P.: A part total cost of ownership model for long life electronic systems. Int. J. Comput. Integr. Manuf. 25(4–5), 384–397 (2010)

Enhancing Programming Competence Through Knowledge-Based Intelligent Learning System

Kanubhai K. Patel and Nilay Vaidya

Abstract Computer programming competence is essential for computer science students. For them, it is challenging to understand computer programming courses easily. It eventually causes a lack of interest in the programming courses. It also impacts on employability of the students. Currently, around 20–30% fresh graduates are only employable and get placed. To overcome the problem, the authors have proposed a framework of Knowledge-based Intelligent Learning System (KILS) which enhance the programming competence of the students and thereby increase their employability. An experiment was conducted on undergraduate and postgraduate computer science students to evaluate the significance of the proposed solution to the problem. The results show that the proposed solution not only enhances their programming competence which ultimately increasing their employability but also improves the students' involvement and attitudes towards programming.

Keywords Knowledge-based learning system · Student learning · Computer programming competence · Student involvement

1 Introduction

A good number of students are taking admission in undergraduate level (such as BCA and B.Sc.(IT)) and postgraduate level (such as MCA and M.Sc.(IT)) computer science programmes. Programming competence is the desired knowledge and skills of students in computer programming-based courses or programmes. For these students, it becomes difficult to understand computer programming courses easily in initial years as it involves the comprehension of theoretical background, practical usage of semantics and syntactic coding and algorithmic skills [1]. It eventually

K. K. Patel (✉) · N. Vaidya
Charotar University of Science and Technology, Changa 388421, Gujarat, India
e-mail: kkpatel7@gmail.com

N. Vaidya
e-mail: vaidyanilay@gmail.com

M. Tuba et al. (eds.), *ICT Systems and Sustainability*,
Advances in Intelligent Systems and Computing 1077,
https://doi.org/10.1007/978-981-15-0936-0_7

causes lack of interest in the programming courses. It also impacts the employability of the students. Currently, around 20–30% fresh graduates are only employable and get placed. To address and overcome the problem, the authors have proposed a framework of Knowledge-based Intelligent Learning System (KILS) which enhances the programming competence of the students and thereby increases their employability. The system includes all three categories of learning, i.e. learning-by-listening, learning-by-watching and learning-by-practicing. It is a web-based solution by which participants learn in each category repeatedly. Following is the list of some of the research questions:

- What is the significance of using this technique to teach various programming courses at undergraduate- and postgraduate-level programmes?
- Which points are to be kept in mind?
- How can we create more interest in the students and thereby make our teaching more effective?
- What are the effects of our system on the academic performance of the students?
- Can we improve the employability of the students?
- Can we improve students' interest in the programming?

In this paper, we have attempted to answer the questions by the experimental research methodology. We have conducted an experiment in teaching technical courses at undergraduate- and postgraduate-level programmes. The objectives of the study are

- To enhance students' involvement in programming course learning,
- To study the effect of the system in augmenting students' involvement in the learning of programming courses and
- Ultimately to improve the students' academic and professional performance.

Section 2 presents the overview of related works while Sect. 3 describes the framework for the intelligent learning system. Section 4 presents experimental study and Sect. 5 presents results. The last section, Sect. 6, concludes the research paper with the future directives.

2 Related Works

A good number of researchers have proposed solutions using knowledge-based learning system [2–6]. Hilles and Abu Naser [3] have proposed a knowledge-based intelligent tutoring system for teaching the MongoDB database. Mona [4] has proposed a multi-agent-based intelligent tutoring system for teaching Arabic grammar.

Simmons et al. [5] proposed a solution to reveal programming competency by using a different approach, that is, game development. Maramis et al. [6] proposed a mobile-video-based learning solution for improving programming competency.

Bashir and Hoque [1] used problem-based learning (PBL) where students engage themselves to self-learn, self-practise and mainly focus on ill-structured problems. Bashir et al. [7] have used problem-based learning (PBL) for E-learning of PHP based on the solutions of real-life problems.

3 Framework of Knowledge-Based Intelligent Learning System (KILS)

Although isolated solutions have been attempted, no integrated solution of improving programming competency of the learner is available to the best of our knowledge. Yet no researcher has given a computation framework to cover all the aspects of the learning process of the learner. Our framework of Knowledge-based Intelligent Learning System (KILS) provides an abstraction of learning by the learner (see Fig. 1). Special emphasis is placed on cognitive aspects and online assessment of perceived knowledge by users. Understanding how learning tasks are constructed is useful in determining how best to improve performance. We should decompose the various tasks of learning in a generic way.

To describe the framework of Knowledge-based Intelligent Learning System (KILS), we divided the learning process into the following three modules (see Fig. 1)

1. Learning,
2. Instructor and
3. Intelligent observer.

3.1 Learning

In this section, an interactive learning environment is provided to the learner. The learner can select the mode of learning as per his choice. Different learning modes are (i) learning-by-listening, (ii) learning-by-watching and (iii) learning-by-practising. The learner can select any of these modes interchangeably. In the learning-by-listening mode, the learner can listen to the audio of the instructor for the

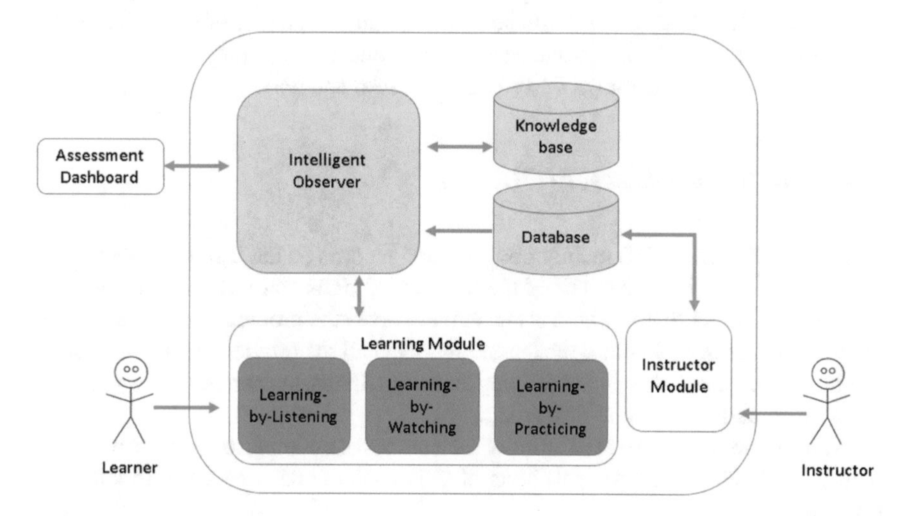

Fig. 1 Schematic diagram of the proposed framework

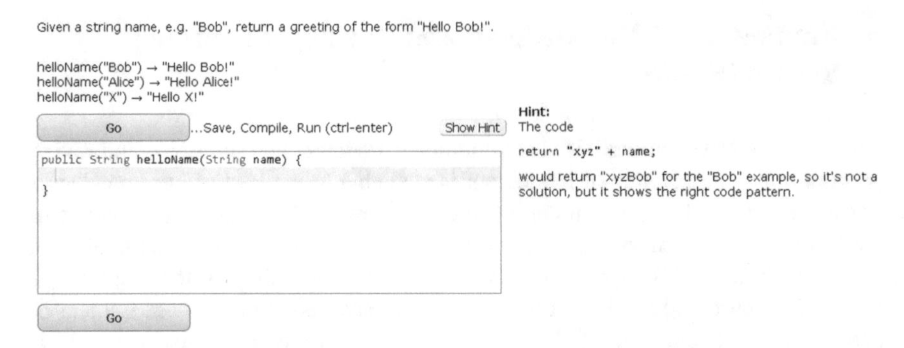

Fig. 2 Learning-by-practicing mode

programming logic. In the learning-by-watching mode, the learner can watch the video of the instructor for the programming. In the video, an instructor shows him how to write a code for a given problem and execute the same.

In the learning-by-practising mode, the learner needs to write a program code for the problem provided by the instructor. The learner writes code and executes the same (see Fig. 2).

3.2 Instructor

By using this module, teachers can frame problems with their solutions. They can set the audio, video and presentation files of each problem. For each problem, the teacher generates audio clips, videos and texts with solutions. She assigns the difficulty level of problems, viz., (i) advance, (ii) general and (iii) simple. She also provides a hint for each problem. She can view the performance and behaviour of the learner in the given environment. The performance and behaviour of the learner will be identified by the Intelligent Observer (as described in the next section).

3.3 Intelligent Observer (OB)

Intelligent Observer (OB) mainly observes and records (i) the learning patterns, (ii) performance and (iii) behaviour of the learners. Various learning patterns might be used by the learner that is whether the learner solves the problem without accessing the video or hints, or whether the learner has used all the provided methods to solve the problem. This helps in identifying better methods to learn the programming course by a particular group of learners.

Performance is based on the terms of the number of attempts to solve the problems with or without taking provided hints of the problem and time taken to solve the problem.

While behaviour is in terms of actions performed by the learner, viz., (i) number of times the learner has used various materials, (ii) flow of use of materials and (iii) time taken by the learner to solve the problem. Learners' jumping pattern is also identified by the OB.

4 Experimental Evaluation

Experimental evaluation has been carried out for the last 2 years (i.e. 2018 and 2017). A total of four batches of undergraduate and four batches of postgraduate-level programme students were given training using our system and the data was collected. During the experiments, Java language programming competency was evaluated in BCA, B.Sc.(IT), MCAL and M.Sc.(IT) students. We have also considered students' result data for the years 2016 and 2015. During these 2 years, training was given using conventional teaching methods only, i.e. without our Knowledge-based Intelligent Learning System (KILS).

5 Results

We have results of the abovementioned four programmes for the years 2015–2018. The following table shows the data on the results of students (see Table 1).

Table 1 Comparative results of students

Programme-year	Percentage of students passed
BCA-2018	96.58
BCA-2017	96.35
BCA-2016	91.44
BCA-2015	91.06
B.Sc.(IT)-2018	97.63
B.Sc.(IT)-2017	97.12
B.Sc.(IT)-2016	92.14
B.Sc.(IT)-2015	91.86
MCAL-2018	98.28
MCAL-2017	99.02
MCAL-2016	93.54
MCAL-2015	93.26
M.Sc.(IT)-2018	91.28
M.Sc.(IT)-2017	92.02
M.Sc.(IT)-2016	87.74
M.Sc.(IT)-2015	88.06

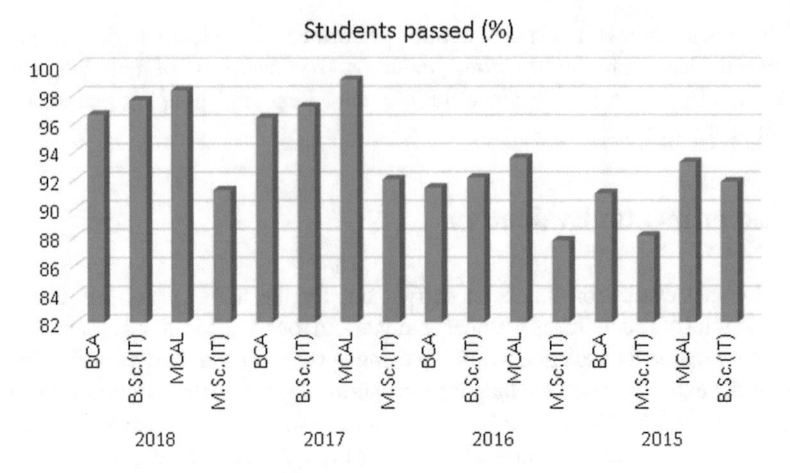

Fig. 3 Comparison of students' results

6 Conclusion and Future Directives

In this paper, we attempt to provide the solution of enhancing the programming competence of computer science students by proposing the framework of the Knowledge-based Intelligent Learning System (KILS). The framework considers mostly all the aspects of improving the students and thereby increasing their employability. The results of the experimental study assure the significance of the framework for the learners as well as higher education institutions. It additionally assures that the solution improves the students' involvement and attitudes towards programming. We would like to consider cognitive factors and carry out the extensive evaluation of the solution with more number of participants as our future works (Fig. 3).

References

1. Bashir, G.M.M., Hoque, A.S.M.L.: An effective learning and teaching model for programming languages. J. Comput. Educ. **3**(4), 413–437 (2016). https://doi.org/10.1007/s40692-016-0073-2
2. Gálvez, J., Guzmán, E., Conejo, R.: A blended E-learning experience in a course of object oriented programming fundamentals. Knowl. Based Syst. **22**(4), 279–286, ISSN 0950-7051 (2009). https://doi.org/10.1016/j.knosys.2009.01.004
3. Hilles, M.M., Naser, S.S.A.: Knowledge-based intelligent tutoring system for teaching mongo database. Eur. Acad. Res. **4**(10) (2017)
4. Mahmoud, M.H.: A multiagents based intelligent tutoring system for teaching Arabic grammar. Int. J. Educ. Learn. Syst. **3**, 52–59 (2018)
5. Simmons, S., Disalvo, B., Guzdial, M.: Using game development to reveal programming competency (2012). https://doi.org/10.1145/2282338.2282359

6. Maramis, G.D.P., Palilingan, V.R., Modeong, M.: Mobile video learning for improving programming competency. In: IOP Conference Series: Materials Science and Engineering, vol. 384 (2018) https://doi.org/10.1088/1757-899x/384/1/012012
7. Bashir, G.M.M., Hoque, A.S.M.L., Nath, B.C.D.: E-learning of PHP based on the solutions of real-life problems, J. Comput. Educ. **3**(1), 105–129 (2016). https://doi.org/10.1007/s40692-015-0050-1

Digidhan Dashboard—Monitoring and Analysis of Digital Payments

Deepak Chandra Misra, Inder Pal Singh Sethi, Om Pradyumana Gupta, Misha Kapoor and Ritesh Kumar Dwivedi

Abstract In today's digital era, Government is trying to digitize most of its public services through its e-initiatives to benefit the common man. Digital transactions are taking place for payment of bills, recharge of metro cards, online shopping, and what not. The need of a teller is diminishing from most of the customer interface points. This has resulted in generation of huge transaction data at every service point. After Hon'ble Finance Minister's announcement of a 2500 crore target of digital transactions in FY 2017–18, many campaigns are being run to promote the growth of digital transactions across the nation. Government is also promoting necessary infrastructure to enable digital transactions in less digitized regions. The Digidhan dashboard is an innovative step taken by the Ministry of Electronics and Information Technology (MeitY) and National Informatics Centre (NIC) to monitor the growth of digital transactions on a central platform. It seamlessly captures the data related with digital payments from various channels like Reserve Bank of India (RBI), National Payments Corporation of India (NPCI), and 110 public sector, private sector, payments, regional rural and foreign banks, ensuring its completeness, accuracy, and relevance. Further, it reveals the potential of data in form of meaningful insights by using data analytics, making it one of most technologically advanced Government portals. The dashboard has paved the way for an advanced mechanism to deal with the volume and variety of digital transactions happening in the nation. It is being further enhanced by development of a chat bot to address user queries. Recently, it

D. C. Misra · I. P. S. Sethi · O. P. Gupta (✉) · M. Kapoor · R. K. Dwivedi
National Informatics Centre, Government of India, Block—A, CGO Complex, Lodhi Road, New Delhi 110003, India
e-mail: op.gupta@nic.in

D. C. Misra
e-mail: dcmisra@nic.in

I. P. S. Sethi
e-mail: sethi@nic.in

M. Kapoor
e-mail: misha.kapoor@nic.in

R. K. Dwivedi
e-mail: ritesh.dwivedi@nic.in

M. Tuba et al. (eds.), *ICT Systems and Sustainability*, Advances in Intelligent Systems and Computing 1077,
https://doi.org/10.1007/978-981-15-0936-0_8

was integrated with the smart city portal to capture and promote digital transactions in 100 smart cities. The dashboard has been widely welcomed by all stakeholders and has a lot of scopes to bring in potential change with respect to Government's idea of digitizing the whole economy. Digidhan dashboard can be accessed at http://digipay.gov.in.

Keywords Digital payments · Digital transactions · Dashboard · Smart city · Data analytics · BI reporting

1 Background

Hon'ble Finance Minister, in his budget speech announced several activities for the promotion of digital payments including a target of 2500 Crore digital payment transactions in FY 2017–18, through Unified Payments Interface (UPI), Unstructured Supplementary Service Data (USSD), Aadhaar Pay, Immediate Payment Service (IMPS) and Debit Cards. Ministry of Electronics and Information Technology (MeitY) is entrusted with the responsibility of leading the initiative on "Promotion of Digital Transactions including Digital Payments."

The benefits of becoming digital are manifold for the citizens as well as for the economy. Government is trying to create awareness about the use of digital modes in rural and less privileged areas and build required infrastructure in these areas for enabling digital payment transactions. MeitY realized the need of a central platform to monitor the growth of these digital transactions as well as absorb the digital payments data and convert it into meaningful insights for the stakeholders for better planning of promotion of digital payments.

MeitY decided to get the Digidhan dashboard developed by National Informatics Centre (NIC) for accurate reporting, monitoring, and analysis of all types of digital payments transactions occurring in the country and enablement of infrastructure through deployment of Physical/Mobile/BHIM Aadhaar PoS devices. The Digidhan dashboard was launched by Shri Ravi Shankar Prasad, Hon'ble Minister of Electronics and Information Technology and Law & Justice on February 13, 2018 in presence of eight State IT Ministers and more than 30 State Secretaries during National Conference of State IT Ministers and IT Secretaries in Vigyaan Bhawan. The home page of the dashboard is shown in Fig. 1.

The dashboard has two levels of access—"General access" that provides details of growth of digital payment and related infrastructure and 'Privileged access' that is given to the stakeholders to review their performance for better implementation and promotion of digital payments. Digidhan dashboard is the first of its kind and is a monitoring tool that accesses and reports data on T + 1 basis from various channels like RBI, NPCI, Banks, etc. The dashboard also employs target setting and monitoring for showcasing top and bottom banks. A composite scorecard is developed to evaluate the performance of banks on assigned targets/parameters and also to create an environment of healthy competition among them. The scorecard

Fig. 1 Home page of Digidhan dashboard

measures each bank on various parameters generating an overall score which qualifies a bank into certain performance category.

The dashboard is also providing digital payments data to Central Govt. portals like eTaal, the e-Transaction aggregation and analysis layer, through API. eTaal provides an aggregated view of e-Transactions performed through e-Governance applications implemented including, but not limited to, the projects of national importance like 31 Mission Mode Projects (MMPs) defined under National e-Governance Plan (NeGP).

Digidhan dashboard is being leveraged to support other Digital India campaigns such as Smart City campaign. The smart city module is integrated with Digidhan dashboard for the purpose of promoting digital payments in 100 smart cities from July 1, 2018 to Oct 31, 2018. The module captures daily performance of these smart cities on the amount collected and the percentage of digital collection. This is a big step toward making the departments realize that a large portion of transactions in the smart city can be transformed into digital payments. Given the Indian landscape, Digidhan dashboard binds the campaigns undertaken in different parts of the country for digitizing the economy and gives aggregated numbers on a single platform.

2 Functionalities

The dashboard offers a number of functionalities including:

- Tracking of total digital payments transactions with mode and bank analysis
- Target setting and monitoring to evaluate performance of banks (top and bottom banks)
- Distribution of digital payments transactions state wise on per capita basis
- Data analysis available through BI-driven dashboards:

  - BHIM transaction and decline analysis
  - Aadhaar and mobile seeding analysis
  - BBPS transaction analysis
  - Closed loop transaction analysis
  - POS deployment and correlation analysis
  - Digital transaction analysis for different banks
  - Digital transaction analysis for different Ministries

- Composite scorecard to evaluate performance of banks on various parameters
- Tracking of POS and BHIM Aadhaar deployment
- Tracking of Mobile/Aadhaar number seeding to bank account
- Bank mode wise report which includes other payment modes (modes that are specific to a particular bank only)
- Smart city reporting portal embedded. The portal automatically consolidates digital payment data from various Departments/Ministries of 100 smart cities and generates reports to track their performance.

3 Key Stakeholders

A number of stakeholders are part of the Digidhan Dashboard ecosystem including:

- Prime Minister's Office (PMO)
- MeitY
- Reserve Bank of India (RBI)
- National Payments Corporation of India (NPCI)
- 110 Banks (public sector, private sector, payments banks, regional rural and foreign banks)
- 100 Smart Cities
- Ministry of Railways
- Department of Posts
- Ministry of Civil Aviation
- Ministry of Road and Transport
- Ministry of Petroleum and Natural Gas
- Ministry of Power
- Department of Telecom
- Ministry of Housing and Urban Affairs
- City Corporations of 100 smart cities.

4 Challenges During Implementation

In the beginning, there was no single secondary source of digital transactions data that could be relied upon for completeness and comprehensiveness of data. Therefore, data was pulled automatically from various agencies like RBI, NPCI, Banks, Closed loop wallets to ensure quality of data. Today, the dashboard is integrated with 24 public sector, 23 private sector, 4 payments, 56 regional rural and 3 foreign banks.

The dashboard was planned to be implemented as a web service initially and pull data from RBI, NPCI, and banks through APIs. Owing to the massive size of transaction data, per day transmission size had reached up to 5 GB. Therefore, a mechanism was devised using the Secure File Transfer Protocol (SFTP) technology for pulling large data from banks and other agencies. SFTP was used for both pulling and pushing files between Bank's SFTP Servers and Dashboard SFTP Servers.

The BHIM app which uses UPI platform was launched on Dec 30, 2017 in New Delhi. The app did not get immediate popularity after launch due to its high transaction failure rate. With the help of Digidhan dashboard, the reason of failure was identified and resolved which helped the app to pick up the volume of successful transactions.

5 Software Environment

5.1 Technology Used

Digidhan Dashboard is developed using Microsoft technologies stack. It is a web application written in C# language, uses SQL Server 2012 database and works on Microsoft .NET Framework 4.5. Windows Communication Foundation (WCF) is used as middleware. The database size is 3 TB and is expected to expand to 6 TB for near future developments. Power BI tool is integrated with the dashboard for showcasing in-depth analysis of digital transactions through data visualization.

5.2 Data Center and Disaster Recovery

The application is hosted on cloud of National Data Centre (NDC), Shastri Park, and New Delhi. The state of the art ICT Infrastructure of NDC includes High-end Blade and Rack Servers, Enterprise level Storage Systems, and Automated Tape Library Systems for data back up and Network & Information Security.

The disaster recovery site is located at Hyderabad. High Availability (HA) option of SQL Server 2012 is configured to make database instances available 24×7 in DC center as well as DR center. As part of the SQL Server Always On offering, Always On Failover Cluster Instances leverages Windows Server Failover Clustering

(WSFC) functionality to provide local high availability through redundancy at the server-instance level—a Failover Cluster Instance (FCI). An FCI is a single instance of SQL Server that is installed across Windows Server Failover Clustering (WSFC) nodes and, possibly, across multiple subnets. On the network, an FCI appears to be an instance of SQL Server running on a single computer, but the FCI provides failover from one WSFC node to another if the current node becomes unavailable.

6 Data Analytics

The digital transaction data has huge potential for data analytics owing to its large volume and variety. Analysis of digital payments is being done and is available on the dashboard in the form of BI reports. Detailed analysis is embedded on the portal with the help of Microsoft Power BI (Table 1).

Table 1 Volume of digital transaction for 16 payment modes integrated with Digidhan dashboard in FY 2018–19

Mode of payment	Volume of digital transactions (in Cr.)	% volume of total digital transaction
PPI	345.91	16.70
Debit card	334.34	16.14
NACH	237.49	11.47
NEFT	194.64	9.40
Others	171.04	8.26
Internet banking	149.35	7.21
Credit card	140.51	6.78
Closed loop wallet	115.09	5.56
IMPS	101.32	4.89
AEPS	98.25	4.74
BHIM	91.30	4.41
Mobile banking	66.21	3.20
NETC	12.68	0.61
RTGS	12.44	0.60
BHIM Aadhaar	0.20	0.01
USSD	0.22	0.01

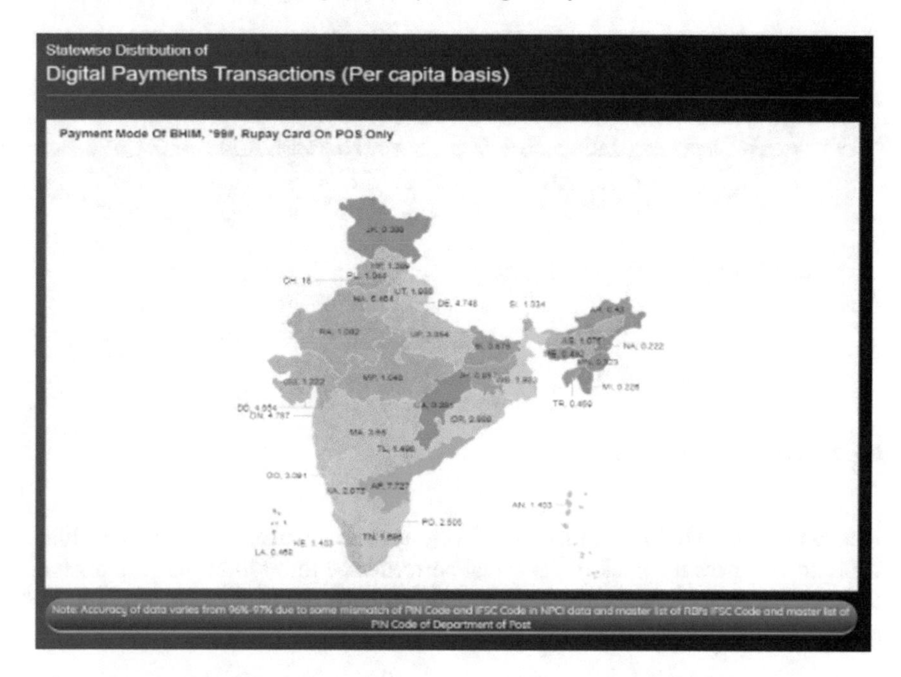

Fig. 2 Distribution of digital payment transactions state wise on per capita basis

6.1 State Wise Distribution of Digital Payment Transactions

A state wise distribution of digital payment transactions is shown on the dashboard with help of a heat map. The visual considers population of the respective State and aggregates digital transactions done through three payment modes, i.e., BHIM, USSD, and Rupay Debit Card on PoS to give an average of total digital transactions per capita basis for each State. The States are marked in red, yellow, and green to signify regions with high medium and low digital payment transactions. This is depicted in Fig. 2.

6.2 Bank Performance Scorecard

Performance of banks is tracked through target monitoring where a target is shared with each bank. The bank is rated on various parameters including target achievement. This rating is used to categorize the bank into a low, medium, and high performer. User can view top and bottom 5 banks on the home page.

A detailed scorecard is also developed to track the performance of banks (Fig. 3). This score card is generated every month and is available to each bank under dashboard login. The scorecard assigns target to each bank on various parameters (KPIs)

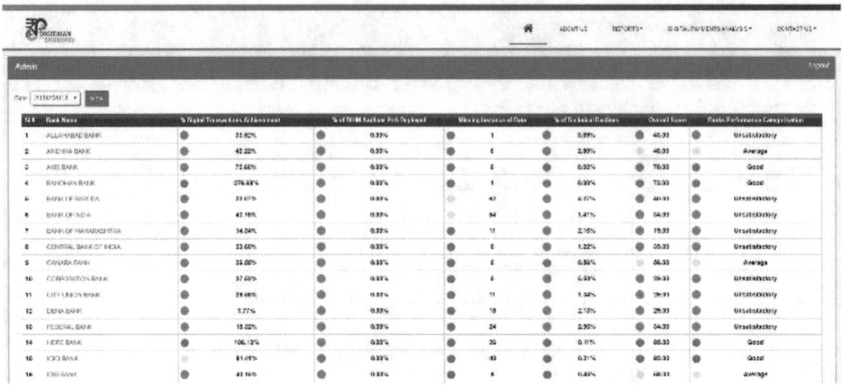

Fig. 3 Bank scorecard summary

like compliance, Digitization Index, UPI System Resilience, Grievance Handling, etc., and evaluates the bank on its actual performance for each of these parameters to arrive at a final score. This score is used to rank the banks in a low, medium, and high performance category.

6.3 Other Reports

A Bank file sharing status report is available under login that gives details of the data shared by each bank for a particular day and indicates the number of days for which a particular bank has shared data in the selected period. Another report lists various payment modes for which a selected bank is providing data. User can view transaction volume for each payment mode as well as the missing days for which data was not received from the bank for that particular payment mode. The missing date information is also compiled bank wise in a separate report for user convenience. Other reports include NPCI Missing data report and NPCI value and volume report.

6.4 Digital Payment Analysis

Data analysis dashboards are built and integrated in the portal through Microsoft Power BI technology. These dashboards are user friendly and provide them with a platform to interact with the data and obtain meaningful insights. Analysis is done for different payment modes, banks, POS deployment, Aadhaar, and mobile number seeding.

The mode wise analysis depicts transaction volume across various digital payment modes taken from RBI, NPCI, and banks. RBI payment modes include RTGS,

NEFT, Credit card, Debit card, and PPI. NPCI payment modes include BHIM, BHIM Aadhaar, IMPS, NACH, USSD, and AePS. Bank payment modes include Mobile banking, Internet banking. Closed loop wallet data is collected from respective metros and Oil Marketing Companies (OMCs). By selecting a particular payment mode from the filter, user can view its monthly transaction volume or do a comparative analysis of all payment modes month wise.

Bank analysis includes a target vs. achievement analysis that depicts banks' achievement against given targets till date. A month-over-month analysis of banks is also available that presents monthly transaction volume for last three financial years for selected bank and payment mode. A comparative analysis drills further into the specific payment modes for each bank and reflects its growth trend for selected time duration.

The POS deployment dashboard (Fig. 4) shows total number of physical/mobile POS devices and BHIM Aadhaar pay POS devices deployed so far, for a selected bank and calculates an achievement percentage for them against set targets. The banks are further ranked on the basis of their achievement into top 5 and bottom 5 banks. A correlation between the total POS devices deployed, total number of transactions and total number of accounts is also available.

The BHIM analysis dashboard gives an overview of the monthly growth of BHIM transactions and the value range in which these transactions are falling. Additionally, it also helps in monitoring the declined transactions by giving the status of technical decline and business decline transactions for a given period. The growth pattern of BHIM transactions is depicted in Fig. 5. It also shows BHIM transaction value analysis where the transactions have been bucketed into different value ranges.

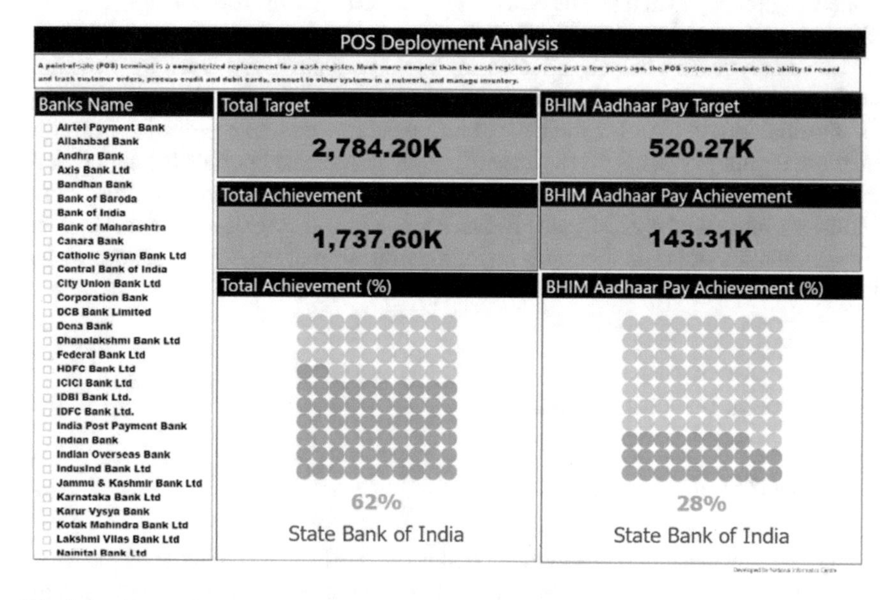

Fig. 4 Dashboard for point-of-sale (POS) device deployment analysis

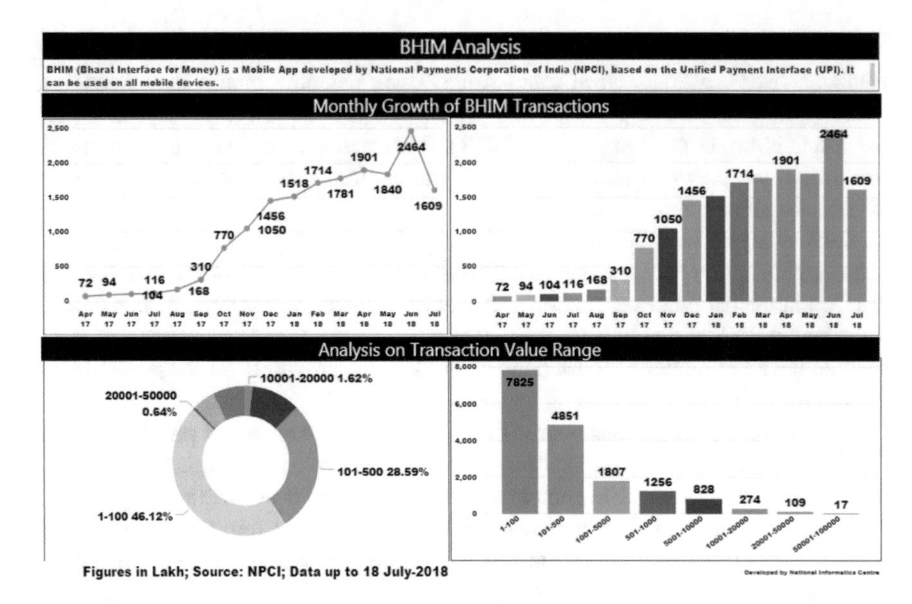

Fig. 5 Dashboard for BHIM transaction analysis

The portal also has a separate analysis dashboard for digital transactions done using the Bharat Bill Payment System (BBPS) interface. It gives a snapshot of the total transactions that are carried out toward payment of various bills like water, gas, landline postpaid, and electricity using the BBPS interface. In addition to aforementioned, digital payment analysis is also done for certain Ministries/Departments and closed loop wallets like IRCTC, DMRC, etc.

Data analytics has helped in increasing the digital payments reach by providing analysis charts and reports to stakeholders, anytime on the go. These are used by all stakeholders to plan for their upcoming meetings and take proactive action to increase digital payments. In the absence of other dedicated portals for digital payment, Digidhan dashboard serves as a chief portal for supporting Hon'ble Prime Minister's agenda of digitizing the Indian economy. Ongoing developments include implementation of Predictive analysis for identification of potential payment modes for next financial year.

7 Current Developments

7.1 Smart City

Being a central platform, the Digidhan dashboard has been integrated with the smart city portal for reporting of digital transaction in 100 smart cities under the smart city

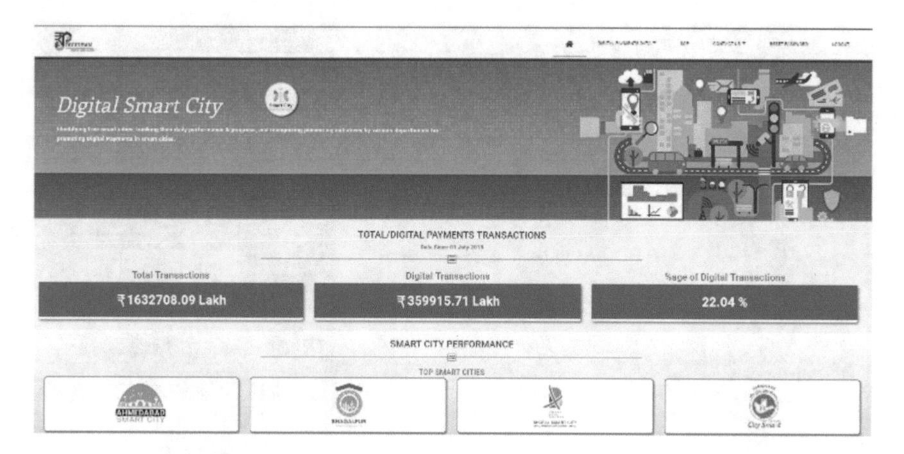

Fig. 6 Smart city portal integrated on Digidhan portal

campaign. The campaign was rolled out in July 2018 and is being led by Ministry of Electronics and Information Technology (MeitY). States and Central Government agencies who are collecting payments from citizens in the smart cities can report on the actual collections and share of digital payments in such collections on the smart city dashboard. This campaign intends to recognize the good initiatives and significant progress made in digital payment enablement and receipts at smart cities. The Smart City Reporting portal home page is displayed in Fig. 6.

Data captured on the smart city portal can reveal the actual status of digital infrastructure enablement in these smart cities and thus help authorities take necessary action toward required growth. For this purpose, several reports are available that consolidates the data submitted by smart city users into meaningful actionable reports. In order to assess the usage of portal, a user activity report is created that lists assigned number of users in each Ministry/Department and the number of users who have logged into portal at least once in a given period. A login report further gives details of all the active users with list of data uploaded. These reports together help in monitoring the data being captured on the portal and encourage inactive smart cities to onboard the platform for the overall success of the campaign.

The smart city data is analyzed and presented through Power BI dashboards. The dashboard provides general insights in form of total digital versus non digital transaction and most digitized Ministries/Departments and smart cities as well as drills down into specific performance of a given smart city. A consolidated report shows percentage of digital versus non digital transactions for each Ministry/Department which drills down to city wise distribution for selected Ministry/Department. The Smart City analysis is shown in Fig. 7.

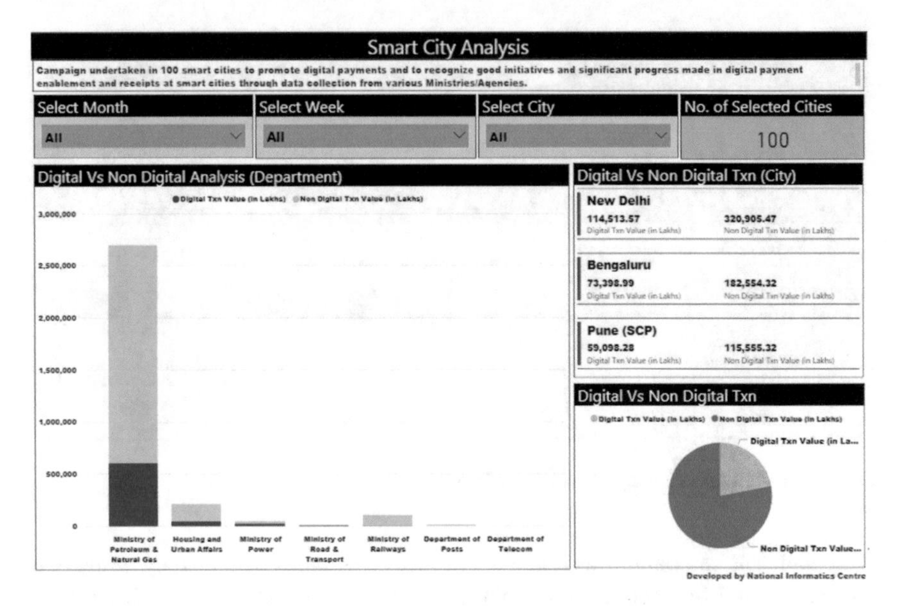

Fig. 7 Smart city analysis on Digidhan dashboard

7.2 *Chat Bot*

Initially, user queries were handled over email. The growth in volume of such queries over time demanded a real-time solution which could eliminate need of a human teller to address such queries. Moreover, it was observed that certain queries were repeating in nature. A chat bot was therefore developed to handle user queries related to use of Digidhan and Smart city portal on real-time basis. The chat bot is built using NLP processing and functions on the concept of self-learning. NLP processing is implemented through Google Dialog Flow. The chat bot window is displayed in Fig. 8 with an illustrative conversation.

8 Green Initiative

Digidhan dashboard is hosted on cloud of National Data Centre (NDC), Shastri Park, and New Delhi. The NDC has been designed to provide full stream of hosting services ranging from physical to shared hosting, dedicated servers with managed hosting solutions to infrastructure services such as Collocation & Bandwidth, Disaster Recovery, etc. NDC is an ISO 27001:2013 & 20000-1:2011 Certified Data Centre.

Benchmarking a Data Centre's energy efficiency is the first key step toward reducing power consumption and minimizing energy costs. Power Usage Effectiveness

Fig. 8 Chat bot window for Digidhan dashboard

(PUE) and its reciprocal Data Center Infrastructure Efficiency (DCiE) are used as benchmarking standards as proposed by the consortium called "The Green Grid."

Following are the additional best practices that were implemented at NDC Shastri Park for improving PUE:

- Enhancing UPS Power Redundancy: UPS Configuration has been converted for both Phase-1 and Phase-2 from $2(N + 1)$ to $(N + N)$ i.e., 100% redundancy
- Providing Blanking Panels in Server racks: This has reduced mixing of cold air with hot air
- Temperature of cold aisle increased from 20 degree to 22 degree
- Continuous monitoring and timely preventive maintenance of all basic infra equipment.

With the above implementations, there has been a substantial improvement in PUE at NDC, which resulted in power saving and operational cost. As per the design of NDC, the desired PUE is 1.7 at 100% load. The basic infra operations team of

NDC has made an effective improvement in this by bringing the PUE to 1.39 even at 60% of current load.

This was achieved by gradually increasing the temperature inside the data center and saving of chilled air by bridging gaps of hot and cold aisle containment, enhancing UPS Power redundancy, continuous monitoring and timely preventive maintenance of all basic infra equipment.

9 Conclusion

Digidhan dashboard has been awarded the Technology Sabha 2018 Award under the category “Analytics/Big Data” by the Indian Express Group at the 24th Edition of Technology Sabha held in Vizag. The dashboard has been appreciated by all stakeholders and is being widely used today by PMO, Finance Ministry, Niti Aayog, and other Ministries/Departments to track growth of digital payment transactions and monitor the performance of banks. Digidhan portal ranking of Jammu and Kashmir bank in achieving targets in digital payment transactions was recently quoted in news.

The dashboard has gained wide popularity and usage among its stakeholders in a short time. It can be marked as a milestone in the definition of a monitoring tool. The dashboard is planned to be further enhanced in technology in future by employing more analytics use cases and prediction reports. Various activities are being conducted to enable predictive analysis of key metrics such as digital payment transactions and its growth. Progressive web apps (PWA) are being planned to be implemented for regions with unstable internet connection and lack of required hardware.

Acknowledgements The authors express their gratitude to Hon’ble Minister of Electronics and Information Technology for supporting this initiative and to Ministry of Electronics &and Information Technology (MeitY) for conceptualizing and leading the project.

References

1. http://digipay.gov.in/
2. http://www.meity.gov.in/digidhan
3. https://datacentres.nic.in/
4. http://informatics.nic.in/uploads/pdfs/dcdef20b_ndc.pdf
5. https://docs.microsoft.com/en-us/sql/database-engine/availability-groups/windows/always-on-availability-groups-sql-server?view=sql-server-2017

Pneumonia Identification Using Chest X-Ray Images with Deep Learning

Nishit Pareshbhai Nakrani, Jay Malnika, Satyam Bajaj, Harihar Prajapati and Vivaksha Jariwala

Abstract Detecting pneumonia is a demanding task which always requires looking at chest X-ray images of patients suffering from it. The normal chest X-ray, reveals clear lungs without any areas of abnormal obscurity or opacification in the image. The Bacterial pneumonia, regularly consists of a focal lobar consolidation (a lung tissue which is filled with liquid instead of air) whereas the viral pneumonia shows distinct and more diffuse 'interstitial' pattern in both lungs. These characteristics can be identified, only with the help of an experienced radiologist. But these characteristics of pneumonia can also be overlapped with other diseases, further complicating the diagnosis. Our objective is to build a Convolutional Neural Network (CNN, or ConvNet) classifier to detect pneumonia in X-ray images of the patient. The dataset for the images is taken from kaggle—a data science learning and competition platform. Convolutional Neural Network (CNN or ConvNet) is a class of deep neural networks that specialises in analysing images and thus is widely used in computer vision applications such as image classification and clustering, object detection and neural style transfer. The above-stated objective will be implemented using Python as a programming language and using concepts of deep learning and neural networks. The dataset consists of 3 folders (train, test, val) and contains subfolders for each image category (Pneumonia/Normal). There are approximately 5000 X-Ray images (JPEG) and 2 categories (Pneumonia/Normal).

N. P. Nakrani (✉) · J. Malnika · S. Bajaj · H. Prajapati · V. Jariwala
Department of Information Technology, Sarvajanik College of Engineering & Technology, Surat 395002, Gujarat, India
e-mail: nakrani.nishit@gmail.com

J. Malnika
e-mail: jaymalnika@gmail.com

S. Bajaj
e-mail: satyambajajd@gmail.com

H. Prajapati
e-mail: beingharihar@gmail.com

V. Jariwala
e-mail: vivaksha.jariwala@scet.ac.in

M. Tuba et al. (eds.), *ICT Systems and Sustainability*,
Advances in Intelligent Systems and Computing 1077,
https://doi.org/10.1007/978-981-15-0936-0_9

Keywords CNN · Pneumonia · Chest X-ray

1 Introduction

More than 10 million cases of patients suffering from pneumonia are diagnosed in India per year and more than 1.8 lacs of patients have died from severity of pneumonia in the year 2017 [1]. The most common and best available method of diagnosing pneumonia currently is through chest X-ray images, which plays a pivotal role in clinical care and control of diseases [2]. Regardless, diagnosing and detecting pneumonia from chest X-ray images is a demanding task that requires availability of skilled and accomplished radiologists. In the given work, we present different models that automatically detect Pneumonia from a given Chest X-ray images.

The models in this work, Pneumonia Identifier, make use of Convolution Neural Networks (CNNs) which expects a Chest X-ray Image as an input and outputs a prediction specifying the presence or absence of pneumonia [3]. We train Pneumonia Identifier on ChestXRay2017 dataset which contains 5863 chest X-ray images (JPEG) categorised as normal/pneumonia and organised into 3 folders (train, test, val) [4]. We use dense connections to optimise such deep neural networks.

Detecting pneumonia can be a challenging task for radiologists [5]. The appearance of pneumonia is often, unclear and can project other diagnoses/diseases. Automated detection of pneumonia from that of chest X-rays at more than average accuracy of practicing radiologists would benefit tremendously in clinical care. It would be also very valuable in areas/populations with inadequate access to diagnostic radiologists.

2 Objectives of the Proposed Approach

Our first objective of proposed approach is to deliver a system that automates the detection of pneumonia with the help of artificial neural networks [6]. In addition to it our approach can be used to achieve accuracy of more than average radiologists. Moreover to that, it is to be helpful to those area/populations which have inadequate access to diagnostic radiologists and to improve costs to predict pneumonia.

3 Proposed Approach

3.1 Flow of Proposed Approach

- Import python libraries: This step involves importing the libraries that are going to be used or will be useful in the project. Some of them are Numpy, Matplotlib, Keras, Pandas, Fast.ai, TensorFlow.
- Input dataset: This step involves including the dataset in the notebook/IDE to be used by the training model.
- Sampling Images: This step involves classifying the images in the dataset as normal or pneumonia i.e. labelling them as 0 (absence of pneumonia) and 1 (presence of pneumonia) (Fig. 1).
- Augmenting training data: The use of augmentation helps us to improve the robustness of the training model to be used. This step involves converting the X-ray images into grayscale form, this leads to easy representation in matrix form and transforming the converted images into resized matrix form.
- Selecting Training Model: Here we select different model like inceptionV3, resent and VGG-16 to learn about layer and get depth knowledge about it [7].
- Build Model: Now on basis of above step and past knowledge of models, we build our own model, then we train our model with dataset we have, test our model with test dataset we have and finally validate our model with dataset which help us to create confusion matrix as well as graphs [8].

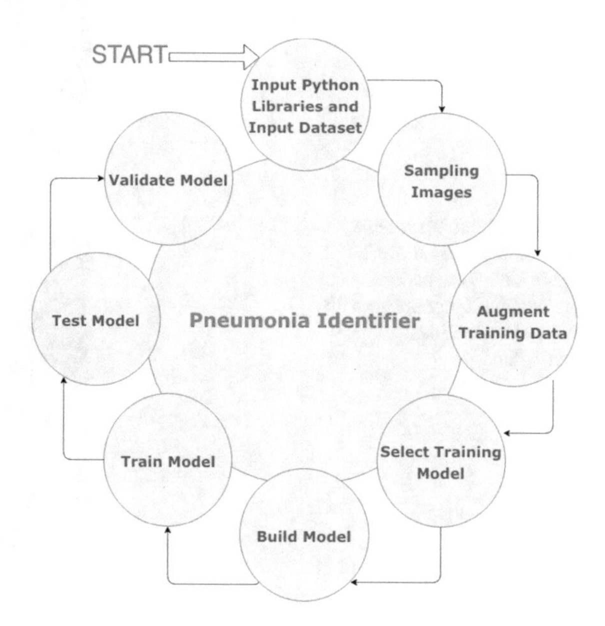

Fig. 1 Steps performed: the above diagram shows the plan of our work which will be apply on dataset we have. Here we going to implement the dataset on our own model and will compare its result with InceptionV3

3.2 SelfLaid Model Architecture

- Our model is 19 layer dense Convolutional Neural Network trained on the chest X-ray dataset [9].
- The output layer in CNN consists of a dense layer that classifies chest X-ray images into 2 categories i.e. normal or pneumonia, it uses sigmoid as an activation function.
- The CNN is trained with an RMSprop optimiser, the model is trained with a batch size of 256 and 6 epochs.
- We set the initial learning rate to 0:005, which is then reduced by a factor of 10 every time the validation loss stagnates/plateaus (changes) after an epoch, and then we save the model with the highest validation accuracy.
- The input images are converted to grayscale and are downscaled to 150 * 150 before being fed to the model (Figs. 2 and 3).

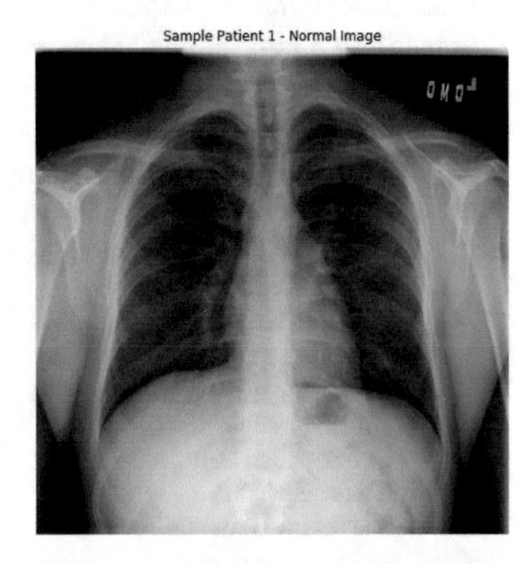

Fig. 2 Patient normal X-ray image: the above figure shows the patient normal X-ray image without any abnormalities

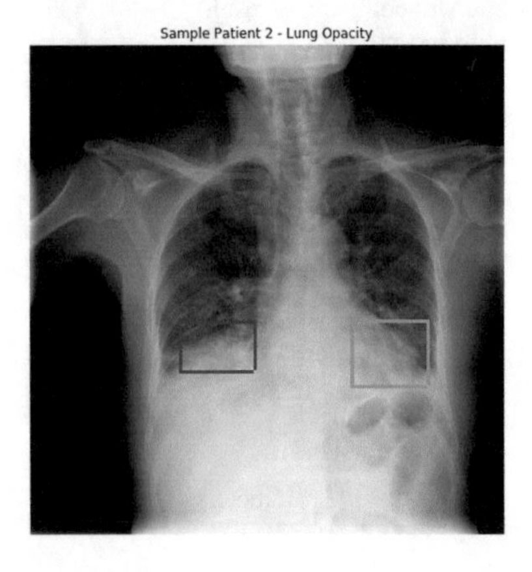

Fig. 3 Patient pneumonia X-ray image: the above figure shows the patient pneumonia X-ray image with some abnormalities highlighted

4 Experiment Setup and Methodology Evaluation

4.1 *Platforms and Tools Used*

- IDE: PyCharm
- OS: macOS (Version 10.13.5)/Windows (Version 10)
- Programming Languages: Python, Javascript, JSON
- Libraries: NumPy, Matplotlib, Keras, Fast.ai, pandas, ScikitLearn, etc.

5 Performance Results

The number of parameters, trainable and non-trainable detected by the 3 models used in the research work. Here we can notice that SelfLaid model used all its parameters for training while other doesn't use some of its parameters for training (Tables 1 and 2).

The epoch wise values of loss, accuracy (for training dataset) and val-loss and val-accuracy (for validation dataset) (Table 3).

The selfLaid network has $f1$-score for 0 (i.e. normal X-rays) of 0.58 and $f1$-score for 1 (i.e. pneumonia X-rays) of 0.84. The inceptionV3 network has $f1$-score for 0 (i.e. normal X-rays) of 0.82 and $f1$-score for 1 (i.e. pneumonia X-rays) of 0.89. The InceptionResnetV2 has $f1$-score for 0 (i.e. normal X-rays) of 0.82 and $f1$-score for 1 (i.e. pneumonia X-rays) of 0.88 (Figs. 4 and 5).

6 Brief Future Aspect and Advantages

6.1 *Advantages*

Easy to use and very valuable in areas/populations with inadequate access to diagnostic radiologists. Also automatic detection of pneumonia from that of chest X-ray images. And also low cost and maintenance.

Table 1 Information regarding parameters used and not used of different models

	SelfLaid model	InceptionV3 model	InceptionResnetV2 model
Total parameters	542,738	22,065,826	54,534,242
Total trainable parameters	542,738	22,031,138	54,473,442
Non-trainable parameters	0	34,688	60,800

Table 2 Information regarding loss, accuracy, validation loss, validation accuracy of different models

SelfLaid model					InceptionV3 model				InceptionResnetV2 model			
Epochs	Loss	Accuracy	Validation loss	Validation accuracy	Loss	Accuracy	Validation loss	Validation accuracy	Loss	Accuracy	Validation loss	Validation accuracy
1	0.5705	0.7421	0.6303	0.625	0.2929	0.88	0.9028	0.758	0.2531	0.89	0.4541	0.859
2	0.4101	0.8168	0.4771	0.7764	0.1582	0.9387	0.6115	0.8237	0.1453	0.9383	0.5541	0.8526
3	0.3186	0.8669	0.4223	0.8077	0.1368	0.9521	0.7507	0.859	0.076	0.9755	0.8413	0.7869
4	0.2763	0.8887	0.6295	0.7188	0.0946	0.9661	1.1412	0.7356	0.0601	0.9783	0.8479	0.7821
5	0.2366	0.9102	0.5048	0.7796	0.0557	0.9835	0.8445	0.7756	0.0581	0.9804	0.8405	0.7853
6	0.2306	0.9099	0.5438	0.7644	0.0348	0.9891	0.8685	0.7724	0.054	0.981	0.8413	0.7853
7	–	–	–	–	0.0333	0.9896	0.8661	0.7724	0.0587	0.9799	0.8394	0.7853
8	–	–	–	–	0.0372	0.9881	0.8698	0.7724	0.0595	0.981	0.8503	0.7837
9	–	–	–	–	0.0326	0.9902	0.8872	0.7724	0.0573	0.9818	0.846	0.7853
10	–	–	–	–	0.0385	0.987	0.8674	0.7756	0.0568	0.9781	0.8371	0.7853

Table 3 Data related to confusion matrix of different models

0-Normal/1-Pneumonia			
True label-predicate label	SelfLaid model	InceptionV3 model	InceptionResnetV2 model
0-0	103	195	201
1-0	16	49	55
0-1	131	39	33
1-1	374	341	335

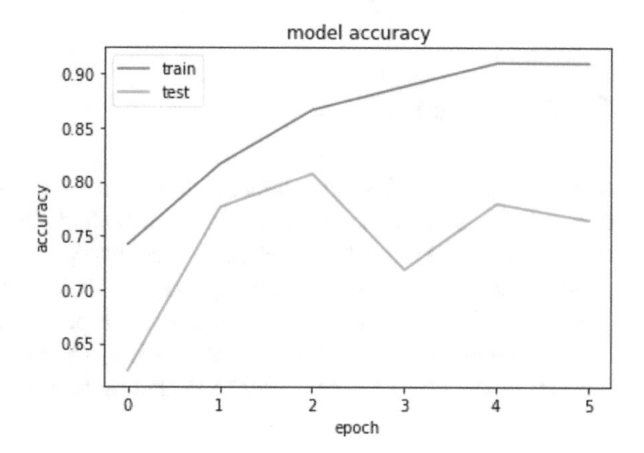

Fig. 4 SelfLaid model accuracy graph: the above graph shows the accuracy of selfLaid model at different epochs for train and test data

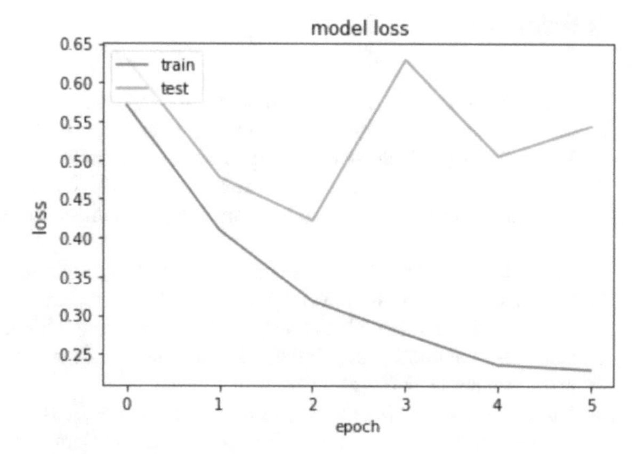

Fig. 5 SelfLaid model loss graph: the above graph shows the loss of selfLaid model at different epochs for train and test data

6.2 *Usefulness with Respect to Existing Solutions*

Current solution to detect pneumonia includes a radiologist to scan and look at the X-ray image, this is an expensive task. Also, human error is inevitable, automated detection of pneumonia can overcome this human error with improved precision and accuracy (recall).

6.3 Scope of Future Work

As the number of scans/images improves in the dataset, precision and accuracy (recall) improve. We can also add different categories of pneumonia if considerable number of images are obtained for different types of pneumonia i.e. bacterial, viral, mycoplasma and others pneumonia. This leads to overcoming human error and reliable detection of pneumonia leading to a fully automated process in the field of differential diagnosis of diseases based on radiological information. We can use similar CNN/model to extend this work on other pulmonary diseases [10].

7 Conclusion

Regardless, diagnosing and detecting pneumonia from chest X-ray images is a demanding task that requires availability of skilled and accomplished radiologists. We trained 3 different CNNs, SelfLaid, InceptionV3, Ince-ResnetV2, which detects pneumonia from frontal-view chest X-ray images. With automated prediction of pneumonia through Chest-X-ray images, we aspire that the work/technology can better the healthcare systems and can provide better healthcare delivery to certain sections of the world where access to skilled radiologists is finite.

References

1. The Economic Times: Available at: https://blogs.economictimes.indiatimes.com/et-commentary/1-8-lakh-deaths-occurred-in-india-due-to-pneumonia/
2. WHO: 2001 available at: http://apps.who.int/iris/handle/10665/66956
3. Aydogdu, M., Ozyilmaz, E., Aksoy, H., Gursel, G., Ekim, N.: Mortality prediction in community-acquired pneumonia requiring mechanical ventilation; values of pneumonia and intensive care unit severity scores. Tuberk Toraks **58**(1), 25–34 (2010)
4. Mendeley: 2017 available at: https://data.mendeley.com/datasets/rscbjbr9sj/2
5. CheXNet: Radiologist-Level on Chest X-Rays, research Mendeley Data, v2 Pneumonia Detection Paper: Available at: https://stanfordmlgroup.github.io/projects/chexnet/
6. Can Artificial Intelligence Reliably Report Chest X-Rays, Research Paper: available at: https://arxiv.org/abs/1807.07455
7. Huang, G., Liu, Z., Weinberger, K.Q., van der Maaten, L.: Densely connected convolutional networks. In: Proceedings of the IEEE Conference on Computer Vision and Pattern Recognition, vol. 1, p. 3 (2017)
8. He, K., Zhang, X., Ren, S., Sun, J.: Deep residual learning for image recognition. In: Proceedings of the IEEE Conference on Computer Vision and Pattern Recognition, pp. 770–778 (2016)
9. Kermany, D., Zhang, K., Goldbaum, M.: Labeled Optical Coherence Tomography (OCT) and Chest X-Ray Images for Classification. https://doi.org/10.17632/rscbjbr9sj.2, Version-3 (2018)
10. Caruana, R., Niculescu-Mizil, A., Crew, G., Ksikes, A.: Ensemble selection from libraries of models. In: Proceedings of the Twenty-First International Conference on Machine Learning, p. 18. ACM (2004)

A Low-Cost IoT-Based Solution for an Integrated Farming Optimization in Nazi BONI University

Téeg-wendé Zougmore, Sadouanouan Malo, Florence Kagembega, Aboubacar Togueyini and Kalifa Coulibaly

Abstract An IoT-based solution for the improvement of the yield in a system of production in which one associates vegetal and animal productions is related in this paper. We deployed a sensor network at the Research on Soil Fertility Laboratory (LERF) and at the Aquaculture and Aquatic Biodiversity Research Unit (UR-ABAQ) of Nazi BONI University. Through this network, we are able to check in real time and continue the pH, the dissolved oxygen and the water temperature of a hatchery of the UR-ABAQ. Also the soil moisture of banana, papaya, and maize is constantly checked, respectively, at UR-ABAQ and LERF sites. A weather station has been deployed to monitor the weather of these research units' site. The equipment deployed is energy-efficient because of its solar panel and has a LoRa radio antenna for communication. The data collected is stored into a cloud platform by two gateways connected to the Internet via 3G MoDem and equipped with a LoRa antenna.

Keywords IoT · Cloud platform · LoRa technology

T. Zougmore (✉) · S. Malo · F. Kagembega · A. Togueyini · K. Coulibaly
Nazi BONI University, 01 BP 1091 Bobo-Dioulasso, 01, Burkina Faso
e-mail: teegwend@gmail.com

S. Malo
e-mail: sadouanouan@yahoo.fr

F. Kagembega
e-mail: florentiakagembega@gmail.com

A. Togueyini
e-mail: togueyinia@yahoo.fr

K. Coulibaly
e-mail: kalifacoul1@yahoo.fr

M. Tuba et al. (eds.), *ICT Systems and Sustainability*,
Advances in Intelligent Systems and Computing 1077,
https://doi.org/10.1007/978-981-15-0936-0_10

1 Introduction

1.1 Study Context

Agriculture and fishing are among the main sectors of Burkina Faso's economy. But Burkina Faso is not traditionally a country of fishing. Its river system covers an area of about 200,000 ha, of which about 72% is the main fish production base, contributing less than 1% of the GDP. The large water surface of the country is therefore not valued in terms of fish production. In fact, the subsector of fish farming is still in its infancy, with an annual production estimated at just over 300 tons.

In addition, with the ever-increasing human pressure on natural resources (deforestation, sand encroachment, overfishing, etc.) and irregular rainfall patterns, there is a drastic reduction in fish populations in natural water bodies. It therefore seems imperative today to develop fish farming in order to fill the gap.

Challenges in boosting the fish farming sector include the availability of good quality fry. UR-ABAQ sets itself the goal of meeting these challenges through fry production and the promotion of fish farming. To make the investments and production efficient, UR-ABAQ is promoting integrated farming. Indeed, the drainage water from fish ponds is a very nurturing source for all types of crop production. It contains the remains of food and fish waste, partly already mineralized and therefore directly accessible and usable by plants (all kinds).

As part of its research activities, UR-ABAQ has a hatchery and a fish farm. Next to the fish farm, the research unit has two experimental fields where it produces banana and papaya. The fields are watered by drainage water from the fish farm. Besides this facility, the Studies and Research on Soil Fertility Laboratory (LERF) experiments the supply of nutrients in the yield of a maize farm.

Our contribution in these research activities consists of the deployment of a sensor network to control the hatchery's water parameters, the soil moisture of the experimental fields and the weather parameters of the experimental site. This deployment was done as part of the WAZIUP project.

WAZIUP[1] is a 3-year EU H2020 funded project started on February 1, 2016. It is a collaborative research project which aims to provide IoT-based solutions at low-cost and low power consumption. It targets many cases that some of them are to enhance the yield of fishing and agriculture.

1.2 Deployment Sites

The sensor network was deployed on two research units' experimental sites of Nazi BONI University. One belongs to the UR-ABAQ research unit and the other is a site of LERF research unit. In UR-ABAQ, experiments are conducted in two buildings.

[1] WAZIUP: www.waziup.eu.

One is dedicated to claria and the other is for tilapia species. There are hatcheries and pre-fattening basins in these buildings. There are also some fattening basins outside. They serve to irrigate some experimental fields on which banana and papayas solos are produced. Among the conducted experiments on UR-ABAQ site, we notice the production of performing clarias larvae and tilapias' super male and artificial reproduction. Also, some nutrition experiments are made for the purpose of finding a low-cost way to feed fishes. There is in addition a watering experimentation applied to production of banana and papaya solos with water coming from the outside fattening basins.

LERF site is the second experimental site. In this site, maize is cultivated on a farm of area 5000 m^2. The experiments performed there are relative to conservation agriculture.[2] It aims to analyze the effects on short, middle and long term of conservation agriculture to the soil fertility and the yield of crops.

1.3 Deployment Process

The objective of this study is to deploy a sensor network to monitor permanently certain physical parameters of the experimental sites. The main parameters considered are PH, dissolved oxygen, and water temperature in UR-ABAQ site. And soil moisture in both research units. To these sensors is added a weather station whose purpose is to measure some parameters of weather (wind direction, temperature, rainfall, atmospheric pressure…).

We would like also through this deployment to make a test of the robustness of the infrastructure. By robustness, we mean the time needed for sensors recalibration—the reliability of measured data—and the energy efficiency of the sensors deployed.

In the process of the rolling-out, we first assembled the sensor devices and the gateways. Second, we deployed them on different sites. Finally we permitted to visualize data measured locally and online.

In this paper, we describe the work we did. The paper is structured as follows: after the introduction, we describe in the second Section how we deployed the sensor network. We present in Sect. 2 the results obtained after the deployment of the sensor network. Finally, in Sect. 5, the perspectives of our work and then the conclusion are presented.

[2]Conservation agriculture: a system in farming that encourages maintenance of a permanent soil cover, minimum soil disturbance, and diversification of plant species. It enhances biodiversity and natural biological processes above and below the ground surface, which contribute to increased water and nutrient use efficiency and to improved and sustained crop production [1].

2 Deployment of the Sensor Network

2.1 Description of the Technology Used

IoT combined with big data techniques can be permitted to collect data on real-time data. In some domains like fish farming and agriculture they can be used to monitor for example water quality and crop status. They can permit also weather predictions, disease detection, and so on. These technologies can be an opportunity for African countries to improve the yield of fish farming and agriculture. But the lack of infrastructure needed by these technologies prevent African countries to enjoy the benefit of IoT and big data. Indeed these technologies require some special hardware which is sometimes expensive and not available in Africa. The deployment we made took into account the cost and the complexity of this technological eco-system. To work around these obstacles, we deal principally with components that are low-cost and low-power consumption to deploy the sensor network.

Sensor devices at low-cost and low power. For our sensor network, we use some devices which are low power and low cost. These devices are assembled by using some components like Arduino boards and LoRa modules. Some physical sensors are also needed. Arduino-like boards are cheap and easy to program. The mainboard used in our work is Arduino pro mini. It is a small form factor board that runs at 8HMZ and powered at 3.3 V. With just 4 AA regular batteries we can power it. It is possible to get it at less than 2 euros from Chinese manufacturers.

LoRa is Semtech's long-range technology. It is a spread spectrum technology that derives from chirp spread spectrum technology [2]. LoRa devices are wireless and cannot emit on long range. They don't need much power. All features make them become the de facto technology for Internet of Things (IoT) networks worldwide [2]. There exist many LoRa modules that can be used to connect to Arduino micro-controller for sending measured data to the gateways. For our use case, HopeRF RFM92 W/95 W and Modtronix inAir9/9B are used. The range of prices to get them is from 7 to 15 euros [3].

All the components used to assemble a fully operational low-cost and low power devices cost less than 15 euros. In addition, the devices can be functional without need of battery replacement for over a year.

Low-cost gateway. A gateway in a sensor network is a middle node. It is between the sensors and the data storage environment. In our deployment, the gateways used are made of Raspberry PI (1B/1B+/2B/3B) that the cost is less than 30 euros. They are also equipped with LoRa modules. Their operating system is Raspbian, a reliable Linux Platform. Many LoRa devices can communicate with one gateway. The gateways can be customized for target applications based on high-level languages such as python [3]. They can store data locally or push them into an IoT cloud platform. One benefit to store data locally is the possibility to consult data even when there is an issue with Internet connection. The maximum power consumption of a Raspberry is 5v. It is then an energy-efficient equipment.

The data collection infrastructure. IoT Cloud and big data platform is the infrastructure used to store data collected. It is built following the Paas paradigm. The IoT part of the platform permits to handle sensors connectivity and to process the received data. By process we mean transform data format, apply some operator and so on in order to store data measured by sensors in a database. It is also possible to manage sensor's information and act on them virtually. For instance we can delete a sensor or modify it but this is virtual. Concerning the big data part of the platform, is its capacity to store the huge volume of data generated by the sensor devices. Also the possibility to make some data analysis. Indeed for the sake of monitoring and actuation, it is possible to extract the most valuable information by filtering, aggregating, and correlating data stored on the platform.

The collection infrastructure is accessible online but the data is also stored locally. This has a benefit to visualize and analyze data even when there is no Internet connection. Indeed data collected is stored in a MongoDB database on the gateway. There is also an installed web server that permits to display the received data in form of graphs.

From the platform, alerts and recommendations can be sent to users by some notifications means like SMS, and social notification (Twitter or Facebook).

2.2 Sensors Deployment

For the assembly and the deployment of the sensors we followed the step described in [4]. The main components we used are Arduino pro mini for soil moisture sensors and ATmega 1284 for water devices. HopeRF RFM 95 W and Modtronix inAir9 are the LoRa modules used to permit the exchange of data between the gateway and the sensor devices. The physical sensor for soil moisture sensor devices used is Fuidino moisture sensor. H-101 PH is the sensor constituting the water sensors device. Soil moisture sensor devices are powered by four AA battery. The water device and weather are equipped with lithium batteries and powered by a solar panel.

The task of assembly consists first of the soldering of header pins on microcontrollers and LoRa modules boards. After we connect physical sensors and LoRa modules to the microcontrollers through soldering or by using Dupont wires, then we upload software into the microcontrollers. We finish by connecting battery to the board and packaging all the components in a box. In Figs. 1 and 2, we show some sensor devices assembled.

Deploying sensors network consists of placing sensor devices in their dedicated places. Also it consists of taking the precaution to place the antenna and solar panel out of the box. Solar panel which covers certain sensor devices box has to be oriented to the south in our case. The antenna of sensor devices and gateways must be in line of sight. Figures 3, 4, 5, and 6 shown below present few deployments of dissolved oxygen PH, weather station and soil moisture sensors.

Fig. 1 Inside view of the weather station

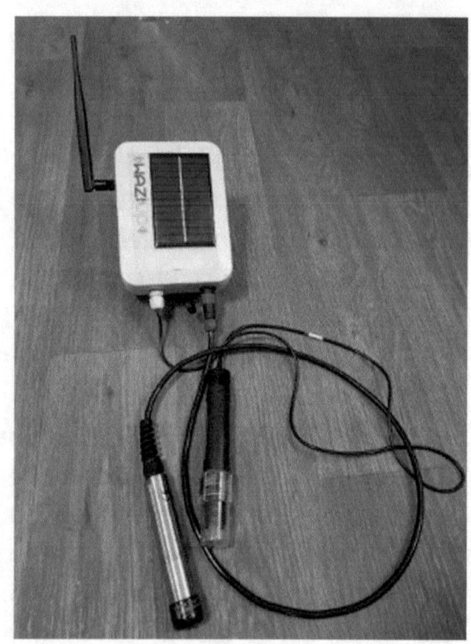

Fig. 2 PH, dissolved oxygen and water temperature measurements device covered by a solar panel

2.3 Gateways Deployment

We place the gateways at some places where their antenna are in a high location possible as recommended in [5]. As gateways equipped with LoRa module, the maximum distance between them and the sensor devices can reach 20 km if they are in the same line of sight. If it is not the case, the distance cannot exceed 2 km. For our case we place the sensor devices at 100 m of the gateways. To the gateways are

Fig. 3 PH and DO sensors inside a hatchery

Fig. 4 Water device equipped with a solar panel fixed outside on a wall

Fig. 5 Weather Station deployed

Fig. 6 Soil moisture sensor attached to a banana plant

plugged 3G USB modems and some configurations must be done such as APN name and service path setting as indicated in [5]. It's possible for the gateways to send SMS to an end-user or a cloud platform. SMS sending is able to serve as backup connection in order to face Internet connectivity problems which may happen in some areas. In Figs. 7 and 8, we show deployed gateways. The antennas are not visible because they are on the roof.

Fig. 7 Gateway fixed on a wall on UR-UBAQ site

Fig. 8 Maize farm gateway installation step

3 Results and Discussions

Physical parameters are measured by deployed sensor devices given some time intervals. For meteorological parameters and soil humidity, values are measured each 10 min. 20 min is the time that separates the measurement of hatchery's water parameters. When the sensor devices measure data, they send them to the gateways. The gateways at their turn send data to IoT cloud platform. On the platform, one can view the last data measured by each sensor or view for each sensor the data measured in a time interval as a graph. This time interval is two days by default but it is possible to indicate the period wished. On the platform in addition to visualization feature, it is possible to edit sensor devices information. Data measured downloading and sensor device removing are also possible actions. We show in Fig. 9 one of a possible visualization view (as graph) of data collected on the platform.

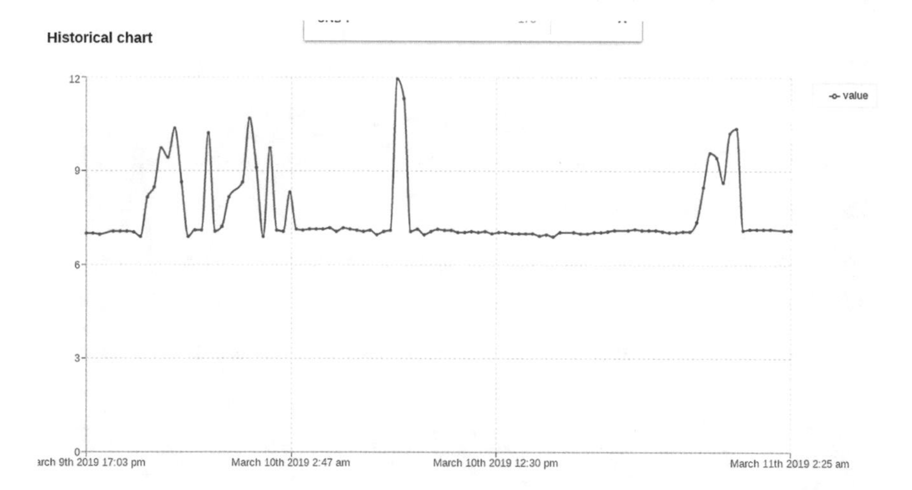

Fig. 9 PH values represented as a graph

Some graphs such as for water temperature (Fig. 10) and PH (Fig. 11) data measured on a period of 01/20/19 to 02/16/19 are shown below. Air temperature and wind speed data collected by the weather station device from 02/20/19 to 03/20/19 are also presented in Figs. 12 and 13.

The air temperature and water temperature are expressed in Celsius degree. The wind speed is expressed in km/h. One can notice in these graphs presented above that the sensor networks are running well without interruption.

Data are sent 24 h/24 into the platform given a respected sending period for the different sensor devices. It has happened sometimes that the gateways stop sending

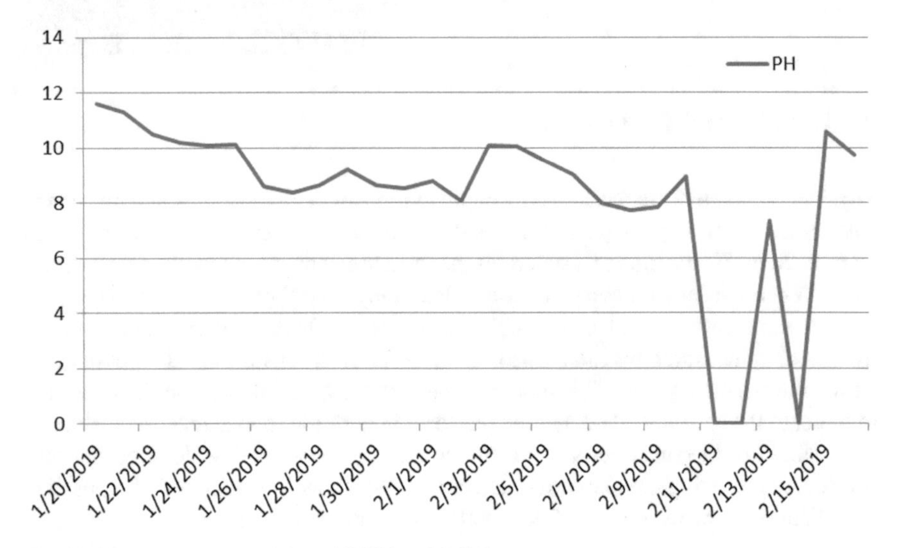

Fig. 10 PH values measured from 1/20/19 to 2/16/19

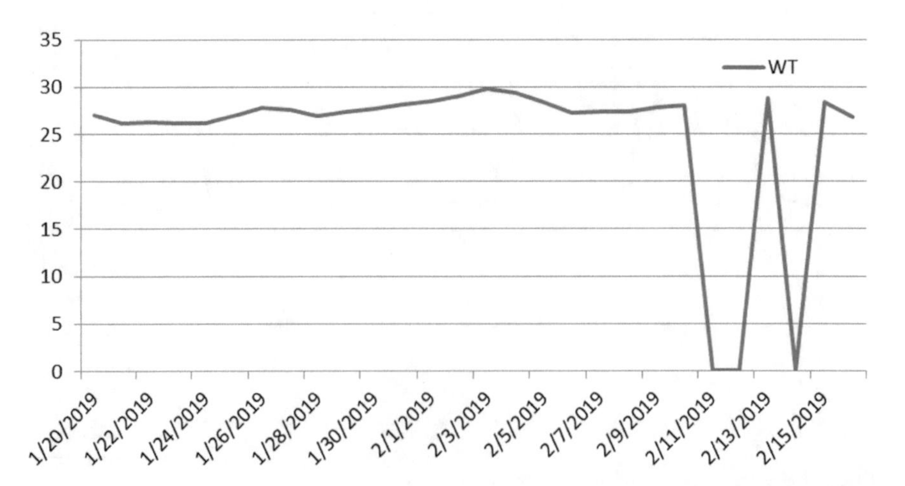

Fig. 11 Water temperature measured from 1/20/19 to 2/16/19

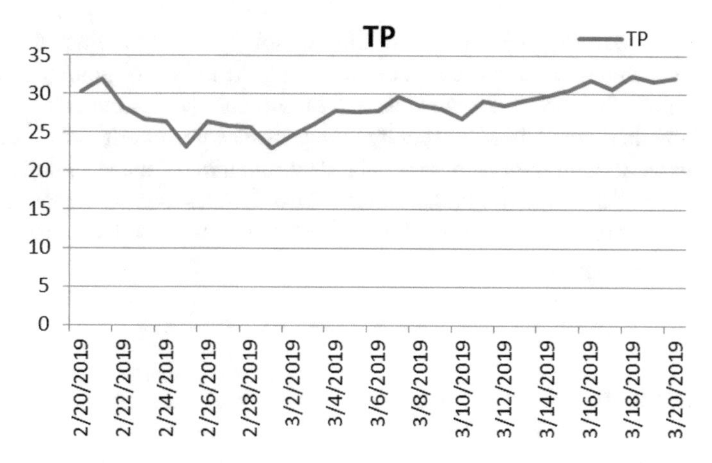

Fig. 12 Air temperature measured from 02/20/19 to 03/20/19

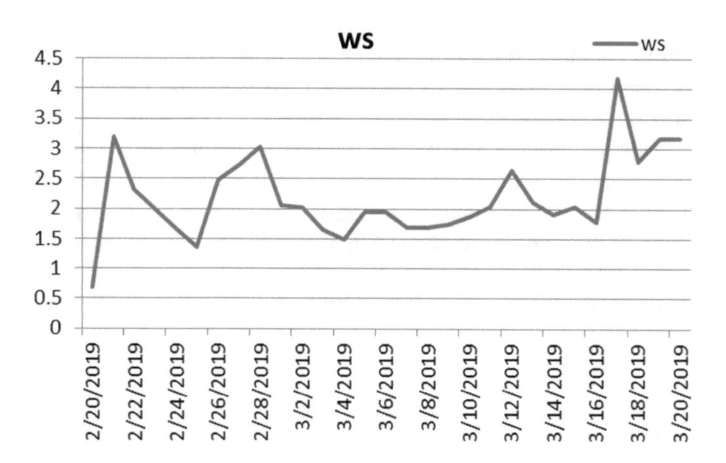

Fig. 13 Wind speed measured from 02/20/19 to 03/20/19

data due to the Internet issue. We also face per moment to battery problems that can prevent sensor devices to emit data. It is important to note that sensors must be calibrated prior to deployment for reliable data. It is therefore necessary to recalibrate them from time to time.

4 Future Works

We will focus on the exploitation of the data collected. The platform must be integrated advanced analytic tools. We will especially:

- evaluate if the maize yield and the fertility of soil in Houet province depends on organic substrates. This will consist of checking the effects of poultry droppings and maggot substrates on maize yield and on soil chemical parameters;
- predict the fish growth in Moussodougou dam: a sensor network will be deployed on Moussoudougou dam to permit—the determination of some biological variables (bio-indicators) from benthic macro-invertebrates and phytoplankton; and the-identification and quantification by machine learning techniques of the exploited main species.

5 Conclusion

In this paper, we have shown that IoT-based solution can be profitable to fish and crop farmers. A sensor network has been deployed and must permit to control the water quality of a hatchery. Also, it is possible with this sensor network to monitor the soil moisture of banana and maize fields. To this sensor network is added a weather station which permits to get the weather parameters of the sites hosting the fields.

The real-time observation of all these parameters on a web platform makes it possible to make the necessary adjustments to improve the yield of the production. Indeed, this sensor network helps fish and agriculture farmers to get better control of fish's production and an optimized watering of agricultural fields. The correlations between the crops production and weather parameters are also able to be determined.

Acknowledgements We thank WAZIUP project which permitted to make this work.

References

1. Food and Agriculture Organization of the United Nations. http://www.fao.org/conservation-agriculture/en/
2. Semtech, What is LoRa? https://www.semtech.com/lora/what-is-lora
3. Pham, C: A DIY low-cost LoRa gateway. http://cpham.perso.univ-pau.fr/LORA/RPIgateway.html
4. Pham, C: Low-cost LoRa IoT device: a step-by-step tutorial. https://raw.githubusercontent.com/CongducPham/tutorials/master/Low-cost-LoRa-IoT-step-by-step.pdf
5. Pham, C: Low-cost LoRa gateway:a step-by-step tutorial. https://raw.githubusercontent.com/CongducPham/tutorials/master/Low-cost-LoRa-IoT-step-by-step.pdf

A Genetic Algorithm for Post-flood Relief Distribution in Kerala, South India

Georg Gutjahr and Haritha Viswanath

Abstract In the summer of 2018, the Indian State of Kerala was affected by a flood disaster. About 1 million people were evacuated to over 3200 relief camps. One of the most severely affected areas was the Chalakudy river basin. Due to its flat topology, large parts of the area were flooded. Also, no proper drainage system was in place during and after the flood, and the area stayed flooded for weeks. In this work, we consider the optimization of the relief distribution by coordinated volunteers in the Chalakudy river area. We are formulating a bi-objective optimization problem of sending volunteers to the various flood relief camps; the two contradicting goals in this objective problem are to minimize the time of the volunteers and the unsatisfied demand. The problem is a variation of a team-orienteering problem with some additional constraints. We present a genetic algorithm to solve the problem efficiently.

Keywords GIS · Disaster relief · Flood modelling · Bi-objective optimization · Tour planning · Team-orienteering problem · Genetic algorithms

1 Introduction

The Indian state of Kerala is located at the southern tip of the subcontinent. Usually, Kerala gets 90% of its rain during the monsoon months. In the beginning of August 2018, heavy rainfall started in Kerala, which resulted in a catastrophic flood. Almost all the dams and reservoirs reached their maximum capacity. The government and the authorities were forced to open all those dams one after the other increasing the disaster. The continuous rain from 14 to 19 August resulted in flooding of 13 of the 14 districts in Kerala. As per the records of India Meteorological Department, the rainfall recorded during the 15–17 August 2018 was comparable only to the situation of the 1924 Kerala flood. The heavy flood resulted in the loss of about 400 people.

G. Gutjahr (✉) · H. Viswanath
Department of Mathematics, Center for Research in Analytics & Technologies for Education (CREATE), Amrita Vishwa Vidyapeetham, Amritapuri, India
e-mail: georgcg@am.amrita.edu

M. Tuba et al. (eds.), *ICT Systems and Sustainability*,
Advances in Intelligent Systems and Computing 1077,
https://doi.org/10.1007/978-981-15-0936-0_11

As per the government estimate, the overall loss is about 8,316 crores in destroyed buildings and infrastructure.

The state of Kerala is divided into three parts: eastern highlands, central midlands and western lowlands. Here we are considering the Chalakudy river area that is located mainly in the western lowlands. The Chalakudy river flows through the Thrissur, Palakkad and Ernakulam districts of Kerala. It is the fifth-longest river (145.5 km) in Kerala. The Chalakudy river basin was severely affected during the flood. Hundreds of people were evacuated to relief camps. Huge destruction was caused to the infrastructure and roads in the area.

During and after the flood, many people in Kerala volunteered to help. Social media was used in many places to coordinate rescue and relief activities. In order to collect relief materials, several non-profit organizations together with the government started collection points at various parts of the state. Through social media, they spread the information regarding the necessities in each place. One such platform was 'Compassionate Keralam', an initiative made by the government under the guidance of Calicut district collector Prashant Nair. The goal was to connect needy people with helping hands. They also connected donors to help affected families.

One of the many non-profit organizations that offered help in association with Compassionate Kerala was the Ayudh (Amrita Yuva Dharma Dhara) youth organization. Ayudh started a helpline on August 16. More than 400 students volunteered to work in three shifts for five days, with 20 phone lines in parallel servicing the Help Line number in order to ensure that every single call was answered. Not only did the volunteers work the phone line, they also fielded help requests via social media. Within five days, the team had fielded more than 25,000 calls, connecting more than 100,000 flood victims across the state with appropriate relief services, including government officials, the Navy, NDRF, Kerala State Police, local fishermen and other volunteering agencies.

Volunteers also ensured that rescued victims were provided with food and clothing, medicine and other basic necessities. They collected relief materials in places like schools and took initiative in distributing it to the needy. After the flood, rehabilitation became a struggle for the authorities as they were in need of sufficient cleaning materials and medicines to prevent the outbreak of epidemics. Ayudh also provided necessary cleaning materials and medicines to various places in the state.

Here we present a model for relief distribution after the flood in the Chalakudy river area. We are formulating a bi-objective optimization problem of sending volunteers to the various flood relief camps; the two contradicting goals in this objective problem are to minimize the time of the volunteers and the unsatisfied demand. The problem is a variation of a team-orienteering problem with some additional constraints. We present a genetic algorithm to solve the problem efficiently.

The paper is organized as follows. Section 2 provides an overview of the literature. Section 3 presents the model for optimal relief distribution. The paper ends with the conclusion in Sect. 4.

2 Literature Review

The demand for humanitarian relief operations has been increased widely around the world. Based on three aspects the papers on logistic attributes can be grouped as a distribution model, inventory model and facility location model [1].

Coherent planning of logistic activities can reduce the number of losses in an affected area. Here we discuss different papers in order to plan the logistics management for a disaster response phase. Some review papers are the following: Review papers from 2010 to recent researches regarding compassionate activities and disaster management [2]. Reference [3] gives a literature survey on recovery planning for disaster management. The overview article by Arenas et al. [4] gives additional detail on dynamic optimization in relief distribution.

Intervention activities have been a topic for researchers since the 1980s. [5] Classification types optimization models for relief distribution are classified according to 5 criteria: modelling types, objective function types, how many objective functions are included, solution approach, routing and optimization method. Modelling includes two types of objective functions which are humanitarian objective function, and combination of cost and humanitarian function. Based on the number of objectives the problems are distinct as multi-objective optimization and single-objective optimization. Moreover the solution methodologies adopted are categorized as heuristics, Metaheuristics and exact methods. The routing of the travel is classified on the basis of the route that needs to be determined by the model and route that presumed to be known. Criteria related to transportation are transportation mode, combined transportation, and availability of vehicles for transportation. There are two modes of transportation; single mode and multimode. Further the capability for combined transportation can be included in the optimization model. Based on the number of available vehicles in the network the classification is done as number of vehicles known and number of vehicles uncertain.

Moreover, types of flow, supply and demand are the three important criteria for flow category. The amount of different types of flows in the network is regarded under flow type, which includes the papers that consider either commodity or injured people and those that consider both commodity and injured people.

In addition, models can be further classified based on the periodicity and commodities. There are two types of models for periodicity: single periodic [6], for example for the relief distribution on the one day: and multi-periodic [7], where the relief distribution is planned multiple days in advance. Further, there are papers that consider single commodity transportation, representing a global demand; but most of the papers consider multi-commodity relief distribution, [8] for commodity such as food, water, medicines and tents.

Finally the models can be solved in different ways: deterministic optimization, stochastic optimization and dynamic optimization.

3 Optimization Model for Relief Distribution

Here we are discussing the problem of finding routes for sending the volunteers around the flood relief camps. There will be a trade-off between the expected time of travel and unsatisfied demand. The problem is modelled as a bi-objective optimization problem to make the trade-off explicit and the aim is to minimize the unsatisfied demand and travel time. The problem becomes a team-orienteering problem.

In order to formally represent the optimization problem, the following notations are introduced. Let N be the number of relief camps and let P be the number of teams of volunteers during the relief distribution. We will use a graph with $N + 1$ vertices, where the first vertex represents the depot and the remaining N vertices represent the relief camps. Commodities are collected at the depot and have to be distributed to the relief camps; therefore, all the tours begin and end at the depot; see [9, 10] for more information about Kerala and the Chalakudy area.

We describe the tours of the volunteers using decision variables in binary form, x_{ijp}, for $i, j = 1, \ldots, N$ and $p = 1, \ldots, P$, with $x_{ijp} = 1$, if in the p-th tour, vertex j is visited after vertex i and 0, otherwise.

We will assume that the p-th team travels in a vehicle that can transport commodities for up to c_p people in relief camps. We will also assume that the lengths of the tours for each team have to be at most $t = 8$ h. Finally, the time taken to travel from vertex i to vertex j is denoted by d_{ij}, the number of commodities that have to be distributed at vertex i is denoted by s_i, and the time that the distribution of commodities at vertex i takes (for $i, j = 1, \ldots, N$) is denoted by v_i.

3.1 Objective Functions

Considering the above-mentioned notations, the two objective functions that explain the flow between the expected time taken for travel and the unsatisfied demand in the campaign can be listed as follows.

Our assumption is that if a team visits a relief camp, all the materials are distributed to all people in this camp so that the demand in a vertex is either completely satisfied or completely unsatisfied. The two objective functions become:

$$\text{Minimize} \quad \sum_{i=1}^{N} s_i - \sum_{p=1}^{P} \sum_{i=1}^{N} \sum_{j=1}^{N} s_i x_{ijp}$$

and

$$\text{Minimize} \quad \sum_{p=1}^{P} \sum_{i=1}^{N} \sum_{j=1}^{N} (v_i + d_{ij}) x_{ijp}$$

Here the objective function written first describes the minimization of the unsatisfied demand and the objective function written second describes minimization of the total work time of the volunteers.

3.2 Constraints

The following constraints are to be satisfied in order for a solution to be feasible. In terms of the decision variables x_{ijp},

$$\sum_{p=1}^{P}\sum_{j=1}^{N} x_{1jp} = \sum_{p=1}^{P}\sum_{i=1}^{N} x_{i1p} = P$$

$$\sum_{p=1}^{P}\sum_{i=1}^{N} x_{ijp} \leq 1, \quad \text{for } j = 2, \ldots, N.$$

$$\sum_{i=1}^{N}\sum_{j=1}^{N} \left(v_i + d_{ij}\right)x_{ijp} \leq t, \quad \text{for } p = 1, \ldots, P.$$

$$\sum_{i=1}^{N} x_{ikp} = \sum_{j=1}^{N} x_{kjp}, \quad \text{for } k = 2, \ldots, N \text{ and all } p.$$

$$s_i \leq u_{ip} \leq c, \quad \text{for } i = 2, \ldots, N \text{ and } p = 1, \ldots, P$$

$$u_{ip} - u_{jp} + cx_{ijp} \leq c - s_j, \quad \text{for } i, j = 2, \ldots, N \text{ and all } p.$$

The first constraint demands that tours begin and end at the depot (vertex 1). The second constraint demands that no relief camp is visited more than once, while the depot can be visited multiple times. The next constraints demand that the lengths of the tours for each team do not exceed t hours and that the tours are closed paths. The last two constraints demand there is no sub tour in the tours and the p-th team of volunteers can transport a maximum of c_p commodities in their vehicle. (this version of the Miller-Tucker-Zemlin constraints are further described in [11, 12].

3.3 Solving the Optimization Problem

To obtain the solution for the optimization problem, we need to get the Pareto front which is the set of all Pareto optimal solutions. Pareto improvement is the movement from one feasible solution to another with at least one of the objective functions to return a better value and also with no other objective function becoming worse off. When no further Pareto improvements can be made, the set of all feasible solutions are said to be Pareto efficient or Pareto optimal.

If the number of camps and volunteers is small, the optimization problem can be solved exactly, using methods such as branch-and-cut or branch-and-price [13]. However, for problem instances of realistic size, heuristic methods have to be used.

In the following, we will describe a genetic algorithm that iteratively approximates the Pareto front closer and closer. Genetic algorithms are a type of evolutionary algorithms that are introduced based on the concept of Darwin's theory of evolution. In genetic algorithms, the set of all feasible solutions are called population. In a sequence of populations, each population is also referred to as a generation. Each solution in a population is called an individual. If we speak of two populations, individuals are also be referred to as children and parents. The goal of genetic algorithms is to obtain a sequence of populations, where each generation provides better and better approximations to the Pareto front. The initial population is generated based on some heuristic. Then, operations such as crossover, mutation, and survival of the fittest are applied to obtain subsequent generations from the previous generations.

Here we are using a particular variant of a genetic algorithm called NSGAII [14], which is optimized for multiple objective optimization problems. The algorithm has also been used for tour planning for the distribution of vaccinations [15]. To adopt the NSGA-II procedure for the problem of relief distribution, we have to find an initial solution and we have to define appropriate operations for selection, crossover, mutation, and survival.

For individual solutions, we are using representations similar to the ones used by Bederina and Hifi for a similar case [16]. In this representation, a solution is a list of tours, and each tour is a list of sites. Crossover procedure is carried out in between randomly selected parents. This procedure randomly selects tours from both parents. In the interim, this may lead to an infeasible solution. Such a solution can violate a few of the constraints in Sect. 4-B, but towards the end of the mutation step tours will always be made feasible. This will be described below and is described in more detail by Bedrina and Hifi [16].

After completing the crossover procedure, mutation is carried out to these children and some local changes are bring off among the individuals. Such changes applied to this problem include:

1. Combining two tours into one
2. Breaking of one tour into two
3. Shuffling of sites in a tour
4. Switching of sub tours between two tours.

There are still chances that the solutions obtained after mutation remain infeasible. In order to make sure that feasible solutions are obtained the following changes are to be done at the end of mutation procedure:

1. Removal of duplicated sites.
2. Removal of vertices that violates capacity c or maximal time t.
3. If more than P tours are contained in the solution set, we randomly remove some of them.
4. If less than P tours are contained in the solution set, we will add empty tours that only contain a loop at vertex 1 ($x_{11p} = 1$ and $x_{ijp} = 0$ for all other values of i and j).

The count of solutions in the population is minimized in a survival step after producing new solutions and applying mutation to them. Survival is done in two steps in such a case: (1) Elimination of dominant solutions; (2) Extra solutions in the jam-packed region get rejected using crowded distance.

Further, the sorting of individuals of the initial population is done into non-dominated fronts; see [8]. Population/crowding distance values are calculated for each of the non-dominated fronts for distinguishing solutions of one and the same non-dominated front. For the Iterative part of the algorithm in each step, child generation is derived from the parent generation. Different individuals from parents are selected for generating the next generation. In NSGA-II the selection of parents depends on the quality of fronts. Within each front, selection is done on the basis of crowding distance. During crossover tours are picked randomly from the parents for generating the children.

The mutation procedure is used for creating n individuals for an offspring population. R_t (a population of size $2n$) is the combined form of population Q_t and P_{t+1}. Now we have a population size of $2n$ and our aim is to reduce the size of the new population P_{t+1} to n, sorting is carried out on R_t. We keep on adding fronts until we ran out of space for additional fronts. Based on the crowding distance, those individuals of the front that cannot be included completely are sorted in descending order, if the population P_{t+1} is not filled completely. Finally, we include the first $n - |P_{t+1}|$ individuals in P_{t+1}, then the loop continues with the next generation.

4 Conclusion

This work discusses the optimization of the relief distribution by coordinated volunteers in the Chalakudy river area. The model takes into account that during the first few days after the flood, considerable uncertainty exists as to which areas are traversable. We therefore use a flood model to estimate safe routes between relief camps.

The paper uses bi-objective optimization problem to describe the different routes for sending the volunteers around the flood relief camps; the two objective functions models the trade-off between the time and number of volunteers and unsatisfied demand. This paper has then presented a genetic algorithm to solve the problem efficiently.

References

1. Manopiniwes, W., Irohara, T.: A review of relief supply chain optimization. Ind. Eng. Manage. Syst. **13**(1), 1–14 (2014)
2. Dascioglu, B.G., Vayvay, O., Kalender, Z.T.: Humanitarian supply chain management: extended literature review. In: *Industrial Engineering in the Big Data Era*, pp. 443–459.

Springer (2019)
3. Ozdamar, L., Ertem, M.A.: Models, solutions and enabling technologies in¨ humanitarian logistics. Eur. J. Oper. Res. **244**(1), 55–65 (2015)
4. Anaya-Arenas, A.M., Renaud, J., Ruiz, A.: Relief distribution networks: asystematic review. Ann. Oper. Res. **223**(1), 53–79 (2014)
5. Barbarosoglu, G., Arda, Y.: A two-stage stochastic programming framework for transportation planning in disaster response. J. Oper. Res. Soc. **55**(1), 43–53 (2004)
6. Rawls, C.G., Turnquist, M.A.: Pre-positioning of emergency supplies fordisaster response. Transp. Res. Part B: Methodol. **44**(4), 521–534 (2010)
7. Lin, Y.-H., Batta, R., Rogerson, P.A., Blatt, A., Flanigan, M.: A logisticsmodel for emergency supply of critical items in the aftermath of a disaster. Socio Econ. Plann. Sci. **45**(4), 132–145 (2011)
8. Haghani, A., Oh, S.-C.: Formulation and solution of a multi-commodity, multimodal network flow model for disaster relief operations. Transp. Res. Part A: Policy Pract. **30**(3), 231–250 (1996)
9. Mohan, A., Gutjahr, G., Pillai, N.M., Erickson, L., Menon, R., Nedungadi, P.: Analysis of school dropouts and impact of digital literacy in girls of the muthuvan tribes. In: 2017 5th IEEE International Conference on MOOCs, Innovation and Technology in Education (MITE), pp. 72–76. IEEE (2017)
10. Pillai, N.M., Mohan, A., Gutjahr, G., Nedungadi, P.: Digital literacy and substance abuse awareness using tablets in indigenous settlements in kerala. In: 2018 IEEE 18th International Conference on Advanced Learning Technologies (ICALT), pp. 84–86. IEEE (2018)
11. Kara, I., Laporte, G., Bektas, T.: A note on the lifted Miller–Tucker–Zemlin subtour elimination constraints for the capacitated vehicle routing problem. Eur. J. Oper. Res. **158**(3), 793–795 (2004)
12. Kara, I.: On the miller-tucker-zemlin based formulations for the distance constrained vehicle routing problems. In: AIP Conference Proceedings, vol. 1309, pp. 551–561. AIP (2010)
13. Gutjahr, G., Krishna, L.C., Nedungadi, P.: Optimal tour planning for measles and rubella vaccination in kochi, south india. In: 2018 International Conference on Advances in Computing, Communications and Informatics (ICACCI), pp. 1366–1370. IEEE (2018)
14. Deb, K., Pratap, A., Agarwal, S., Meyarivan, T.: A fast and elitist multiobjective genetic algorithm: Nsga-II. IEEE Trans. Evol. Comput. **6**(2), 182–197 (2002)
15. Gutjahr, G., Kamala, K.A., Nedungadi, P.: Genetic algorithms for vaccination tour planning in tribal areas in Kerala. In: 2018 International Conference on Advances in Computing, Communications and Informatics (ICACCI), pp. 938–942. IEEE (2018)
16. Bederina, H., Hifi, M.: A hybrid multi-objective evolutionary optimization approach for the robust vehicle routing problem. Appl. Soft Comput. **71**, 980–993 (2018)

Classification of Cancer for Type 2 Diabetes Using Machine Learning Algorithm

Ashrita Kannan, P. Vigneshwaran, R. Sindhuja and D. Gopikanjali

Abstract Cancer prognosis is crucial to control the suffering and death of diabetic patients. Diabetes is a prolonged disease caused by the deficiency in the amount of insulin generated by the pancreas. Type 2 diabetes occurs due to the inability of the cells to respond to the production of insulin that results in increased concentration of glucose in the blood. We have attempted to bring in novelty to predict and classify cancer types such as breast, liver, and colon cancer for Type 2 diabetic patients through this paper. We have analyzed key common parameters like triglycerides, age, menopause age, number of pregnancies, etc. for Type 2 diabetes and cancer patients. We then gathered values for these parameters and used them to train and validate the Random Forest model that we have used for classification. We have been able to predict and classify the type of cancer using our model and achieve an accuracy level of 86%.

Keywords Liver cancer · Colon cancer · Breast cancer · Type 2 diabetes · Random forest

---

A. Kannan (✉) · R. Sindhuja · D. Gopikanjali
Department of Information Technology, Rajalakshmi Institute of Technology, Chennai, Tamil Nadu, India
e-mail: ashrita.kannan.2015.it@ritchennai.edu.in

R. Sindhuja
e-mail: sindhuja.r.2015.it@ritchennai.edu.in

D. Gopikanjali
e-mail: gopikanjali.d.2015.it@ritchennai.edu.in

P. Vigneshwaran
Department of CSE, School of Engineering and Technology, Jain University, Bangalore, Karnataka, India
e-mail: vigneshwaran05@gmail.com

M. Tuba et al. (eds.), *ICT Systems and Sustainability*,
Advances in Intelligent Systems and Computing 1077,
https://doi.org/10.1007/978-981-15-0936-0_12

1 Introduction

Diabetes Mellitus is a relentless pancreatic disease caused due to deficiency in the generation of insulin. This increases the concentration of glucose within the blood, which in turn affects various body frameworks, in specific the blood vessels and nerves. Non-Insulin Dependent Diabetes Mellitus (NIDDM) is very common and accounts for around 90% of all cases of diabetes around the world. Type 2 diabetes and cancer have many common risk factors but there is an incomplete understanding of potential biologic connections between the two diseases. Type 2 diabetes in vast majority of cases is largely related to old age and being overweight having a family history of this occurrence.

1.1 Diabetes with Breast Cancer

During the post-menopause period, the insulin sensitivity decreases and changes in hormone level can trigger fluctuations in blood sugar level. Due to this, the hormone estrogen and progesterone are affected based on the reaction of body cells to insulin. The common factors associated with breast cancer are family history of breast cancer, post-menopausal age, obesity, women with no pregnancy or pregnancy after 35 years of age, early menstruation (before age 12), old age (after age 55), a lifestyle of physical inactivity and consumption of alcohol. Cancer that is present within the breast is known as non-invasive breast cancer and its ability to spread to the outer surface of the breast is referred to as breast cancer.

1.2 Diabetes with Liver Cancer

Liver cancer is associated with weight and NAFLD, a condition in which the liver accumulates surplus fat, and risked by diabetes. The major factors that cause liver cancer are obesity, high insulin, triglycerides, cholesterol, and blood pressure level. Patients with liver cancer experience large changes in the signaling of the cancer cell that boosts the severity of tumors due the prevalence of cancer. Understanding the working mechanism of the cancer cells and diabetes can provide better treatments for the affected patients. In a recent study about 85% of Type 2 diabetes died due to liver cancer. The study of incident glucose tolerance and liver cirrhosis in patients, helps to plan the treatment for these patients.

1.3 Diabetes with Colon Cancer

Diabetes is associated with high insulin production also known as hyperinsulinemia, which is thought to increase the risk of colon cancer. Common factors that encourage the development of colon cancer include age greater than 50 years, family history of colon cancer, history of colon polyps, obesity, smoking, and diet. Colon cancer is related to the rectum where the digestion process is affected due to unhealthy food habits. Thus diabetic affected patients have chances of developing colon cancer that reduces the breaking down of the food leading to indigestion. Due to this, cells within the gland grow rapidly and if not diagnosed at an early stage will start growing in the muscles and will spread to other parts. High blood pressure accelerates colon cancer in older women. From a study, it is inferred that 141,000 people were identified with cancer and at least 50,000 deaths were caused due to rectum cancer in 2011. Consuming food with high fiber and low-fat can help prevent colon cancer. Colon cancer is called colorectal cancer.

2 Related Work

Due to the recent advent in technology, Machine learning has been widely used to predict if diabetic patients could develop cancer. There are many algorithms and the most common are Artificial Neural Networks (ANN), K Nearest Neighbor (KNN), Naïve Bayes, SVM, and Random Forest used for classification problems [1]. The paper identifies cancer and polygenic disorder discrimination ANFIS by an adaptive group-based KNN. By comparing various algorithms or methods with the same dataset taken from Pima Indian were discussed. ANN and ANFIS combining with Adaptive Group-based KNN will improve the accuracy. 80% of accuracy has been achieved by them [2].

C. Kalaiselvi proposed a model to predict heart diseases and cancer in diabetic patients. Heart disease, cancer, and diabetes are based on some common factors that majorly cause these diseases, which include obesity, age, gender, etc. In the first stage, the attributes are extracted by the Particle Swarm Optimization algorithm. In the second stage, the Adaptive Neuro-Fuzzy Inference System with AGKNN algorithmic program has been used to classify the heart diseases and cancer in Type 2 diabetic datasets [3]. The author proposed a technique by using Linear Regression to predict pancreatic and the author concluded the Linear Regression model more accurately predicted cancer than the artificial neural network model. They consider attributes such as insulin level, hyperglycemia, hyperinsulinemia, and inflammation [4]. This study infers that patients with Type 2 diabetic lead to developing colorectal cancer at higher risk. They consider parameters such as Age group, gender, occupation, insulin, metformin, hyperglycemia, stroke, and obesity [5]. This paper infers that data related to diabetes are converted into valuable knowledgeable data used for prediction of many diseases using SVM [6]. The author introduces a technique to identify the hidden patterns for non-communicable diseases such as diabetes and cancer in different perspectives using various data mining algorithms [7]. This study infers that diabetes increases the risk of death from cancer because lifestyle and food habits play a vital role for causing cancer in diabetic patients [8].

3 Classification of Cancer Using Random Forest

3.1 Training

The process of training involves providing data to an algorithm to learn from the input data. To train our Random Forest Classifier we have used the diabetic dataset. The dataset consists of 15 key parameters that is used to train the Random Forest classifier. This model iteratively identifies the best weight for the input and then fit to do classification. Once the training phase is complete, the predict method is applied to classify the input to return result as whether a diabetic patient is likely to get cancer. Further, it classifies the types of cancer such as breast, liver, and colon. The Random Forest algorithm gives a training accuracy of 99%.

3.2 Random Forest

Random Forest is a supervised machine learning algorithm. A Random Forest model fits a number of decision trees from various samples drawn from the dataset and uses weight averaging to improve the prediction accuracy and control overfitting. In a decision tree, the input is a whole dataset which is given at the top as the root node and which then branches off into various smaller sets of decision trees. It builds multiple decision trees and merges them together to form the final prediction. It uses the median for handling missing numerical values and modes for handling missing categorical values. The various decision trees in the Random forest are run in parallel. While building the trees there is no interaction between these trees. In this method, a massive amount of decision trees are created. Every decision tree is fed by each observation results. The most frequent outcome for all examination is returned as the final output.

3.3 Testing

The diabetics dataset is split into train and test data in the ratio of 80:20. We have used 200 entries with 15 attributes as test data. Predictions have been made by considering the set of tuples as well as the individual predictions. The predicted results where zero is the possibility of not getting cancer, one being the possibility of getting the liver cancer, two being the possibility of getting breast cancer and three being the possibility of getting colon cancer. We got the testing accuracy as 86%.

3.4 Confusion Matrix

The confusion matrix is used to relate the performance of a classifier on the test data for which the true positives are known. It allows us to picturize the performance of the Random Forest algorithm (Fig. 1).

It allows for easy identification of confusion between the four class labels. A confusion matrix may be an outline of prediction results on a classification. The number of correct and incorrect predictions is identified with count values and the datasets are broken down by 4 class. It shows how our classification model is confused when it makes a prediction with the three different types of cancer. It gives us details about the types of errors that are being made.

4 Results and Discussion

See Figs. 2, 3, 4.

Fig. 1 Confusion matrix

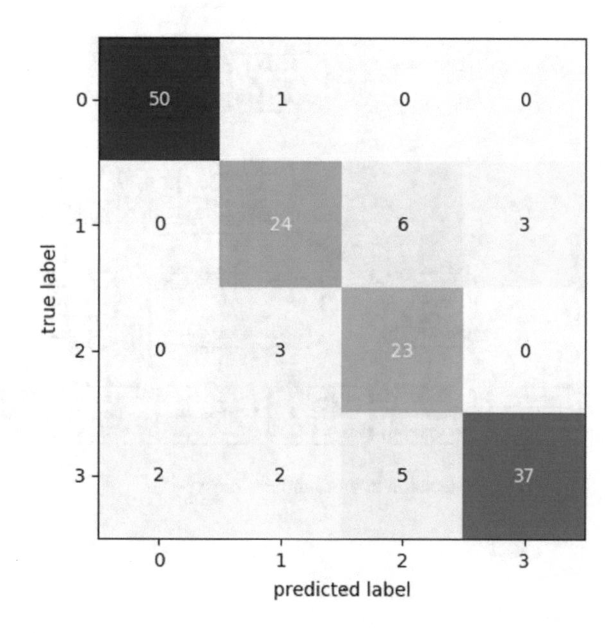

```
The shape of our features is: (1048, 16)
The shape of our features is: (156, 16)
Training Features Shape: (838, 15)
Training Labels Shape: (838,)
Testing Features Shape: (156, 15)
Testing Labels Shape: (156,)
```

Fig. 2 Splitting of dataset into train and test

```
Test Accuracy  ::  86.0 %.
classification report
              precision    recall  f1-score   support

           0       0.96      0.98      0.97        51
           1       0.80      0.73      0.76        33
           2       0.68      0.88      0.77        26
           3       0.93      0.80      0.86        46

   micro avg       0.86      0.86      0.86       156
   macro avg       0.84      0.85      0.84       156
weighted avg       0.87      0.86      0.86       156
```

Fig. 3 Accuracy using random forest algorithm

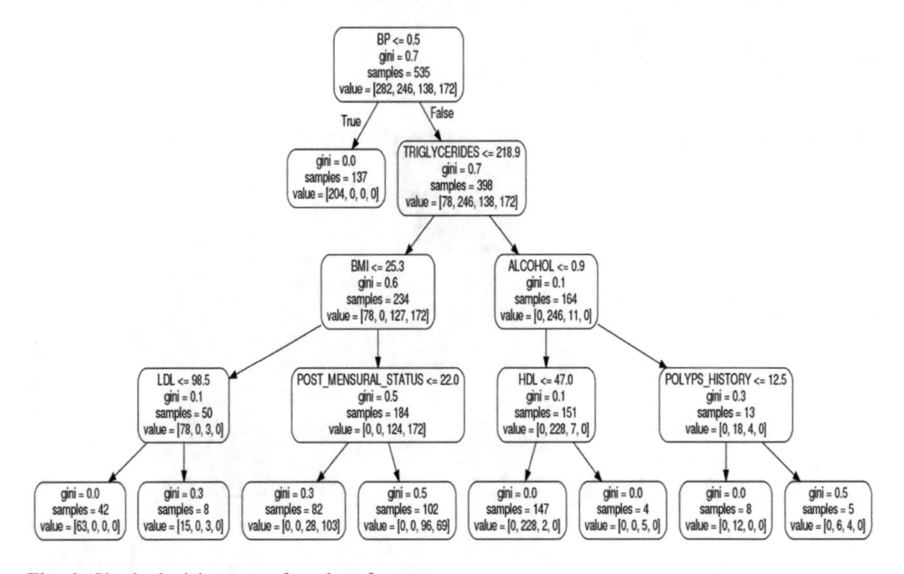

Fig. 4 Single decision tree of random forest

5 Conclusion and Future Work

Patients with a record of Type 2 diabetes have a higher chance of getting cancer than those without diabetes. The Random Forest model has been used effectively to predict whether a Type 2 diabetic patient is vulnerable to cancer or not. If the values as specified in Table 1 are met, then it is possible to narrow down and classify a particular type of cancer such as breast, liver, and colon cancer. We have achieved accuracy of 86%. In future work, cancer can be predicted and classified using deep neural network. Hence, the accuracy of this model can be further improved by testing at odds with a larger dataset.

Table 1 List of parameters and their range

Parameters	Values (in range)
BMI Body mass index	Both male and female
	Normal BMI –18.5 to –24.9
	Abnormal range <18.5 & >24.9
Liver cancer	
Triglycerides	Normal level: less than 150 mg/dL
	Border level: 150–199 mg/dL
	High: 200–499 mg/dL
	Very high: 500 mg/dL or more
HDL High-density Lipoproteins	High risk: less than 40 mg/dL
	Desirable: 60 mg/dL and above
LDL Low-density Lipoproteins	Optimal: less than 100 mg/dL
	Near optimal above optimal: 100–129 mg/dL
	Border line/high: 130–159 mg/dL
	High: 160–189 mg/dL
Insulin	Normal: 80–100 mg/dL
	High risk: >160 mg/dL
BP Blood Pressure	Normal: <120/<80
	Elevated: 120–129/<80
	Hypertension: 130–180/90–120
	Severe: Hypertensive crisis >180/Hypertensive crisis >120
Colon cancer	
Age	Age in years

(continued)

Table 1 (continued)

Parameters	Values (in range)
Smoking	Yes/No
Polyps history	Polyps diameter ≥10 mm
Breast cancer	
Post-menstrual status	High risk: >50 years of a risk age
Family history of breast cancer	Yes/No
No. of pregnancy	Normal: >5
	Risk: 1–5
	High risk: 0
Age of 1st delivery	High risk: >30 years of age
Menstruation age	High risk: <12 years of age
Alcohol	High risk for women: >0.0188 liters per day

References

1. Rau, H.H., Hsu, C.Y., Lin, Y.A., Atique, S., Fuad, A., Wei, L.M., Hsu, M.H.: Development of a web-based liver cancer prediction model for type II diabetes patients by using an artificial neural network. Comput. Methods Programs Biomed. (2015)
2. Kalaiselvi, C., Nasira, G.M.: A novel approach for the diagnosis of diabetes and liver cancer using ANFIS and improved KNN. Res. J. Appl. Sci. Eng. Technol. **8**(2), 243–250 (2014)
3. Kalaiselvi, C., Nasira, G.M.: Prediction of heart diseases and cancer in diabetic patients using data mining techniques. Indian. J. Sci. Technol. **8**(2) (2015)
4. Hsieh, M.H., Sun, L.M., Lin, C.L., Hsieh, M.J., Hsu, C.Y., Kao, C.H.: Development of a prediction model for pancreatic cancer in patients with type 2 diabetes using logistic regression and artificial neural network models. Cancer Manage. Res. (2018)
5. Hsieh, N.H., Sun, L.M., Lin, C.L., Hsieh, M.J., Sun, K., Hsu, C.Y., Chou, A.K., Kao, C.H.: Development of a prediction model for colorectal cancer among patients with type 2 diabetes mellitus using a deep neural network. J. Clin. Med. (2018)
6. Kavakiotis, I., Tsave, O., Salifoglou, A., Maglaveras, N., Vlahavas, I., Chouvarda, I.: Machine learning and data mining methods in diabetes research. Comput. Struct. Biotechnol. J. **15** (2017)
7. Jeewandara, N., Asanka, P.P.G.D.: Data mining techniques in prevention and diagnosis of non communicable diseases. Int. J. Res. Comput. Appl. Robot. ISSN 2320-7345 (2017)
8. Chen, Y.: Association between type 2 diabetes and risk of cancer mortality: a pooled analysis of over 771,000 individuals in the Asia Cohort Consortium. Diabetologia (2017)
9. Wang, M., Hu, R.Y., Wu, H.B., Pan, J., Gong, W.W., Guo, L.H., Zhong, J.M., Fei, F.R., Yu, M.: Cancer risk among patients with type 2 diabetes mellitus—a population-based prospective study in China. Sci. Rep. **5**, 11503 (2015)
10. Jee, S.H., Ohrr, H., Sull, J.W., Yun, J.E., Ji, M., Samet, J.M.: Fasting serum glucose level and cancer risk in Korean men and women. JAMA (2005)
11. Renehan, A.G., Zwahlen, M., Minder, C., O'Dwyer, S.T., Shalet, S.M., Egger, M.: Insulin-like growth factor (IGF)-I, IGF bindingprotein-3, and cancer risk-Systematic review and meta-regression analysis. Lancet (2004)
12. Chiu, C.C., Huang, C.C., Chen, Y.C.: Increased risk of gastrointestinal malignancy in patients with diabetes mellitus and correlations with anti diabetes drugs. A nationwide population-based study in Taiwan. Intern. Med (2005)

13. Kumara Kumar, J., Agilan, S.: Liver cancer prediction for type-2 diabetes using classification algorithm. Int. J. Adv. Res. Comput. Res. **9**(2) (2018)
14. Hippisley-Cox, J., Coupland, C.: Development and validation of risk prediction algorithms to estimate future risk of common cancers in men and women: prospective cohort study. Bmj Open (2015)

ICT in Education for Sustainable Development in Higher Education

Shalini Menon, M. Suresh and S. Lakshmi Priyadarsini

Abstract The purpose of this paper is to explore and encapsulate the ICT enablers that can facilitate education for sustainability in higher education and to understand the interrelationship between these enablers. This study contributes to the literature survey of the developments so far in the successful incorporation of ICT in sustainable education. This paper discusses the process of shifting to the new paradigm learning from some of the successful ICT initiatives implemented by universities across the globe. The results of this study will assist the policymakers and management of universities and colleges in understanding important ICT enablers that can facilitate sustainability in higher education.

Keywords ICT for sustainable education · Sustainability · Higher education · Sustainable development · Sustainable education

1 Introduction

The past decade has seen growth in the number of publications on the topic Sustainability in higher education which can be attributed to the increase in the number of journals that are committed to the cause of creating and disseminating knowledge on sustainability and the general concern toward the issue that if not redressed now will become irretrievable.

Higher education has a social responsibility and is empowered to educate and influence thoughts and actions of people. Sustainability is a value-laden concept that makes it difficult for universities to teach as the content on sustainability alone

S. Menon (✉) · M. Suresh
Amrita School of Business, Amrita Vishwa Vidyapeetham, Coimbatore, India
e-mail: shalini.binu@hotmail.com

M. Suresh
e-mail: drsureshcontact@gmail.com

S. Lakshmi Priyadarsini
Department of Zoology, Government Victoria College, Palakkad, India
e-mail: lakshpriyadarsini@gmail.com

M. Tuba et al. (eds.), *ICT Systems and Sustainability*,
Advances in Intelligent Systems and Computing 1077,
https://doi.org/10.1007/978-981-15-0936-0_13

would not be able to change learners' prior assumptions. Learning that comes from students experiencing real-life issues related to sustainability can be instrumental in deconstructing the frame of references and assumptions from prior experiences leading to construction of new knowledge and new values and belief that are in favor of the society.

2 Sustainability in Education

Superfluous changes to the curriculum might not be sufficient to bring about the change [1]. Interlocking the content and activities can prove to be synergistic in promoting education for sustainability. Engaging students in different ways throughout the learning process [2, 3] can catalyze the transformation of the whole institute empowering students with multiple learning outcomes. Studies on incorporating sustainability in higher education have emphasized the holistic, and interdisciplinary approach that hones systemic, critical, and analytical thinking [4] in students for which the authors believe that the students should be subjected to active learning [5, 6]. Providing students with the opportunity to experience the real issues [7] would enable them to understand the problems from perspective of different stakeholders and in the process reflect on their own values and behavior constructing new knowledge.

Some of the pedagogical methods used across the globe for teaching sustainability are problem-based methodology [8], inquiry-based learning [9, 10], field trips [11], case study [12, 13], etc. [14] reflected on the importance of ICT suggesting higher education institutes focus on the three concepts ICT, Curriculum, and Pedagogy. According to [15–17] interactive platforms such as ICT and e-learning enhance communication and collaborations. Universities should utilize the potential of E-learning and Social media in promoting and communicating environmental and other issues related to sustainability [18, 19]. Reference [18] argued that these tools alone cannot serve as a substitute for face-to-face teaching–learning process, they need to be blended with other experiential learning methods. Reference [20] supported distance education, online teaching–learning as it can reduce the consumption of energy. Reference [21] pointed using educational simulated games as one of the tools that can be experimented with in communicating sustainability issues.

3 ICT and Sustainability

In the world where technology has made inroads into the lives of people, the way we learn has evolved. Participative and social web technology such as social networking sites, wikis, blogs, video sharing platforms (YouTube), and web applications have made information available at click of a mouse and has facilitated collaborative, social learning [22]. Universities lately have been actively floating courses in asynchronous and synchronous mode through e-learning platforms; open-source educational systems, massive online open courses, and virtual learning environments [23].

The universities face the challenge of how to capitalize on the reliance and dependence of the millennial generation on technology in imparting education in general and education for sustainability in particular [24, 25]. Though there have been efforts by universities in supporting traditional classroom teaching with ICT-enabled teaching methodologies, how far these methods have been successful in teaching sustainability is still blurring.

The literature review revealed two aspects of ICT and Sustainability. The first aspect is incorporating the concept and principles of Sustainability in ICT/computing program enabling students to design and develop products and processes that are pro-environmental while the other aspect is the ICT applications in facilitating education for sustainability reducing the impact of other unsustainable actions on the society.

4 Sustainability in Computing Program

The growing dependence on ICT infrastructure in developed countries has been the reason for increasing carbon footprint, issues with e-waste disposal because of the short life of e-devices and the obsolescence of the software have impacted the environment in a negative way [26]. In order to mitigate the negative impact of ICT on environment, universities are making efforts to include sustainability in designing and developing ICT goods and services that consume less energy and material and are sustainable throughout their life cycle. This is achieved by following sustainability through design and sustainability in design also called green ICT [27–29].

References [30, 31] suggested three ways of integrating sustainability in computing programs: in a centralized manner equivalent to a course on sustainability, distributive way by including modules on sustainability in courses or in a blended way as reflected across programs and operations of institutions. Reference [32] implemented sustainability concept in the form of a course in Media technology program. Though the students were able to relate to the concept, they failed in connecting sustainability with their field of work which demonstrates the need for bridging this gap by weaving sustainability across all the programs and activities in campuses. Reference [33] suggested enhancing system systems thinking competency along with computational in order to deal with problems of social and environmental sustainability. Reference [34] recommended some of the topics that could motivate students pursuing computing programs; incorporation of positive and negative impact of ICT on society, ways of substituting virtual by physical to reduce the unsustainable acts, educating on the ways of reducing the carbon footprint caused by ICT and disposal of e-waste.

5 ICT in Facilitating Sustainability Education

Literature review revealed that universities today realizing the importance of ICT in education have incorporated it in different ways. Universities across the globe have been using Learning Management system (Moodle) for uploading course materials, assignments and assessments [35]. Tools like Moodle saves time of lecturers and are easy to manage. Apart from LMS, the other technology-enhanced learning adopted by universities are cloud learning services, social networking applications, m-learning applications, virtual and augmented reality interventions, etc. Reference [36]. How many of these tools and applications are put to use by universities in promoting sustainability is intriguing.

Reference [37] reflected on two key elements that are important for promoting sustainability one being education and the other networking. Teaching sustainability should provide students with active and experiential learning. E-learning is one such platform that can facilitate collaborations of different stakeholders working toward global goal of combating unsustainable practices. Universities have the option of going totally online by offering courses only in the distance education mode or can opt for blended learning where face-to-face learning is complemented with online learning. Portuguese distance learning university is committed to offering innovative and quality assured courses at undergraduate and postgraduate levels on environmental and sustainability science only in the distance mode. A platform like has benefited full time professional looking for professional development in ESD as the university has experienced opting for more courses [38].

Reference [18] studied the development and evaluation of the portal EDUCATE and m-learning app EDUCATE IN SITU offered in the curricula of architecture and recommended that for effective promotion of a multidisciplinary concept like sustainability a blend of face to face and online is critical, situated learning supported by e-learning and m-learning interface that are student-friendly and engaging can contribute in this case in developing sustainable designs.

Keeping the flexibility and choices that an online course can offer to students and professionals, a group comprising of a faculty member, specialist in engineering education and a specialist in pedagogy at the University of Nottingham started online courses on sustainability-Nottingham Open Online Course (NOOC) open to students and staff of different courses and university campus and Massive Open Online Course (MOOC) open to all developing content, pedagogy and process that develops holistic sustainability understanding among the engineering students [39].

Universities have also experimented with more than one ICT application like at the University of Jaen, Spain a new tool called the renewable energy video sharing tool collates information from different sources by using techniques of media sharing, podcast, and videocast. The videos can be accessed from anywhere at any time, the university by doing so has cut down on their visits to the sites to understand the process of generating energy from renewable sources. The videos are made available to students through YouTube channel and web portal. Apart from these Facebook and Twitter are used to communicate about the latest video updates [40].

Reference [15] identified four developmental areas for achieving literacy on sustainability (conceptual development, identity development, skill development, and the development of confidence) and suggested ways of achieving these with the support of e-learning:

- Sustainability concept can be explained with the help of documentaries, debating on issues, blogging, and using games that have proved to evoke emotions in students.
- Dialoging within the learning community, communication between the student and the staff through e-mail and other managed learning environments, developing e-portfolio or learning logs of personal experiences, debating on international online forums would help students in self-identification.
- Assigning simulated problem blending the face-to-face mode with e-learning can motivate students with internet research, collecting data, communicating with experts and stakeholders enhancing their critical and analytic skills.
- Making students accountable for their work and feedback on their work as they proceed can boost their confidence.

Social media can play a crucial role in facilitating sustainability. According to [25] social media platforms can influence opinions, invite debating, and can prove instrumental in changing behavior of people toward sustainability.

Reference [19] discussed the role of social media in educating society on environmental sustainability. According to [41] sustainability leaders can disseminate information about the institute's policies and campus activities through social media. This can boost the morale of students directly or indirectly involved in campus sustainability activities encouraging others also to be a part of the community by adopting and involving in sustainable campus initiatives. Recruitment of volunteers to take part in events is another purpose for which sustainable leaders use social media.

All said and done for a university to include these tools it is crucial to understand whether the government supports such innovations. In the case of Sub-Saharan African region the initial connectivity and the development of data networks led to creation of research and education networks that helped in collaboration with the US scientists on the issues of climate change [42].

6 Social Implications

Higher education institutes are hubs of knowledge creation and innovations. Institutes hold the social responsibility of disseminating this knowledge not only among the students and staff but also with other stakeholders. The growing impetus on making ICT services available through digital community centers in rural and isolated areas by way of Government legislations and programs [43] has increased the opportunity for higher education institutes in reaching out to students in remote and isolated places. Higher education institutes can capitalize on ICT-enabled methodologies such as e-class, e-learning, distance learning, web-based learning [44, 45] and other platforms like outreach activities and projects in promoting education on sustainability among students who live in remote places. Universities can also design and develop programs delivered through distance mode for the working population empowering them with decision-making skills that promote sustainable practices in their workplaces.

7 Conclusion

The literature on the use of ICT tools in promoting education for sustainable development is very limited. This could be as the author [16] mentioned in his study because of the lack of understanding of the concept sustainability, lack of knowhow of using ICT in the teaching–learning process, resistance from employees toward incorporating new technology, or lack of organizational culture that supports ICT-enabled practices. Although a number of institutions have incorporated ICT in their pedagogy, the use of ICT applications and tools in imparting education for sustainable development needs to be explored. To grow and to be successful in this competitive world it is important to keep pace with technology and transform the pedagogy to incorporate more such platforms. Universities need to explore more options of incorporating use of ICT applications and tools in imparting sustainability literacy.

References

1. Sterling, S., Thomas, I.: Education for sustainability: the role of capabilities in guiding university curricula. Int. J. Innov. Sustain. Dev. **1**(4), 349–370 (2006)
2. Mintz, K., Tal, T.: The place of content and pedagogy in shaping sustainability learning outcomes in higher education. Environ. Educ. Res. **24**(2), 207–229 (2018)
3. Beringer, A., Adomßent, M.: Sustainable university research and development: inspecting sustainability in higher education research. Environ. Educ. Res. **14**(6), 607–623 (2008)
4. Sherren, K.: A history of the future of higher education for sustainable development. Environ. Educ. Res. **14**(3), 238–256 (2008)
5. Kalamas Hedden, M., Worthy, R., Akins, E., Slinger-Friedman, V., Paul, R.: Teaching sustainability using an active learning constructivist approach: discipline-specific case studies in higher education. Sustainability **9**(8), 1320 (2017)
6. Shephard, K.: Higher education for sustainability: seeking affective learning outcomes. Int. J. Sustain. High. Educ. **9**(1), 87–98 (2008)
7. Kolb, A.Y., Kolb, D.A.: Learning styles and learning spaces: enhancing experiential learning in higher education. Acad. Manage. Learn. Educ. **4**(2), 193–212 (2005)
8. El-adaway, I., Pierrakos, O., Truax, D.: Sustainable construction education using problem-based learning and service learning pedagogies. J. Prof. Issues Eng. Educ. Pract. **141**(1), 05014002 (2014)
9. Aditomo, A., Goodyear, P., Bliuc, A.M., Ellis, R.A.: Inquiry-based learning in higher education: principal forms, educational objectives, and disciplinary variations. Stud. High. Educ. **38**(9), 1239–1258 (2013)

10. Dewoolkar, M.M., George, L., Hayden, N.J., Neumann, M.: Hands-on undergraduate geotechnical engineering modules in the context of effective learning pedagogies, ABET outcomes, and our curricular reform. J. Prof. Issues Eng. Educ. Pract. **135**(4), 161–175 (2009)
11. Cusick, J.: Study abroad in support of education for sustainability: a New Zealand case study. Environ. Dev. Sustain. **11**(4), 801–813 (2009)
12. Montiel, I., Antolin-Lopez, R., Gallo, P.J.: Emotions and Sustainability: A Literary Genre-Based Framework for Environmental Sustainability Management Education. Acad. Manage. Learn. Educ. **17**(2), 155–183 (2018)
13. Steiner, G., Laws, D.: How appropriate are two established concepts from higher education for solving complex real-world problems? a comparison of the Harvard and the ETH case study approach. Int. J. Sustain. High. Educ. **7**(3), 322–340 (2006)
14. Makrakis, V., Kostoulas-Makrakis, N.: Course curricular design and development of the M. Sc. programme in the field of ICT in education for sustainable development. J. Teach. Educ. Sustain. **14**(2), 5–40 (2012)
15. Diamond, S., Irwin, B.: Using e-learning for student sustainability literacy: framework and review. Int. J. Sustain. High. Educ. **14**(4), 338–348 (2013)
16. Jääskelä, P., Häkkinen, P., Rasku-Puttonen, H.: Teacher beliefs regarding learning, pedagogy, and the use of technology in higher education. J. Res. Technol. Educ. **49**(3–4), 198–211 (2017)
17. Din, N., Haron, S., Ahmad, H.: The level of awareness on the green ICT concept and self directed learning among Malaysian Facebook users. Procedia-Social Behav. Sci. **85**, 464–473 (2013)
18. Altomonte, S., Logan, B., Feisst, M., Rutherford, P., Wilson, R.: Interactive and situated learning in education for sustainability. Int. J. Sustain. High. Educ. **17**(3), 417–443 (2016)
19. Hamid, S., Ijab, M. T., Sulaiman, H., Md. Anwar, R., Norman, A. A.: Social media for environmental sustainability awareness in higher education. Int. J. Sustain. High. Educ. **18**(4), 474–491 (2017)
20. Caird, S., Lane, A., Swithenby, E., Roy, R., Potter, S.: Design of higher education teaching models and carbon impacts. Int. J. Sustain. High. Educ. **16**(1), 96–111 (2015)
21. Mercer, T.G., Kythreotis, A.P., Robinson, Z.P., Stolte, T., George, S.M., Haywood, S.K.: The use of educational game design and play in higher education to influence sustainable behaviour. Int. J. Sustain. High. Educ. **18**(3), 359–384 (2017)
22. Stepanyan, K., Littlejohn, A., Margaryan, A.: Sustainable eLearning in a Changing Landscape: A Scoping Study (SeLScope). UK Higher Education Academy (2010)
23. Bush, M.D., Mott, J.D.: The transformation of learning with technology: Learner-centricity, content and tool malleability, and network effects. Educ. Technol. **49**(1), 3–20 (2009)
24. Henry, B.C.: ICT for sustainable development. Sci. Technol. **2**(5), 142–145 (2012)
25. Bodt, T.: Role of the media in achieving a sustainable society. Centre of Bhutan Studies, pp. 459–500 (2007). http://crossasia-repository.ub.uni-heidelberg.de/358/. Accessed 2 May 2019
26. Hilty, L.M., Hercheui, M.D.: ICT and sustainable development. In: What Kind of Information Society? Governance, Virtuality, Surveillance, Sustainability, Resilience, pp. 227–235. Springer, Berlin, Heidelberg (2010)
27. Hilty, L.M., Aebischer, B.: ICT for sustainability: an emerging research field. In: ICT Innovations for Sustainability, pp. 3–36. Springer, Cham (2015)
28. Suryawanshi, K., Narkhede, S.: Green ICT for sustainable development: a higher education perspective. Procedia Comput. Sci. **70**, 701–707 (2015)
29. Klimova, A., Rondeau, E., Andersson, K., Porras, J., Rybin, A., Zaslavsky, A.: An international Master's program in green ICT as a contribution to sustainable development. J. Clean. Prod. **135**, 223–239 (2016)
30. Mann, S., Muller, L., Davis, J., Roda, C., Young, A.: Computing and sustainability: evaluating resources for educators. ACM SIGCSE Bull. **41**(4), 144–155 (2010)
31. Cai, Y.: Integrating sustainability into undergraduate computing education. In: Proceedings of the 41st ACM Technical Symposium on Computer Science Education, pp. 524–528. ACM (2010)

32. Eriksson, E., Pargman, D.: ICT4S reaching out: Making sustainability relevant in higher education. In: ICT for Sustainability 2014 (ICT4S-14). Atlantis Press (2014)
33. Easterbrook, S.: From computational thinking to systems thinking: a conceptual toolkit for sustainability computing. In: ICT for Sustainability 2014 (ICT4S-14). Atlantis Press (2014)
34. Hilty, L.M., Huber, P.: Motivating students on ICT-related study programs to engage with the subject of sustainable development. Int. J. Sustain. High. Educ. **19**(3), 642–656 (2018)
35. Isaias, P., Issa, T.: E-learning and sustainability in higher education: an international case study. Int. J. Learn. High. Educ. **20**(4), 77–90 (2013)
36. Daniela, L., Visvizi, A., Gutiérrez-Braojos, C., Lytras, M.: Sustainable higher education and technology-enhanced learning (TEL). Sustainability **10**(11), 3883 (2018)
37. Bordbar, M., Allahyari, M.S., Solouki, M.: E-learning as a new technology for sustainable development. ARPN J. Eng. Appl. Sci. **7**(3), 332–337 (2012)
38. Azeiteiro, U.M., Bacelar-Nicolau, P., Caetano, F.J., Caeiro, S.: Education for sustainable development through e-learning in higher education: experiences from Portugal. J. Clean. Prod. **106**, 308–319 (2015)
39. Sivapalan, S., Clifford, M.J., Speight, S.: Engineering education for sustainable development: using online learning to support the new paradigms. Australas. J. Eng. Educ. **21**(2), 61–73 (2016)
40. Torres-Ramírez, M., García-Domingo, B., Aguilera, J., De La Casa, J.: Video-sharing educational tool applied to the teaching in renewable energy subjects. Comput. Educ. **73**, 160–177 (2014)
41. Carpenter, S.E.R.E.N.A., Takahashi, B., Cunningham, C., Lertpratchya, A.P.: The roles of social media in promoting sustainability in higher education. Int. J. Commun. **10**, 4863–4881 (2016)
42. Bothun, G.D.: Data networks and sustainability education in African universities: a case study for Sub-Saharan Africa. Int. J. Sustain. High. Educ. **17**(2), 246–268 (2016)
43. Armenta, Á., Serrano, A., Cabrera, M., Conte, R.: The new digital divide: the confluence of broadband penetration, sustainable development, technology adoption and community participation. Inform. Technol. Dev. **18**(4), 345–353 (2012)
44. Bordoloi, R.: Transforming and empowering higher education through Open and distance learning in India. Asian Assoc. Open Univ. J. **13**(1), 24–36 (2018)
45. Andreopoulou, Z.: Green Informatics: ICT for green and sustainability. Agrárinformatika/J. Agric. Inform. **3**(2), 1–8 (2012)

Edge Detection and Enhancement of Color Images Based on Bilateral Filtering Method Using K-Means Clustering Algorithm

S. Sai Satyanarayana Reddy and Ashwani Kumar

Abstract In the recent, object recognition and classification has become an emerging area of research in robotics. In many image processing applications accurate edge detection is very crucial and plays an important role. The continuous and connected edges detection of color images is important in many applications such as satellite imagery and discover cancers in medical images, etc. The detection of these edges is very difficult and most of the edge detection algorithms do not perform well against broken and thick edges in color images. This paper proposed an edge detection and enhancement technique using bilateral method to decrease the broken edges of an optimization model in order to detect the edges. Combining bilateral filtering with convolution mask show all edges that are necessary by analyzing window one-by-one without overlapping. The proposed scheme is applied over the color image for manipulating the pixels to produce better output. The simulation results are performed using both noisy images and noise-free images. For producing the experimental results Standard deviation, Arithmetic mean are calculated. With the use of these parameters' quality assessment of corrupted and noisy images and the effectiveness of the proposed approach is evaluated.

Keywords K-Means algorithm · Bilateral filter · Edge detection · Convolution mask · Standard deviation

1 Introduction

In most of the systems of computer vision, intensity information and orientation of edges in images are useful and act as inputs for further processing in order to detect objects. For the success of such systems, precise information about edges is vital. Information about edges is broadly used in image classification, pattern recognition,

---

S. Sai Satyanarayana Reddy · A. Kumar (✉)
Vardhaman College of Engineering, Hyderabad, India
e-mail: ashwanikumarcse@gmail.com

S. Sai Satyanarayana Reddy
e-mail: saisn90@gmail.com

M. Tuba et al. (eds.), *ICT Systems and Sustainability*,
Advances in Intelligent Systems and Computing 1077,
https://doi.org/10.1007/978-981-15-0936-0_14

image registration, and image segmentation [1]. Continuous contours of the object boundaries can also be provided by edge detection algorithm from an application-level view [2]. However, the computations required for the establishment of these continuous data would be complex and take much time. These areas are responsible for shaping the contours which serve as the boundary of objects. Even though there have been a large number of solutions that are recommended in order to improvise the precision of the edges [3]. Even in the algorithms that are existent the most prominent obstacle in the detection of continuous edges is noise phenomena [4]. This result causing different values for pixels and minimizes the efficiency of the scheme in case of noisy environment. Illumination phenomenon is another significant barrier which complicates the operation of edge detection [5]. The threshold-based technique can be utilized by various edge detection algorithms to perform better edge detection [6]. Edges serve as the boundaries between distinct textures of image and termed as an edge [7]. Image reconstruction, data compression, and image segmentation can be done with the help of detection of edges for an image [8]. In particular, weighted average of pixel values in the neighborhood can be computed by Gaussian low-pass filtering in which the weights get reduced with distance from the center of the neighborhood. The noise values that can cause the corruption of the pixels nearby are mutually very less correlated than the actual data [9]. According to the work done by Sobel, the operator i.e., Sobel operator (Sobel filter) which focuses on edge enhancement which is a part of edge detection with little smoothening [10]. A convolution mask of 3×3 is applied over the image for manipulating the pixels to produce better output. Because of this, it has its own drawbacks such as sensitivity to noise and directional because many small local maxima will be produced by noise [11]. The response to a step edge is across several pixels, so a post-processing is needed for edge thinning.

The prime aim of this research work is to implement edge detection and enhancement operation in a simple and easy way so that it can be understandable to everyone by eliminating the drawbacks of existing algorithms to some extent. In Sect. 1, the introduction of edge detection and enhancement especially for color images are shown. In Sect. 2 the author demonstrates related work. In Sect. 3, the proposed approach for implementation of edge detection and enhancement based on bilateral filtering method is shown. The implementation result is shown in Sect. 4. Finally, Sect. 5 concludes the research paper.

2 Related Work

The implementation of edge detector is a non-trivial problem and this work done by Djemel Ziou and Salvatore Tabbone on edge detection technique [4]. The result of theoretical analysis and other requirements designed by edge detectors are required to get a running program. Here the case of multi-scale edge detection is considered; the increase of the scale must not drastically affect the computation time it is one of the requirements, and also the convolution mask method is considered as it is sensitive to scale increase. In addition, they added that a detector must preserve the properties of the resulting image from theoretical analysis. In fact, when edge detection involves a convolution operation of the image and a continuous filter, the implementation requires sampling of this filter in its discrete form. This method has multiple advantages. No filter cutoff is necessary. The measurement of spatial gradient of an image operates on a 2-D performance and this work is done by Sobel. At each point the above gradient finds inside "*I*" says an input grayscale image. A pair of 3×3 convolution masks is used by the Sobel edge detector, here one gradient forms in x-direction and another in y-direction this is used for estimation. The mask over the image is modified with square pixels at a time that provides as a result where value changes into original image that shifts one pixel to form right and continues till it finds its end of the row and again automatically starts its next row, it contains first and last row that cannot be manipulated by 3×3 masks.

The use of the multistage algorithm to detect edges in images in a wide range is proposed by Canny [3]. This canny scheme is a complex edge detector that takes a long time to get results. First image runs through a Gaussian blur to avoid noise. Here the angle and magnitude are applied when the algorithm is used to determine the portion of the edges. In 1986 optimal smoothing filter is the solution of canny and is considered by mathematical problem of detection, minimizing and localization of multiple resources to get single edge. Four exponential terms that were shown by filters in the form of first deviation of Gaussian [12]. He also introduced non-maximum suppression, where the edge gradient direction is pointed. Frequency and orientation based on Gabor filter for edge detection that represents and is similar to human perception in texture representation and discrimination [13, 14]. The sinusoidal plane wave is modulated by Gabor filters and is connected to Gabor wavelets. They are designed by the number of rotation and dilations. The Gabor filters are convolved with the signal, that result is called Gabor space, and the advantage is a better fit in receptive fields of weight functions, it is very important in image processing and edge detection [15, 16]. Cartoon style rendering of 3D scenes was introduced by Decaudin [17]. Buffering is used to determine discontinuity in edge spaces uses in these techniques. We get results in cartoon style rendering in edge enhanced process. The texture mapping hardware used in the real time was in cartoon style rendering [18]. The algorithm was presented at the border, silhouette and crease edges. The screen space information that contains 2-D texture in edge intensities that are available in edge map. They describe a fast and new implementation based on the convolution with an FFT and the kernel, the color image filtering and cross bilateral filtering to be demonstrated.

3 Proposed Method

The proposed approach focuses on dealing with the important features of the color image based on RGB color scheme. It points to the edges in the image dynamically and points them out accordingly. The area of concentration changes according to the image. The algorithm does not depend on certain constraints and also acts dynamically. The algorithm aims to show all edges that are necessary by analyzing window one by one without overlapping. Our proposed scheme consists five important blocks namely: pre-processing, edge detection, Bilateral Filtering, and k-means algorithm shown in Fig. 1. Input color image $I(x)$ is pre-processed by de-noisy technique to produce noise-free image $N(x)$. Then the detection process of edge is carried on for the image which generates a transformed image $P(x)$. This converted image suffers by applying bilateral filtering with k-means algorithm yielding the filtered image $Q(x)$. Further edge enhancement by using segmentation and threshold is performed on these enhancement images to get final edge map $E(x)$. The entire proposed approach can be understood by a block diagram given below.

In our proposed approach the broken edges of the image are reconstructed by obtaining the edge map using a convolution mask of 3×3 that is applied over the image for manipulating the pixels to produce better output of the image when noise is present. Firstly, the input image $I(x)$ is pre-processed for normal image processing which produce an image $N(x)$. Thenceforth edge detection is performed which uses bilateral filter for noise removal and k-means algorithm for sharpening the edges we get $P(x)$ image. Finally edge enhancement is performed on $P(x)$ image by applying direct convolution on threshold value we obtained $Q(x)$ from this image the reconstructed image $E(x)$ is generated.

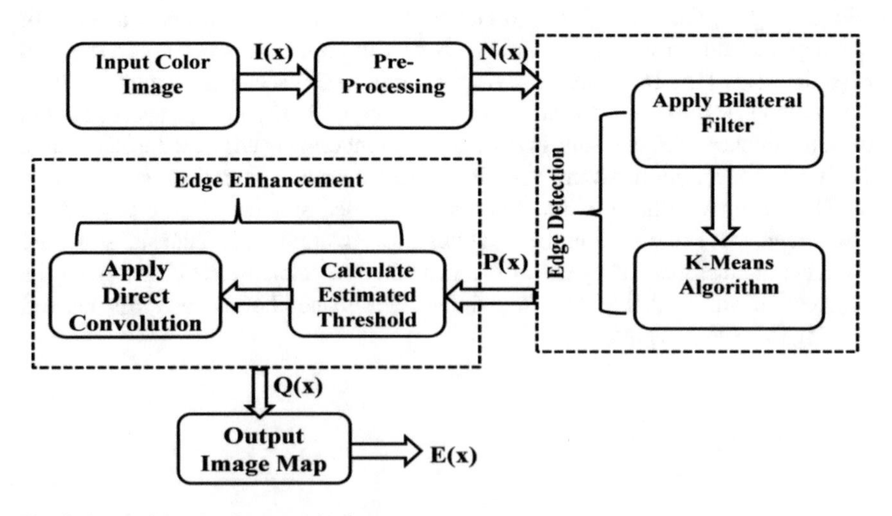

Fig. 1 Block diagram of proposed scheme

3.1 Bilateral Filtering

Bilateral Filtering is highly effective in noise removal and keeps the edges sharp because it only considered nearby pixels. It contains methods such as non-iterative, simple and local. These methods can be applied on color-based or grayscale-based images both are in geometric closeness and this will prefer the values like range and domain. Tomasi and Manduchi coined in (1998) the "bilateral filter" [19–22].

3.2 NumPy

One of the highly optimized libraries is NumPy for the numerical operation. Array structure and OpenCV are changed to NumPy arrays. Provides several amounts of image for the future. Other libraries like SciPy, Matplotlib supports libraries to be used for the numerical operations with arrays. We must write code in OpenCV-python this is a mathematical operation like standard deviation, mean, etc. To measure the quality of the amount of variation in standard deviation has a set of data values. The data points are close to mean to deviate the low standard points of the set, whereas high standard deviation spread over the wide range of values of data points that easily spread out. In our algorithm we have used the standard deviation by considering the window size of 3 × 3 and replace the resultant in the center position. The standard deviation is calculated by using Eq. 1.

$$\sigma = \frac{\sqrt{\sum_{i=1}^{n}(x_i - \bar{x})2}}{n - 1} \tag{1}$$

where: n = number of data points, x_i = Value of data, $\bar{x}$ = Mean of the data

3.3 K-Means Algorithm

The aim of the data points into k clusters as partitions. Each point is near to the mean assigning to the ṅ data points. The mean of the cluster called "center." Applying of k-means yields k which supports the original n data points. These are more similar to each other than that which belong to the other clusters. This technique is used for dynamic clustering.

$$J(v) = \sum_{i=1}^{c}\sum_{j=1}^{c_i}\left(\|x_i - v_j\|\right)^2 \tag{2}$$

4 Result Analysis

To check the performance of the proposed scheme noisy and noise-free color images are taken as input. The quality assessment of these images is done by calculating the parameters like Standard deviation (NumPy), Arithmetic mean and convolution mask. In this section, for producing our result we have taken different original images as input and detecting the edges at various levels like first we have analyzed Sobel operator method then we have detected the edges by bilateral filtering method and finally we get enhanced images. We can conclude that our proposed scheme reduces the noise from noisy images as shown below. As mentioned above the proposed technique for edge detection is to overcome the problems raised by the existing algorithms. For instance, we have taken Baboon, Man, Bird, Boat, Eline, and Lena images of size 512 × 512 color images.

Figures 2a–d, 3, 4, 5, 6, 7a–d demonstrate the original image, images obtained by applying Sobel operator, edge detection, and edge enhancement, respectively. Our results are compared with Sobel operator which can be seen in Figs. 2b, 3, 4, 5, 6, 7b. From the above results, we can conclude that our proposed approach reduces the noise which is shown in Figs. 2d, 3, 4, 5, 6, 7d. As mentioned above the scheme uses a method for edge detection to conquer the problems raised by the previously published algorithms. In this scheme bilateral filtering with k-means algorithm makes use to analyze the performance of the scheme. Each pixel (x) of the original image is replaced with correlated normalized neighbor pixel by using the mathematical equations. Salt and pepper noise can be demonstrated as a bright and dark spot into the image.

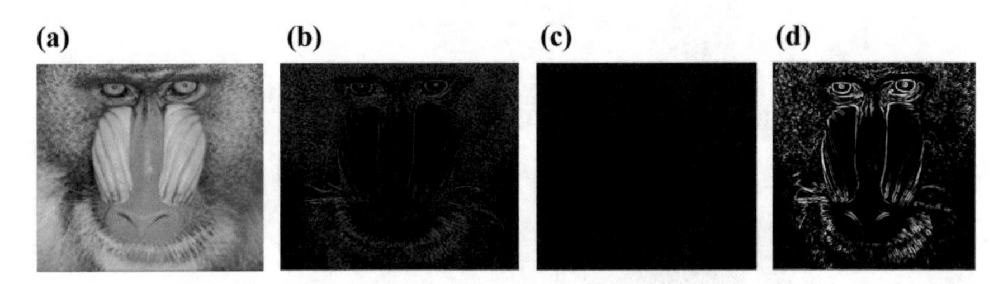

Fig. 2 **a** Baboon image **b** Sobel operator **c** edge detection **d** edge enhancement

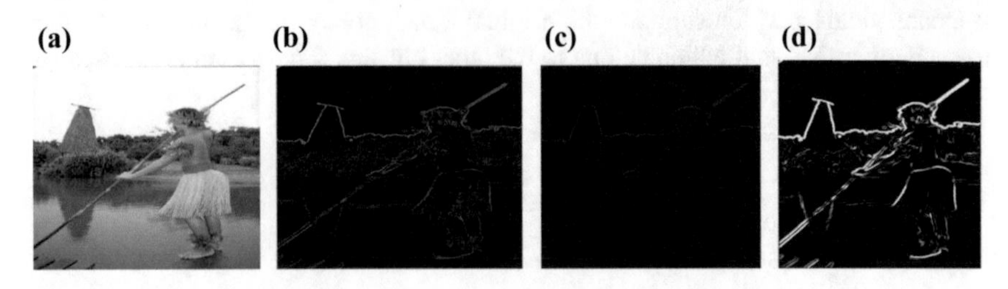

Fig. 3 **a** Man Image **b** Sobel operator **c** edge detection **d** edge enhancement

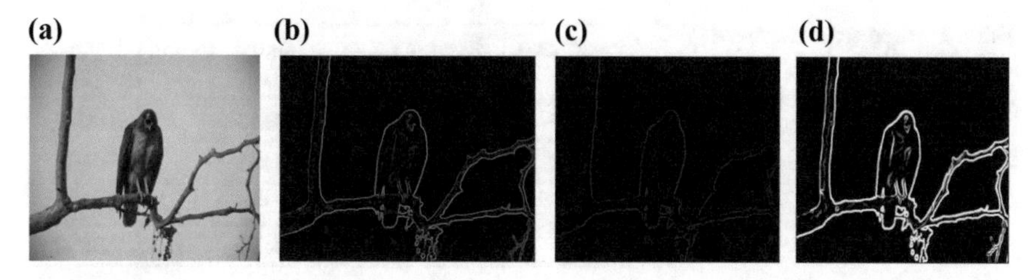

Fig. 4 **a** Bird Image **b** Sobel operator **c** edge detection **d** edge enhancement

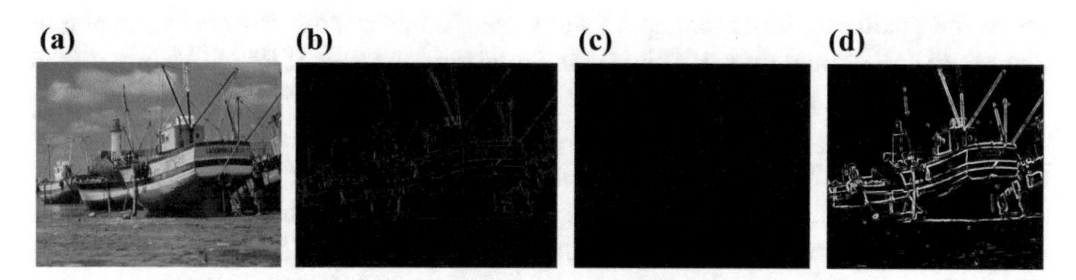

Fig. 5 **a** Boat image **b** Sobel operator **c** edge detection **d** edge enhancement

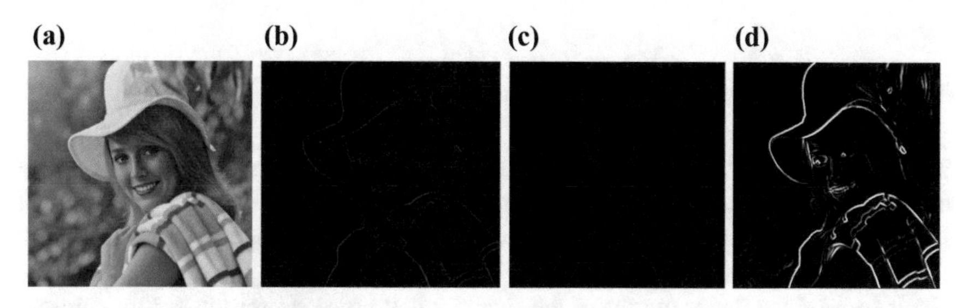

Fig. 6 **a** Eline image **b** Sobel operator **c** edge detection **d** edge enhancement

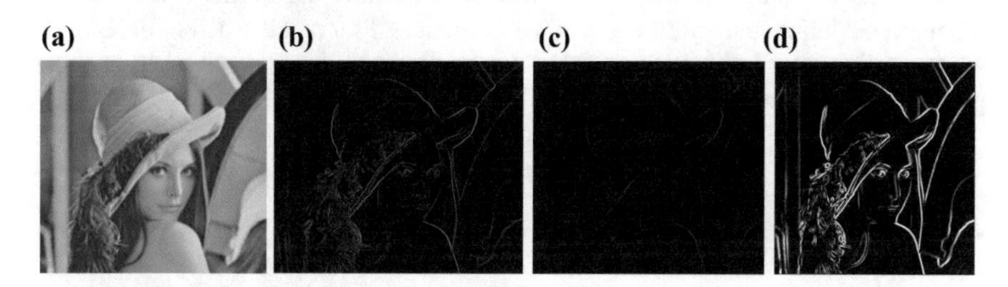

Fig. 7 **a** Lena image **b** Sobel operator **c** edge detection **d** edge enhancement

During the edge detection process this noise corrupts the edges of the image. Edges contain most of the important information of the images if there are sharp changes in the intensity these edges represent points. Hence, edges are more prone

Table 1 Represents the mean values for different edge detection and enhancement technique in case of Lena image

Noise variance (%)	Sobel operator (6)	Edge detection (13)	Proposed approach
3	0.0169	0.0133	0.0099
6	0.0183	0.0168	0.0107
9	0.0198	0.0175	0.0116
12	0.0223	0.0194	0.0130

to noise detection if noise present in the image leads us to detect false edges. To assess the quality of the edge map, mean is used to find out the error between output image and the original image. This is calculated with the help of Eq. 2. We have taken salt and pepper noise with density $(\sigma) = (3\text{–}12)\%$. For representing the performance comparison of the proposed approach, we have used Sobel operator. By considering Figs. 1a–d, 2, 3, 4, 5, 6, 7a–d the output images of the proposed scheme clearly show the better quality in edge detection over Sobel operator. This operator did not work well in noisy environment because it misses some true edges. Table 1 represents the mean values for different edge detection and enhancement technique in case of Lena image with the presence of salt and pepper noise with different variance.

5 Conclusion

In this work, we have used a new approach for detecting edges of color images by using a bilateral filter with convolution that are applied for enhancing edge map. Here, the broken edges of image are reconstructed by obtaining the edge map using convolution mask of 3×3 which is applied over the image for manipulating the pixels to produce better output of the image when noise is present. By making the use of bilateral filtering along with k-means clustering increased the performance. To maintain the quality of original color images we have used RGB color scheme. Standard deviation and arithmetic mean are calculated for reconstructing the broken edges of the color images. Simulation results indicate that the proposed scheme performs well in noisy environment.

References

1. Agaian, S.S., Baran, T.A., Panetta, K.A.: Transform-based image compression by noise reduction and spatial modification using Boolean minimization. In: IEEE Workshop on Statistical Signal Processing (2003)
2. Baker, S., Nayar, S.K.: Pattern rejection. In: Proceedings of IEEE Conference Computer Vision and Pattern Recognition, pp. 544–549 (1996)
3. Canny, J.: A computational approach to edge detection. IEEE Trans. Pattern Anal. Mach. Intell. (1986)

4. Ziou, D., Tabbone, S.: Edge detection technique an overview. Int. J. Pattern Recognit. Image Anal. (1998)
5. Kumar, A., Ghrera, S.P., Tyagi, V.: An ID-based secure and flexible buyer-seller watermarking protocol for copyright protection. Pertanika J. Sci. Technol. **25**(1), 57–76 (2017)
6. Sharifi, M., Fathy, M., Mahmoudi, M.T.: A classified and comparative study of edge detection algorithms. In: Proceedings of the International Conference on Information Technology: Coding and Computing, pp. 117–220 (2002)
7. Folorunso, O., Vincent, O.R., Dansu, B.M.: Image edge detection: A knowledge management technique for visual scene analysis. Inf. Manage. Comput. Secur. **15**(1), 23–32 (2007)
8. Kumar, A., et al.: A lightweight buyer-seller watermarking protocol based on time-stamping and composite signal representation. Int. J. Eng. Technol. **7**(4.6), 39–41 (2018)
9. Gonzalez, R., Woods, R.: Digital Image Processing, 2nd edn. Prentice-Hall Inc, pp. 567–612 (2002)
10. Shin, M.C., Goldgof, D., Bowyer, K.W.: Comparison of edge detector performance through use in an object recognition task. Comput. Vis. Image Underst. **84**(1), 160–178 (2001)
11. Kumar, A., Ansari, D., Ali, J., Kumar, K.: A new buyer-seller watermarking protocol with discrete cosine transform. In: International Conference on Advances in Communication, Network, and Computing, pp. 468–471 (2011)
12. Peli, T., Malah, D.: A study of edge detection algorithms. Comput. Graph. Image Process. **20**,1–21 (1982)
13. Pei, S., et al.: The generalized radial Hilbert transform and its applications to 2-D edge detection (any direction or specified direction). In: Proceedings of the International Conference on Acoustics, Speech and Signal Processing, pp. 357–360 (2003)
14. Torre, V., Poggio, T.A.: On edge detection. IEEE Trans. Pattern Anal. Mach. Intell. **PAMI-8**(2), 187–163 (1986)
15. Yang, T., Qiu, Y.: Improvement and implementation for Canny edge detection algorithm. In: Proceedings of the SPIE (2015)
16. Saito, T., Takahashi, T.: Comprehensible rendering of 3D shape. In: Computer Graphics (Proceedings of SIGGRAPH'90), **24**(4), 197–206 (1990)
17. Decaudin, P.: Cartoon looking rendering of 3D scenes. Research Report 2919, INRIA (1996)
18. Lake, A., Marshall, C., Harris, M., Blackstein, M.: Stylized rendering techniques for scalable real-time 3D animation. In: NPAR'00: Proceedings of the 1st International Symposium on Non-Photorealistic Animation and Rendering. ACM, New York, NY, USA, pp. 13–20
19. Kumar, A., Ghrera, S.P., Tyagi, V.: Modified buyer seller watermarking protocol based on discrete wavelet transform and principal component analysis. Indian J. Sci. Technol. **8**(35), 1–9 (2015)
20. Kumar, A., Ghrera, S.P., Tyagi, V.: A new and efficient buyer-seller digital watermarking protocol using identity based technique for copyright protection. In: Third International Conference on Image Information Processing (ICIIP). IEEE (2015)
21. Kumar, A., Gupta, S.: A secure technique of image fusion using cloud based copyright protection for data distribution. In: 2018 IEEE 8th International Advance Computing Conference (IACC). IEEE (2019)
22. Kumar, A.: Design of secure image fusion technique using cloud for privacy-preserving and copyright protection. Int. J. Cloud Appl. Comput. (IJCAC) **9**(3), 22–36 (2019)

Drainage Toxic Gas Detection System Using IoT

Swapnil Manikrao Alure, Rahul Vishnupant Tonpe, Aniket Dhanaraj Jadhav, Supriya Tryambakeshwar Sambare and Jayshri D. Pagare

Abstract Sensing technology has been widely investigated and utilized for gas detection. Due to the different applicability and inherent limitations of different gas sensing technologies, researchers have been working on different scenarios with enhanced gas sensor calibration. We developed an ammonia (NH_3) sensor using the PANi solution. It was observed that the sensor gives a quick response to it, which is less than 100 s. This sensor is operable at room temperature. Interfacing of the sensor is done with Arduino and Android application is used to display the results. This monitoring application can be used as a safety precaution for laborers.

Keywords Sensor · NH_3 sensor · Android · Arduino · IoT

1 Introduction

In India recently many drainage workers have lost their lives due to the toxic gases produced in drainage. If these gases were detected in earlier stages then it will help to reduce the rate of deaths.

In the twentieth century, there was a famous technique for the detection of toxic gases in coal mines. This method is called a Canary Bird technique. Miners took this bird with them in mines. A canary in a coal mine is an advanced warning of some danger. Their anatomy requires more oxygen and makes them more sensitive to toxic gases such as methane and carbon monoxide both of which have no color, odor, or

S. M. Alure (✉) · R. V. Tonpe · A. D. Jadhav · S. T. Sambare · J. D. Pagare
MGM's Jawaharlal Nehru Engineering College, Aurangabad, MH, India
e-mail: swapnilalure93@gmail.com

R. V. Tonpe
e-mail: rahul.v.tonpe@gmail.com

S. T. Sambare
e-mail: supriyasambare@gmail.com

J. D. Pagare
e-mail: Jaydp2002@yahoo.co.in

M. Tuba et al. (eds.), *ICT Systems and Sustainability*,
Advances in Intelligent Systems and Computing 1077,
https://doi.org/10.1007/978-981-15-0936-0_15

taste. Signs of distress from the bird indicated to the miners that the condition was unsafe.

Ammonia (NH_3) is one of the most common toxic gas found in drainage. It is also very harmful to the human body and it may cause the death of the person after a certain level. If we are able to detect the presence of this gas in the earlier stage it will be very helpful.

2 Literature Survey

Gupta et al. [1] have discussed FET sensor which detects the ammonia gas at room temperature on low concentration. It also shows less sensitivity toward Carbon monoxide, carbon dioxide, and hydrogen.

Knittel et al. [2] have discussed new readout circuit for FET-based gas sensors that are designed. It is designed in such a manner that it is able to detect ammonia and hydrogen gases with one device.

Hakimi et al. [3] have discussed ammonia sensor which is fabricated using PANi and nitrogen-doped graphene quantum dots and they used silver (Ag) and aluminum (Al) as electrodes to sensing film.

Pawar et al. [4] have discussed in this paper oxidative chemical polymerization and sol gel methods that are used to synthesize PANi-TiO_2.

Vaghela et al. [5] have discussed the agarose-guar gum-polyaniline films that were synthesized by situ synthesis process and discussed its resistance to ammonia and air.

3 Materials and Reagents

PANi solution is best for ammonia detection. It shows good results for ammonia at room temperature. The materials required are H_2SO_4 (1 M) and aniline (0.5 M).

4 Experimental Methods

Device development consists of three modules

(1) Sensor development
(2) Interfacing with Arduino
(3) Application layer.

4.1 Sensor Development

To create the PANi solution we added 1.8626 ml of aniline and 3.92 ml of H_2SO_4 in 40 ml of deionized water and kept it under magnetic stirring for 90 min.

After creating PANi Solution using chemical synthesis method fabrication of material is done by the electrochemical process [6]. The material formed from the process is deposited on the copper clad chip and kept them to dry for 24 h. After that, we checked the resistance of each chip and tested it under a controlled environment by exposing it to ammonia gas. Figure 1 shows the resistance of the sensor chip which is shown around 11 Ω.

We tested its sensitivity by providing parameters as 3 min baseline, 2 min response time and 1 min recovery time. The results are shown in Fig. 2 graph.

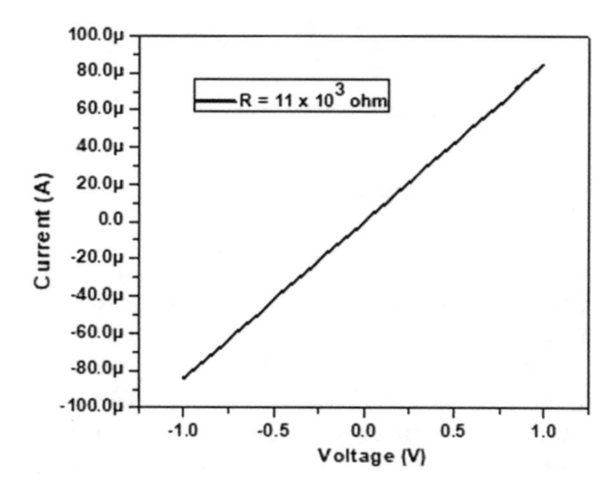

Fig. 1 Resistance of sensor

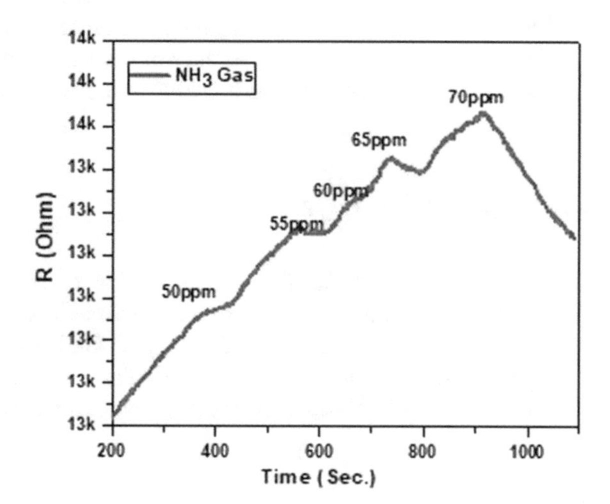

Fig. 2 NH_3 gas sensing result

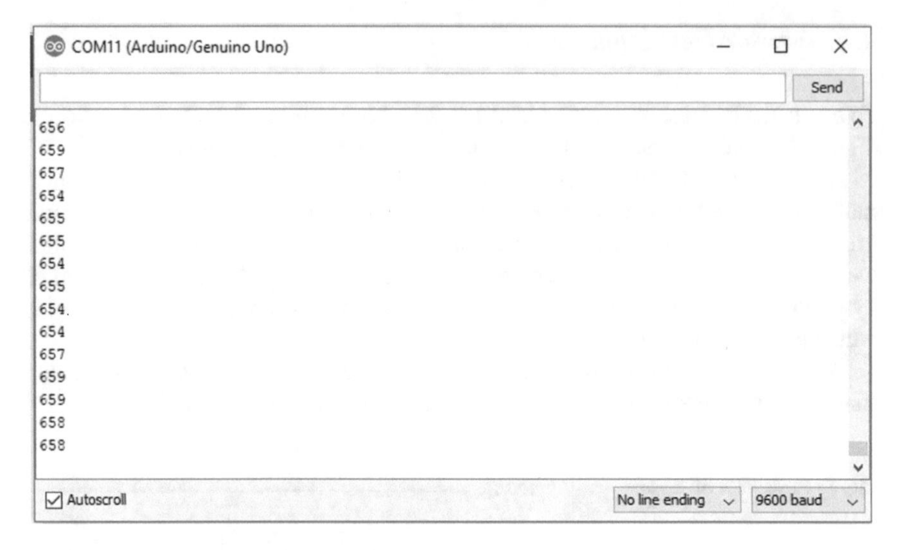

Fig. 3 Original resistance before exposing the gas

4.2 Interfacing with Arduino

To read the data sensed by our sensor we used Arduino with the Wi-Fi module. The data from the sensor is in the form of a change in resistance. When the gas is exposed to the sensor it shows the increase in resistance.

The sensor has its own resistance already and when we connect it to the Arduino it will show its original resistance on the serial monitor of Arduino IDE as shown in Fig. 3 (the original resistance of sensor is 658) and when we expose the gas against the sensor, it shows the increase in resistance as shown in Fig. 4 (the changed resistance of the sensor is above 900). The following screenshots show the value of sensor resistance before and after exposing the gas, respectively.

Figure 3 shows the original resistance of the sensor which is present between 655 and 660. When we expose the gas it increases suddenly above 900 as shown in Fig. 4.

4.3 Application Layer

The last layer of this device is the application layer. Here we used Arduino with the Wi-Fi module to send data to the firebase database. This database refreshes the value of resistance as it changes. Figure 5 shows the GUI of the firebase database.

Using an android application we fetch those values from the database. In the Android application, it will show the original resistance of the sensor on the top left

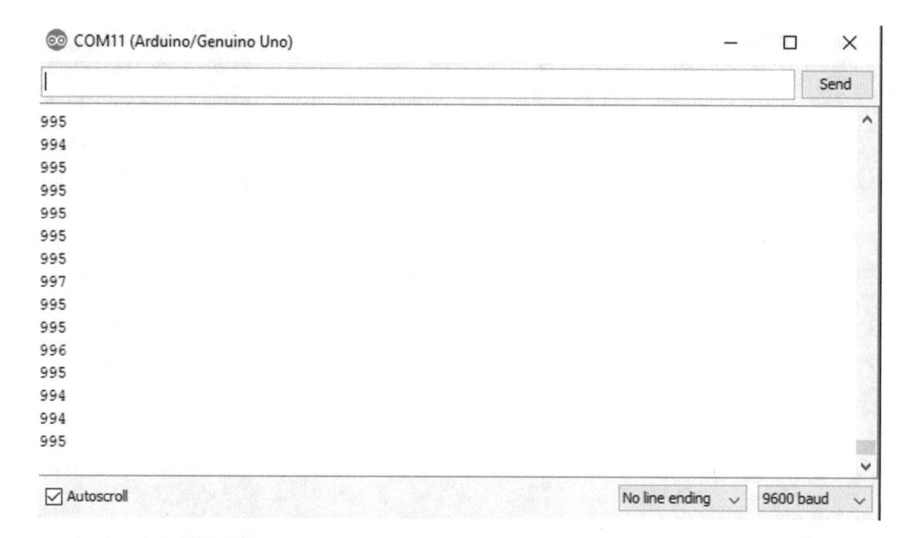

Fig. 4 Change in resistance after exposing the gas

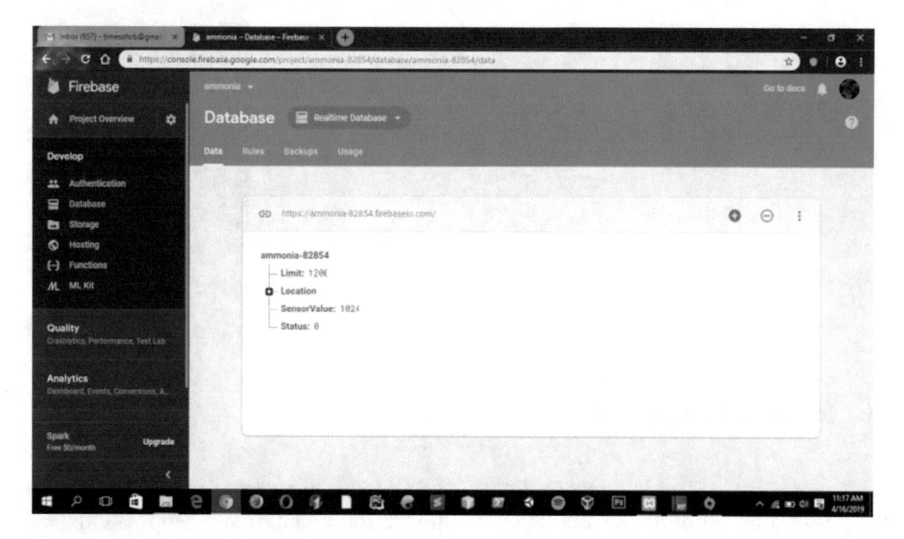

Fig. 5 Firebase database GUI

corner. We can set the level of the resistance above which we want to turn on the alert. So the limit is set in accordance with resistance observed before exposure of gas to the sensor. The application shows a green icon until it crosses the limit we set as shown in Fig. 6. When it crosses the limit, the icon turns red as shown in Fig. 7.

Fig. 6 Android app (safe)

Fig. 7 Android app (unsafe)

5 Results and Discussion

Our sensor has a very less response time of 83 s but it has some limitations also. Its response time is not fixed sometimes it will take too much time when it is exposed to a large amount of gas.

As we are using 8-bit Arduino for interfacing it will return value until 1024. So if the value exceeds 1024 then it will return only 1024 constantly.

6 Conclusion

This device will sense the presence of ammonia at a very low concentration of (<50 ppm). Ammonia sensor works at room temperature. Due to GUI representation, it will be easy to detect the level of harmful gas. Since the Wi-Fi module has been used this data can be accessed from everywhere.

Acknowledgements We are thankful to Dr. Pagare J. D. (Assistant Prof. at Jawaharlal Nehru Engineering College, Aurangabad), Dr. Mahendra Shirsath Director of RUSA (Rashtriya Uchchattar Shiksha Abhiyan) Centre for Advanced Sensor Technology for allowing us to use their laboratories and other facilities. We are also thankful to Mrs. Manasi Mahadik, Mr. Nikesh Ingale, Mr. Gajanan Bodkhe, Mr. Sayyad Pasha for their valuable guidance in this process.

References

1. Gupta, R.P., Gergintschew, Z., Schipanski, D., Vyas, P.D.: A New Room Temperature FET-Ammonia Sensor. IEEE (2002)
2. Knittel, T., Burgmair, M., Freitag, G., Zimmer, M., Eisele, I.: Combined ammonia and hydrogen gas sensor. Sensors. IEEE (2003)
3. Hakimi, M., Salehi, A., Boroumand, F.A., Mosleh, N.: Fabrication of a room temperature ammonia gas sensor based on polyaniline with N-doped grapheme quantum dots. IEEE Sens. J. (2018)
4. Pawar, S.G., Patil, S.L., Chougule, M.A., Raut, B.T., Godase, P.R., Mulic, R.N., Sen, S., Patil, V.B.: New method for fabrication of CSA doped PANi-TiO_2 thin-film ammonia sensor. IEEE Sens. J. (2011)
5. Vaghela, C., Kulkarni, M., Haram, S., Karve, M., Aiyer, R.: Biopolymer-Polyaniline composite for wide range ammonia sensor. IEEE Sens. J. (2016)
6. Huang, C.K., Liao, C.Y., Lin, S.C., Han, Y.K.: The Electrochemical Synthesis of Polyaniline as a Hole Transport Layer for Polymer LED. IEEE (2011)

An Efficient Approach for Job Recommendation System Based on Collaborative Filtering

Ranjana Patel and Santosh K. Vishwakarma

Abstract Managing huge measure of enlisting data on the web, a job seekers dependably invests hours to find helpful ones. To decrease this relentless work, we structure and actualize a recommendation system for online job-seeking job recommender systems are wanted to achieve an uncommon state of precision while making the rating predicts which are significant to the client, as it turns into a repetitive assignment to review a huge number of jobs, posted on the web for instance LinkedIn, fresherworld.com, naukri.com and so on intermittently. In spite of the fact that a great deal of job recommender systems exist that utilization various techniques, here undertaking have been put to make the job recommendations based on applicants profile coordinating just as safeguarding applicants job conduct or inclinations. The collaborating filtering contains a list of rating that the previous user has already given for an item. This paper shows a concise review of collaborative filtering rating prediction based job recommender system and their execution utilizing RapidMiner.

Keywords Recommendation system · Collaborating filtering · Rating prediction

1 Introduction

In recent years, the volume of information present online has developed exponentially. A segment of this information is identified with web based various stages. The assessment of such information or potentially the extraction of data is troublesome because of its vast volume. Hiring the correct ability is a test looked by all organizations. This challenge is intensified by the high volume of candidates if the business is work escalated, developing and faces high wearing down rates. In this way, a job posting for a Java software engineer can without much of a stretch pull in

R. Patel (✉)
Gyan Ganga Institute of Technology & Sciences, Jabalpur, India
e-mail: ranjanap191@gmail.com

S. K. Vishwakarma
Manipal University Jaipur, Jaipur, India
e-mail: santosh.kumar@jaipur.manipal.edu

M. Tuba et al. (eds.), *ICT Systems and Sustainability*,
Advances in Intelligent Systems and Computing 1077,
https://doi.org/10.1007/978-981-15-0936-0_16

a huge number of thousands of utilizations in half a month. Most IT Services organizations are immersed with countless candidates. The job openings are promoted through different ways like online job portal, paper ads, and so forth. Competitors who are interested to apply for the job opening transfer their profile through an assigned site. The site commonly gives an online structure where the hopeful enters insights concerning their application like individual information [1], instruction and experience subtleties, aptitudes, and so forth. The goal of enabling the contender to enter metadata in an online structure is to catch the data in an increasingly organized organization to encourage robotized investigation. This screening procedure [2] is pivotal as it legitimately influences consumption quality and subsequently benefits from the organization.

During the screening the top few applicants, who are shortlisted, experience further assessments meetings, composed tests, bunch talks and so forth. The input from these assessment forms is utilized to settle on the last employing choice.

Recommender Systems (RS) give a significant and effective answer for this issue. Recommender Systems investigate the client profile/conduct [3] and propose items/administrations identify with the premiums of the client. The issue of prescribing jobs to clients varies in a general sense from traditional recommendation system issues, for example, suggesting books, items, or movies to clients. While the general goal of the above is to amplify the client's commitment rate, just a single key distinction is that a job presenting is commonly implied on contract just one or a couple of workers, though a huge number of clients could conceivably be suggested for utilization for a similar book, item, or movies [4]. Perfect job recommendation system would need to prescribe the most pertinent jobs to clients.

Collective filtering filters data utilizing other individuals' recommendations. It is assemble the information that individuals who have chosen in their assessment of specific things in the past will logical to choose again later on. For instance, an individual looking for a job ask for recommendations from online job portal and news commercial. Explicit job portal recommendations with comparative consolation are more reliable than recommendations from others. This strategy gives client explicit recommendations to items that are predicated on rating designs without the requirement for outside data about either items or users [5]. The users are asked to describe the rate from items for example 1–5 star measure.

Collaborative Filtering systems can be sorted as User based and Item-based. Depending on the degree of similar designated, Pearson or Cosine may be further sub-categorized among each of the user-based and item-based collaborative tactic, In user based and item based, the exertion is to anticipate client's appraisals for items that has not rated and by then propose to him the item(s) with the highest rating(s).The design of Recommender systems in RapidMiner has been in addition improved over the Recommender Extension. We use Rapidminer Extensions operator that takes over an association's needs and the abilities of applicants.

2 Review of Literature

Belsare and Deshmukh [2] in their paper "Employment Recommendation System utilizing coordinating Collaborative filtering and Content Based Recommendation" demonstrates the activity seeking process well-ordered and characterize the most ideal path for employment looking with coordinating occupation profile. The work done by Zhang et al. in their paper "A Research of job Recommendation System Based on Collaborative Filtering" also gives significance on the important aspects of applying collaborative filtering approach on job recommendation. They justify the difference between CBF and CF and describe the Method of Similarity calculation with numeric formulas.

Zheng et al. [6] center for web Intelligence "Matrix Factorization in Recommendation System". Their approach to characterize completely precise expectation utilizing factor wise learning matrix factorization. The method research incorporates business related to estimation of notoriety of worker for ideal hiring decisions, just as business related to positioning and importance parts of occupation coordinating in labor commercial centers.

Wang et al. [7] examined job commercial center as a two-sided coordinating business sector utilizing locally stable coordinating calculations for taking care of the issue of getting another line of job utilizing social contacts. The work done by Tool et al. [5] state that Collaborative filtering stands up to troubles of versatility and besides recommendation exactness. They propose a cross breed user model to remove a portion of its drawbacks. The recommender system in light of this model not simply holds the advantage of proposal precision in memory-based procedure, moreover has the adaptability in a similar class as model-based method.

3 Recommendation Algorithm

In content-based method [8], features of items are theoretical and contrast and a profile of the user's preference. In other word, this algorithm attempts to recommend items that are similar to those that a user likes in the past. It is broadly connected in data retrieval (IR). Anyway it performs seriously in media field, for example, music or movie recommendation since it is difficult to extricate items traits and acquire user's affection at times.

Another method known as collaborative filtering, is a well-known recommendation algorithm that puts together its prediction and recommendations with respect to the rating or behavior of different users in the system [9]. There are two types of it known as user based and item based collaborative filtering. In user based CF, we find other users whose past rating behavior is like that of the current user and utilize their ratings on different items to predict what the current user will like. While in Item-based CF, it similitude's between the rating examples of item [10]. Since

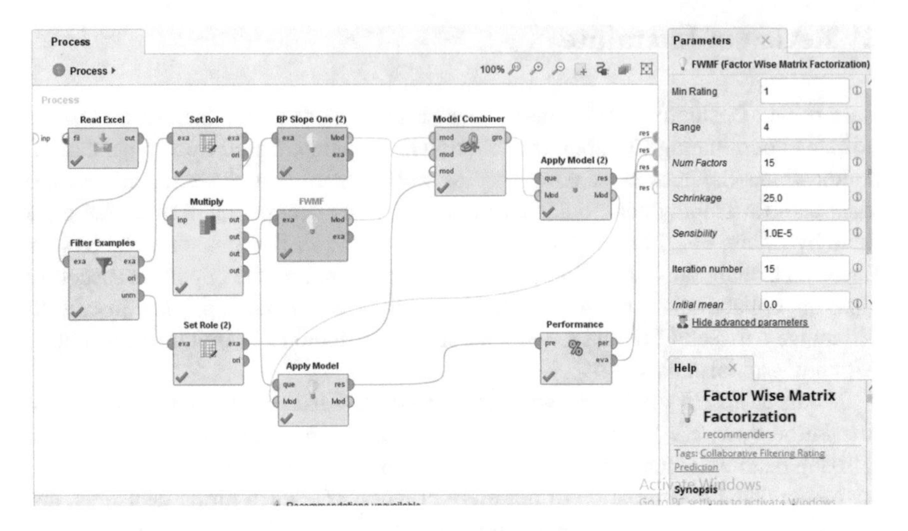

Fig. 1 Rating prediction to put on multiple models to a set of data

finding comparable item is simpler than finding comparable users, and characteristics of items are increasingly steady that user's inclination, item-based methods are reasonable for off-line computing [11].

We have design and implemented the recommendation system with the Rapidminer data mining tool. The collaborative filtering is associated in parallel and their result pooled in a weighted way to assemble a model for recommender. For this purpose we have used twofold operators multiply and Model Combiner (Rating Prediction). Here also uses BP slope one and FWMF within value of number factor 15 (Fig. 1).

On progressively running this multiple model procedure, with defaulting parameter, there is a minor improvement in the execution of a recommender system. The outcome is appeared in

And finally we have got more accuracy using by the matrix factorization with factor wise learning. This model Improve Large Accuracy of Recommender Systems. We need the matrix factorization in recommendation system for decrease dimension. Matrix factorization algorithms work by decomposing the user-item interaction matrix l into the result of two lower dimensionality rectangular matrices.

4 Result

We have used user K-NN, slope one, bi polar slope method as a single model and compared the results with various standard evaluation parameters such as RMSE, MAE, NMAE. We have also used minimum of 15 iterations for all algorithms (Table 1).

Table 1 Comparison of models for ratings

Operator	RMSE	MAE	NMAE
Model combiner	0.434	0.329	0.082
Bipolar slope one	0.607	0.374	0.093
Slope one	0.895	0.67	0.168
User KNN	1.025	0.764	0.191

Comparison of RMSE (Root Mean Square Error), MAE (Mean Absolute Error) and NMAE (Normalize MAE) for different operators of input dataset for Collaborative item rating prediction accuracy.

Now we are showing chart of these errors performance in graph (Fig. 2).

The RMSE performance for all operator is shown in graph for decreasing the errors and get the more accurate prediction for unrated websites (Fig. 3).

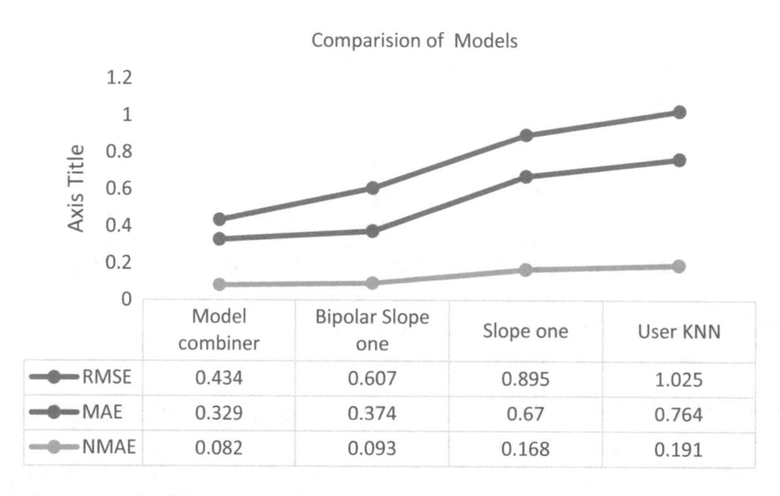

Fig. 2 Comparing RMSE, MAE, NMAE for all recommend operator

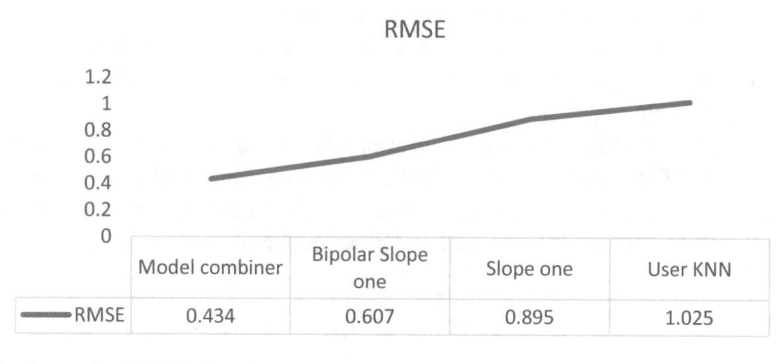

Fig. 3 Comparing RMSE, for all recommend operator

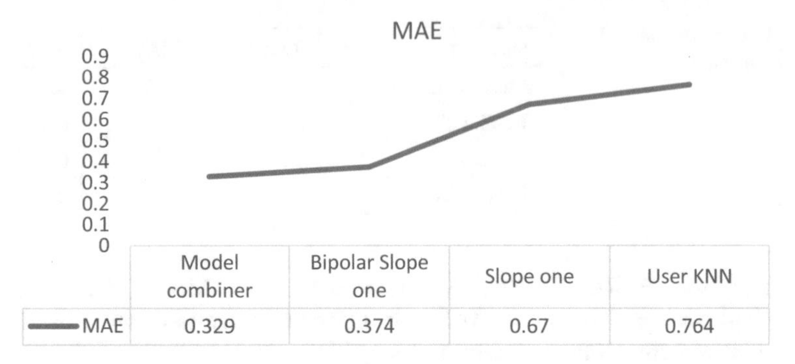

Fig. 4 Comparing MAE, for all recommend operator

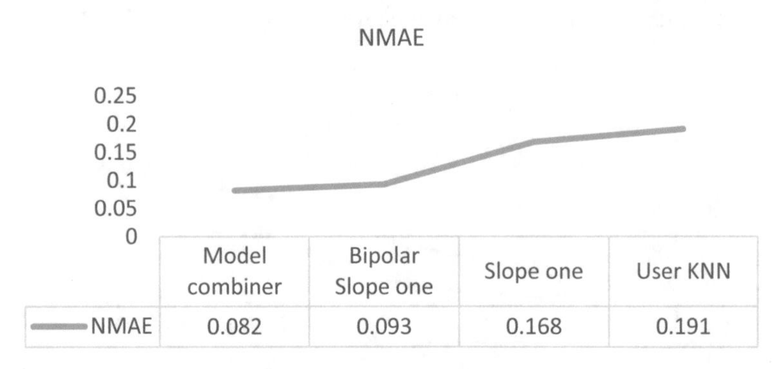

Fig. 5 Comparing NMAE, for all recommend operator

The MAE performance for all operator is shown in graph for decreasing the errors and get the more accurate prediction for unrated websites (Fig. 4).

The NMAE performances for all operators are shown in graph for decreasing the errors and get the more accurate prediction for unrated websites (Fig. 5).

The Matrix Factorization with factor wise learning is a matrix factorization based recommendation model that utilizes expanding number factor and iteration number of factors combined with collaborative information to make recommendations for some user. By using the FWMF, we increase the performance from other models appear in the Table 1.

We have also evaluated the factor wise matrix method with different number of parameters and the results are listed in Table 2. As the number of factor increases, the value of RMSE, MAE, NMAE decreases and its gives the optimum errors for more than 100 factors. This method seems to be prominent with compare to the remaining methods.

Table 2 Factor wise matrix factorization

Number of factors	RMSE	MAE	NMAE
10	0.493	0.420	0.203
20	0.267	0.228	0.057
30	0.153	0.130	0.032
40	0.088	0.074	0.019
50	0.052	0.044	0.011
60	0.031	0.026	0.006
70	0.019	0.016	0.004
80	0.011	0.009	0.002
90	0.006	0.005	0.001
100	0.004	0.003	0.001
110	0.002	0.002	0.000

5 Conclusion

This paper discuss the various methods of collaborative recommendation system used for job recommendation system. The experiments have been carried out using RapidMiner machine learning tool. Different similarity measures are used to compute how much similar all the items are to each other in the matrix. Different algorithms for recommendation such as KNN, Slope one, Bi-polar and Factor based matrix factorization method has been implemented and compared as various levels. Based on the results, it has been observed that the recommendations which has been generated might not be that useful to the jobseekers as they are not personalized. It has been concluded that factor wise matrix factorization using different increasing number of factors and iteration produces the optimum results.

References

1. Ramezani, M., Bergman, L., Thompson, R., Burke, R., Mobasher, B.: Selecting and applying recommendation technology. In: Proceedings of International Workshop on Recommendation and Collaboration in Conjunction with International ACM on Intelligence User Interface (2008)
2. Rafter, R., Bradley, K., Smyth, B.: Automated collaborative filtering applications for online recruitment services. In: Adaptive Hypermedia and Adaptive Web-Based Systems. Lecture Notes in Computer Science, vol. 1892, pp. 363–368 (2000)
3. Balabanovic, M., Shoham, Y.: Conent–based collaborative recommendation. Commun. ACM **40**(3), 66–72 (1997)
4. Ha-Thuc, V., Xu, Y., Kanduri, S.P., Wu, X., Dialani, V., Yan, Y., Gupta, A., Sinha, S.: Search by Ideal Candidates: Next Generation of Talent Search at LinkedIn (2016). https://doi.org/10.1145/2872518.2890549
5. Hayes, C., Cunningham, P.: Smart radio—community based music radio. Knowl. Based Syst. **14** (2001)

6. Belsare, R.G., Deshmukh, V.M.: Employment Recommendation System using Matching Collaborative Filtering and Content Based Recommendation
7. Wang, Q., Yuan, X., Sun, M.: Collaborative Filtering Recommendation Algorithm based on Hybrid User Model. FSKD (2010)
8. Pazzani, M.J., Billsus, D.: Content-based recommendation systems. The Adaptive Web. Springer Berlin Heidelberg, pp. 325–341 (2007)
9. Adomavicius, G., Tuzhilin, A.: Toward the next generation of recommender systems: a survey of the state-of-the-art and possible extensions. IEEE Trans. Knowl. Data Eng. **17**(6), 734–749 (2005). Jacobs, I.S., Bean, C.P.: Fine particles, thin films and exchange anisotropy. In: Rado, G.T., Suhl, H. (eds.) Magnetism, vol. III. Academic, New York, pp. 271–350 (1963)
10. Schafer, J.B., Frankowski, D., Herlocker, J., et al.: Collaborative filtering recommender systems. The Adaptive Web. Springer Berlin Heidelberg (2007)
11. Sarwar, B., Karypis, G., Konstan, J., et al.: Item-based collaborative filtering recommendation algorithms. In: Proceedings of the 10th International Conference on World Wide Web. ACM (2001)
12. Wei, K., Huang, J., Fu, S.: A survey of e-commerce recommender systems. In: International Conference on Service Systems and Service Management, pp. 1–5, June 2007
13. Zhang, C., Cheng, X.: An ensemble method for job recommender systems. In: Recommender Systems Challenge'16, Boston, MA, USA 2016 ACM, Sept. 2016
14. Jain, A., Vishwakarma, S.K.: Collaborating filtering for movie recommendation using RapidMiner. Int. J. Comput. Appl. **169** (2017)
15. Sarwar, B., Karypis, G., Konstan, J.A., Riedl, J.: ItemBased collaborative filtering recommendation algorithms. In: Proceedings of the 10th International Conference of World Wide Web, pp. 285–295 (2001)
16. De Pessemier, T., Vanhecke, K., Martens, L.: A scalable, high-performance algorithm for hybrid job recommendations. In: Proceedings of the Recommender Systems Challenge (RecSys Challenge'16). ACM, New York, NY, USA, Article 5, 4 pp. (2016)
17. Zhang, Y., Yang, C., Niu, Z.: A research of job recommendation system based on collaborative filtering. In: International Symposium on Computational Intelligence and Design (2014)
18. Miheleie, M., Antulov-Fantulin, N., Bosnjak, M., Smuc, T.: e-LICO: An e-Laboratory for Interdisciplinary Collaborative Research in Data Mining and Data—Intensive Science by the European Community 7th Framework ICT (2007)

Constraint-Driven IoT-Based Smart Agriculture for Better e-Governance

Pankaj Kumar Dalela, Saurabh Basu, Sabyasachi Majumdar, Smriti Sachdev, Niraj Kant Kushwaha, Arun Yadav and Vipin Tyagi

Abstract Agriculture, with its allied sectors, is one of the largest livelihood providers in India as well as other developing countries. It also contributes a significant figure to the Gross Domestic Product (GDP), 23% in India in 2016. Several research works have used IoT to automate and revolutionize different aspects of agriculture but have not been able to achieve a significant level of automation in India. These solutions are not able to have ground-level impact due to challenges faced in deploying wireless sensor networks, developing of application and incorporating diversity of crop, soil type, and climate parameters in a single solution. Other practical limitations include absence of funds with farmers to deploy such sophisticated solutions, lack of technical knowledge, and erratic power supply in remote areas. In this paper, we propose a solution to the above challenges faced in field deployment in the form of a Geo-Intelligent Platform with inherent monitoring of soil and crop irrigation requirements using IoT nodes. Resource-based analytics are developed by integrating data from the field nodes, weather forecasts, and Agromet Advisory from Agriculture department. The proposed solution is aimed to provide a direct knowledge sharing channel between farmer and government agencies to facilitate farmers

P. K. Dalela · S. Basu (✉) · S. Majumdar · S. Sachdev · N. K. Kushwaha · A. Yadav · V. Tyagi
Centre for Development of Telematics, Mehrauli, New Delhi, India
e-mail: saurabh.basu.cs@gmail.com

P. K. Dalela
e-mail: pdalela@gmail.com

S. Majumdar
e-mail: sabyasachi3.cse@gmail.com

S. Sachdev
e-mail: smriti359@gmail.com

N. K. Kushwaha
e-mail: niraj.kushwaha@cdot.in

A. Yadav
e-mail: arun.yadav@cdot.in

V. Tyagi
e-mail: vipin@cdot.in

M. Tuba et al. (eds.), *ICT Systems and Sustainability*,
Advances in Intelligent Systems and Computing 1077,
https://doi.org/10.1007/978-981-15-0936-0_18

in proper crop management and government agencies in policy formation, subsidy finalization and issuing of advisories as part of e-Governance initiatives adopted worldwide.

Keywords Geo-Intelligence · Ad hoc-WSN · GIS · oneM2M · Agriculture · LoRa · e-Governance

1 Introduction

Smart Agriculture with IoT (discussed in [1, 2]) enables the application of the right amount of water to the crop at the right time and ensure its uniform distribution in the field. In absence of proper irrigation and with erratic rainfall distribution in India, crops wither in the harsh tropical climate. In irrigated areas, chemical degradation of land is a subsequent development on account of long duration waterlogging. Hence, irrigation needs to be done as per the general climate in the area, weather forecast, and type of crop and soil for a healthy harvest.

Another problem faced by farmers of India is the unreliable supply of electricity in rural areas. Farmers are dependent on the timing of electricity availability to manually control motors to irrigate the field, also generally motors are located in locations remote from their residences. This manual control of irrigation systems must be automated for the health of crops, conservation of water as well as ease of living for the farmers. The data inherently collected by this process can be useful for further analysis not only by farmers but also by government agencies for policy formation, subsidy finalization and issuing of advisories [3].

This paper proposes an IoT-based smart agriculture system using a wireless ad hoc network. It allows placement (and replacement) of sensors with laying of new crop patterns and as per farmer's needs. It allows monitoring of power status at irrigation motor and its automated control according to soil moisture status. This data is analyzed and presented on ISRO Bhuvan GIS [4] integrated with census data and village boundaries for utilization by farmers and government bodies. one M2M platform is used to bind the various logical entities of the network with the data from the large no. of physical devices and provide a secured, scalable and interoperable M2M network.

The paper is organized as follows: Sect. 2 gives a brief background on the technologies used in the proposed solution. Sections 3 and 4 describe the system architecture and challenges overcome in the field deployment, respectively. Section 6 concludes the paper.

2 Background

LoRa [5] is a long-range low power modulation technique for IoT applications developed by Semtech. Owing to the long range and low power consumption of LoRa, large fields can be covered with few gateways and battery-powered nodes with low investment as compared to other IoT communication technologies such as ZigBee [6]. Though cellular technologies exhibit large coverage required by large scale agricultural applications as proposed by this paper, however, cellular network is generally weak in rural areas of India and hence not suitable. Hence, LoRa has been chosen for wireless connectivity in this smart agriculture solution.

The network architecture for LoRa is shown in Fig. 1. Data from large no. of devices is aggregated by LoRa gateways which forward the information toward the server.

The proprietary MAC layer was chosen for LoRa instead of the open standard of LoRaWAN as when such large numbers of sensor devices are interconnected in the IoT network; there arises concern pertaining to security of data, control and management of devices, scalability of network, and interoperability between devices. Though LoRaWAN has some of the above capabilities, it only caters to LoRa nodes.

oneM2M [7] is one of the standards of IoT that defines the architecture of connected devices enabled by any technology and provides management and security features for a heterogeneous network. oneM2M Architecture and network configurations are shown in Fig. 2.

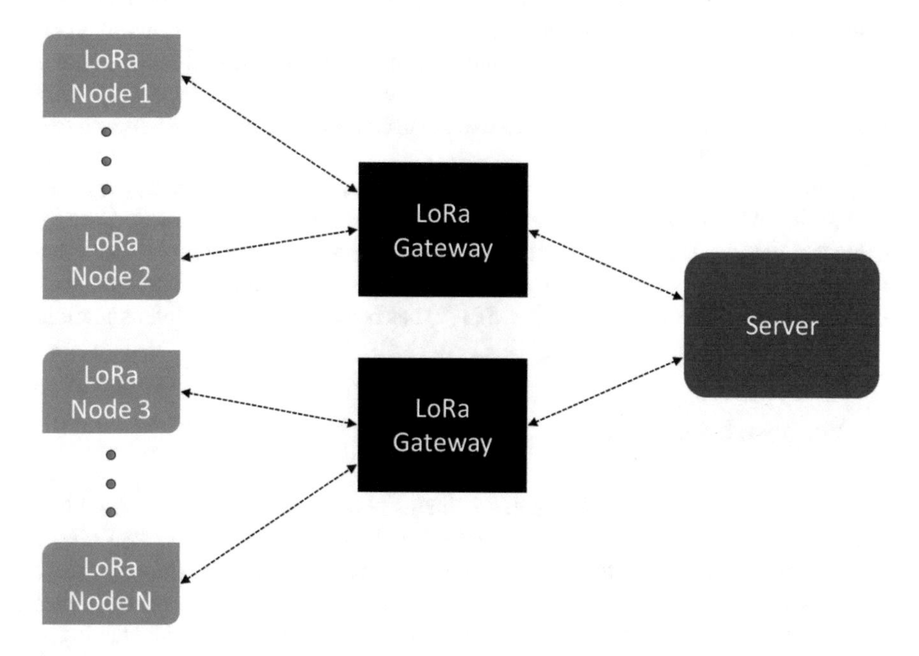

Fig. 1 Network architecture of LoRa

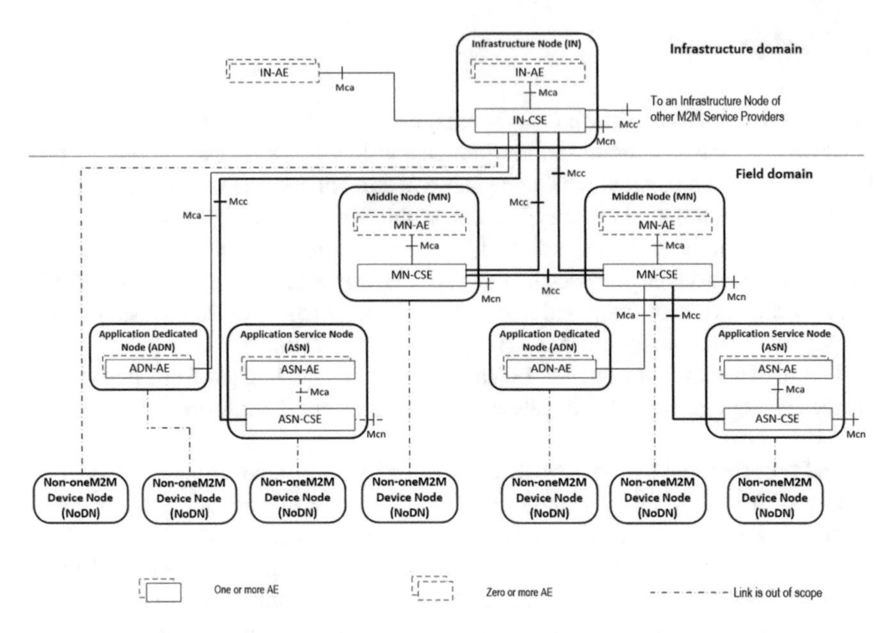

Fig. 2 oneM2M network architecture

oneM2M network [8] consists of Application Dedicated Node (ADN), Application Service Node (ASN), Middle Node (MN), Infrastructure Node (IN), and NononeM2M Device Node (NoDN). Application Entity (AE) modules mainly interact with end systems like smart light and Common Services Entity (CSE) modules contain Common Services Functions (CSFs) which are used to cater to requests from AE. Mcc and Mca are reference points that are used for communication between two CSEs and AE–CSE, respectively.

To cover the large fields, groups of sensor devices need to be formed as per M2M networking which will be aggregated into a single device (i.e., Gateway) through LoRa technology. The Gateway will communicate with application server through a common service layer of oneM2M. For the proposed solution, oneM2M compliant C-DOT Common Services Platform (CCSP) [9] has been used for the implementation.

3 Proposed Solution

This paper proposes a fully automated irrigation system based on wireless ad hoc sensor network deployed in the agricultural field. It proposes LoRa-based sensor devices for real-time monitoring of soil moisture, temperature, and pH because of LoRa's low cost, long range and long battery life. Devices can be deployed by farmer without any technical training as per crop sowing and ridge pattern, without having to log the position of the devices. These devices can be removed prior to harvesting and can be charged prior to next deployment due to their long battery life. The proposed solution also involves installation of LoRa device for power supply monitoring and automatic control of irrigation motor. LoRa gateway acts as aggregator for the soil sensor devices and power monitoring devices and relays the data to the application server. Figure 3 depicts the network architecture of proposed solution.

The application server maintains database of the region-wise soil type parameters and corresponding irrigation requirements. The proposed method uses GIS-based geo-fencing technique wherein a geo-fence corresponding to an irrigation motor is created. Data from sensors with GPS coordinates in the same geo-fence result in activation/de-activation of that motor only. The application server takes the decision to switch ON/OFF irrigation motor taking into account the general climate in the area, weather forecast, type of crop, type of soil, and real-time soil parameters. Our application server interfaces with oneM2M for access to the sensor data and motor control device through oneM2M compliant LoRa Gateway. This common service layer allows for easy integration with different applications of different government organizations. The system also enables remote management, customization

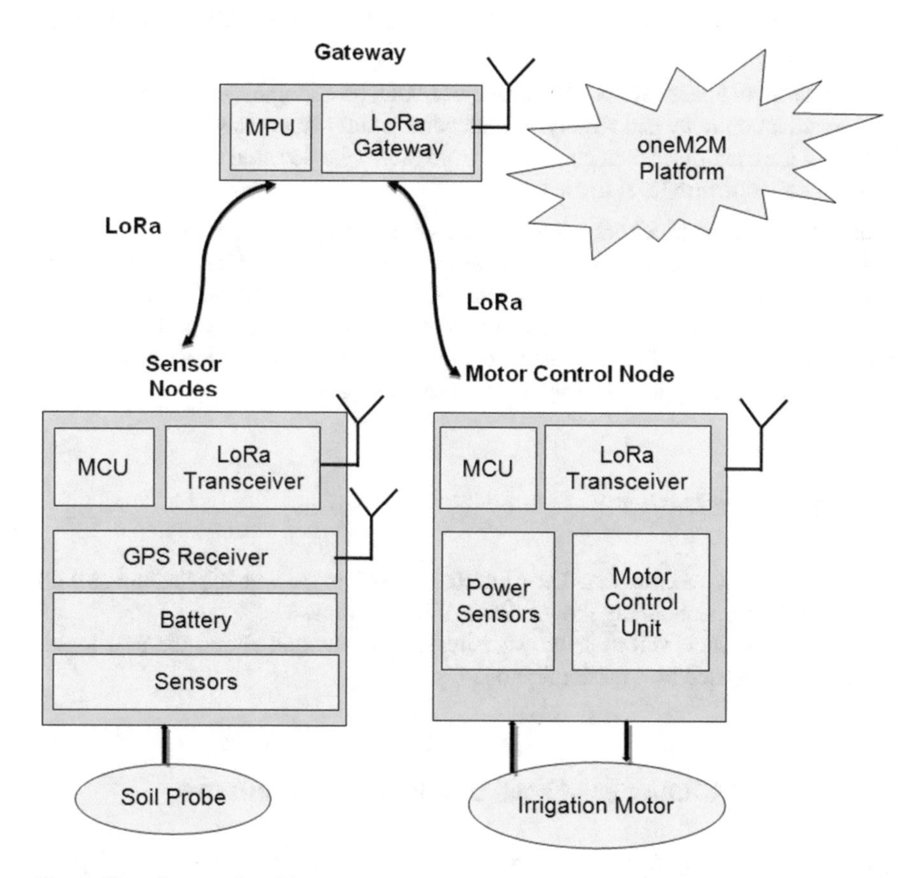

Fig. 3 Phase 2 network architecture

of automation and viewing of live and historic data is possible with web application and android application. This data is analyzed and presented on ISRO Bhuvan with administrative boundaries and Census data.

3.1 Hardware Design

Sensor Node: In the proposed solution, sensor devices are placed in the field at regular intervals of distance and approximate locations are noted. These devices are to be removed prior to harvesting and charged prior to the next deployment. Sensor data is sent to the LoRa gateway periodically using LoRa technology which is then forwarded to server. Each sensor device consists of Microcontroller Unit (MCU), LoRa Transceiver, Sensors, and Battery.

- Soil moisture sensor and pH sensor employ probes that are dug into the soil at appropriate depth.
- MCU coordinates sensor data acquisition and wireless communication with gateway.

Motor Control Node (MCN): It consists of MCU, power monitoring circuitry [10], motor activation relay and battery. A high current solid-state relay is used to switch ON/OFF the motor based on commands from the server decision being made on the basis of soil moisture level in the area.

Gateway: Smart Agriculture gateway consists of Microprocessor Unit (MPU) and LoRa Gateway. The Gateway acts as an aggregator/concentrator for all LoRa nodes. Sensor data received by the LoRa gateway is forwarded to the ADN-AE residing in the MPU. The information then travels to the application server in the standardized format of oneM2M. Also control commands sent by the application server are translated to LoRa by the gateway and sent to the targeted MCN.

3.2 Software Design

The user will have a Graphical User Interface (GUI) in the web application as well as in mobile app, containing the locations of installation sites.

The GUI presents soil moisture statistics graphically and allows the user to see soil moisture sensed by a particular node Fig. 4.

4 Practical Challenges Faced and Proposed Solutions

Different Soil and Crop Types: Different parts of the field have different irrigation requirements based on crop growing in that area. Uniform soil moisture threshold for

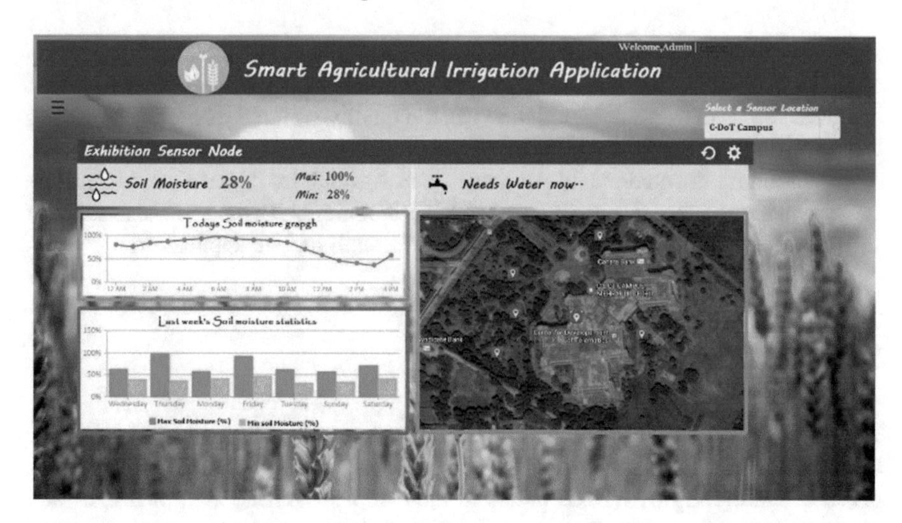

Fig. 4 Phase 1 desktop application GUI

activation/de-activating motor for the entire field resulted in overwatering of some regions and under-watering of others with impact on crop quality. In the proposed solution, different upper and lower thresholds were set for different soil and crop types based on Agromet data. This customization ensured that irrigation is most effective for each crop.

Deployment Constraints: Deployment and recording of location required skilled manpower which adds to the overall cost of the solution. Farmers have good knowledge of crop sowing and ridge pattern; hence, this expertise could be useful in deployment to get effective monitoring and irrigation. Hence, a need was felt to develop nodes that did not require additional technically skilled manpower and were easy to install. In the proposed solution, each device was fitted with GPS receiver to obtain exact location information. This information is transmitted by node at the time of deployment for automatic registration to server with location information.

Device Battery Life: The frequency for sending soil data was set at once every 10 min to closely monitor the soil status in near real time. However, this high frequency had a drastic impact on battery life. A hybrid approach has been implemented to tackle this issue, frequency was reduced to once per hour for general operation and frequency was set to once per 10 min while pump remains on.

Erratic Power Supply: When soil moisture level went outside the allowed range a control message was sent to the irrigation motor. In rural India, erratic power supply is common, due to which, the control message could not be executed and the fields were not irrigated properly. Hence, there is a need to monitor power supply of irrigation motor to ensure irrigation takes place whenever electricity becomes available.

Along with the above-mentioned enhancements, users can view the soil conditions of different sites and conditions of irrigation motors using intuitive GUI which has been designed in multiple languages supporting various vernacular languages of India, keeping in mind the needs of its intended users. Weather forecasts were integrated so as to prevent degradation in crop quality suffered due to overwatering caused by heavy rains subsequent to irrigation.

Operational information and Advisory from Agriculture department can also be forwarded to the concerned users through SMS in regional languages. Users can also give crop-related inputs/feedback to platform through mobile application.

5 Field Trial

The field implementation and trial of this proposed system were conducted at C-DOT Campus in New Delhi, India for both Phases. It is being used for soil moisture and pH level monitoring of various plants of the campus covering area of 3 Km^2 and automated irrigation since December 5, 2017. A drive test was also carried out for verifying the range of LoRa gateway placed at G+3 of the building as shown in Fig. 5.

From GUI user can view the soil conditions of different sites and conditions of irrigation motors as shown in Fig. 5. Here each field is monitored and irrigated independently to ensure efficient irrigation. The real-time status of soil moisture, irrigation motor, and power is shown on GIS Table 1.

Fig. 5 Phase 2 with geo-fencing

Table 1 Field wise moisture level and irrigation motor status

Field ID	Moister level	Motor status	Motor power status
Field-1	Normal	OFF	Not available
Field-2	Need watering	OFF	Available
Field-3	Increasing	ON	Available
Field-4	Normal	OFF	Available
Field-5	Decreasing	OFF	Not available
Field-6	Normal	OFF	Available

Flag (Depicts Soil Sensor Node) Color:

- Black—Sufficient moisture. No action required.
- Green—Currently being watered by an irrigation pump.
- Orange—Soil drying up. Will need water soon.
- Red—Soil is very dry, crop affected. Requires immediate watering.

Irrigation Motor Color:

- Blue—Motor is ON.
- Green—Motor is OFF and Power available.
- Red—Motor is OFF and Power unavailable.

6 Conclusion

In this paper, oneM2M-based Smart Agriculture sensor node, irrigation motor control node with LoRa gateway design has been presented. It is integrated with Geo-Intelligent analytics based on Weather forecasting and Agromet database. The field trial has been done for monitoring, receiving information and advisories about the plants of C-DOT campus, New Delhi, India to test such a system in practical scenarios. Further, Geo-Intelligent platform has been integrated with ISRO Bhuvan map with Census and village boundary data. Various government agencies such as Electricity department, Agricultural Department, and allied ministries can interface with the proposed platform to use the data gathered pertaining to location-wise availability of electricity, yield, and type of crop per hectare, Village/District/State wise Electricity, water consumption, and feedback from the farmers. Thus, the proposed solution provides a two-way communication channel between farmer and government agencies to facilitate farmers in proper crop management and government agencies in policy formation, subsidy finalization, and issuing of advisories as part of e-Governance initiatives adopted worldwide.

References

1. Wasson, T., Choudhury, T., Sharma, S., Kumar, P.: Integration of RFID and sensor in agriculture using IOT. In: 2017 International Conference on Smart Technologies for Smart Nation (SmartTechCon), pp. 217–222 (2017)
2. Chardy, M., et al.: Agriculture internet of things: AG-IoT. In: 27th International Telecommunication Networks and Applications Conference (ITNAC) (2017)
3. https://farmer.gov.in. Accessed 10 Jan 2019
4. http://bhuvan.nrsc.gov.in/map/bhuvan/bhuvan2d.php. Accessed 10 Jan 2019
5. https://www.lora-alliance.org/. Accessed 10 Jan 2019
6. Maia, R.F., Netto, I., Tran, A.L.H.: Precision agriculture using remote monitoring systems in Brazil. In: IEEE Global Humanitarian Technology Conference (GHTC) (2017)
7. www.onem2m.org. Accessed 10 Jan 2019
8. oneM2M TS-0001 Functional Architecture. http://www.onem2m.org
9. Dalela, P.K., Basu, S., Sabyasachi, S., Majumdar, S., Kushwaha, N.K., Tyagi, V.: Surveillance enabled smart light with oneM2M based IoT networks. In: 2017 Computational Intelligence, Communications, and Business Analytics, pp. 296–307
10. Dalela, P.K., Basu, S., Majumdar, S. Saldhi, A., Tyagi, V.: Real time monitoring of power resources with surveillance based on M2M communication 2016. In: IEEE International Conference on Advanced Networks and Telecommunications Systems (ANTS) 6–9 Nov 2016. 978-1-5090-2193-2

A Network Intrusion Detection System with Hybrid Dimensionality Reduction and Neural Network Based Classifier

V. Jyothsna, A. N. Sreedhar, D. Mukesh and A. Ragini

Abstract In the recent years, there is a lot of developments in the technology where lots and lots of information are crawling in the network. As the technology is increasing the threats and cyber attacks are also gradually increasing while is leading to evolve new security mechanisms. There are many classical security models such as firewalls, encryption, and authentication schemes. But these techniques are not able to secure today's computers and networks from attacks. One of the best solutions for today's network attacks is Intrusion Detection System (IDS). IDS are used to monitor and analyze the behavior of the network traffic. In this work, a novel network Intrusion Detection System with Hybrid Dimensionality Reduction and Neural Network Based Classifier is proposed. In this, Information Gain (IG) and Principal Component Analysis (PCA) are used for dimensionality reduction and multilayer perception technique is used to classify the data. The performance of this proposed method is estimated on benchmark dataset of network Intrusion Detection System i.e., NSL-KDD. The experimental results exhibit that the model designed has provided an improvement in accuracy and also provides less computational time and minimal false alarm rate.

Keywords Network IDS · Information Gain (IG) · Principal Component Analysis (PCA) · Multilayer perception · Neural networks · False alarm · Performance · Accuracy

1 Introduction

As the speed of broadband Internet is growing, social applications became more attractive source of exchanging the information. So, the threats and cyber attacks are also gradually increasing which is becoming a critical factor for maintaining the

V. Jyothsna (✉)
Sree Vidyanikethan Engineering College, Tirupati, India
e-mail: jyothsna1684@gmail.com

A. N. Sreedhar · D. Mukesh · A. Ragini
Chadalawada Ramanamma Engineering College, Tirupati, India

M. Tuba et al. (eds.), *ICT Systems and Sustainability*,
Advances in Intelligent Systems and Computing 1077,
https://doi.org/10.1007/978-981-15-0936-0_19

network protected from malicious activities. Researchers have contributed a wide range of contributions for further improving the security of the network, in the form of firewalls, anti-viruses, encryption and authentication schemes, and also security standards [1]. Firewalls are used to monitor the packets going in and out of the system or the network. Based on the predefined rules in the database of the firewall it will check whether the incoming or outgoing transaction is normal or attack. As they depend on static rules enabled in the firewall it cannot detect the complex attack scenarios. Anti-virus software relies on know patterns of attacks. If a new pattern arises, it will not be able to detect. One of the best solutions for today's network attacks is Intrusion Detection System (IDS). IDS are used to monitor and analyze the behavior of the network traffic. IDS are of two types: host-based and network-based IDS. A host-based Intrusion Detection System evaluates the events actions mainly associated with OS like process identifiers and system calls. Further, a network IDS examines the activities related to network i.e., traffic volume, IP addresses, service ports, protocol usage, etc., misuse IDS or anomaly IDS are the classifications of network-based IDS [2]. The misuse detection uses well-defined patterns of the attack that exploits the weakness in the system. Misuse detection IDS are also called as signature-based systems. They classify the records as normal or attack based on the specific signature stored in the system. In Anomaly-based detection behavior of the system is analyzed. When a new record comes its behavior is compared with the existing normal behavior. Whenever the deviation between the normal behaviors exceeds a predefined threshold then the record is treated as attack otherwise treated as normal. Many machine learning approaches have evolved to analyze the network traffic. Recently, Neural Network Based IDS are evolving. A neural network has the benefit of easier illustration of nonlinear relationship between i/p and o/p with inherent computational speed.

The organization of the rest of the paper is as follows. The Sect. 2 presents the literature survey of the proposed system. Section 3 specifies the overall process of the proposed methodology. An empirical result of the proposed model is presented in Sect. 4. Finally, in Sect. 5 includes conclusion.

2 Related Work

An efficient hybrid model is proposed by Shadi Aljawarneh et al. [3] to calculate intrusion scope threshold degree on the optimal features during the training. The optimal features are extracted using Vote algorithm with Information Gain. The results exposed for this model states that there is a significant decrease in the computational and time complexity in calculating the feature association impact scale.

Srinivas Mukkamala et al. [4] describe an approach for detecting the anomalies using neural networks and support vector machines. The main aim is to discover useful attributes that analyzes the user behavior in the network. The relevant attributes are used to build classifier that can recognize intrusions. Experimental results state that the proposed model is more efficient and accurate classifiers in detecting the attacks.

A prototype of Intrusion Detection System suitable for TCP/IP networks is proposed by Mauricio et al. [5]. The prototype first captures the packets. The captured packets are further fed to neural network to analyze and identify the malicious behavior. The identification of the behavior is based on the past known profiles of the network. In this method, the system has the capable of adapting to the network, because when new patterns are analyzed neural network retrained them and can be added to the database.

With the hypothesis that audit records of the system usage can be used to identify the abnormal patterns in the network, Denning [6] proposed a model. The behavior of the user in the profiles are been represented in terms of metrics and statistical models. Based on these metrics and the model rules are been framed to detect the anomalous behavior. The advantage of the system proposed is that, this model is a general-purpose intrusion detection expert system which is independent of any particular system, application environment, system vulnerability, or type of intrusion.

To improve the performance of Intrusion Detection System, Lippmann et al. [7] proposed a model using keyword selection and neural networks. This model is mainly designed to detect user-to-root (U2R) attacks. The improved keyword system model is able to detect unknown as well as known attacks with the same computation time. The experimental results state that, it reduced the false alarm rate.

3 Proposed Methodology

In this paper, a novel network Intrusion Detection System with Hybrid Dimensionality Reduction and Neural Network Based Classifier is proposed. In this, Information Gain (IG) and Principal Component Analysis (PCA) are used for dimensionality reduction and multilayer perception technique is used to classify the data. Figure 1 represents the framework of proposed model. It consists of training and testing phase.

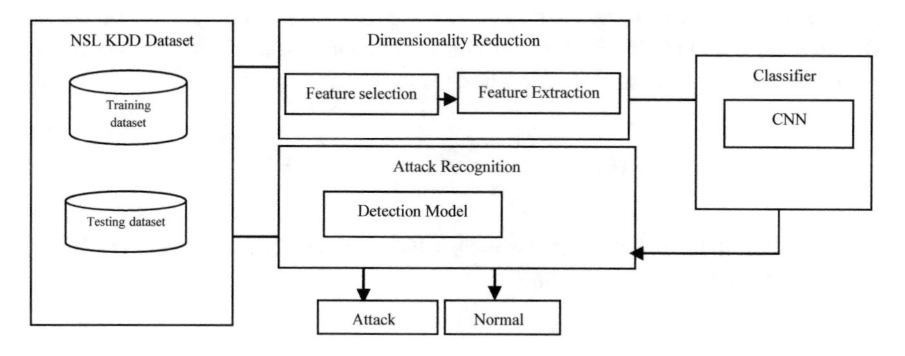

Fig. 1 Framework for the proposed model

In the training phase, the dataset with the class field is taken into consideration. Initially, the features are preprocessed to reduce the redundant and missing value records in the dataset. Then, the data is normalized to convert the all features value into one range. Further, the features undergo feature selection and feature extraction phase. Most useful and relevant features are selected from available data is selected in feature extraction phase and Existing features are combined to develop more useful ones in feature selection phase. Finally the optimal features are given to the multilayer perceptron to extract the rules. In the testing phase, the records without the class files are given to the detection model. The detection model classifies the record as normal or attack based the rules or patterns obtained from the training set. The main advantage of this model is that the system has the capable of adapting to the network. If the new pattern is analyzed, neural network retrained them and can add to the database such that further such patterns arising will be able to detect them. This model is used to detect both known and unknown attacks.

3.1 Data Preprocessing

Usually, the dataset will not be organized for directly applying to any of the data mining or machine learning or any statistical technique. So, the inconsistent, incomplete data should be removed before applying to any technique to avoid generation of misclassification patterns. This process is called Data preprocessing. And then the nonnumerical attributes should be represented in numerical.

3.2 Data Normalization

Normalization is a method frequently applied as part of data preparation for machine learning. Due to diversity of the values of the attributes, the classifier patterns may have some errors. To avoid such issues, normalization is the process of convert the all features value into one range, without distorting divergence in the ranges of values. One of such technique is *Min-Max normalization* where data will fit in a predefined boundary. The formula for Min-Max normalization is as follows:

$$f(x) = \frac{(x - x_{\min})}{(x_{\max} - x_{\min})}.(d - c) + c \tag{1}$$

where the minimum and maximum values of attribute x are $x_{\min}$ and $x_{\max}$ respectively, $[c, d]$ is the interval range, i.e., [0, 1].

3.3 Dimensionality Reduction

The features undergo feature selection and feature extraction phase for reducing the dimensionality for removing irrelevant and redundant data which helps in increasing learning accuracy, and reduces the false alarm rate. In this paper, two-dimensionality reduction techniques are used. Initially, most useful and relevant features are selected from available data is selected in feature extraction phase and in the secondly, Existing features are combined to develop more useful ones in feature felection phase. Dimensionality reduction helps us in reducing the process complexity and further helps in decreasing the time taken for training the dataset.

Algorithm for Hybrid Dimensionality Reduction

Input:

Input dataset *X*, where (X includes n instances *with its corresponding T features*).

Output: Dimensionally reduced dataset

procedure Compute_IG (*X*):

1. *Calculate estimated information required to categorize a given instance*
2. *while* $1 \leq i \leq n$ *do*
3. *Compute entropy for attribute Ti*
4. *Compute information gain for attribute Ti*
5. *Y* $\leftarrow$ *the k attributes with the highest scores*
6. *return Y*

procedure Compute_PCA (*Y*):

7. *Compute the covariance matrix of Y*
8. *Compute eigenvector (q1,…, qi) and eigenvalues (λ1,…, λi) of the covariance matrix above*
9 *Z* $\leftarrow$ *the k eigenvectors with largest eigenvalues*
10 *return Z the new k-dimensional feature space*

3.3.1 Feature Selection

All the features in the dataset will not be useful for discriminating the classes. The redundant and irrelevant data will lead to misclassification of the data, which affects the performance of the overall system. So, feature selection helps us to combine the features to develop more useful ones which reduce the process complexity. In this paper, Information Gain (IG) is used for feature selection. Initially, entropy for each attribute is calculated. Then Information Gain is calculated for the attribute. The values will be sorted with the Information Gain. The features with highest Information Gain score are taken into consideration. After the feature selection process using Information Gain, the features are reduced from 41 to 20.

FEATURE SELECTION ALGORITHM USING INFORMATION GAIN (IG)
Input*: *Original Dataset with n dimensions
Output*: *Reduced Dataset with r dimensions

1. *Begin*
2. *For i = l to N do*
3. *Gain[i] = information[i]*
4. *Sort Gain[i] in ascending order*
5. *End*

3.3.2 Feature Extraction

The features selected in the feature selection process are given as an input to feature extraction method. The feature extraction method further selects the relevant features from available data, which can be helpful in reducing the time for processing and also reduces the process complexity. The Principle Component Analysis (PCA) is used for feature extraction. The PCA calculates the covariance between the attributes.

The formula for covariance matrix C of the dataset is defined as follows:

$$C = \frac{1}{M}\sum_{i=1}^{M} \emptyset_i \emptyset_i \quad (2)$$

The eigen values $(\lambda_1, \lambda_2, \ldots \lambda_N)$ and eigenvectors $(u_1, u_2, \ldots u_N)$ of a covariance matrix C is calculated. The K eigenvectors with largest eigen values are selected as the new k—dimensional feature space.

The criteria used to determine the dimensionality of subspace K is as follows:

$$\frac{\sum_{i=1}^{k} \lambda_i}{\sum_{i=1}^{\mathrm{N}} \lambda_i} > \text{threshold} \quad (3)$$

After the feature extraction process using Principle Component Analysis (PCA), the features are reduced from 20 to 13.

3.4 MultiLayer Perceptron

A multilayer perceptron is a type of feed-forward neural network, consisting of at least three layers such as an input layer, a hidden layer, and an output layer. An MLP can be analyzed as a logistic regression classifier where the input is transformed using a learnt nonlinear transformation which will pass it via weighted connections to the hidden layer. These computations are passed to the succeeding layers, until a finalized vector is calculated in the output layer. The connection weights of the net_i

are determined as the network function that maps the input vector onto the output vector. Each neuron i in the network computes its activation s_i as follows:

$$\text{net}_i = \sum_{j \in \text{pred}(i)} s_j w_{ij} - \theta_i, \tag{4}$$

where, set of predecessors of unit i is denoted as pred(i).

Connection weight from unit i to unit j is denoted by w_{ij}.

s_i is calculated by feeding the net input through a nonlinear activation function.

Typically, the sigmoid logistic function is the computable derivative with respect to its incoming excitation is called as net input represented as net_i

$$s_i = f_{\log(\text{net}_i)} = \frac{1}{1 + e^{-\text{net}_i}} \tag{6}$$

$$\frac{\partial s_i}{\partial \text{net}_i} = f_{\log}^{i}(\text{net}_i) = s_i * (1 - s_i) \tag{7}$$

Algorithm for MultiLayer Perceptron is as follows

Step 1: Initialize weights
Step 2: Do
Step 3: For every tuple x part of the training dataset
Step 4: Assign O ← neural-net output and later forward the acquired outcome to the subsequent
Step 5: Assign T ← outcome for any given tuple x
Step 6: Compute error ← [T, O) at each unit
Step 7: Calculate the weights w, from hidden layers to output layer
Step 8: Adjust the weights for all neurons in the MLP network until the process is complete or exit criteria is met
Step 9: Output MLP network

4 Empirical Analysis of the Proposed Model

A novel network Intrusion Detection System with Hybrid Dimensionality Reduction and Neural Network Based Classifier is proposed to classify the intrusions using a benchmark dataset NSL–KDD [9]. If the magnitude of input data is very large then it faces difficulties to interpret the association among the transactions and features. 20% of training dataset of NSL–KDD contains 41 features and more than 125,973 records. The records may contain redundant data and false connections which hamper the process of detecting intrusions. The features undergo feature selection and feature extraction phase for reducing the dimensionality for removing irrelevant and redundant data which helps in increasing learning accuracy, and reduces the false

```
> print(f)
result ~ src_bytes + dst_bytes + flag + diff_srv + same_srv +
    dst_ho_s + d_h_s_r_r + logged_in + d_h_s_r_v + dst1_host +
    dst_host2 + count + serror_rte + srv_serror + dst_host1 +
    dst_host + dhssp + srv_diff + srv_count + dst_hst4
<environment: 0x0000000020b31eb8>
> |
```

Fig. 2 Attributes obtained after information gain

```
> summary(pc)
Importance of components%s:
                         PC1    PC2    PC3     PC4     PC5     PC6     PC7     PC8     PC9
Standard deviation     2.734 1.6381 1.3609 1.27912 1.05135 1.00389 0.99388 0.96979 0.87267
Proportion of Variance 0.356 0.1278 0.0882 0.07791 0.05264 0.04799 0.04704 0.04479 0.03626
Cumulative Proportion  0.356 0.4838 0.5719 0.64986 0.70249 0.75048 0.79752 0.84230 0.87857
                         PC10    PC11    PC12    PC13    PC14    PC15    PC16   PC17    PC18
Standard deviation     0.7818 0.68052 0.65382 0.62865 0.52731 0.41680 0.26296 0.2290 0.18425
Proportion of Variance 0.0291 0.02205 0.02036 0.01882 0.01324 0.00827 0.00329 0.0025 0.00162
Cumulative Proportion  0.9077 0.92973 0.95008 0.96890 0.98214 0.99041 0.99371 0.9962 0.99782
                          PC19    PC20    PC21
Standard deviation     0.15973 0.12033 0.07596
Proportion of Variance 0.00121 0.00069 0.00027
Cumulative Proportion  0.99904 0.99973 1.00000
> |
```

Fig. 3 Output for PCA algorithm

alarm rate. In this work, a novel network Intrusion Detection System with Hybrid Dimensionality Reduction and Neural Network Based Classifier is proposed. In this, Information Gain (IG) and Principal Component Analysis (PCA) are used for feature reduction and multilayer perception technique is used to classify the data. The performance of this work was evaluated based on NSL–KDD dataset (Fig. 2).

Dimensionality Reduction

Feature Selection

On the Normalized data, after applying Information Gain, most important 20 features are selected. (Fig. 3).

Feature Extraction

Features obtained after Information Gain are given as an input to Principle Component Analysis. These are the features extracted from Principle Component Analysis. It is further given as input to classification (Fig. 4).

Classifiers: **MultiLayer Perception**

5 Conclusion

Recent Literature reveals that an efficient feature selection methodology is the main factor to support the classification method in the detection process. The features undergo feature selection and feature extraction phase for reducing the dimensionality for removing irrelevant and redundant data which helps in increasing learning

```
Confusion Matrix and Statistics

          Reference
Prediction   0   1
         0 408  12
         1  10 518

               Accuracy : 0.9768
                 95% CI : (0.9651, 0.9854)
    No Information Rate : 0.5591
    P-Value [Acc > NIR] : <2e-16

                  Kappa : 0.953
 Mcnemar's Test P-Value : 0.8312

            Sensitivity : 0.9761
            Specificity : 0.9774
         Pos Pred Value : 0.9714
         Neg Pred Value : 0.9811
             Prevalence : 0.4409
         Detection Rate : 0.4304
   Detection Prevalence : 0.4430
      Balanced Accuracy : 0.9767

       'Positive' Class : 0
```

Fig. 4 Output for MLP

accuracy, and reduces the false alarm rate. In this paper, In this work, a novel network Intrusion Detection System with Hybrid Dimensionality Reduction and Neural Network Based Classifier is proposed. In this, Information Gain (IG) and Principal Component Analysis (PCA) are used for attribute reduction and multilayer perception technique is used to classify the data. The Experimental result shows that, the proposed system is accurate to the level of 97.68%. The failure percentage is about 2.3%, which is significant and arises due to diversity of the feature values. The successful rate of intrusion recognition is representing by the metric called sensitivity, which is acknowledged as around 0.9761. The successful rate of fair operation finding is represented by the metric called specificity that identified as 0.9774, which is not significant in this experiment due to the negligible count of fair transactions given as test input. Hence, the proposed model is designed to identify and categorize the anomalies with more accuracy and less training time.

References

1. Jyothsna, V., Rama Prasad, V.V.: Anomaly based network intrusion detection through assessing feature association impact scale. Int. J. Inform. Comput. Secur. (IJICS), **8**(3), 241–257 (2016)
2. Jyothsna, V., Rama Prasad, V.V.: FCAAIS: anomaly based network intrusion detection through feature correlation analysis and association impact scale. J. Inform. Commun. Technol. Express **2**(3), 103–116 (2016)
3. Aljawarneh, S., Aldwairi, M., Yassein, M.B.: Model. J. Comput. Sci. **25**, 152–160 (2018)
4. Mukkamala, S., Janoski, G., Sung, A.: Intrusion detection using neural networks and support vector machines. In: Proceedings of the 2002 International Joint Conference on Neural Networks, pp. 1702–1707 (2002)
5. Bonifkio Jr, J.M., Cansian, A.M., de Carvalho, A.C.P.L.F., Moreira, E.S.: Neural networks applied in intrusion detection systems. In: IEEE International Joint Conference on Neural Networks Proceedings. IEEE World Congress on Computational Intelligence, pp. 205–210 (1998)
6. Denning, D.E.: An intrusion-detection model. IEEE Trans. Softw. Eng. pp. 222–232 (1987)
7. Lippmann, P., Cunningham, R.K.: Improving intrusion detection performance using keyword selection and neural networks. Comput. Netw. **34**, 597–603 (2000)
8. Saloa, F., Nassif, A.B.: Dimensionality reduction with IG-PCA and ensemble classifier for network intrusion detection. Comput. Netw. **148**, 164–175 (2019)
9. Dhanabal, L., Shantharajah, S.P.: A study on NSL-KDD dataset for intrusion detection system based on classification algorithms. Int. J. Adv. Res. Comput. Commun. Eng. **4**(6), (2015)

Rule-Based Approach to Sentence Simplification of Factual Articles

Aryan Dhar and Ambuja Salgaonkar

Abstract A model for automatic text simplification by splitting the longer sentences from a given text has been presented. The paper's proof-of-concept model demonstrated the following capabilities: (i) the average number of characters per sentence in the output text is reduced to 54% of that of the input text, (ii) 90% of the newly generated sentences are comparable with those generated by a human expert. The Flesch-Kincaid Grade Level (FKGL) of text complexity used in this experiment is 16.5, suitable for graduate students. The FKGL of the processed text is 9.7, suited for tenth grade students. Sentence simplification has various applications in natural language processing, including automatic question generating systems. This work is aimed at simplification so as to enhance accessibility of text and its comprehension by learning disabled children.

Keywords Automatic sentence simplification · Rule-based approach · ICT for accessibility enhancement

1 Introduction

There has been a growing interest in the integration of technology and education, but it must be noted that technology is not a replacement for pedagogy [1]. Rather, the purpose of technology is to help with the achievement of goals such as expanding access, enhancing the quality of education, and improving internal efficiency [2]. Arguably, two fundamental purposes of the integration of technology in education are to increase access to those with learning disabilities like dyslexia, and to augment the internal efficiency of question generation by paving the way for implementing automatic question generators (AQG) which serve as key motivations for the work.

A. Dhar (✉)
Podar International School, Mumbai, India
e-mail: aryan.dhar@hotmail.com

A. Salgaonkar
Department of Computer Science, University of Mumbai, Mumbai, India
e-mail: ambujas@udcs.mu.ac.in

M. Tuba et al. (eds.), *ICT Systems and Sustainability*,
Advances in Intelligent Systems and Computing 1077,
https://doi.org/10.1007/978-981-15-0936-0_20

Sentence simplification has been identified as a hurdle for achieving higher success rates while generating "well-formed" questions using AQGs [3]. Moreover, splitting complex sentences into simpler ones enables greater comprehension by those who are less-skilled in reading [4]. Potential beneficiaries include persons with dyslexia, aphasia, and deafness, which are learning disorders that inhibit literacy [5].

This paper proposes a novel scoring model and a rule-based approach to sentence simplification. It demonstrates exceptional accuracy as well as a reduction in complexity using the framework outlined in this paper. The paper has treated sentence simplification as a "splitting operation" out of three possibilities: splitting, deletion, and paraphrasing [6], and has employed a soft-computing approach.

A brief literature survey has been given below, and the novelty and contributions of the research have been presented. Section 3 provides the details of the model and Sect. 4, the results and discussions. Both the achievements and shortcomings of the work are discussed. The last section summarizes the key findings of this research.

2 Previous Research

The first sentence-splitting rules, proposed in 1994 by Chandrasekar [7], identified phrases in complex sentences that could then be simplified manually. However, the rules failed to detect relative clauses. This was then rectified in 1996 in the dependency-based model of Chandrasekar et al. [8]. It used linear pattern matching rules, similar to what this paper proposes. The novelty of this paper is that it employs a second set of heuristic and general rules that rectifies the errors caused by linear pattern matching rules, thereby achieving better accuracy.

More recent approaches include Siddharthan [10], which applies transformation rules to typed dependencies in an effort to reduce parsing errors. This paper's contribution is not the reduction of such errors, but rather, the improvement in penetrability of the corpus. Glavaš and Štajner [11] used a novel approach that relied on event extraction to simplify text in news articles. This was achieved by the incorporation of a neural component.

Several subsequent papers including Narayan et al. [12] employ neural components in their framework as well. Kriz et al. [13] employ neural components to offer multiple simplifications and then rank them to enhance fluency, adequacy, and simplicity. The present paper contributes to making the text more readable by reducing average sentence length. This paper has employed a soft-computing approach that is relatively less costly and more effective when a satisfactory answer is called for, instead of the one and only answer. This is a significant novelty of this work. Another motivation of the work was to make the sentence simplification process more accessible to educators.

3 Methodology

The database consists of two hundred consecutive sentences selected from news/wiki articles. The first and the last word of each sentence was Part of Speech (PoS) tagged.

The distribution of various PoS in the first and the last word of each of the sentences is listed in Table 1.

We noticed that if two successive words separated by a comma, semicolon, or hyphen had an amalgamated score of greater than 0.2, then the punctuation mark could be converted into a full stop and two grammatically correct sentences could usually be generated from the result.

The algorithm for the scoring model is given below:

```
Input corpus
character in corpus == "," or ";" or "-"; =>
Get POS of word before character
ender =POS of word before character
Get POS of word after character
starter =POS of word after character
score = probability of occurrence of ender at end+probability
of occurrence of starter at start
ender in ("CD", "JJ", "JJS", "JJR") OR starter in ("JJ", "JJS",
"JJR") =>score=0
ender in ("NNS", "NN", "NNP") AND starter is "DT" => score=0
score>0.2 => character = "."
```

A set of rules is devised to cater to common exceptions in factual articles. For example, when an article cites an argument for or against a proposition, the sentence can be converted into two or more sentences. Additionally, some of these rules cater to sentence-splitting opportunities that the scoring model may have missed while generating correct sentences from them, e.g., conjunctions in the middle of a sentence without a preceding comma.

Next, when a sentence starts with a verb due to splitting an original sentence at commas, this error is rectified by starting the sentence with the subject as identified

Table 1 Probability of occurrence of a certain PoS at start of the sentence

PoS	Probability (score) of occurrence at the start of sentence	Probability (score) of occurrence at the end of sentence
NN	–	0.41
NNS	–	0.15
RB	0.08	0.08
NNP	0.10	0.07
PRP	0.24	0.06
DT	0.13	–
IN	0.11	–
CC	0.08	–

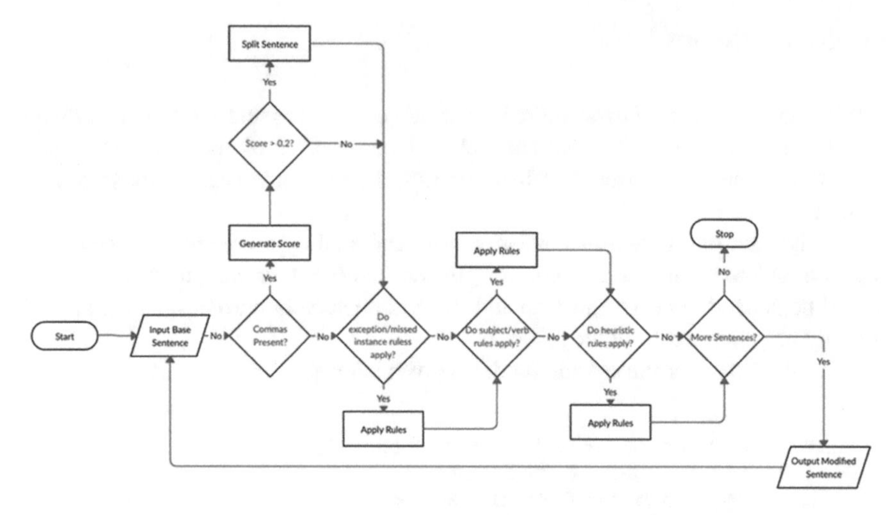

Fig. 1 Flowchart of the automatic sentence simplification process

by the Spacy NLP toolkit, or with the general phrase "The subject(s)". Verbs after unchanged commas or adverbs are also split into independent sentences using this method. In the case of past participles, a subject is added with "was/were". In the case of present participles, a subject is added and the verb is converted into its past tense form. Constructs following a comma, like singular noun and a verb, or "has/had" and a verb with a subject, are locations where a sentence could be split. In such a situation, a comma needs to be replaced by a full stop. The task is to identify appositives from commas and generate correct new sentences from them. Conjunctions, interjections, and connectives following a comma are to be metamorphosed so as to generate separate sentences.

These changes give rise to a few mistakes in sentence simplification. These were analyzed and heuristic rules to rectify them were developed. Essentially, this involves the elimination of connectives, prepositions, or injunctions when they come at the beginning of a new sentence.

Often the base corpus sentence contains an injunction to add information about the subject in the main clause, but later on, the main clause is made into an independent sentence by the previous rules, resulting in a meaningless clause starting with an injunction/preposition (Fig. 1, Table 2). Sample heuristic rules are given below.

3.1 Experimental Setup

An input of 50 sentences from Wiki articles about Mahatma Gandhi (15 sentences), Dubai (14 sentences), Guardian articles about the Sudan Coup (11 sentences), and Brexit (10 sentences not part of training set) was considered. The output was then analyzed to determine the accuracy and identify the shortcomings of the algorithm.

Table 2 Sample heuristic rules

Rule number	Algorithm
R1(or)	'or' is present in sentence => 'or' is used to suggest an alternative term for => 'This is alternatively known as' else 'Or' is used to give an alternative course of action => "Another option is"
R2(on)	'on' is present in sentence => 'on' refers to present location OR present position => 'It is on' else 'on' refers to event in the past =>'It was on'
R3(at)	'at' in present sentence => 'There was' + Appropriate Article
R4(because)	'because' in present sentence => 'because' is related to past event => 'The aforementioned was because' else 'The aforementioned is because'
R5(with)	'with' is present in sentence => 'with' refers to a simultaneous event => 'This occurred as' else 'with' is used to add information about past event => 'This was with' else 'Based on'

4 Results and Discussion

4.1 Sample Correct Transformations

Consider the following sentence: "Adult non-Muslims are allowed to consume alcohol in licensed venues, typically within hotels, or at home with the possession of an alcohol license." The algorithm first computes the scores at each comma. The first comma is a combination of NNS + RB (0.23) and the second is of NNS + CC (0.23). Therefore, the sentence is split to give: "Adult non-Muslims are allowed to consume alcohol in licensed venues. Typically within hotels. Or at home with the possession of an alcohol license." The heuristic rules (R1) recognizes the "or" and its purpose, changing it to "Another option is". Additionally, the other rules identify the incorrect beginning with "typically" and modify it to give: "Adult non-Muslims are allowed to consume alcohol in licensed venues. This is typically within hotels. Another option is at home with the possession of an alcohol license (Table 3)." A shortcoming in this example is discussed in Sect. 4.4 below.

Table 3 Other sample correct sentence transformations

Old sentence	New sentence
Ahmed Awad Ibn Auf, who is also an army general, said political detainees would be released but that a state of emergency would continue for three months and that a curfew from 10 p.m. to 4 a.m. would be enforced for at least a month	Ahmed Awad Ibn Auf, who is also an army general, said political detainees would be released. Nevertheless, a state of emergency would continue for three months. A curfew from 10 p.m. to 4 a.m. would be enforced for at least a month
Gelder then composed and released an interview summary, cabled it to the mainstream press, that announced sudden concessions Gandhi was willing to make, comments that shocked his countrymen, the Congress workers, and even Gandhi	Gelder then composed and released an interview summary. Gelder cabled it to the mainstream press. It announced sudden concessions Gandhi was willing to make. These were comments that shocked his countrymen, the Congress workers, and even Gandhi
On the southeast coast of the Persian Gulf, it is the capital of the Emirate of Dubai, one of the seven emirates that make up the country	It is on the southeast coast of the Persian Gulf. It is the capital of the Emirate of Dubai. It is one of the seven emirates that make up the country
The "great storm" of 1908 struck the pearling boats of Dubai and the coastal emirates toward the end of the pearling season that year, resulting in the loss of a dozen boats and over 100 men	The "great storm" of 1908 struck the pearling boats of Dubai and the coastal emirates toward the end of the pearling season that year. The storm resulted in the loss of a dozen boats and over 100 men
Assuming leadership of the Indian National Congress in 1921, Gandhi led nationwide campaigns for various social causes and for achieving Swaraj or self-rule	Gandhi assumed leadership of the Indian National Congress in 1921. Gandhi led nationwide campaigns for various social causes and for achieving Swaraj or self-rule
However, the option to hold a referendum in all circumstances is likely to be strongly opposed by Unite, a powerful ally of Corbyn on the committee, and the CWU, the Communication Workers Union, which last weekend confirmed union policy to oppose a second referendum carried by more than 90% of delegates	However, the option to hold a referendum in all circumstances is likely to be strongly opposed by Unite. Unite is a powerful ally of Corbyn on the committee, and the CWU. CWU is the Communication Workers Union, which last weekend confirmed union policy to oppose a second referendum carried by more than 90% of delegates

4.2 Summary Statistics

By employing the rule-base discussed above, 36 sentences were simplified to generate 80 new sentences. The Flesch-Kincaid Grade Level (FKGL) is an extensively used test that rates the readability of the text (Fig. 2). (The FKGL score is roughly equivalent to the number of years of schooling required to understand the text.) The value is computed using the following formula [14]:

$$\text{FKGL} = 0.39\left(\frac{\text{total words}}{\text{total sentences}}\right) + 11.8\left(\frac{\text{total syllables}}{\text{total words}}\right) - 15.59 \quad (1)$$

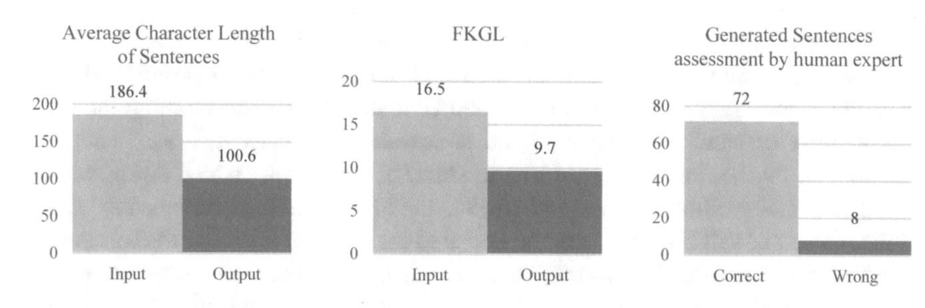

Fig. 2 Histograms summarizing results

Table 4 Rule distribution table

Rule type	Correct instances	Incorrect instances	Accuracy (%)
Scoring model	12	2	85.7
Heuristic rules	11	0	100
Subject/verb/noun rules	19	4	82.6
Rules for "exceptions" and "missed instances"	16	2	88.9

It is noted that the purpose of sentence simplification for those with reading disorders is not to simplify the standard of language—this makes further penetrability untenable. Therefore, the reduction in FKGL to half of its original is significant.

4.3 *Rule Distribution Table*

This gives a mean accuracy of 89.3% with standard error of 3.79%. We use a 90% confidence interval where critical probability is 0.95 with three degrees of freedom Table 4. This gives a T-score of 2.35 which imputes a marginal error of 8.9%; therefore the 90% confidence interval of accuracy is 89.3 ± 8.9%.

4.4 *Discussion of Results and Errors*

The most common error (two occurrences) was not due to the rule-based algorithm but due to an incorrect subject identification by the Spacy toolkit. A good example of this is: "In 1896, fire broke out in Dubai, a disastrous occurrence in a town where many family homes were still constructed from barasti-palm fronds." was converted to "The year was 1896. Fire broke out in Dubai. Dubai is a disastrous occurrence in a town where many family homes were still constructed from barasti-palm fronds." In the third sentence, "Dubai" rather than "fire" is identified.

Other errors included failure to assign an article to the subject identified (one occurrence), tense not rectified in a sentence (two occurrences), and one final error was due to incorrect punctuation in the source text. Another shortcoming included the inability of handle complicated and punctuated sentences such as: "This was highlighted by the famous Salt March to Dandi from 12 March to 6 April, where, together with 78 volunteers, he marched 388 km (241 mi) from Ahmedabad to Dandi, Gujarat to make salt himself, with the declared intention of breaking the salt laws."

Moreover, sometimes, information is lost in intermediary sentences as seen in the first example in 4.1. A final shortcoming is that many of the rules rely on "exceptions" that occur in the writing styles typical of factual articles and therefore the model may not work as effectively for fictional texts. An approach to overcome these shortcomings would include mechanisms to identify the subject/object in a sentence with higher accuracy. The training set would need to be expanded to develop more rules.

Nevertheless, it is noted that this paper achieves an appreciable mean accuracy of 89.3% while simultaneously avoiding simplification of the vocabulary, which makes it ideal in educational environments where language standards must be preserved. Ultimately, by producing a more comprehensible output, the algorithm possesses the ability to level the playing field for students with learning disorders.

5 Conclusion

This study shows that it is possible to simplify sentences and enhance readability in factual texts using a rule-based approach. The proposed model enabled us to achieve a high accuracy of 90% for the task under consideration. Almost 50% reduction in the FKGL, i.e., complexity of the text, results in a greatly enhanced penetrability of factual texts. It is hoped that this would interest many educators to employ this model.

References

1. Jhurree, V.: Technology integration in education in developing countries: Guidelines to policy makers. Int. Educ. J. **6**(4), 469–483 (2005)
2. Haddad, W.D., Jurich, S.: ICT for Education: prerequisites and constraints. In: Haddad, W.D., Draxler, A. (eds.) Technologies for Education: Potential, Parameters, and Prospects, p. 47. For UNESCO by Knowledge Enterprise (2002)
3. Divate, M., Salgaonkar, A.: Ranking model with a reduced feature set for an automated question generation system. In: Sharma, D.S, Sangal, R., Sherly, E. (eds.) Proceedings of the 12th International Conference on Natural Language Processing, pp. 384–393 (2015)
4. Mason, J., Kendall, J.: Facilitating reading comprehension through text structure manipulation. Alberta J. Med. Psychol **24**, 68–76 (1979)
5. Siddharthan, A.: A survey of research on text simplification. Int. J. Appl. Linguist. **165**(2), 259–298 (2014)

6. Feng, L.: Text simplification: a survey. In: The City University of New York, Technical Report (2008)
7. Chandrasekar, R.: A hybrid approach to machine translation using man machine communication. Ph.D. Thesis, Tata Institute of Fundamental Research/University of Bombay, Bombay (1994)
8. Chandrasekar, R., Doran, C., Srinivas, B.: Motivations and method for sentence simplification. In: Proceedings of COLING'96, pp. 1041–1044 (1994)
9. Siddharthan, A.: Syntactic simplification and text cohesion. Technical Report 597, University of Cambridge (2004)
10. Siddhathan, A.: Text simplification using typed dependencies: a comparison of the robustness of different generation strategies. In: Proceedings of the 13th European Workshop on Natural Language Generation, pp. 2–11 (2011)
11. Glavaš, G., Štajner S.: Event-centred simplification of news stories. In: Proceedings of the Student Research Workshop associated with RANLP, pp. 71–78 (2013)
12. Narayan, S., Gardent, C., Cohen, S.B., Shimorina, A.: Split and rephrase. In: Proceedings of the EMNLP, pp. 617–627 (2017)
13. Sulem, E., Abend, O., Rappoport, A.: Simple and effective text simplification using semantic and neural methods. In: Proceedings of the 56th Annual Meeting of the Association for Computational Linguistics (Long Papers), pp. 162–173 (2018)
14. Kriz, R. et al.: Complexity-weighted loss and diverse reranking for sentence simplification. In: North American Association of Computational Linguistics (NAACL) (2019)
15. Kincaid, J.P., Fishburne Jr., R.P., Rogers, R.L., Chissom, B.S.: Derivation of new readability formulas (Automated Readability Index, Fog Count and Flesch Reading Ease Formula) for Navy enlisted personnel. Research Branch Report 8–75, Millington, TN: Naval Technical Training, U. S. Naval Air Station, Memphis, TN

Motion Intervene Surveillance Control Using IoT

Annapurna Kai, Mohamad Zikriya, Jayashree D. Mallapur and Shridevi C. Hiremath

Abstract Safety and security have always become a basic necessity for the people. With the growing concern over physical security, surveillance cameras have been installed outdoors more and more in every industry sector and public space from towns, airports and railroad stations to offices, retailers, healthcare providers and more. Installing cameras under the open sky with expensive outdoor housings became essential to monitor and detect the unusual activity outside the home/offices, Installation of a CCTV based surveillance for home is costly as it requires additional components like hard disk for storage, a monitor display and a 24 × 7 connected Wi-Fi router for surveillance of home over internet. The other drawback of using CCTV as for surveillance is that, it does not provide any alerts or notification to the user whenever an event occurs, and thus there is no preventive option given in the CCTV Surveillance. To solve this problem, we came across an idea of Smart home surveillance using IoT and also we tend to provide an affordable Smart home surveillance keeping the minimal cost that can be used by a family/person having the average source of income. The overall working of the device is handled by a controller board, interfaced along with a camera that would trigger an alert message referring the motion in front of the camera. Another advantage of our device is, it tends to work on a fully autonomous mode by capturing image of moving body (i.e. Human being, Animal).

Keywords Surveillance · Home security · CCTV · IoT

A. Kai (✉) · M. Zikriya · J. D. Mallapur · S. C. Hiremath
Department of Electronics and Communication, Basaveshwar Engineering College, Bagalkot 587102, India
e-mail: annapurna731@gmail.com

M. Zikriya
e-mail: zakriyat@gmail.com

J. D. Mallapur
e-mail: bdmallapur@yahoo.co.in

S. C. Hiremath
e-mail: shiremath837@gmail.com

M. Tuba et al. (eds.), *ICT Systems and Sustainability*,
Advances in Intelligent Systems and Computing 1077,
https://doi.org/10.1007/978-981-15-0936-0_21

1 Introduction

Home security has become the basic necessity for the people. In current scenario, the working environment consists of long working hours wherein majority members in a nuclear family adhere to work day and night long hours without ensuring a proper security surveillance to their home, thus this is leading to an increase in the threats causing unusual event to occur in the absence of the family members during their work hours. The present techniques to safeguard home and kids in the absence of their parents do not provide an efficient and reliable monitoring of home. The present technology consists of bulky cameras, and would require an additional memory support for storage of the footage taken during the surveillance, thus due to these constrains the traditional method (CCTV) opted for home surveillance in turn increases the cost, and also the traditional methodology used does not provide an instant action/alert to the family member during the occurrence of an unusual activity at the home premises.

Keeping these all drawbacks of the CCTV approach for home surveillance, we have come up with a solution that would surpass all drawback of CCTV based home surveillance and would ensure reliable and efficient security of home during the absence of family members. The proposed solution would also tend to send an instant picture of unusual activity that occurs at the home premises.

The Hardware used for the application to work consist of a development board i.e. raspberry pi 3 B+ interfaced along with a pi-camera, whereas software side completely rely on RASPIAN LINUX operating system that would completely handle and ensure proper working of the motion based surveillance application that is made to run on RASPIAN OS.

This running application on the development board would trigger an E-mail alert along with the captured image as an attachment whenever there is a motion in front of the camera, upon getting an E-mail alert the Person/Family member could also be able to the live stream status of the home in his android mobile via an android application and IoT Interface.

Our device helps to take action immediately at that unusual activity and with this we make sure to overcome all the above said drawbacks of traditional CCTV based surveillance system, the key parameters unlike cost memory requirement and energy consumption of our device are completely taken into the considerations and accordingly implemented in order to provide better efficiency and high reliability. The paper is organized into sections. The section two covers related work, section three covers proposed work, section four covers conclusion.

2 Litrature Survey

- In the paper [1] the components used are camera, raspberry pi, LCD display IR for night vision and USB is used for storage. When the motion is detected in camera, it uses the image processing to detect where exactly the motion is detected. System will now transmits the images captured in the camera over IoT, so the user can view online.

- In the paper [2] they have used Machine Learning for Theft detection. The detection of theft occurrence is captured in the camera using the Machine Learning. With the help of the convolution neural networks system will detect motion and send captured images and message to owner. It has option to call the police.
- In the paper [3] they have used PIR module that will be continuously monitoring home or work space. The PIR module detects an intruder and sends signal to microcontroller. The controller is connected to alarm system, system will send signal to cloud in turn cloud will send an alert signal to users mobile phone. It is also having thumb print reader for controlling of the opening and closing of the door for home security.
- The paper [4] the designed system will detect an intruder at home. System will send alert message through GSM module to the user when the intruder is detected. The user can get message of intrusion from anywhere on internet enabled device. It uses an IP address of the installed IP webcam of mobile.
- In the paper [5] they have focused on individual face recognition. raspberry pi and IoT is used to get push notice through e-mail server if individual is identified. And then the video/Photos are sent to cloud server.
- In the paper [6] the captured images are transmitted via 3G dongle and then sent to smart phone. Raspberry pi controls detectors and continuous video recording is stored in it. It will also stream live video for surveillance. They can also be used for future playback.
- The paper [7] uses camera, PIR sensor and GPS tracker. PIR sensor will detect the motion and camera will capture the image and live video. Raspberry pi will stores it for future playback. GPS tracker will track the location of the intruder.
- The paper [8] will use an web camera, camera will capture live images in the area it is placed if any object is moving. If movement is found computer starts to record the video and buzz the alarm and it will send SMS to user.

These are some of the existing technologies, where images and video are recorded using different applications. Our paper is focused only on the instant camera capturing and send to E-mail. By doing like this memory and time will be saved. User can also take an instant action.

3 Proposed Work

The technology usage in a day today life, make life easy safe and comfortable. We have come with a technological help for home security to help husband and wife working together and keeping home locked. The following block diagram represents the overall working of our proposed work.

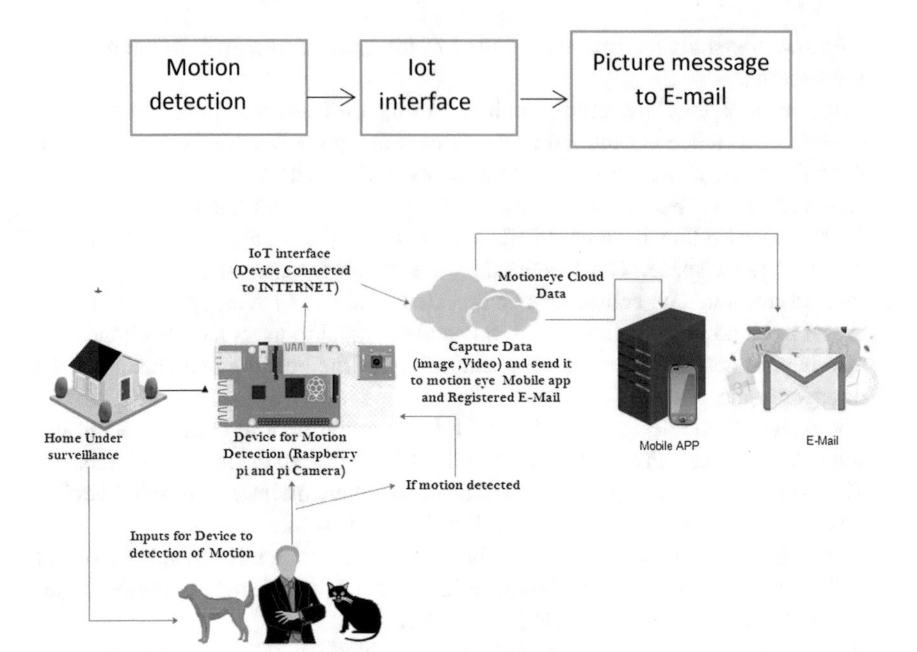

3.1 Motion Detection

Whenever any motion occurs in front of the camera interfaced to a controller board, an image of that instant is captured by the controller and an immediate alert is sent to the person via IoT interface.

3.2 IoT Interface

As mentioned earlier our project is meant to provide instant alerts to the user if there is a motion detected by the hardware (raspberry pi and pi camera). These alerts are handled via the aid of internet without any restrictions over the distance. Thus project utilizes IoT platform in order to send captured images by device as an email, the device requires an internet connection which is provided either via Ethernet or using a mobile hotspot.

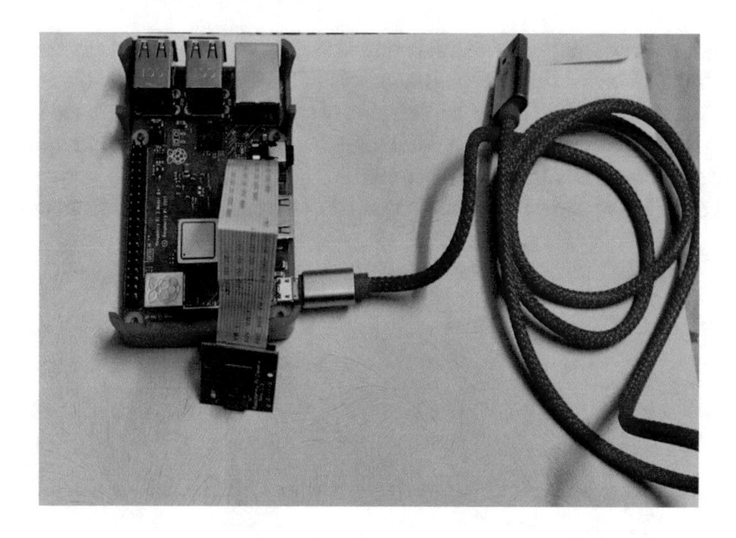

3.3 *Experimental Set Up*

The above picture shows experimental set up. USB cable used for power supply, camera for capturing the image and memory card for storage purpose.

4 Experimental Results

The below figures shows the experimental results of the project. Project tends to give more importance for the safety home while the family members are out during there working hours. The existing technology does not provide instant action.

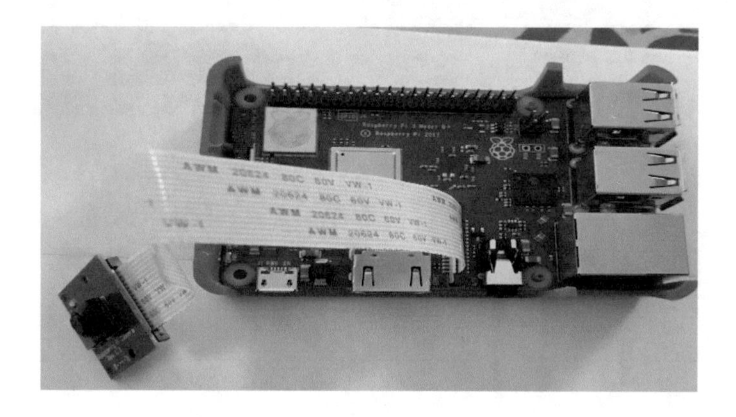

The below graph shows a detailed comparison between traditionally used CCTV and raspberry pi based home surveillance system and as we look below in the categories done for both the CCTV and raspberry pi, The Home surveillance based on raspberry pi based system is more efficient in terms of Cost, Memory requirements, Image size, Video Size, Instance response.

The below figure shows the results consisting of received E-mail from the gadget.

COMPARISION OF RASPBERRY PI MOTION SURVELLIANCE VS TRADITIONAL CCTV

	COST	MEMORY REQUIREMENTS (interms of GB)	POWER CONSUMPTION (In terms of Voltage)	IMAGE SIZE (In terms of MB)	VIDEO SIZE (In terms of MB)	INSTANCE RESPONSE (In terms of Seconds)
CCTV Camera	15000	1000	12	5	10	0
RASPBERRY PI	3500	16	5	1.5	5	5

RASPBERRY PI ■ CCTV Camera

52230
Camera1
2019-05-04
05:45:07

Camera1
2019-05-04

5 Conclusion

The paper mainly focuses on the low cost implementation and also low memory consumption for the safety monitoring of home premises. The above said system ensures security of the home using advanced technologies unlike the IoT and hardware like raspberry pi and pi camera. The proposed system will also give valid proof of introducer as picture message to owner's mobile phone for further process. This system will make working people tension free for their locked home.

References

1. Anjum, U., Babu, B.: Iot based theft detection using raspberry pi. Int. J. Adv. Res. Ideas Innov. Technol. **3**(6), 131–134 (2017)
2. Kushwaha, A., Mishra, A., Kamble, K., Janbhare, R.: Theft detection using machine learning. Int Conf. Innov. Adv. Technol. Eng. **8**, 67–71 (2018). ISSN (e): 2250-3021, ISSN (p): 2278-8719
3. Safa, H., Priyanka, N.S., Priya, S,V.G, Vishnupriya, S., Boobalan, T.: Iot based theft premption and security system. Int. J. Innov. Res. Sci. Eng. Technol. **5**(3), 4312–4317 (2016)
4. Patil, A., Mondhe, S., Ahire, T., Sonar, G.: Auto-theft detection using raspberry pi and android app. Int. J. Res. Eng. Appl. Manage. (IJREAM) **02**(07), 15–17 (2016). ISSN: 2494-9150
5. Nandurkar, A.A., Nathaney, G.: An internet of things approach for security surveillance using raspberry-pi. Int. J. Innov. Res. Comput. Commun. Eng. **5**(4), 1–4 (2017)
6. Prasad, S., Mahalakshmi, P., Sunder, A.J.C., Swathi, R.: Smart surveillance monitoring system using raspberry pi and PIR sensor. Int. J. Comput. Sci. Inform. Technol. 5(6), 7107–7109 (2014). ISSN-0975-9646
7. Madhuri, R., Mayuri, P., Ashabai, W., Rupali, M., Ankita, M.: Study of theft detection and tracking using raspberry pi and PIR sensor. Int. J. Latest Trends Eng. Technol. **6**(1), 290–294 (2015)
8. Upasana, A., Manisha, B., Mohini, G., Pradnya, K.: Real time security system using human motion detection. Int. J. Comput. Sci. Mob. Comput. **4**(11), 245–250 (2015)

Broadband Power-Line Communication in Self-Reliant Microgrid Distribution Network-an Overview

Z. A. Jaffery, Ibraheem and Mukesh Kumar Varma

Abstract Broadband communication has become an essential part of modern life. The broadband communication over power line is an emerging technology that conveys information through the prevailing power-line structure. The self-reliant distribution network is independent flexible cells operating with distributed energy resource and local loads at low-voltage distribution network. A microgrid is an ultimate power system to use renewable energy sources for electric energy generation. Generation using renewable energy sources can decrease greenhouse effects; though, it cannot meet the request of loads strongly. In this paper, power-line communication techniques that have been in use for microgrid are discussed and summarized. The paper also incorporates the various modeling aspects of power-line communication channel of the self-reliant microgrid distribution network. The simulation study is also carried out and the results for Quadrature binary shift keying (QPSK) are presented. The simulation results obtained for SNR based on the threshold BER compared with those calculated theoretically. These results are found to be comparable.

Keywords Broadband power-line communication · Channel modeling · Channel characteristics · Microgrid

---

Z. A. Jaffery (✉) · Ibraheem
Electrical Engineering Department, Faculty of Engineering & Technology, Jamia Millia Islamia, New Delhi, India
e-mail: zjaffery@jmi.ac.in

Ibraheem
e-mail: ibraheem@jmi.ac.in

M. K. Varma
Department of Electrical Engineering, FET, J. M. I., New Delhi, India
e-mail: mkvsuni@gmail.com

Department of Electrical Engineering, G. B. Pant Polytechnic, New Delhi, India

M. Tuba et al. (eds.), *ICT Systems and Sustainability*,
Advances in Intelligent Systems and Computing 1077,
https://doi.org/10.1007/978-981-15-0936-0_22

1 Introduction

The demand for broadband communication has considerably grown worldwide. Communications over power lines have a history of more than a hundred years [1, 2]. The fast growth of communications technology recently made it likely to practice the power-line network for high-frequency data transmission. The broadband communication over power line is an emerging technology that uses the prevailing power-line structure of low-voltage- and medium-voltage distribution networks as a communication medium for data, voice, and video transmission [3, 4]. A microgrid is a relatively small group of Distributed Generators (DG), Energy Storage Systems (ESS) and flexible loads that are electrically interconnected. The microgrid is functioned, organized, and accomplished in a unlike way from traditional grids, primarily in that it depend on supplementary data collection and communication. The high-speed communication is vibrant for management and control of microgrid [4, 5].

A microgrid efficiently combines the transmission of energy and data communication. Actual information gathering, suitable and consistent data transmission, and effective processing and intellectual investigation of multilevel information are the main essential requirements for appropriateness, precision, and generality of data. The main feature which makes a microgrid smarter is two-way Broadband communication through the existing power-line network [4]. The Indian government aims to explore the possibilities to reach this broadband communication technology to interior urban, suburban, and rural areas throughout India. This paper presents an overview of broadband power-line communication for the microgrid distribution network and their applications.

The paper is organized as follows. In Sect. 2 introduces, the architecture of microgrid communication system. Section 3 discusses the microgrid monitoring communication network and power-line channel characteristics. Section 4, describes the overviews of the power-line communication system. In Sect. 5, the channel characteristics and analysis for the proposed channel, are presented. Analysis and Error Performance evaluation of the channel model is given in Sect. 6. Finally, in Sect. 7, discussion on conclusions drawn from the study is reported.

2 The Architecture of Microgrid Communication System

The microgrid is a part of the electrical power distribution network with single or multiple distributed energy resources, and variety of loads. The microgrid distribution system consisting of is a rural residential network operating at low voltage and is primarily powered by photovoltaic (PV) and wind power station [4]. These renewable resources provide clean power and high-speed internet to the consumer. A microgrid is one of the most effective and efficient ways of integrating renewable energy resources into the utility grid. The microgrid has AC as well as DC generation, transmission, and distribution based generation. The overall system is synchronized to the central existing grid having conventional resources with bidirectional flow of energy. The basic model of a microgrid system is presented in Fig. 1. The various distributed generations which are considered in this model are, solar PV generator, the wind generators (WECS), fuel cells, micro-hydel, diesel generators, and reciprocating engine, AC and DC buses are interlinked to each other by using suitable coupling converters. A microgrid typically operates within two steady-state operational modes, namely grid-connected mode and islanding mode [5]. Microgrid in islanded mode can be a result for providing alternative power for communications, lighting, and other applications. In grid-connected mode, local grouping of electricity resources and loads are coupled to and synchronized with the centralized utility grid supply. The word self-reliant microgrid distribution network will ensure the security, quality of power, communication with better control and management performance.

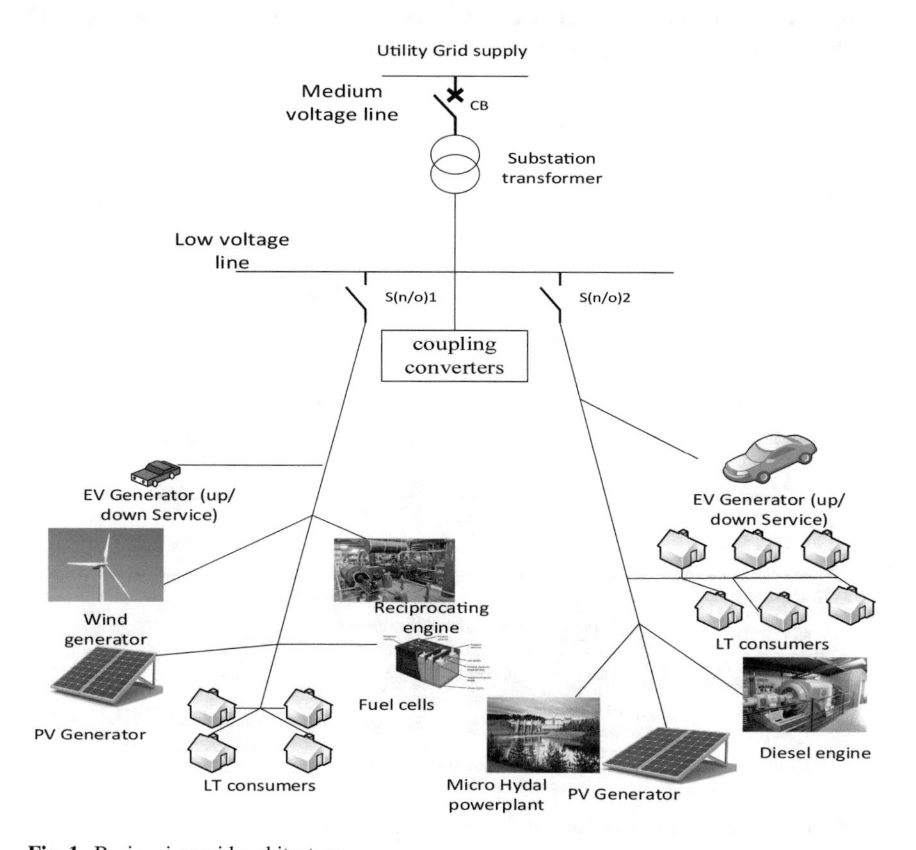

Fig. 1 Basic microgrid architecture

3 Microgrid Monitoring Communication Network

The communication arrangement of a microgrid is basically used for the generation, transmission, and distribution of power and can be separated into two parts based on the following possible applications:

(1) The monitoring communication system for the microgrid networks and
(2) The communication link between the main data server of the microgrid and the delivery system.

Figure 2 describes the structural design of the microgrid. With power-line communication machineries, the system works for energy dispatch, online monitoring of power apparatus, managing of site operation, energy data collection, and anti-theft of outside services. The broadband power-line communication system incorporates computer network tools, control machinery, detecting, and metering technology, and can join the microgrid with power-line communication networks.

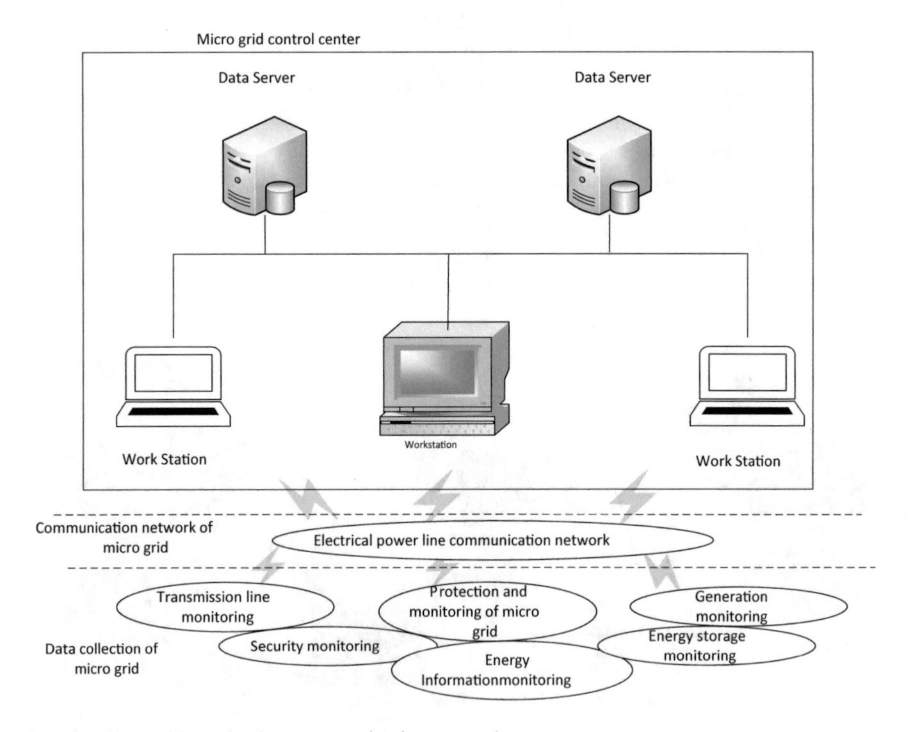

Fig. 2 Microgrid monitoring communication network

4 Power-line Communication Overview

The term broadband power-line communication (BPLC) is used to describe the technology for carrying data through the standing power lines and cables which are the primary means to carry electric powers. The power-line skills can be grouped into narrowband power-line communication, operating usually below 500 kHz while broadband power-line communication, operates usually at frequencies above 2 MHz [2].

One of the main novel features of power-line communication over other options is that it can be deployed at the low cost, due to the usage of the prevailing power lines as a communication channel. In addition, power-line communication is the only type of communication medium that is integrally connected to the topology of the power distribution network. As such, knowledge of the power-line communication grid topology can be simply used to extract information about power system connectivity. Power-line communication also offers essential physical safety aspects which not all resolutions offer. Power-line communication channels suffer from attenuation owing to reflection and frequency fading, as well as introduction to unalike types of noise. However, power-line communication modeling is considered as the main challenge in designing power-line communication networks due to the dynamic characteristics of the power-line communication channel.

The microgrid communication system is planned for bidirectional transmission, observation and control, and data communication. The term broadband power-line communication (BPLC) is a favorable broadband communication network. In this high-frequency data, voice, and video, are transferred through the prevailing low-voltage power lines and cables through the coupling device as shown in Fig. 3, the coupling device consists of a coupling capacitor, a coupling filter, and the line trap. The coupling capacitor in combination with the coupling filter will provide a low impedance path for PLC signal and high impedance path to the power signal. The line

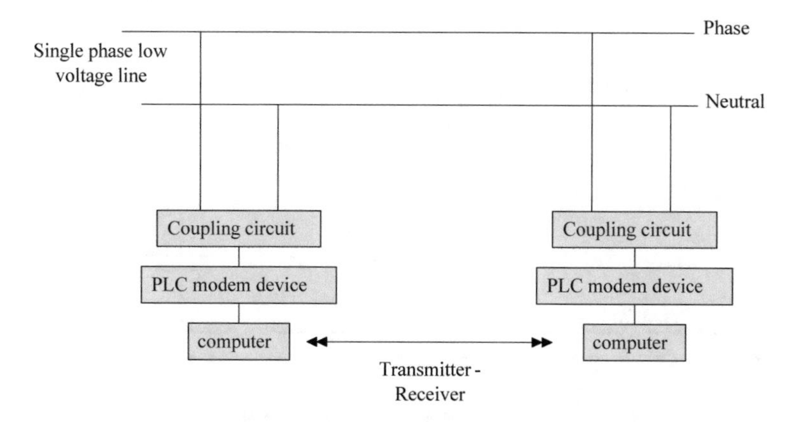

Fig. 3 Microgrid power line communication structure

trap is responsible for preventing the PLC signal from being short-circuited by the substation. The matching transformer offers the impedance equal to the 50 Ω or the 75 Ω coaxial cable and the characteristic impedance of the power line (150–500 Ω). However, broadband power-line communication channel suffers from attenuation due to reflection and frequency fading, as well as experience to diverse types of noise.

5 Channel Characteristics and Its Analysis

The basic traditional distribution network, the power line composed of a pair of conductor, i.e., two-conductor with an identical cross section that is parallel to each other. According to the electromagnetic theory, the power line is regarded as the two-parallel wires. Figure 4 shows the equivalent circuit for, the two-conductor line transmission channel [2].

In this Fig. 4, R, L, G and C are per unit length resistance (Ω/m), inductance (H/m), conductance (S/m) and capacitance (F/m) respectively. The electric quantities are dependent by the geometric and associated parameters.

$$Z_c = \sqrt{\frac{(R + j\omega L)}{(G + j\omega C)}} \tag{1}$$

And

$$\gamma = \sqrt{(R + j\omega L)(G + j\omega C)} = \alpha + j\beta \tag{2}$$

where, the α real part is are the attenuation constant (in Np/m) and β-imaginary part is and phase constant (in rad/m) of the propagation constant.

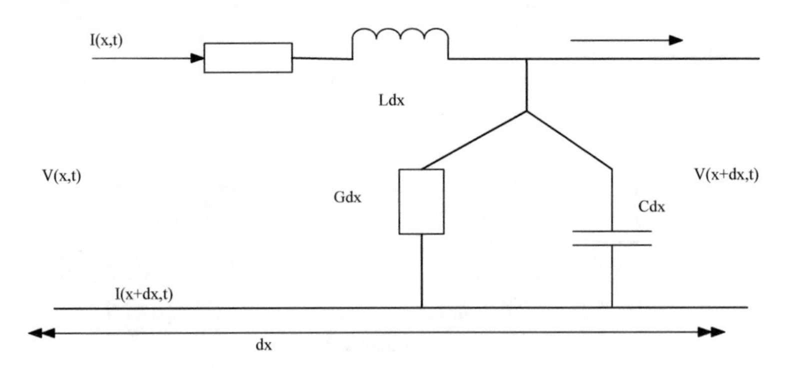

Fig. 4 The equivalent circuit for a two-conductor transmission line

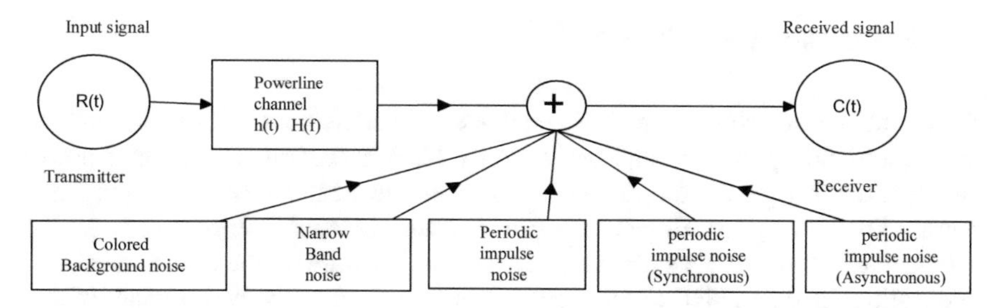

Fig. 5 The channel model

$$H(f) = \frac{V(x = d_i)}{V(x = 0)} = e^{-\gamma d_i} = e^{-\propto(f)d_i} e^{-j\beta(f)d_i} \tag{3}$$

The channel surroundings of the microgrid are inferior than the usual power-line grid network. The grid assembly and transferring of a large number of energy terminals in the microgrid will alter the impedance characteristics of the powerline network and thus disturb the transmission characteristics of the power line. Also it makes the power grid noisier and more complex. The channel is also subjected to various additive noises in the system. Power-line noise can be considered into numerous classes reliant on its origin, level, and the time domain signature [2], as the colored back-ground noise, narrowband noise, periodic impulsive noise (synchronous or asynchronous to the power supply frequency), besides the aperiodic impulsive noise. The mixed hybrid channel model for the broadband power-line communication system is now given as shown in Fig. 5.

The arrangement of the power line is consisting of various branches and unequal impedance causes multiple reflections. When the power line is used for high frequency communication, the data signal does not fallow the straight path between the transmitter and receiver, which makes, multipath signal propagation. Relating the multipath signal propagation, communication frequency and length of the line, the channel transfer function indicated as [1].

$$H(f) = \sum_{i=1}^{N} g_i \times e^{[-(a_0 + a_1 f^K)d_i]} \times e^{\left[-2\pi f\left(\frac{d_i}{v_p}\right)\right]} \tag{4}$$

where N is the number of multi-paths, the g_i is weighting factor, d_i is path length of Ith path, a_0 and a_1 is attenuation parameters, and v_p phase velocity, τ_i is path delay, f-frequency and k-exponent of attenuation factor (0.2–1.0). Part of equation $e^{[-(a_0+a_1 f^K)d_i]}$ is an attenuation portion and the part $e^{\left[-2\pi f\left(\frac{d_i}{v_p}\right)\right]}$ is known as delay portion. The attenuation parameters a_0, a_1 are attenuation constants, and K is a constant (valued 0.5–1), that can be initiate with level of frequency response.

6 Performance Evaluation

Renewable energy sources for the generation of electric power have acquired the attention by utility industry and academic world. A microgrid is an isolated power structure that has generators and loads. For a remote area, microgrid is an ultimate power system for generating electrical power with renewable energy sources like solar and wind energy. This paper uses MATLAB software for the channel analysis, attenuation, BER and SNR of the proposed channel model. The performance of any communication channel depends primarily on the signal-to-noise ratio (SNR) at the receiving end. The propagation of the PLC signal is governed by the attenuation and noise levels by the following equation:

$$\mathrm{SNR} = 10\log\frac{P_{\text{received}}}{P_{\text{Noise}}} \tag{5}$$

It can be seen that more attenuation in the PLC signal along the path will result in less signal strength (SNR). In addition, more accumulated noise will decrease SNR, or in other words, will make the signal sustain less attenuation. In this section, attenuation and noise will be discussed separately. When a signal travels through various components of the system, it will be subjected to attenuation due to the loss incurred on each system component. This attenuation (loss) can be measured in decibels, which allows the losses to be added regardless of the change in impedance [4, 5]. The attenuation is calculated as follows:

$$\text{Attenuation}\,(dB) = 10\log\frac{P_{\text{Transmitted}}}{P_{\text{Received}}} \tag{6}$$

If Z_l is the load impedance and as we calculated the characteristic impedance Z_c as per Eq. (1), the attenuation in powerline communication signal can be calculated as

$$\text{Attenuation}\,(dB) = 20\log\frac{Z_c + Z_l}{2\sqrt{Z_c}Z_l} \tag{7}$$

Data transmission above a carrier by means of different digital modulation schemes. This digital modulation is Amplitude Shift Keying (ASK), Frequency Shift Keying (FSK) and Phase Shift Keying (PSK). The Phase Shift Keying modulation system binary Phase Shift Keying (BPSK) and Quadrature Phase Shift Keying (QPSK), those are least effected by noise. Quadrature Phase Shift Keying modulation is the commonly used M-array PSK system among others since it does not suffer from Bit Error Rate (BER) degradation while the bandwidth efficiency is improved. The results shown in Fig. 6, for the Error performance analysis using quadrature binary shift keying (QPSK) modulation techniques of different SNR values based on

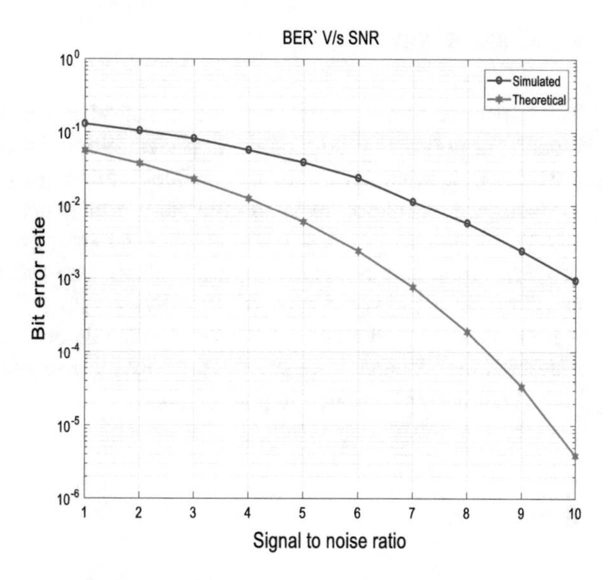

Fig. 6 BER verses SNR plot

the threshold BER of simulated and theoretical values. Figure 7, shows the plot of BER versus SNR, for QPSK modulation in Gaussian environment, on comparison, the results, that used Gaussian filter is good then non-Gaussian filter schemes.

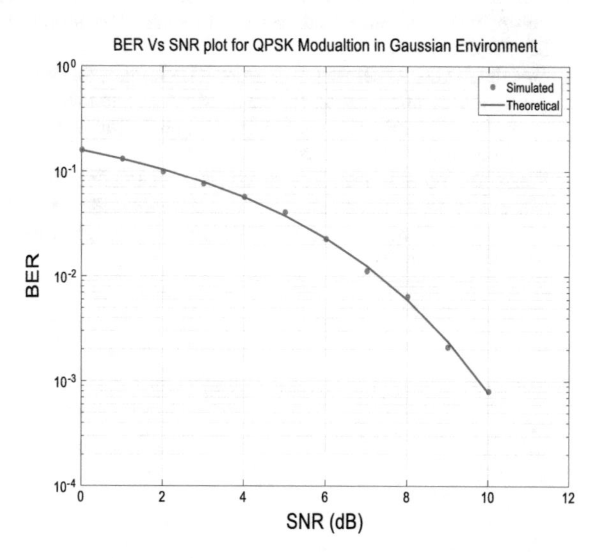

Fig. 7 BER verses SNR in Gaussian environment

7 Conclusions

This paper presents an overview of the microgrid, power-line channel characteristics, broadband power-line communication technology, its application, etc. This paper provides a detailed review of power-line communication technologies as applied to microgrid. A microgrid is an ultimate power structure to use renewable energy sources for electric power generation that can reduce greenhouse effects. The results of QPSK modulation techniques are evaluated on the basis of SNR, which were calculated for a threshold value of BER. Results show that SNR of proposed method is high for simulated value as compared to the theoretical value. We hope the ideas presented in this paper have smooth the way for new research information and have helped out current research in perception.

References

1. Zimmermann, M., Klaus, D.: A multipath model for the power line channel. IEEE Trans. Commun. **50**(4), 553–559 (2002)
2. Meng H., Chen, S., Guan, Y.L., Law, C.L., So, P.L., Gunawan, E., Lie, T.T.: Modeling of transfer characteristics for the broadband power line communication channel. IEEE Trans. Power Delivery **19**(3), 1057–1064 (2004)
3. Kumar, S., Chinnamuthan, P., Krishnasamy, V.K.: Study on renewable distributed generation, power controller and islanding management in hybrid microgrid system. J. Green Eng. **8**(1), 37–70 (2018)
4. Giustina, D.D., Andersson, L., Casirati, C., Zanini, S., Cremaschini, L.: Testing the broadband power line communication for the distribution grid management in a real operational environment. In: IEEE International Symposium on Power Electronics, Electrical Drives, Automation and Motion, pp. 785–789 (2012)
5. Wang, D., Yang, J., Lin, S.: Analysis on transmission characteristics of broadband power line carrier and power frequency integrated communication for new energy microgrid. In: 7th IEEE International Conference on Electronics Information and Emergency Communication (ICEIEC), pp. 300–303 (2017)

An Experimental Study of Mobile Phone Impact on College Students

Maitri Vaghela, Kalyan Sasidhar and Alka Parikh

Abstract Smartphones have revolutionized the pervasive computing research domain. With various sensors embedded in them, they are being applied as experimental platforms for sensing, computing, and communication. These devices are utilized to sense certain physical quantity like motion, sound, location, etc., to infer user's physical activity levels, location/context, and mobility patterns. The devices are also being used to understand user application usage styles and correlate them to users' addiction to mobile phones, mental states, personality types, depression, emotion, and stress levels. However, such studies have not been attempted in India. The Indian research on mobile overuse includes studies conducted purely through surveys and personal interaction. In this work, we attempt at using the mobile sensing mechanism to study how smartphones impact the academic, social, and mental lives of students.

Keywords Mobile sensing · Smartphones

1 Introduction

Impact of mobile technology on human behavior is an emerging area of study and the work in this paper seeks to contribute to that knowledge base. Mobile phones have changed the way people shop, play, do business, or even socialize. They are rightly called a revolution because they changed the entire behaviour of the society. However, concerns are expressed that people are becoming less social, communication is more superficial, computer games have replaced physical sports, etc., through excessive

---

M. Vaghela · K. Sasidhar (✉) · A. Parikh
Dhirubhai Ambani Insititute of Information and Communication Technology, Gandhinagar 382007, Gujarat, India
e-mail: pathapks@gmail.com

M. Vaghela
e-mail: maitrivaghela5@gmail.com

A. Parikh
e-mail: alka2chat@gmail.com

M. Tuba et al. (eds.), *ICT Systems and Sustainability*,
Advances in Intelligent Systems and Computing 1077,
https://doi.org/10.1007/978-981-15-0936-0_23

usage of these portable devices. Recent statistics from one survey show that India has the world's second highest mobile phone users with the figure 750 million [1] and another study by TCS [2] found that 70% of sampled high school students in 14 cities of India today own smart phones with a larger user base in smaller cities than the metropolitan cities. A study by Lilavati Hospital, found that 79% of the population between the age group of 18 and 44 have their cell phones with them almost all the time, with only two hours of their waking day spent without their cell phone in hand [3].

The studies assessing mobile phone addiction that have been undertaken in India used the manual techniques of data collection primarily through questionnaire based surveys [4]. Smartphones today, offer rich and powerful sensing, processing and communication capabilities. A new paradigm termed mobile sensing, offers an unobtrusive way of passively sensing personal, group, and community scale activities pertaining to health, environment, and transport to name a few [5]. The recent years has seen an increased usage of myriad built-in sensors in smartphones such as the microphone, accelerometer, compass, GPS, camera, light, proximity and radio technologies like Wi-Fi and Bluetooth, to monitor human movements and activities (sitting, standing, walking, etc.).

Western nations have attempted at studying the impact of smartphone technology on college students through mobile phone sensors [6]. However, with vast differences in culture, living styles, and education systems between the western nations and India, these studies do not give us insights into the conditions faced by the Indian students. Hence, this paper aims to understand smartphone addiction, its association with student academic life and mental health in an Indian student cohort. There is no work in the Indian context that has attempted to explore correlations between continuous, automatic sensing data from smartphones and qualitative responses to several different questionnaires. In that sense, this is the first effort to use automatic and continuous smartphone sensing to assess mental health and behavioral trends of a student body.

2 Related Work

Dixit et al. [7] studied how a group of 200 students felt when there was no mobile connection, reducing talk time balance or running out of battery. 20% of the students responded as becoming stressed in such situations. Aggarwal et al. [8] studied 192 resident doctors in North India and found that about half of the participants reported feeling anxious when switching off the mobile phones. Sonu Subba et al. [9] studied 336 medical students and found that 85.3% checked their phones up to 10 times. In case of network inaccessibility or a phone malfunction, 39% of the students said they would be very upset or extremely upset. The proportion of students with the symptom of ring anxiety was 34.6% in this study.

Davey et al. [10] compiled six research studies related to smartphone's addiction across 1304 participants. Meta-analysis was performed to calculate smartphone addiction magnitude in India. Results ranged from 39 to 44%. Bisen Shilpa et al. [4] designed a questionnaire related to usage of various types of apps. Response from 100 students revealed that male students are more prone to addiction than female. Akhil Mathur et al. [11], conducted a study through an Android app to analyze smartphone behavior. The analysis was complemented by a survey with 55 users and semi-structured interviews with 26 users. Results revealed that (a) smartphone usage among Indian users was most prominent during late night hours (12 a.m.–4 a.m.), (b) nearly 75% of the notifications were viewed in less than one minute and (c) Indian users are extremely battery conscious.

One work closely similar to ours is by Wang et al. [6] who developed a mobile application to study student behavior. They found that the activity levels, social interaction levels, mobile phone usage, and others showed an increasing trend during normal days followed by a dip during exams. To conclude this section, we find that almost all studies conducted in India are based only on qualitative methods. Also, these studies do not talk about excessive phone use and its impact on mental health and academic performance. In contrast, we propose a framework with custom developed mobile app that uses the sensors for inferences rather than depending only on user response. Moreover, we look at academic performance included as one of the variables related with excessive smartphone use.

3 Methodology

Fifty undergraduate engineering students of Dhirubhai Ambani Institute of Information and Communication Technology were chosen for this case study. The main means of collecting data was through a mobile app developed for automatic, continuous sensing (through the embedded built-in sensors). The app was installed on the students' smart- phones and data was captured for 24 h each day for 45 days. From the sensor data, we extracted the following information:

1. Physical activity (Idle, Moving): The accelerometer senses any movement made on the device. If there is no movement, we assume the device to be idle and the person using the phone also to be idle.
2. Indoor mobility: By capturing Wi-Fi access point information, we extracted the location and the duration spent at each location.
3. Sleep duration: By monitoring the idle status of the device through the accelerometer, whether the light sensor is ON or OFF, the duration for which the screen remained locked, we extracted the sleep hours.
4. Conversation levels: We used the microphone to capture the sound data and inferred the number of conversations and duration of each conversation per day. Neither the content nor the speaker was identified.
5. Social Interaction: By recording the Bluetooth IDs we look at how many IDs are regularly being scanned by a particular phone and combine this data with the conversation levels to extract the social interaction hours.

6. App usage: Our data collection app recorded the app usage only if the application was active on the screen. Background applications were not considered.

The mobile data was supported by daily surveys on certain parameters like quality of sleep, number of hours slept, mood, and the activities undertaken during the day (at-tended classes, went for gym or for a movie, etc.). Such records captured the behavior pattern of the students, generally referred to as the EMA (ecological momentary assessment). The EMA was used to measure the parameters like stress levels, feelings of depression and this data was triangulated with the data collected from the smartphone.

4 Findings

4.1 Utility of Mobile Sensors

The first thing that we found was that the difference between what the students believe they do versus what the mobile app shows they actually did was significant. The students were asked in the daily survey, how many hours did you use the mobile phone today? The same information was examined from the mobile sensor records. The Pearson correlation coefficient between the reported usage hours and the actual usage hours turned out to be a mere −0.103. That means that qualitative surveys gave results that are not very reliable. Figure 1 shows the result.

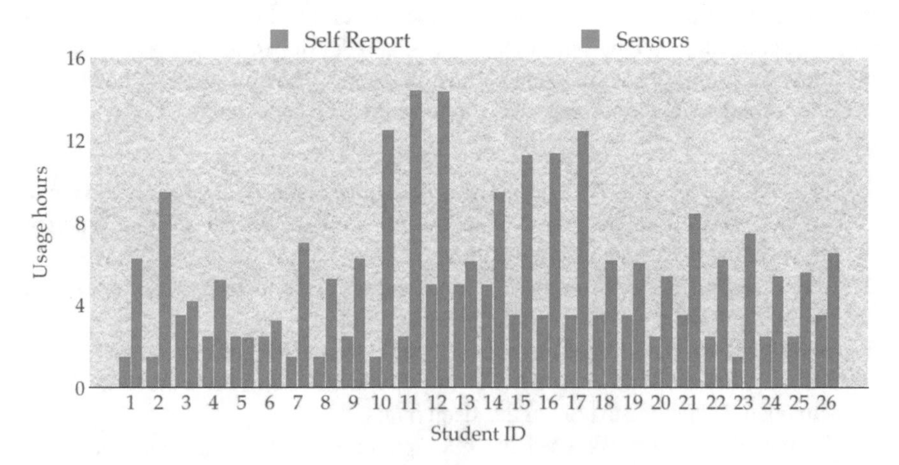

Fig. 1 Utility of sensor data: the graphs show that there is a significant difference between re-ported and measured number of hours for which the students used the phone

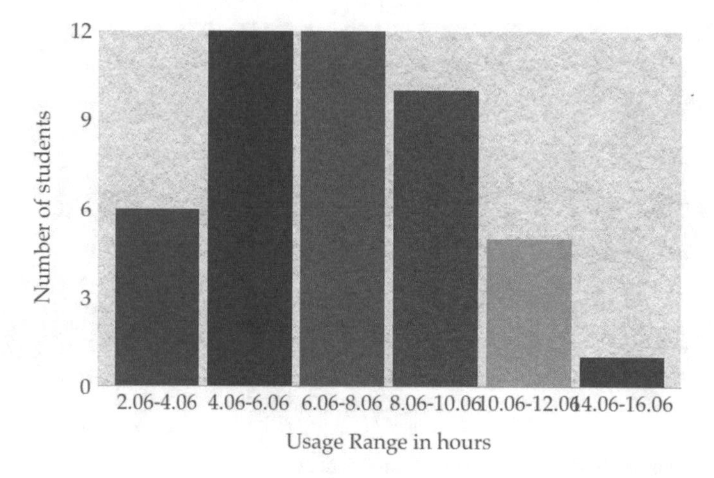

Fig. 2 Distribution of students for mobile phone usage

4.2 Mobile Usage Hours

The average mobile usage among the students was as high as 6.87 h per day. However, the curve is skewed towards the higher side showing that more students used the phone for more than the average 7 h compared to the ones who used it for lesser hours. Figure 2 shows that 12 students used the mobile for 4–6 h and 6–8 h each, 10 students used it for 8–10 h, 5 students used it for 10–12 h and 1 student used the phone for 14 h each day. Thus the phone usage is indeed heavy.

4.3 Application Popularity

Our app also recorded how much time a student spends on different apps. From this data, we extracted three most used apps by each student. We found that for 33 out of 47 students, the most used app was Whatsapp. Instagram was a close second with 27 students. However, most students were using both apps. They would spend on an average 2 hours a day on each of these apps. We found the evidence of Facebook declining in its popularity—only seven students were using it. YouTube was an app to spend a lot of time for 11 students, although we suspect this number must be more—the students must be using it more on computer than on mobile phones.

The general belief that students spend a lot of time playing games on mobile also turned out to be somewhat true. 70% students were spending about 4 h each day on social media but 53% students (25 out of 47) were found to be spending 1–5 h on an average on games. 10 students were found to be spending hours playing PubG. The average hours spent on games by addicts was 3 h, which is substantial.

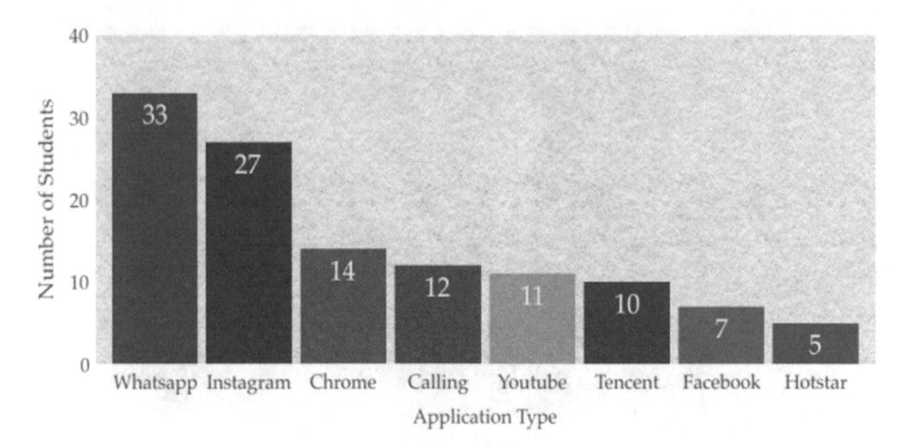

Fig. 3 Distribution of students for different applications

4.4 Socialization

Whatsapp and Messenger are modes of socialization too but here we limit ourselves to conservative meaning of socialization—talking on phone and talking personally to people. We found that the usage of mobile phone for talking was not very high. 45% of the students talked on the phone only for 15 min or less. 30% talked for 15–30 min a day. Thus an overwhelming number—75% of the students do not talk on phone for more than 30 min daily. Only 12% students talk on the phone for more than an hour. Interaction over phone seems to have declined in importance for students (Fig. 3).

To clarify our conclusion on the personal/social interaction, we calculated the number of hours of voice conversations a particular student over the course of a day. We did not capture the actual content of the conversation but only captured the amplitude levels. Figure 4 illustrates the result. The overall average interaction hours across the student body was around 4.5 h, which we consider healthy.

4.5 Physical Exercise

From the sensor data, we extracted the idle and non-idle hours spent by a student. Idle included sitting at a particular location and using his/her mobile phone. We do not consider sitting in lecture halls. Figure 5 illustrates the average hours students moved or remained idle. On an average, the idle hours were found to be 13 h/day and non-idle were 4 h/day.

The result indicated lesser hours of active movement by the students. We coupled this finding by asking them their interest in exercising and if interested, how many hours/week they wanted to exercise and the actual hours they exercised. Out of 1558 days for which we have information, only 10% of the days the students were

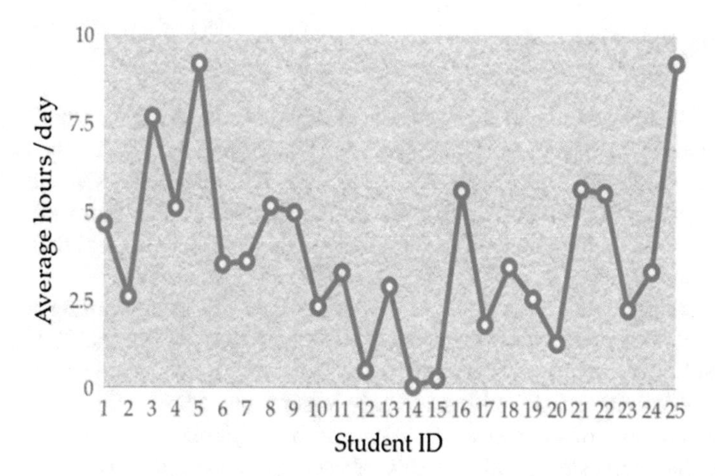

Fig. 4 Social interaction hours: average hours of vocal interaction hours/student for 45 days

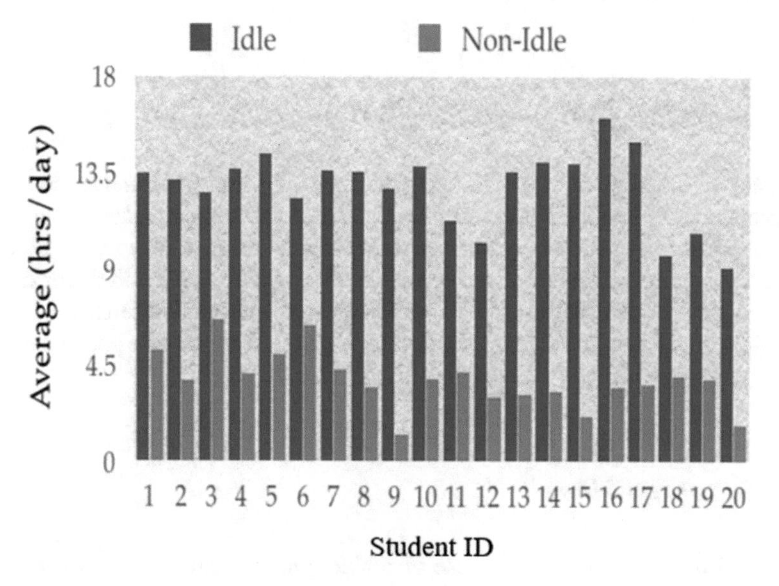

Fig. 5 Physical activity hours: average idle and non-idle hours per day

found to be exercising. If we see it student-wise, only 3 out of 47 students exercised every day. 18 out of 47 students (38%) exercised once in three days (that means not very regularly). The remaining 56% did not do much physically. Thus the general belief that the young generation is more sedentary seems to be true.

4.6 Impact on Sleep

DAIICT is a residential college so students live in hostels. Given the usually charged atmosphere in hostels, most students seem to be sleeping for 5–6 h only. As can be seen from the figure, the curve is skewed towards left; that means students sleeping for lesser than 6 h are many more compared to students sleeping for more hours (6 and above). We found that on 40% of the days, students slept for less than 6 h. This is worrisome because it means that the students are sleep starved.

Interestingly, when we asked them whether they felt sleepy during lectures or while doing other tasks during the day, the response was always "no"!! That means that their bodies do not yet feel the stress of less sleep. Students are able to carry on their daily activities easily with seemingly lesser sleep. May be this has something to do with quality of sleep also. For 83% of the sample days, the students reported that they had good sleep. Only on 16% of the days, some difficulty in sleeping was reported. This shows that if the sleep is sound for even 5 h, the person can manage to remain fresh (Figs. 6 and 7).

4.7 Impact on Academic Performance

The data on academic results is stored in the DAIICT portal and can be accessed by the professors. We looked at the results of each of the sample students and found that the correlation between mobile use and academic result is very weak at −0.18. Some heavy users of mobile were found to be having CPI of 8.5 and 9, considered as very high according to the DAIICT standards. Similarly, students making less use of mobile phones often ended up with average grades. Thus mobile phone use does not seem to negatively affect academic performance.

4.8 Impact on Mood

We found that in spite of heavy use of mobile phones, only for 4% of the total 1558 days, the students reported being somewhat or highly depressed. If we do not consider "somewhat depressed" as not serious then just for 1% of the days were the students found to be depressed. The correlation coefficient was found to be 0.28, again statistically insignificant. Thus in general, the mental state of the students was found to be normal. Their moods were not negatively affected by either mobile phone use or any other factor (Fig. 8).

We had also asked in the survey whether the mental condition was "happy". 64% of the days, the students reported being happy and 25% of the days, they were neutral. Thus for 89% of the days, the mental state of the students was in equilibrium or positive. This indicates that the student population is generally in healthy condition.

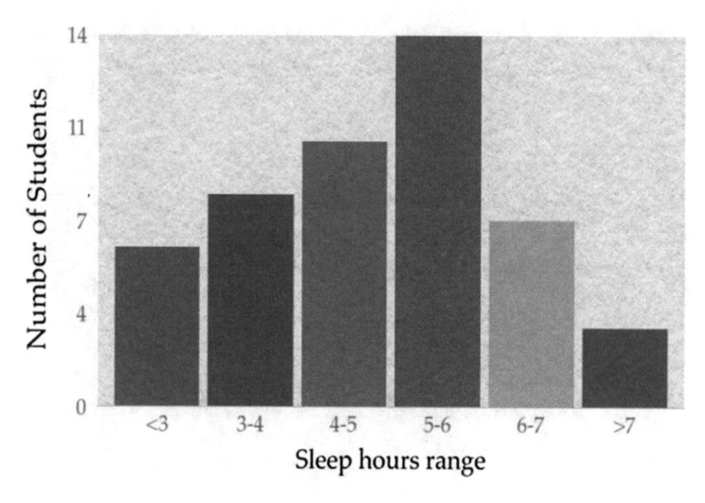

Fig. 6 Sleep hours: average sleep hours extracted from sensor data

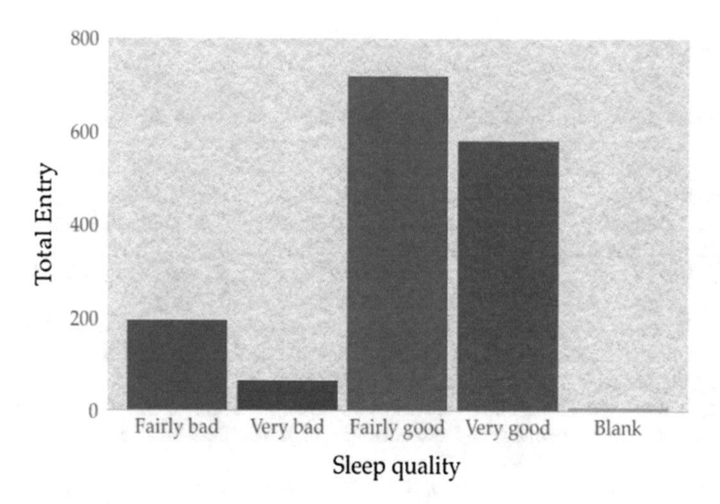

Fig. 7 Student perceived sleep quality

5 Conclusion

This study has collected accurate data round the clock on many parameters of student behavior with the help of mobile sensors. The study finds that the sample college students studied was using the mobile phone for about 7 hours on an average in a day, which is indeed a long duration. They were also found to be sleeping for 5–6 h an average in the night, far below the recommended eight hours. However we find that their academic performance, alertness or moods are not affected negatively because of these habits.

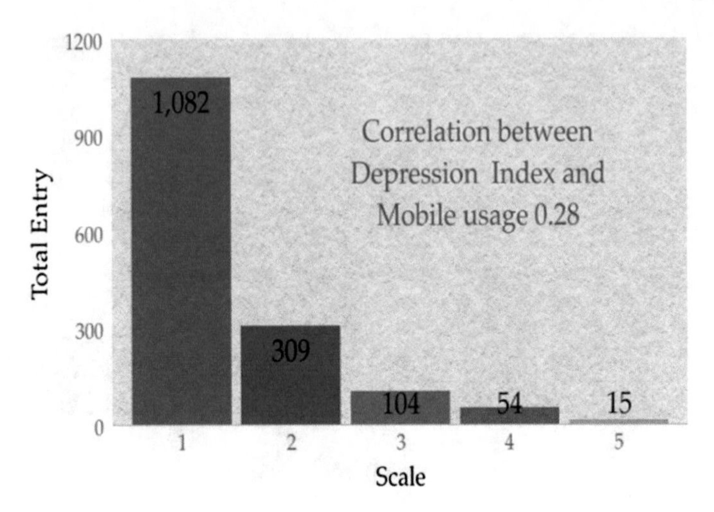

Fig. 8 Distribution of students for mobile phone usage

References

1. http://statista.com/statistics/274658/forecast-of-mobile-phone-users-in-india
2. http://tcs.com/digital-divide-closes-students-owning-smart-phones
3. https://gadgets.ndtv.com/mobiles/news/70-percent-students-use-smartphone-in-india-survey-381068/
4. Bisen, S., Yogesh, D.: An analytical study of smartphone addiction among engineering students: gender differences. Int. J. Indian Psychol. **4**(1), 70–83 (2016)
5. Lane, N.D., Miluzzo, E., Lu, H., Peebles, D., Choudhury, T., Campbell, A.T.: A survey of mobile phone sensing. IEEE Commun. Mag. **48**(9), 140–150 (2010)
6. Wang, R., Chen, F., Chen, Z., Li, T., Harari, G., Tignor, S., Zhou, X., Dror, B.-Z., Campbell, A.T.: Student life: assessing mental health, academic performance and behavioral trends of college students using smartphones. ACM UbiComp, 3–14 (2014)
7. Dixit, S., Shukla, H., Bhagwat, A. et al.: A study to evaluate mobile phone dependence among students of a medical college and associated hospital of central India. Indian J. Community Med. **35**(2), 339–341 (2010)
8. Aggarwal, Munish, Grover, Sandeep, Basu, Debasish: Mobile phone use by resident doctors: Tendency to addiction-like behavior. German J Psychiatry **15**(2), 50–55 (2012)
9. Subba, S.H., Mandelia, C., Pathak, V.M., Reddy, D., Goel, A., Tayal, A., Nair, S., Nagaraj, K.: Ringxiety and the Mobile Phone usage pattern among the students of a Medical College in South India. J. Clin. Diagn. Res. JCDR **7**(2), 205–209 (2013)
10. Davey, S., Davey, A.: Assessment of smartphone addiction in indian adolescents: a mixed method study by systematic-review and meta-analysis approach. Int. J. Prev. Med. **5**(12), 1500–1511 (2014)
11. Mathur, A., Kalanadhabhatta, L.M., Majethia, R., Kawsar, F.: Moving beyond market research: demystifying Smartphone user behavior in India. Proc. ACM Interact. Mobile, Wearable Ubiquit. Technol. **1**(3), 82 (2017)

Air Pollution Monitoring Through Arduino Uno

Meenakshi Malhotra, Inderdeep Kaur Aulakh, Navpreet Kaur and Navneet Singh Aulakh

Abstract Air pollution is a serious problem which is affecting the population worldwide. To measure the level of adversities, one should measure the concentration of pollutants available in the atmosphere. The conventional methods were expensive and were not user-friendly to visualize the measured data. The new technology gives the advantage of measuring data with the help of sensors and deploying that data on the cloud. The data can then be viewed from anywhere anytime. To measure the level of pollutant in the environment we used Arduino Uno as hardware with some sensors and visualized the measure data with the help of graphs.

Keywords IoT · Cloud · Arduino uno · Raspberry Pi · Pollution · Gases

1 Introduction

Earth is made up of three natural resources that are Land, Water, and Air. These three resources are the source for living but nowadays, technology is affecting these resources and affecting the lives too. We are focusing on smartness of everything instead of improving it. The main cause to any problem is the increasing number of population which directly or indirectly increases the demand. The increasing demand affects the resources in such a way mentioned below

---

M. Malhotra (✉) · I. K. Aulakh · N. Kaur
University Institute of Engineering and Technology, Panjab University, Chandigarh, India
e-mail: meenakshi.malhotra@outlook.com

I. K. Aulakh
e-mail: ikaulakh@yahoo.com

N. Kaur
e-mail: nvprtkr7@gmail.com

N. S. Aulakh
Central Scientific Instruments Organization, Chandigarh, India
e-mail: aulakh@csio.res.in

M. Tuba et al. (eds.), *ICT Systems and Sustainability*,
Advances in Intelligent Systems and Computing 1077,
https://doi.org/10.1007/978-981-15-0936-0_24

Land Resources: As the land requirement for living increases, increases the requirement for home and infrastructure, due to which forest coverage is being cleared which leads to degradation of environment.
Water Resources: As the population is increased in number, the production of food grains increases. Formers need more pesticides which is one of the main causes polluting the ground water also imposing more irrigation facilities, would further means construction of damns which intern destroy the eco-system of the water.
Air Resources: Increasing population increases the demand for transportation and that increasing vehicles emits the toxic gases. These gases when reaching to the upper atmosphere, react with the already available gases and causes the acid rain. Harmful gases emission from factories and tree cutting is also the causes of air pollution. Formers burn crop which also lead to the air pollution.

Pollution affects everything: human lives, Agriculture industry and so on. Human life due to air pollution is affected in such a way that one can face heart disease, lung disease, can affect the brain functionality, eye irritation, and throat infection, and can affect the child within the womb of a lady and many more. On the other hand, Agriculture department faces the same problem due to the polluted gases available in the environment. These toxic gases can harm the farming. As to prevent from these problems, farmers use pesticides to secure the crop but actually it affects the nutritious content of food. There are lots of gases available in the atmosphere which affects the living like PM 10, PM 2.5, Carbon, Monoxide, Methane, Nitrogen Dioxide, Ozone, Sulfur Dioxide, Lead, Smoke Gas, Butane, etc.

The first and focusing point should be to measure the level of toxic gases presence in the atmosphere (Indoor or Outdoor environment) so that the strict action can be taken to mitigate the problem.

Monitoring with intelligent system can increase the accuracy and lead to develop the effective technique to reduce the problem related to pollution. Nowadays IoT (Internet of Things) is broadly acceptable by many industries and even by the academician to do the projects as well as for the research. IoT makes it possible for every device to communicate to each other easily. An effective monitoring system can help the industry to take further action to reduce the affect of toxic gases if any.

This paper is organized as Sect. 2 presents the work done so far in the field of air quality monitoring systems. Section 3 reports the experimental setup with the outcomes and Sect. 4 concluded the idea behind the paper.

2 Literature Review

In [1], the authors monitored the real-time weather data by deploying sensors in different location. The data collected by sensors was send to the cloud and Google spreadsheet is used for the analysis. The authors used Raspberry Pi with IoT shield to execute the monitoring process. For the unexpected level of toxic gases, an alarm based system was considered. For the smooth communication between Raspberry Pi and Google server, API was used. In [2], the authors aim was to measure the indoor temperature at different time interval. They used Net Zero Energy building concept. To do the work, the experimental setup was made up with wireless SMD SHT11 (measure temperature and Humidity) and IoT gateway C4EBOX. A classroom with sensors at different location was used and placed to the counter the readings. One week of reading was taken and only two days out of seven were noticed with high temperature and low temperature. In [3], the authors collected the weather data making the Raspberry Pi as workstation and XBEE as the communication protocol. A user from anyplace of the world can access the data via Ethernet or from internet. The authors focused on each step from collecting the data and to sending the data to the cloud. In [4], the authors in the paper focused on real-time weather data collection and alarming system. The experimental setup was created with Raspberry Pi and IoT gateway. For the collected data to be communicated to the cloud/server, Thingspeak (open source cloud service) was used. HTTP protocol was used for the sending and receiving of the data. In [5], authors used Raspberry Pi as the measurement of the toxic gases available in the environment. Wi-Fi module was used for the communication. To display the data over the webpage, MEAN stack was used in the system. In [6], the experimental setup was created with Arduino mega, Raspberry Pi and with various sensors. To see the collected data by the Arduino mega, Arduino IDE was used to give power supply to the Arduino mega, Raspberry Pi was used with the help of Ruby code. The data was then sent to the server with the help of Ruby on Rail framework which interns implements web socket. This will create the communication channel between the server and Raspberry Pi in full duplex model. In [7], authors focused on giving the idea about Internet of Things architecture keeping in mind the key elements while designing an IoT based application or project. How an IoT-based application well communicates to the device or servers also explained in terms of focusing on cloud platform. In [8], the authors have given the overview related to IoT. The architecture, prototype and the application where IoT was successfully deployed and still gaining popularity, was explained. The authors point out each and every single element to be considered before jumping to IoT projects or application. In [9], the authors surveyed about Internet of Things. They focused on some issues related to IoT like addressing issues when starting with the IoT application as well as the networking issues while maintain the communication between IoT application and the cloud server. The security related issues were also considered for the hassle free communication between devices. In [10], the authors collected weather data through IoT-WSMP (Internet of Things enables Wireless Sensing and Monitoring Platform). The focusing parameters were temperature, Humidity, and Light. The monitored and recorded data was analyzed by LabVIEW. The whole communication was done through mobile application via LabVIEW. In [11], the authors surveyed the various IoT-based cloud platform for the communication between IoT-based application and to the server. A comparison between each and every cloud platforms was done perfectly. The paper also given the idea about the challenges related to IoT cloud based platform for further research purpose.

3 Experimental Setup and Results

The smart monitoring is done with the help of sensors where sensors collect the level of gases available within the range of those sensors. The collected data by the sensors are then transferred to the base station. The base station with the help of the IoT platform or the protocols send the data to the cloud where the data is then analyzed. The experimental setup is made up of Hardware Module, Sensor Module, and the Software Module.

3.1 Hardware Module (Arduino UNO)

The Arduino UNO is a microcontroller which is open source in property. It acts like a controller. The data collected by sensors is converted from analog to digital form with the help of Arduino Uno. Arduino IDE sends the textual data from the board to the cloud [12].

3.2 Sensor Module

- DHT11—Temperature and Humidity Sensor:
  The DHT11 is used to measure temperature and humidity. Without the need of analog pin, it measures the air (within the range of the sensor) and splits out the digital form of the signal to the pin. It measures the data in every 2 s interval [13].
- MQ2—Smoke Sensor:
  MQ-2 senor is used where the possibility of gas leakage is high like in home or in the factory. It is appropriate to measure the level of butane, propane gas, methane gas, alcohol, hydrogen, smoke, etc.
- MQ9—Carbon-Monoxide Sensor
  This is also a gas sensor but it is used to measure the level of LPG, CO, and CH_4 gases present in the air.

3.3 Software Module

ThingSpeak is an open source software based on IoT. It is used for storing and retrieving the data from the hardware using HTTP protocol over LAN or Internet. Thingspeak is suitable for the application like location tracking, sensor logging with the feature of status update [14].

It gives the benefit of analysing and visualizing the uploaded data using MATLAB from anywhere (Fig. 1).

Fig. 1 Physical view of hardware setup

To use the Thingspeak, first you have to register yourself in the Thingspeak clouds which intern creates your channel. With the help of these channels the analysis will be done.

Here, we can read temperature, humidity, smoke, and carbon-monoxide data from DHT11, MQ2, and MQ9 sensors and upload it to a ThingSpeak cloud using Arduino Uno and the .csv file of the data collected. Arduino Uno is MCU, it fetches the data from the sensors and processes it and then we create a .csv file. It can then be transferred to the IOT cloud.

Arduino is powered using USB cable and the three sensors are connected to the Arduino board. The USB cable not only gives power to the Arduino board but also transfers the data from the Arduino board to computer. Those data can be analyzed and result can be shown in the computer. Connections are shown above in Fig. 2.

The objective of the work are mentioned as

- To use sensors DHT 11, MQ2, and MQ 9 for acquisition of data regarding the temperature, humidity, smoke, and carbon monoxide values of the environment.
- To transmit the data from the sensors to the cloud by converting the data to .csv file.
- To use the data acquired for further analysis using ThingSpeak.

Testing

Test Case-1

The Arduino Uno was interfaced with only a single sensor, i.e., DHT11 sensor. The real-time values of temperature and humidity were seen on the computer screen. The code worked fine.

Status—Passed

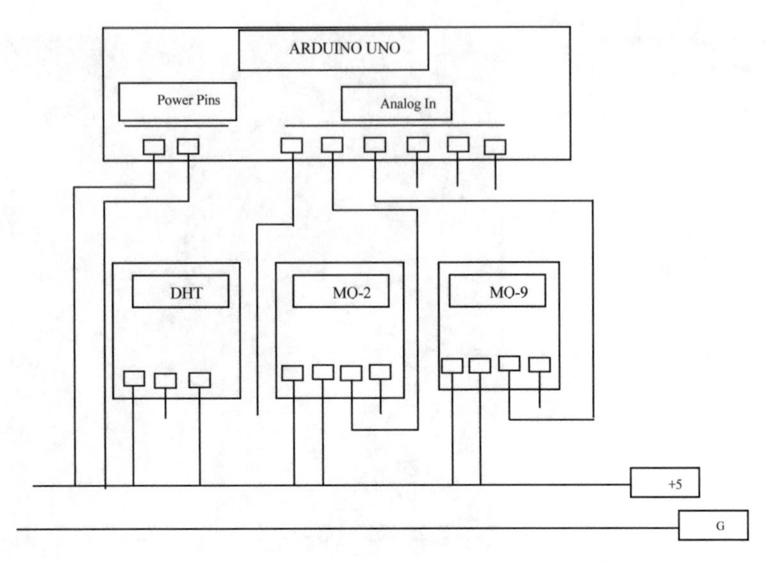

Fig. 2 Circuit diagram of connection between Arduino Uno and sensors

Test Case-2
All the sensors were interfaced with the Arduino but the data of the Smoke sensor was not displayed. The connections and the code were rechecked to remove the bug.

Status—Error in Code

Debugging—The code was corrected and all the data was displayed as needed.

Test Case-3
Now a Wi-Fi module esp8266 was interfaced with the Arduino to directly send the data to the cloud. Proper connections with the cloud could not be made.

Status—Problem in data Analysis due to improper connection to cloud

Debugging—A different approach was followed, the data recorded was converted into a .csv file which was then uploaded on the cloud for analysis.

All the tests performed helped in improving the working of our system and in making the project bug-free (Figs. 3, 4, 5 and 6).

4 Conclusion

This paper with the help of Arduino Uno monitors the gases available in the atmosphere with the help of sensors attached to the hardware and the information of the air pollutants are represented using graphs which user can access from anywhere through the cloud. The temperature and humidity sensors examine the changes occurring in the climate where on the other hand CO gas and smoke sensors provides the variations of the pollutant concentration in the environment. The monitoring platform gives the

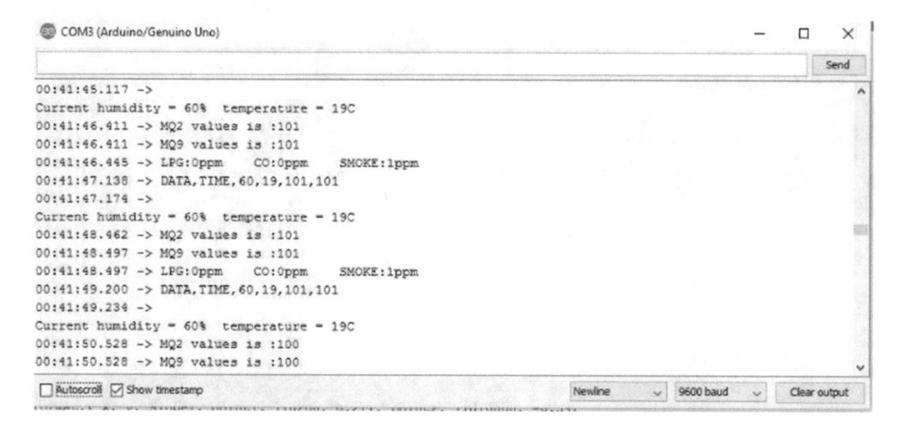

Fig. 3 Initial values of CO, smoke and LPG

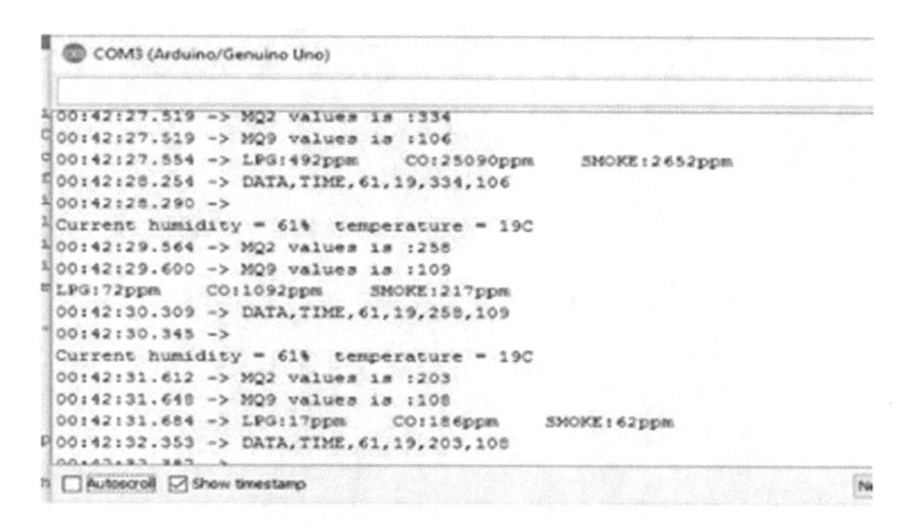

Fig. 4 Match stick burnt close to sensors

information which helps to keep the cloud value up to date. This paper focused on the Arduino Uno as the hardware platform as it is cheap to maintain and easy to use for the small projects handling. An alarm can be added to the circuit to notify the user in case of excess smoke conditions, i.e., Smoke alarm and an SMS can be sent to clients notifying them with the temperature/humidity/smoke/CO parameters, can be consider in the future work.

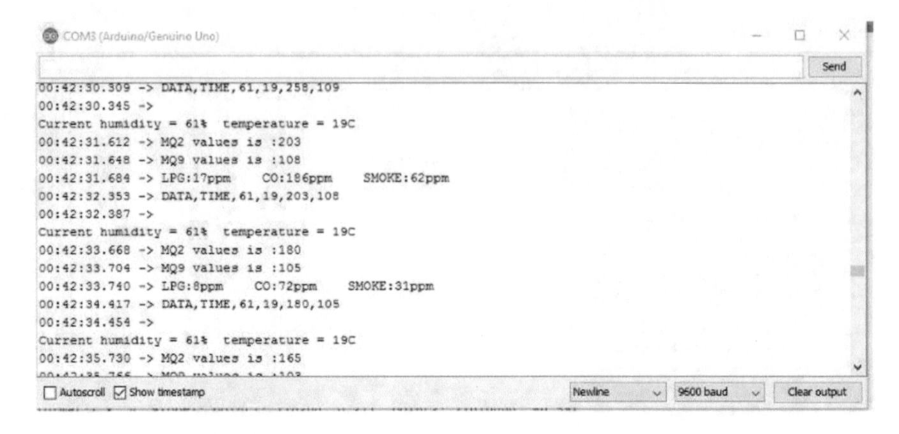

Fig. 5 When the match stick is placed at a distance

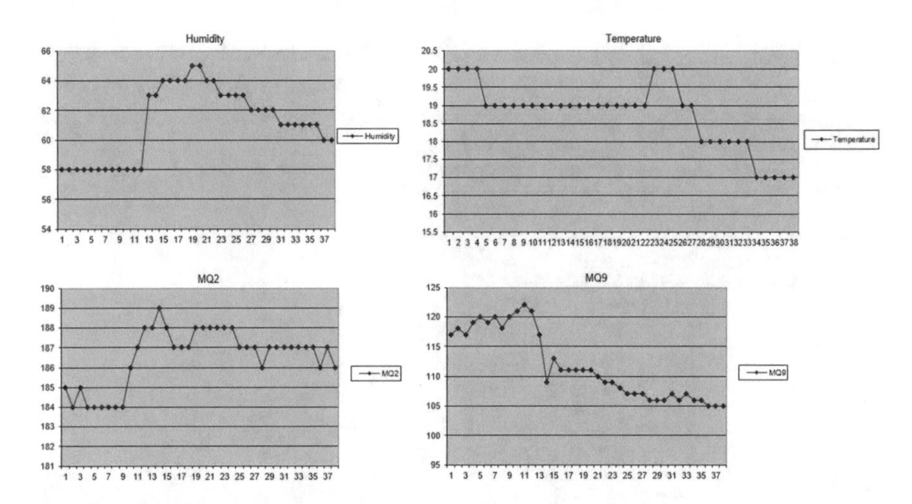

Fig. 6 Humidity graph, temperature graph, MQ 2 graph and MQ 9 graph

References

1. Jose, D.: IoT based pollution monitoring system using raspberry-pi. Int. J. Pure Appl. Math. **118**(24), (2018)
2. Irulegi, O., Serra, A., Hernández, R.: Data on records of indoor temperature and relative humidity in a University building. Data Brief **13**, 248–252 (2017)
3. Jaladi, A.R., Khithani, K., Pawar, P., Malvi, K. and Sahoo, G.: Environmental monitoring using wireless sensor networks (WSN) based on IOT. Int. Res. J. Eng. Technol. **4**(1), (2017)
4. Balasubramaniyan, C., Manivannan, D.: IoT enabled air quality monitoring system (AQMS) using raspberry pi. Indian J. Sci. Technol. **9**(39), 1–6 (2016)
5. Parmar, G., Lakhani, S., Chattopadhyay, M.K.: An IoT based low cost air pollution monitoring system. In: 2017 International Conference on Recent Innovations in Signal processing and Embedded Systems (RISE) (pp. 524–528). IEEE 2017

6. Kirthima, A.M., Raghunath, A.: Air quality monitoring system using Raspberry Pi and web socket. Int. J. Comput. Appl. **975**, 8887 (2017)
7. Gubbi, J., Buyya, R., Marusic, S., Palaniswami, M.: Internet of things (IoT): a vision, architectural elements, and future directions. Future Gener. Comput. Syst. **29**(7), 1645–1660 (2013)
8. Sethi, P., Sarangi, S.R.: Internet of things: architectures, protocols, and applications. J. Electr. Comput. Eng. (2017)
9. Atzori, L., Iera, A., Morabito, G.: The internet of things: a survey. Comput. Netw. **54**(15), 2787–2805 (2010)
10. Shah, J., Mishra, B.: Customized IoT enabled wireless sensing and monitoring platform for smart buildings. Procedia Technol **23**, 256–263 (2016)
11. Ray, P.P.: A survey of IoT cloud platforms. Future Comput. Inform. J. **1**(1–2), 35–46 (2016)
12. Arduino—Education. https://www.arduino.cc/en/Main/Education
13. Dey, N., Mukherjee, A.: Embedded Systems and Robotics with Open Source Tools. CRC Press (2018)
14. Thingspeak IoT: https://thingspeak.com/pages/learn_more

Revamping of Telecom ICT in India

Anil Kumar Sinha, Vivek Gupta, Gargi Bhakta, Archna Bhusri and Ramya Rajamanickam

Abstract There has been a huge change in the entire chain of ecosystem including system, business processes, and functions related to handling of Unsolicited Commercial Communications (UCC). Thus, there is an immense need for re-engineering the architecture which is robust, non-repudiative, and confidential. A framework which handles huge scalable data and keeps it distributed as well as is secure on the agenda. Telecom Regulatory Authority of India (TRAI) is cause-driven and constantly moving forward in curbing UCC. With the growth of the technology and innovations, ways of spamming the consumer has also increased. It has become very crucial to detect and prevent the various ways of UCC through the new regulations of the Telecom Commercial Communications Customer Preference Regulations (TCCCP), 2018. This move came as a result of the increasing commercial calls even after all precautionary and regulatory measure taken by the Authority. Every Service Provider has to amend Codes of Practice for Unsolicited Commercial Communications Detection before allowing any commercial communication through its networks. TRAI also faces various set of concerns like making consumers choose their preference of UCC messages they wish to receive, escalation/resolution of complaints within stipulated time frame and communication to consumers regarding their complain status, etc. This paper studies the current system and illustrates a system to cater to the rising demands of an ICT fueled environment and provides solution to

A. K. Sinha (✉) · V. Gupta · G. Bhakta · A. Bhusri · R. Rajamanickam
National Informatics Centre, Lodhi Road, New Delhi 110003, India
e-mail: anilksinha@nic.in

V. Gupta
e-mail: vivek.gupta@nic.in

G. Bhakta
e-mail: gbakhta@gov.in

A. Bhusri
e-mail: abhusri@gov.in

R. Rajamanickam
e-mail: ramya.rajamanickam@nic.in

M. Tuba et al. (eds.), *ICT Systems and Sustainability*,
Advances in Intelligent Systems and Computing 1077,
https://doi.org/10.1007/978-981-15-0936-0_25

handle the challenges and support the government in relieving the consumers from hassle of UCC.

Keywords Unsolicited commercial communication UCC · Blockchain · Smart contracts · TRAI · Telecom commercial communications customer preference regulations (TCCCP)

1 Introduction

The Telecom Regulatory Authority of India (TRAI) has come up with the idea of using Blockchain concepts in the new regulation—Telecom Commercial Communications Customer Preference Regulations (TCCCP) 2018 to handle and regulate Unsolicited Commercial Communications. The various Service Providers like Airtel, Idea-Vodafone, BSNL, JIO, etc., will connect to the public blockchain of the system. The wide picture includes the private Distributed Ledger Technology, Smart Contracts, Artificial Intelligence, Machine Learning, and Cloud Services.

Governments and related public sector bodies around the world are reforming their public administration organizations for delivery of more efficient and cost-effective services, as well as better information and knowledge sharing with their stakeholders. The only aspect which has not been spelt out in detail is e-Governance and related pillars of ICT enabled solutions, without which dream of smart city implementation cannot be completed. TRAI always has a keen eye towards using the new trends in technology for the benefits of the consumers.

It has obligated upon the Service Providers to develop and maintain an ecosystem which inter-alia includes detection, identification and action against senders of Commercial Communication who are not registered. Authority has also mandated every Service Provider to undertake and develop Codes of Practice (CoP) for Unsolicited Commercial Communications Detection before allowing any commercial communication through its networks. Designing a robust system which adheres to the regulations and meets up the ICT standards amidst all technical setbacks that lay ahead is a challenge. This paper studies the current system and takes in mind the individual Service Provider as a use case and how they can approach the system.

2 Findings in the Current System

Various problem areas which are related to effectiveness and efficiency of the current regulatory framework are

i. The preference chosen by the consumer goes through a strict process and takes a considerable time before becoming effective.
ii. The considerable time taken for effectiveness is exploited by the unregistered telemarketer and ends up in becoming a burden to the consumer.

iii. In spite of strict and adverse actions taken, violators find numerous and innovative ways for making UCC.
iv. Plenty of complaints were observed where a non-telemarketer was targeted, resulting in disconnection of their numbers.
v. A platform-free mobile application was in demand to meet all the requirements of the consumers and to be readily available for them in just a click away.
vi. Multiple cases of concern has been reported, where consumer is not aware of how to take back the consent given by him for transactional messages, which were later misused.
vii. Entities meant for solely sending transactional messages were coaxed to send UCC which was forbidden.
viii. Numerous entities were not traced to the violator unregistered telemarketer and was forced to halt at middle level of trace.
ix. With growth in technology, numerous new and innovative methods has emerged for making UCC which were new to consumers and were a big hassle to handle.

3 Research and Analysis Strategies

After series of analysis and research and comparing it with other similar environments, following steps details the process of how to go with the new design. This is the outcome of new regulation and has evolved with the joint initiative &concurrence of all the Service Providers. This paper discusses in detail how the Service Provider can build a system which can effectively connect to TRAI's system and is in line with the regulations.

i. Provide for various Entities internally within TSP or delegated to a partner entity who would perform the respective functions assigned as per TCCCP Regulation.
ii. Cover process for registration of Senders and their obligations, who will be assigned and allocated the headers (SMS or voice CLIs).
iii. Provide conditions for Network system functioning including SLAs and architecture
iv. Provide minimum set of information which will be put on DLT system for sharing with different Entities and among TSPs.
v. From the implementation date of the TCCCP Regulations, in case any Originating Access Provider (OAP) is not ready with the systems and processes and has not published its CoPs (prepared under TRAI's TCCCP Regulation, 2018), the Terminating Access provider (TAP) may block commercial communication to terminate on its network from such OAP, provided that the TAP shall not restrict any commercial communication from OAP for reasons owing to its own systems and processes not being ready in accordance with the TCCCP Regulations 2018.

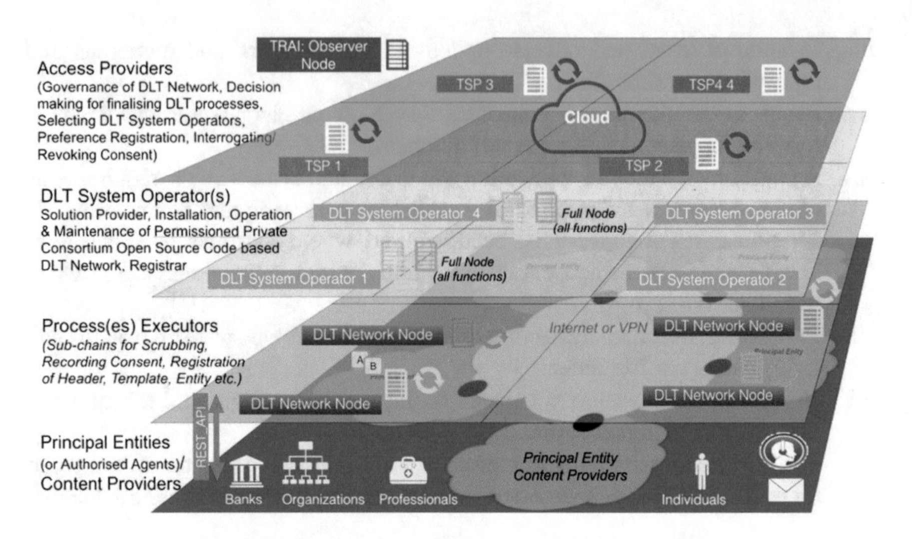

Fig. 1 Implementation of distributed ledger technology by TSPs and entities

vi. The Service Provider, at its sole discretion, may choose to perform the functions of Entity by its own or may also choose to delegate the same to a third party. If delegated to a third party, the TSP should ensure proper authorization of such Legal entity through legal agreements, which should at least contain strict conditions for safety, security, and confidentiality of the information being made available over the DLT (Distributed Ledger Technology) system (Fig. 1).

vii. A private consortium which has permission for DLT is the right plan for the UCC framework. It benefits all the stakeholders and accounts for better and quick performance.

4 Case Study: Proposed Detection Methods by Tsps

It has become very crucial to detect and prevent the various ways of Unsolicited Commercial Communications to regulate the system framework. With the growth of the technology and innovations, innovative way of spamming has also increased. Following are the major proposed detection methods for TSP which are broadly categorized as follows:

4.1 Auto-Dialer

a. Ratio of Silent Calls to total attempted calls for a registered entity exceeding 3% over a period of 24 h by an entity using Auto-Dialer for Commercial Communications Calls.
b. Ratio of silent calls to total attempted calls for a registered entity exceeding 1% over a period of 24 h by an entity using Auto-Dialer for Commercial Communications Calls.
c. If a sender intends to use Auto-Dialer to make commercial communication, it shall notify in writing to the originating access provider, before usage.
d. Notify entities to whom voice resources for telemarketing have been allocated that "No Sender registered for making commercial communication shall initiate calls with an Auto-Dialer that may result in silent or abandoned calls".
e. Undertake that use of Auto-Dialer for making commercial communication should not result in silent or abandoned calls exceeding the limit specified in this regulation. In case of failure to meet the limits, the concerned sender shall be liable for any consequences arising from the TCCCPR, 2018.
f. If any Sender/Entity is found to be using Auto-Dialer resulting into Silent or Abandoned calls, without informing the TSP in advance, TSP should issue warning notice to the said Sender/Entity for first instance. In case of subsequent repeat instance, the TSP should withdraw the telecom resources from the said Sender/Entity.
g. If there are more than 10 customer complaints w.r.t. receipt of call from Auto-Dialer from a MSISDN allocated to a registered Sender/Entity (who has not informed the TSP about usage of AutoDialer), originating TSP should send warning notice to the said Sender/Entity and seek compliance of the Points.
h. The complaint received from a same customer number but against the different Sender ID's or number of such Sender/Entity; the same shall be treated as distinct complaint.

4.2 Signature Solutions and Enhancements

a. All Access Providers need to identify Signatures, keywords and phrases and ensure that no SMS, having similar signature, from any source or number originating more than a specific count of SMS per hour is delivered through its network. This will only be applicable in case of P2P SMS except where specific exemption has been provided by the TSP on case to case basis. For the avoidance of doubt, Signature meanscontentsofcommercialcommunicationshavingsameorsimilarcharactersorstringsorvariants thereof, but does not include subscriber related information.

b. All Signatures detected from the DL-complaints/reports to be captured periodically after due diligence and applying a threshold decided by TSP from time to time.
c. All Signatures or new patterns detected or learned by one Service Provider to be shared amongst the Access Providers on regular basis through DLT system or as per the agreed mode. The list/database maintaining the signatures are continuously updated.
d. Improved Signature Solutions on an ongoing basis for detection and upgradation.

4.3 Deploying Honeypots and Using Information Collected by It

a. Honeypot is a mechanism which has virtual numbers configured in the system, with capability of receiving voice calls and SMS. Such system to record voice and SMS received by it, calling number, date, time, duration, etc., and generate reports.
b. For deploying Honeypots, TSPs need to identify five MSISDN from any series (for each LSA) which are not recycled.
c. The output of Honeypots, i.e. CDR analysis and content recorded to be used for investigation and establishment of complaints on best effort basis.
d. The details of Honeypot solution would be available once IT system and solution is finalized by TSPs. If the solution being implemented by TSPs has configurations contrary to above, then they should inform the TRAI.

4.4 Information from Other Network/IT Elements

a. Use of information like IMEI check (usage of more than a specified number of SIMs through a single IMEI in a given period), age on network, for detection of UCC or for utilizing the same during complaint resolution, to be examined and internal guidelines are formulated and evolved from time to time.
b. Usage pattern analysis of reported number.
c. Dictionary attacks: Entity-Scrubber to ensure identification of Dictionary attacks/telephone number harvesting software on a best effort basis, where the Sender/Entity/Telemarketer is originating SMS/calls for termination on MSISDNs in series or particular pattern. This can be done on a best effort basis only.
d. In case instances of UCC/promotional calls/SMSs are observed by terminating TSP, originated by unregistered Telemarketers (i.e., using normal 10-digit MSISDN) of same TSP or any other TSP, the terminating TSP may temporarily

block calls/SMSs from such unregistered telemarketers from terminating on its subscribers. Such numbers to be put on DLT system, so that it could be useful to all TSPs (particularly originating TSP) while investigating complaints and taking action against the same.

5 Smart Contracts Used by Individual TSPs

Smart contracts can be encoded into the blocks to carry out instructions about the secured data. The ownership data is kept in the Blockchain to remain trustfully traceable and irreversible. Smart contracts are conversion of an agreement, the terms and conditions into a scripting language. The program runs the rules that are validated and executed. Smart contract script is loaded into the chain to validate. After the initial steps the contract is immediately executed. A third party does not have any role in smart contract, the code handles it. The Service Providers contracts can be encoded into the blocks to carry out instructions about the secured data. These smart contracts will be activated by events that of the Blockchain can read from another source. Blockchain can provide secure proof of the ownership by storing hash value of the digital data in a time-stamped transaction. The ownership data are kept in the Blockchain to remain trustfully traceable and irreversible

- Smart Contract is written in a scripting language including all the set of logic and rules for executing the contract.
- Contract is uploaded in blockchain and executed via distributed network.
- Any number of conditions or rules can be written, and it is up to the users to choose the conditions before execution of the contract.
- The mining concept comes into picture at this stage. In a blockchain, any user with internet facility can become a client and can create contract and upload the same for any number of specific cases.

5.1 Smart Contracts Using Ethereum

There are several implementations of smart contracts. In this paper let us see about Ethereum client, geth.Genesis.jsonfile is used in initialization of the private Ethereum blockchain. This shall be used by the individual Service Providers for private blockchain. The blockchain is initialized and started. The necessary accounts by TSPs are added in the blockchain after its start up. Private Ethereum blockchain requires mining to build the block and then blocks are added. Now, the contract has to be compiled to bytecode and deployed. After which the contract interacts with the public variables and the required functions are called for execution. Thus executing the contract in blockchain (Fig. 2).

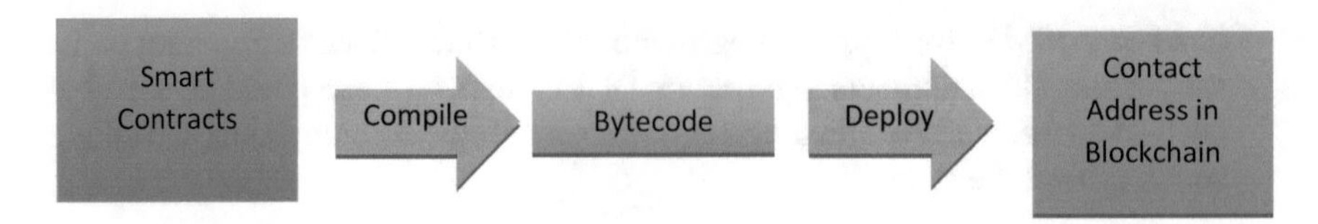

Fig. 2 Smart Contracts execution in Blockchain

In our UCC framework, the role of TSP plays a major impact. Considering the use case of any individual TSP in connecting with the TRAI's public interface via, private blockchain is the agenda.

5.2 TSP Role Based Access

A design architecture should suffix the needs of the Service Provider also by amending the regulations and considering the highly scalable system. Usage of apt technology shall give way the same. Considering the individual TSP use case model, the design is briefed in depth. The below scenario shows, how the TSP admin roles can be created, deleted, and executed by using smart contracts inside the private DLT of TSP. As in the case of using the system across the country with number of users from the same TSP, it is mandated to have TSP admin role to restrict and secure the data (Fig. 3).

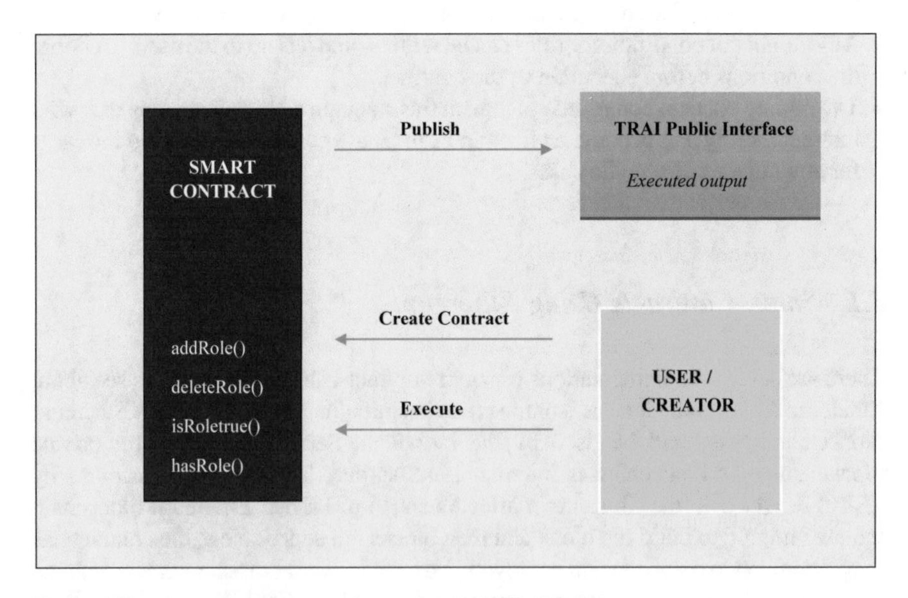

Fig. 3 TSP smart contract framework overview with TRAI interface

5.3 Algorithm for Execution of TSP Roles

Given below is a simple algorithm of how TSP roles are created, deleted and executed with a sample TSP admin role. This is further developed according to required roles of the TSP while connecting to TRAI's DLT and performing the necessary activities on a large scale (Fig. 4).

```
contract TSPbasedRole
{
 address creator;
 mapping(address => mapping(string => bool)) roles;

 function RoleBasedAcl ()
{
   creator = msg.sender;
 }

 function addRole (address service_provider, string role) hasRole('TSPadmin')
 {
   roles[entity][role] = true;
 }

 function deleteRole (address service_provider, string role) hasRole('TSPadmin')
{
   roles[entity][role] = false;
 }

 function isRoletrue (address service_provider, string role) returns (bool)
 {
   return roles[entity][role];
 }

 modifier hasRole (string role)
 {
   if (!roles[msg.sender][role] && msg.sender != creator)
   {
     throw exception;
   }
.......
.......
   }
}
```

Fig. 4 Sample algorithm of TSP role

6 Benefits of Implementation

i. Effective and timely handling and resolution of UCC complaints raised by customers.
ii. Comply with the TCCCP regulation, 2018.
iii. Cover process and modes for registration of complaints by customers. Complaints raised within 3 days of UCC event, will be treated as valid.
iv. Process for complaint handling, verification and resolution, including necessary action on UCC made by RTM (registered telemarketers) and UTM (unregistered telemarketers).
v. Provide Network system functioning conditions including Service Level Agreements and architecture.
vi. Provide minimum set of information which will be put on DLT system for sharing with different Entities and in between TSPs.

7 Conclusion

The paper discusses about how Blockchain concept along with smart contracts could be used in UCC framework. A use case of individual TSP is considered and solution is provided for the overall system to connect with TRAI's DLT. Going forward in this digital era this system takes care of nonrepudiation and confidentiality as smart contracts play a major role in it. This E-Governance design not only forms a bridge between government services and people but also enriches their lives in a way never before. It provides G2C (for Indian subscribers), G2B (for Telemarketers) and G2G (Telecom Regulatory Authority of India) solution.

References

1. Mishev, A., Karafiloski, E.: Blockchain Solutions for Big Data Challenges. IEEE Eurocon 2017, Ohrid, R. Macedonia (2017)
2. Sinha, A.K.: A New Digital Infrastructure for IndianTelecom. ICEGOV, Delhi (2019)
3. Gupta, R.: E-Transformation in Indian Telecom Sector through m-Governance. ICEGOV, Delhi (2017)
4. Lee, J.-H.: BIDaaS-Blockchain Based ID As a Service. IEEE Access in Special Section on Intelligent Systems for the Internet of Things (2017)
5. Karamitsos, I., Papadaki, M., Al Barghuthi, N.B.: Design of the blockchain smart contract: a use case for real estate. J. Inform. Secur. **9**, 177–190 (2018)
6. https://blockgeeks.com/guides/smart-contract-development/
7. https://medium.com/existek/what-is-smart-contracts-blockchain-and-its-use-cases-in-business-271a6a23cdda
8. https://www.thehindubusinessline.com/info-tech/trai-proposes-use-of-blockchain-technology-to-curb-pesky-calls-sms/article24022511.ece
9. https://trai.gov.in/sites/default/files/RegulationUcc19072018.pdf

The Mordor Shaper—The Warsaw Participatory Experiment Using Gamification

Robert Olszewski and Agnieszka Turek

Abstract The article promotes the idea of using the ICT and AR technologies, geoinformation as well as the gamification method to develop so-called a 'serious game' and support the process of creating (geo)information societies in the smart city. The aim of developing this game is to optimise the issue of spatial development and public transport through the gamification stimulated 'carpooling', on the one hand, and to conduct a participatory experiment and study the ethical issues of using the ICT in an urban society, on the other. The test field used to carry out such an experiment is the office district of Warsaw called Mordor, where over one hundred thousand people are employed. They use mobile devices and the ICT technology on a daily basis and are impressionable to social gamification.

Keywords Geoparticipation · ICT · Smart city · Serious game · Gamification · Geoinformation · Mordor · Spatial data mining

1 Introduction

Smart city is a term meaning effective integration of physical, digital and social systems in urban space, in order to ensure a sustainable, prosperous living environment for citizens (BSI PAS 180: 2014) [1]. Creating a modern, people-oriented city, so-called human smart city, however, requires a much more sophisticated use of modern technology. Smart city should be a city in which public issues are solved using information and communication technologies (ICT) with the involvement of various types of stakeholders working in partnership with the city authorities [2]. To achieve rational spatial planning in the smart city, as well as the development of participatory democracy and an open geoinformation society, the ability to capitalise on the potential of available spatial data is just as crucial as the ability to stimulate the

R. Olszewski (✉) · A. Turek
Faculty of Geodesy and Cartography, Warsaw University of Technology, Warsaw, Poland
e-mail: robert.olszewski@pw.edu.pl

A. Turek
e-mail: agnieszka.turek@pw.edu.pl

M. Tuba et al. (eds.), *ICT Systems and Sustainability*,
Advances in Intelligent Systems and Computing 1077,
https://doi.org/10.1007/978-981-15-0936-0_26

local community by means of innovative geoinformation technologies. Capitalising on spatially localised data retrieved from social media, geosurveys or gamification using mobile devices and spatial information, makes it possible to accumulate and transform a huge volume of geographic data.

For the modern information society, shaped in an age of civilisational transformation, games and gamification represents not only virtual entertainment platforms, but more often also serves as a tool for the participatory shaping of reality. Civilisational transformation, as defined by Pink [3], is associated with moving from the Information Age to the Conceptual Age. During the Agricultural Age, progress was a result of farming innovations; the Industrial Age focused on perfecting tools and machinery; in the Information Age, the main emphasis was on gathering data, while in the Conceptual Age, the primary concern is knowledge extraction. One of the most effective—and also most morally debatable—ways of gathering information from users in the smart city era is the use of ICT, gamification techniques [4–9] and so-called serious games; these include persuasive games, which are not only aimed at gathering information from players, but also at simultaneously changing their attitudes [10–15].

Efficient analysis of participatory planning in a smart city usually requires not only the use of reference data (e.g. official topographic databases) and open source data (e.g. Open Street Map), but also the analysis of geolocalised social media data (e.g. GeoTwitts) or even the collection of social data by means of dedicated tools, such as surveys, games and web applications for mobile devices using GPS navigation and augmented reality (AR) technology. Using this approach allows (through the use of data mining techniques) to convert 'raw' data into useful information and spatial knowledge used in the process of geoparticipation and creation of human smart city (see Fig. 1).

The use of the ICT as well as digital geoinformation plays a significant role in the process of shaping the smart city. The expansion of technology, stimulating the development of an information society also fosters, indirectly, the development of a civil society, which leads to the democratisation of decision-making processes. The feedback between the development of democracy (civic society) and the technological revolution (information society) is very visible in the area of geoinformation. According to the EU Directorate-General for Information Society, over 50% of public information in the EU that has economic value is geoinformation, whereas the Federal Geographic Data Committee of the USA estimates that about 80% of public data contains a spatial component. Therefore, we witness (and are active participants!) shaping a geoinformation society, which widely uses geoinformation obtained by means of commonly available geoinformation facilities.

The motivation for the authors is the development of an innovative methodology to activate the inhabitants of the smart city in the process of social geoparticulation using gamification. An important contribution is the achievement of the synergy effect between the methodology used in the applied social sciences and the possibilities offered by the use of geoinformation technologies and social gamification.

The article presents a prototype of a decision-making 'serious game' for the selected business district of Warsaw, Poland. The first section contains a description

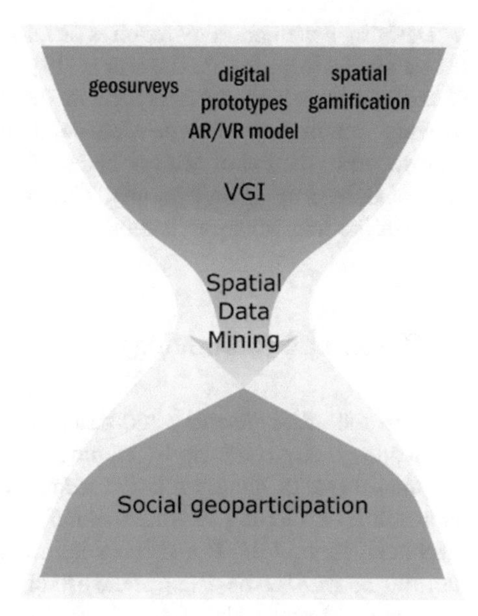

Fig. 1 Geoparticipation hourglass in human smart city

of the area of study, including the specificity of the area. Next, the main assumptions of developing the methodology are presented. The final part includes the description of the conducted Warsaw Participatory Experiment.

2 Research Area

The research area is restricted to approximately 6 km^2 in Służewiec, part of the Mokotów district of Warsaw, in the south-western part of the city. With Domaniewska street at its centre, this is now the biggest office district of Warsaw. According to various estimations, 83,000–100,000 people commute here daily to work in almost 100 buildings that cover over 1 million square metres of floor space. It is the largest concentration of corporations in Warsaw, comprising global and national companies from various industries, as well as trade and services. When most employees choose to commute by car, a true transport apocalypse takes place. In terms of transport, this area is one of the most poorly planned places in the city. During rush hours, people are spilling out of tram stops and it takes an hour to travel the busiest 100 m of the route. All of this is why the residents of Warsaw have nicknamed the area surrounding Domaniewska 'Mordor', a reference to the place where evil forces accumulated in J.R.R. Tolkien's Lord of the Rings trilogy.

Additionally, in the preliminary stage of research, the authors identified the main problems occurring in the area by processing answers from the dedicated questionnaire among responders, who are professionally involved with this area, using spatial data mining methodology [16].

Devising a methodology requires developing a prototype of a 'serious game' for the Varsovian 'Mordor'. In the article [17], there is a description of the first elements of the process which constitute the methodical background to the optimisation of workers' commuting and a possible solution to traffic congestion in the smart city. Warsaw office district of Mordor and social geoparticipation in urban spatial development is held up as an example. The game and its pilot implementation are the topics of the following publication.

3 Research Methodology

To solve the urban planning and transport problems of the smart city, it is crucial to encourage city residents by tapping into the potentials of gamification and the common usage of geoinformation technologies, ICT and AR. At the core of this approach is developing an attractive mobile serious game that utilises digital maps, GPS navigation, AR/VR and other information technologies to make the solution more attractive. The next step is the use of data mining and KDD techniques for the exploration of spatial data (see Fig. 2) [18–23].

The authors developed a specially designed decision-making game (*Mordor Shaper*) for mobile devices and personal computers that enables users to highlight 'problematic' places and structures in need of revitalisation, development, alteration or maintenance on a background map. This game also enables a simulation of the process of carpooling—sharing your car with other passengers or travelling in someone

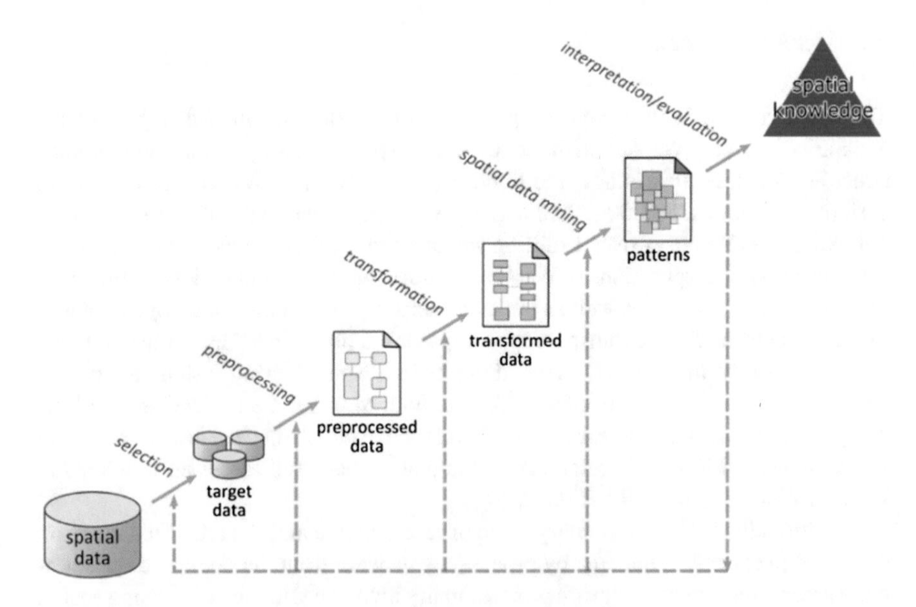

Fig. 2 The process of transforming spatial data into spatial knowledge

else's vehicle. What is particular about *Mordor Shaper* is the moral aspect present in the game mechanics and narration. The established set of rules allows for almost boundless influence on urban space, assessing its qualities and making both 'rational' and 'controversial' decisions, resulting in changes to the cityscape.

Mordor Shaper may be an example of serious games which, according to the authors, can also be used to raise awareness of the role of social responsibility and the consequences of undertaking amoral activities in the process of shaping and using space. It is an increasingly popular approach among researchers and game developers, as evidenced by projects such as the 'Games for Change' initiative (www.gamesforchange.org), aimed at supporting social impact games. An important element of creating such games is being aware of the moral implications inherent in the gamification system, which rewards certain behaviours and attitudes. This feedback mechanism between the game and the player, based on reward and punishment (positive or negative stimuli from the game system, as defined by behavioural psychology), constitutes the basis of players' learning. The rules of assessment appear objective to the user, even though the game developer established them arbitrarily. Hence, the developer has the ethical responsibility of being aware what patterns of behaviour and assessment the application shapes, and whether it might unintentionally and adversely manipulate the player.

Generally, the developed game *Mordor Shaper* allows—through serious games analytic [24]—for the conversion of individualised unit data to spatial knowledge relevant for the development of the human smart city [25].

4 Warsaw Participatory Experiment

By using gamification techniques to assess and shape the vision of urban development, the Warsaw Participatory Experiment (WPE) consciously refers to the famed Stanford Prison Experiment, conducted by Philip Zimbardo in 1971 [26]. The prototype of the *Mordor Shaper* game was developed as an application dedicated to mobile devices with the Android system. During the implementation of the prototype, the GPS technology was also used, which allows for the game to be carried out directly outdoors and through the Augmented Reality enable the participants to visualise the prizes, badges and other symbols. At the initial stage of the game, players are asked about their attitude to the idea of carpooling as a social cooperation solution to transport problems and urban congestion. At the second stage of the game, the players are allocated at random to one of the groups: orcs, elves or hobbits. Like in the classic SPE experiment, the players of the *Mordor Shaper* game are assigned to their roles at random (see Fig. 3). The aim of the 'elves' is to create common open spaces, for example, parks, museums, etc. For the 'orcs', the most important thing is to 'ban' common open spaces, claim new territory and capture other 'race' representatives. The aim of 'hobbits' is to mark places, where restaurants, fast food ones, bars, etc. can be set up. Despite specific aims, every game player can fully enjoy freedom

Fig. 3 *Mordor Shaper* menu and 'orcs' activity (smartphone version)

to build new alliances (even with members of other races) and fashion the Warsaw district of Mordor into a dweller-friendly smart city.

The initial study was carried out during the 'City of Dreams' geoinformation hackathon organised in 2017 by the authors of this article. Extensive research was conducted in 2018 and 2019, involving employees of Warsaw Mordor and students of the Warsaw University of Technology within the Warsaw Participatory Experiment. 337 respondents working in Mordor took part in the survey.

126 people participated in various variants of the game dedicated to mobile devices. The findings prove that employees of the city district suffer from a shortage of common open spaces. The actual experiment was carried out with the help of land management students, the Faculty of Geodesy and Cartography at the Warsaw University of Technology, under the supervision of researchers from the Warsaw University of Technology and the Institute of Applied Social Sciences of the University of Warsaw. While the set of rules and the game mechanics remained identical, the narration was conducted differently. In one instance, the introduction was neutral and limited to explaining the rules and technicalities of the game; in the second case, there was a focus on drawing attention to the moral aspect of the decisions made by players; in the third case, the narrator emphasised the role of optimising investment gains and the necessity for modernising the city. The results of the experiment indicate the importance of ethical issues in the process of social participation, as well

as the use of gamification tools in social education, spatial planning and shaping the vision of urban development. *Mordor Shaper* is an example of a 'serious game', whose basic aim is not only to entertain. 'Serious games' serve to gain, develop and consolidate particular skills as well as to solve problems. According to the authors of this article, such games might be used to raise awareness of social responsibility and consequences of taking moral and immoral actions in the process of transport optimisation as well as developing and using spaces. The developed set of rules and game mechanics enable us to have an unlimited influence on other participants of the carpooling process and on shaping the urban space. It provides the opportunity to make unrestrained assessment of the process and to take both 'rational' and 'controversial' decisions which change the city scenery.

Due to the peculiarity of the city district—dubbed Warsaw Mordor—the plot and the rules of the serious game refer to some elements of Tolkien's prose. According to the survey, almost all of the one hundred thousand employees of Warsaw Mordor call themselves 'orcs' and consistently use terms from the fantasy world in their working lives. This has allowed for the development of a serious game, in which the ICT and AR are used in the context of the Warsaw office district set in the Mordor reality. The game resources are symbolic goods (virtual badges, points, ranking, etc.). The participants of the gamification are not only particular employees, but also companies (for the sake of the game called 'clans'). The institution whose employees will score the highest number of points in carpooling or presenting ways of a spatial development in Mordor receives the title of 'the Lord of Mordor', which, thanks to the AR technology, is visualised through relevant symbols, flags and trophies on the buildings of the particular companies. Despite the virtual character of both the prizes and the way of their visualisation that requires a mobile application and the AR technology, over 81 percent of the respondents in the survey expressed their interest in the game. Therefore, the use of this attitude allowed for a cost free optimisation of the public transport and reduction of congestion. The majority of respondents would be willing to play an active part in carpooling (as a driver, car owner or passenger) solely because of the exciting gamification set in the fantasy world using state-of-the-art ICT solutions.

5 Discussion and Summary

In the game developed by the authors, the moral aspect of the gamification process is conveyed in the narration of the game leader as well as in the scoring system, which classifies some behaviour as desirable and others as 'improper'. The user seems to find the assessment rules objective, although they are freely set up by the designer.

Depending on the narration of the game leader and the type of characters that the game player picked at random, the final results differed significantly. The members of the orc clan tended virtually to create closed spaces in the game environment. Participants randomly assigned to a group of elves easily played the role of builders of a modern art gallery, city parks, a botanic garden or got involved in drilling a new metro line to improve the public transport for all the 'clans'. What is interesting is that both the field tests that used mobile applications, were fostered by GPS and conducted in an urban space, as well as small tests that used digital Google Maps and Open Street Map brought similar results. The gamers quite easily yielded to stimulation and were keen to take on the roles of elves and orcs. In the first case they aimed at claiming as big area of Mordor as possible, creating enclosures or virtually hindering the other participants of the serious game from functioning in the urban space. This way of behaviour is equally typical of those who, at the beginning of the game, were against carpooling and of those who supported the idea. In the case of the elves, the participants also easily (even those socially uninvolved on a daily basis) aimed at 'opening' the city and developing it in the spirit of a human smart city.

Due to the interdisciplinary aspect of the conducted research, to assess its results it is essential to apply not only the tools of the spatial data mining and the ICT, but also a sociological analysis of the shaping processes in a geoinformation society of the smart city, of both 'open' and 'withdrawn' type. The conducted research showed that the use of the ICT and a serious game with a well-chosen narration can lead to both shaping participatory democracy in the smart city as well as a technocratic city of high technology and social relationships fashioned in an authoritarian way.

The authors of the Warsaw Participatory Experiment and *Mordor Shaper* allow users to freely shape the surrounding space. Changes introduced in the virtual game space are visualised using high-quality cartographic charts, thematic maps and point tables. Other players, as well as the game narrator evaluate (support, ignore or challenge) the proposed actions, making users aware of the social consequences of fulfilling their ambitions.

The creators of the WPE believe that using well-defined rules and gamification mechanics may not only enable large-scale public consultations but, perhaps more importantly, may also contribute to an awareness of the social consequences of the actions involved in this process. While the authors of the WPE wish to research the moral aspects that stimulate gamification processes, the emphasis is on modifying *Mordor Shaper* on the basis of the results. The game, with its obvious educational value, strengthens the participatory activity of its users with regard to creating a vision of urban development and supporting an open (geo)information society. Social participation for sustainable development—as a free exchange of opinions on shaping the surrounding space—may be an important element of public discourse in the information society and can, for example, be conducted using properly designed gamification tools.

Using modern technologies in participatory processes requires an appropriate method to be selected and the entire process to be planned, including the final evaluation. The research conducted by the authors proves the effectiveness of applying the analysis of spatially localised data from VGI sources, geosurveys and gamification. A holistic approach to the problem of processing and harmonising spatial data retrieved from many sources collectively, makes it possible to achieve synergy between them and to find interesting spatial relationships.

References

1. BSI Group: PAS 181 Smart city framework: https://www.bsigroup.com/en-GB/smart-cities/Smart-Cities-Standards-and-Publication/PAS-181-smart-cities-framework. Last accessed 28 April 2019
2. Manville, C.: Mapping Smart Cities in the EU (2014)
3. Pink, D.H.: A Whole New Mind: Why Right-brainers Will Rule the Future. Riverhead Books (2005)
4. Kapp, K.: The Gamification of Learning and Instruction. Pfeiffer/Wiley (2012)
5. Kim, T.W.: gamification ethics: exploitation and manipulation. In: Proceedings of the ACM SIGCHI Gamifying Research Workshop (2015)
6. Olszewski, R., Wieszaczewska, A.: The application of modern geoinformation technologies in social geoparticipation. In: Gotlib, D., Olszewski, R. (eds.) Smart City. Spatial Information in Smart Cities Management. PWN, Warszawa (2016)
7. Salen, K., Zimmermann, G.: Rules of Play. MIT Press (2006)
8. Werbach, K.: (Re)Defining Gamification: A Process Approach. Persuasive Technology. Lecture Notes in Computer Science, vol. 8462, pp. 266–272 (2014)
9. Lattemann S., Robra-Bissantz C., Zarnekow S., Brockmann R., Stieglitz T. (eds.): Gamification. Using Game Elements in Serious Contexts. Springer International Publishing (2017)
10. Abt, C.: Serious Games. University Press of America (1970)
11. Adams, E., Dormans J.: Game Mechanics. Advanced Game Design. New Readers/Pearson Education (2012)
12. Aldrich, C.: The Complete Guide to Simulations and Serious Games. Pfeiffer/Wiley (2009)
13. Bartle, R.: Designing Virtual Worlds. New Riders (2003)
14. Bogost, I.: Persuasive Games. MIT Press (2007)
15. Kapp, K.M: The Gamification of Learning and Instruction: Game-Based Methods and Strategies for Training and Education, 1st edn. Pfeiffer, San Francisco, CA, USA, p. 336. (2012)
16. Olszewski R., Turek A.: Application of the spatial data mining methodology and gamification for the optimisation of solving the transport issues of the "Varsovian Mordor". In: Tan, Y., Shi, Y. (eds.) Proceedings of Data Mining and Big Data. First International Conference, DMBD 2016 Lecture Notes in Computer Science, vol. 9714, pp. 103–114. Springer International Publishing, https://doi.org/10.1007/978-3-319-40973-3_10 (2016)
17. Olszewski, R., Pałka, P., Turek, A.: Solving smart city transport problems by designing car-pooling gamification schemes with multi-agent systems: the case of the so-called "Mordor of Warsaw". Sensors **18**, 1–25 (2018). https://doi.org/10.3390/s18010141
18. Berry, M.J.A.; Linoff, G.S.: Mastering data mining. In: The Art and Science of Customer Relationship Management. Wiley, New York, NY, USA (2000)
19. Fayyad, U.M.; Piatetsky-Shapiro, G.; Smyth, P.: From data mining to knowledge discovery in databases. In: AI Magazine; American Association for Artificial Intelligence: Menlo Park, CA, USA, vol. 17, pp. 37–54 (1996)
20. Miller, H.J., Han, J.: Geographic Data Mining and Knowledge Discovery. Taylor & Francis, London, UK (2001)
21. Lu, W.; Han, J.; Ooi, B.C.: Discovery of general knowledge in large spatial databases. In: Proceedings of the Far East Workshop on GIS, Singapore, pp. 275–289 (1993)
22. McGraw, K.L., Harbison-Briggs, K.: Knowledge Acquisition: Principles and Guidelines. Prentice Hall, Englewood Cliffs, NJ, USA (1989)
23. Han, J., Kamber, M., Pei, J.: Data Mining: Concepts and Techniques. Elsevier Science & Technology, Saint Louis (2011)
24. Loh, C.S., Yanyan, S., Ifenthaler, D.: Serious Games Analytics. Methodologies for Performance Measurement, Assessment, and Improvement. Springer (2015)

25. Olszewski, R., Turek, A.: Using fuzzy geoparticipation methods to optimize the spatial development process in a smart city. In: Proceedings: 4th IEEE International Conference on Collaboration and Internet Computing. CIC 2018, pp. 430–437. https://doi.org/10.1109/cic.2018.00065 (2018)
26. Zimbardo, P.G., Maslach, C., Haney, C.: Reflections on the stanford prison experiment: genesis, transformations, consequences. In: T. Blass (eds.) Obedience to authority: Current Perspectives on the Milgram Paradigm pp. 193–237. Lawrence Erlbaum Associates, Mahwah, N.J. (2000)

A Hybrid Deployment Approach for Grid-Based Wireless Sensor Network

Sukhkirandeep Kaur and Roohie Naaz Mir

Abstract In this paper, we propose a load balancing clustering approach for grid-based deterministic deployment in Wireless Sensor Networks (WSN). A hybrid deployment is considered where deterministic deployment is performed at the edges of the grid and random deployment is performed in the whole network. Heterogeneity in terms of energy is provided to the nodes at the edges. These nodes will be more powerful than normal nodes and will act as Cluster Heads (CHs). Uneven clustering is performed in the network. Using uneven clustering and heterogeneity, the network lifetime of the hybrid deployment in grid-based network can be improved when large number of nodes are deployed. The network operates for 800 s more in case of 200 nodes.

Keywords Wireless sensor network · Uneven clustering · Load balancing · Heterogeneity

1 Introduction

As a large number of sensor nodes are deployed in WSN applications, scalability is also an important parameter that needs to be considered while designing protocols for WSN. Protocols developed for WSN should be energy efficient, more scalable, less prone to failures, etc. WSN nodes can be deployed in a deterministic manner or a random manner. Here, we propose a hybrid deployment for grid topology. Both deterministic and random deployments are performed for the nodes and load balancing is performed by maintaining optimal cluster sizes. A novel clustering approach is implemented considering the node density of grids while forming clusters. Grids that

S. Kaur (✉)
Lovely professional University, Jalandhar, India
e-mail: sukhkirandeep.23328@lpu.co.in

R. N. Mir
National Institute of Technology, Srinagar, India
e-mail: naaz310@nitsri.net

M. Tuba et al. (eds.), *ICT Systems and Sustainability*,
Advances in Intelligent Systems and Computing 1077,
https://doi.org/10.1007/978-981-15-0936-0_27

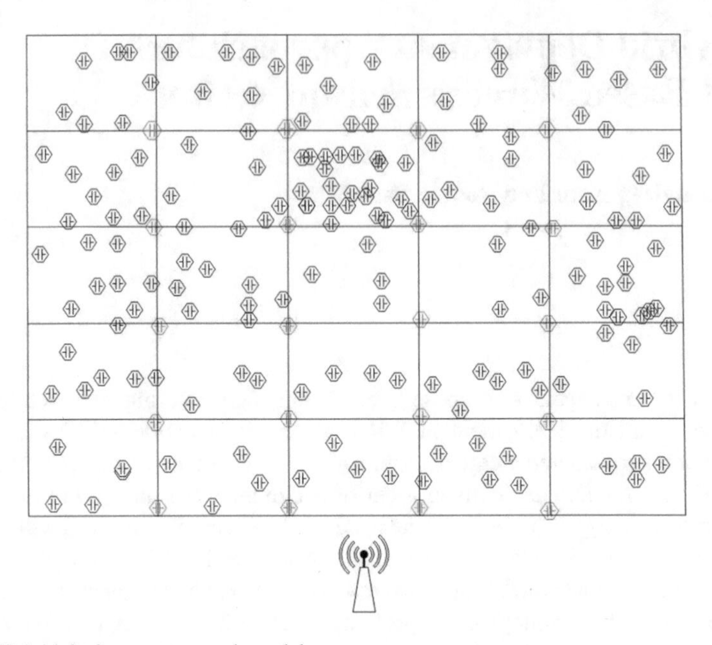

Fig. 1 Hybrid deployment network model

have a low node density combines with other grids to form clusters. Merging of grids is decided on the basis of Inter threshold Gap (ITG). CHs are selected from the nodes that are deployed in a deterministic manner and for data transmission, the next hop is chosen based on the distance to BS and residual energy. The proposed approach can find its applicability in various WSN scenarios. The best scenario we can consider for such deployment is farmland applications in precision agriculture where fields are partitioned and on partitioning points, we deploy heterogeneous nodes whereas, in fields, random deployment is followed. In Fig. 1, nodes at the edges are represented by green nodes and are more energy-efficient than normal nodes that are distributed randomly in the network.

A lot of research work has been done on clustering in WSN. Survey on clustering has been discussed in [1–3]. Clustering Algorithms are further divided into centralized [4–6], distributed [7, 8], and hybrid algorithms [9–11]. In clustering, heterogeneity can be provided to some nodes to increase the performance of the network. Heterogeneity can be described in terms of energy, link, and computational resources. Most of the algorithms describe heterogeneity in terms of energy where some nodes are more energy- efficient and they act as CHs. Some of the techniques where heterogeneity is discussed in terms of energy are [12–15].

1.1 Proposed Network Model

The performance of a network can be improved by deploying a minimum number of heterogeneous nodes in the network. Square topology is considered and grids of equal size dimension are formed. For example, for 500*500 m topology, 25 grids of 100 m are formed. BS is placed at the lower central position. The main motive here is to balance the load of the network by distributing the load of the highly-dense area. In deterministic deployment, there is a minimum overhead of forming and maintaining clusters but when random deployment is involved, selecting large number of CHs will impose more overhead. Here we propose a load balancing approach for a hybrid network that includes the benefits of both the uneven clustering and heterogeneity.

1. **Classifying the grids**

After deployment, a number of nodes in each grid are calculated and the threshold is defined. If a number of nodes are less than the threshold called lower threshold (LT), the grid is classified as a sparse grid if the nodes in a grid are more than the upper threshold (UT), the grid is classified as a dense grid.

2. **Merging and splitting zones**

Merging and splitting of the grids in a network are done based on the following classification. If two adjoining grids in a row are sparse, they are merged until the number reaches the upper threshold. If both adjoining grids are in a row are dense, then the operation is performed along the column. If one grid is sparse and the other is dense, split the dense grid and combine with sparse until it reaches the upper threshold.

Figure 2 depicts the splitting and merging of zones where similar color represents the merged zones. Zones that are not merged send data to a CH that is located at lower-left corner and merged zones send data to the CH at the center while the CH of 2nd merged zone is put to sleep mode. Dark green color represents the active CHs in the current scenario and the other shows the CHs that are put to sleep mode.

3. **Forming clusters**

Once grids are formed they send a request message to the green node at the edge of a grid. Only the green node acts as a CH and once that node receives the request message, it forms TDMA schedule of the nodes. Nodes send data to their respective CH in their time slot. If a node at the edge of a grid does not receive any request message, it enters into sleep mode.

4. **Data transmission**

Data is collected by CHs from their cluster members and they send collected data to BS. Data transmission takes place in a single- hop communication within a cluster as the clusters are formed from nearby nodes and the CHs send data to BS via multi-hop communication through other CHs.

The algorithm describing various steps of proposed approach is given as follows:

Algorithm 1 Load Balancing - Hybrid Deployment

Step 1: Input
Area: Square area topology(A = M $*$ N) where M = N
Sensor nodes: N
Base Station: BS
GM: Grid Members
LT: Lower Threshold
UT:Upper Threshold.

Step 2: Initialization
Divide the network into equal sized grids
Distribute green nodes on grid points,
Distribute normal randomly in the whole network.

Step 3: Categorize grids
Compare the grids in horizontal fashion, i.e., row-by-row
if (Number of GM $\leq$ LT) **then** set it as sparse grid
end if
if (Number of GM $\geq$ UT) **then** set it as dense grid
end if

Step 4: Combine or split grids
Compare adjoining grids row-by-row.
if (CurrentGrid = Sparse && NeighborGrid = Sparse) **then**
Merge CurrentGrid & NeighborGrid until
GM $\leq$ UT
end if
if (CurrentGrid = Sparse && NeighborGrid = Dense) **then**
Merge with GM of NeighborGrid until
GM $\leq$ UT
end if
if (CurrentGrid = Dense && NeighborGrid = Dense) **then**
Move along the column and compare the values.
end if

Algorithm 2 Load Balancing - Hybrid Deployment

Step 5: Data Transmission phase
if (node $\in CH_{set}$) **then**
Transmit data to BS via multi-hop communication.
else
  if (node $\in MZ_{set}$) **then**
Transmit data to CH in one hop communication.
  end if
end if

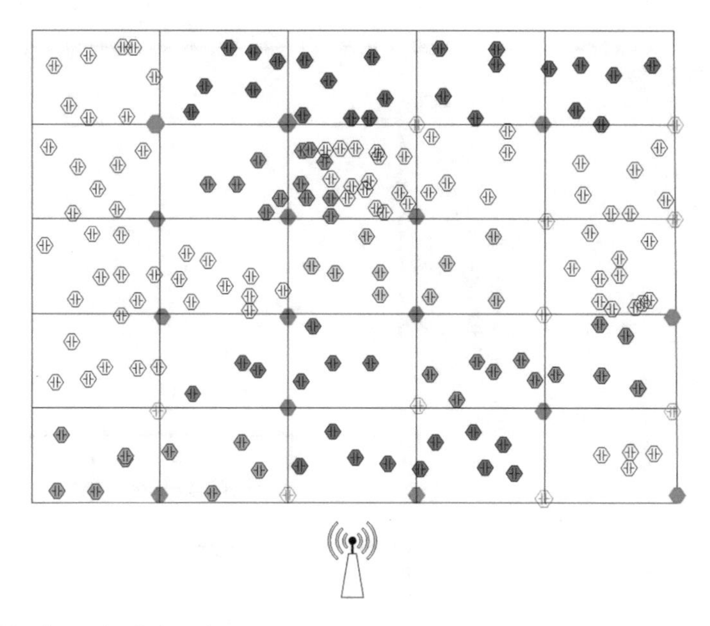

Fig. 2 Merging and splitting of zones

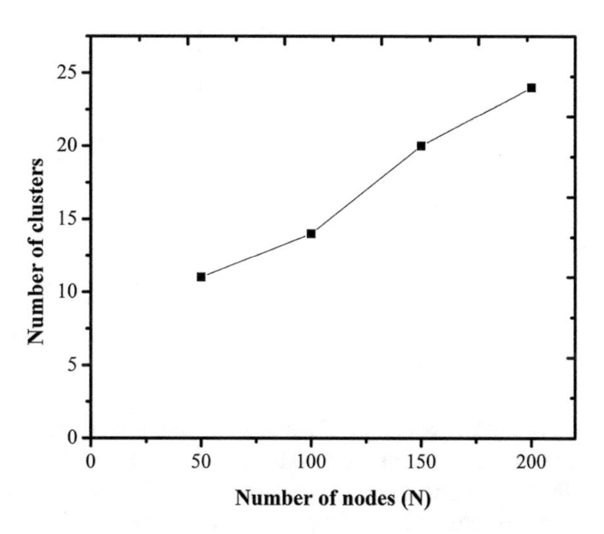

Fig. 3 Number of nodes versus clusters formed

2 Results

In Fig. 3, number of clusters formed for different number of nodes are evaluated. In this approach using heterogeneity and load balancing, the number of clusters is reduced. On average, with 15% of CHs, required connectivity is achieved.

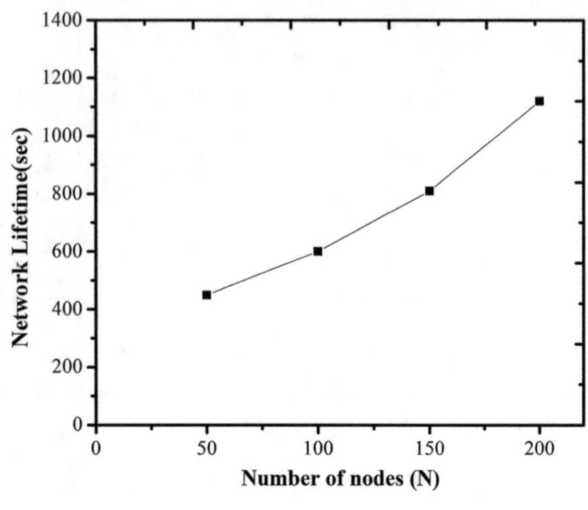

Fig. 4 Network lifetime with heterogeneity

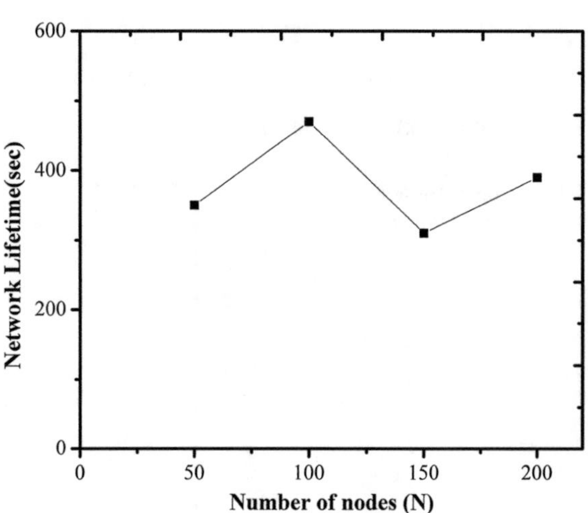

Fig. 5 Network lifetime without heterogeneity

Compared to earlier CH number estimation, here using merging, splitting of zones, and by putting some CHs to sleep mode, we have improved the network performance. The results obtained are based on the average number of simulations.

Network lifetime of the proposed approach is compared in Figs. 4 and 5. CHs in this network are provided heterogeneity in terms of infinite energy. So, lifetime in Fig. 4 depends upon energy consumption of sensor nodes. Lifetime increases with an increase in the number of nodes because for more number of nodes depending upon threshold value, the rate of merging of zones is low and nodes transmit at shorter distances that in turn increases their lifetime. In Fig. 5, in approach without heterogeneity, maximum lifetime is achieved with 100 nodes and it decreases for more number of nodes. Therefore, scalability can be achieved in the network using heterogeneity.

Here, from the results, we can deduce that implementing heterogeneity in clustering can increase the lifetime of the network with less number of CHs.

3 Conclusion

This paper proposed a load balancing clustering approach for grid-based deterministic deployment. Proposed approach is simple and results show that significant energy can be saved by incorporating uneven clustering and by using heterogeneity in terms of energy. We evaluate the performance of our approach through simulations for hybrid deployment. Using uneven clustering and heterogeneity, the network lifetime of the hybrid deployment in grid-based network can be improved when large number of nodes are deployed. The network operates for 800 s more in case of 200 nodes.

References

1. Boyinbode, O., Le, H., Mbogho, A., Takizawa, M., Poliah, R.: A survey on clustering algorithms for wireless sensor networks. In: 2010 13th International Conference on Network-Based Information Systems, pp. 358–364 (2010)
2. Jadidoleslamy, H.: An introduction to various basic concepts of clustering techniques on wireless sensor networks. Int. J. Mob. Netw. Commun. Telemat. (IJMNCT) **3**, 1–17 (2013)
3. Liu, X.: A survey on clustering routing protocols in wireless sensor networks. Sensors, 12(8), 11113–11153 (2012)
4. Kuila, P., Jana, P.K.: Energy efficient load-balanced clustering algorithm for wireless sensor networks. Procedia Technol. 6, 771–777 (2012)
5. Low, C.P., Fang, C., Ng, J.M., Ang, Y.H.: Efficient load-balanced clustering algorithms for wireless sensor networks. Comput. Commun. **31**(4), 750–759 (2008)
6. Gao, T., Jin, R.: A regional centralized-clustering routing algorithm for wireless sensor networks. In: 4th International Conference on Wireless Communications, Networking and Mobile Computing, pp. 1–4 (2008)
7. Ding, P., Holliday, J., Celik, A.: Distributed energy-efficient hierarchical clustering for wireless sensor networks. In: International Conference on Distributed Computing in Sensor Systems, pp. 322–339. Springer, Berlin, Heidelberg (2005)
8. Jin, Y., Wang, L., Kim, Y., Yang, X.: EEMC: an energy-efficient multi-level clustering algorithm for large-scale wireless sensor networks. Comput. Netw. **52**(3), 542–562 (2008)
9. Zhu, J., Lung, C.H., Srivastava, V.: A hybrid clustering technique using quantitative and qualitative data for wireless sensor networks. Ad Hoc Netw. **25**, 38–53 (2015)
10. Wang, Z., Lou, W., Wang, Z., Ma, J., Chen, H.: A hybrid cluster-based target tracking protocol for wireless sensor networks. Int. J. Distrib. Sens. Netw. **9**(3), 494863 (2013)
11. Hajiaghajani, F., Naderan, M., Pedram, H., Dehghan, M.: HCMTT: hybrid clustering for multi-target tracking in wireless sensor networks. In: 2012 IEEE International Conference on Pervasive Computing and Communications Workshops, pp. 889–894 (2012)
12. Gupta, G., Younis, M.: Load-balanced clustering of wireless sensor networks. In: IEEE International Conference on Communications, vol. 3, pp. 1848–1852. IEEE (2003)

13. Wajgi, D., Thakur, N.V.: Load balancing based approach to improve lifetime of wireless sensor network. Int. J. Wirel. Mob. Netw. **4**, 155 (2012)
14. Kumar, D., Aseri, T.C., Patel, R.B.: EEHC: Energy efficient heterogeneous clustered scheme for wireless sensor networks. Comput. Commun. **34**(4), 662–667 (2009)
15. Duan, C., Fan, H.: A distributed energy balance clustering protocol for heterogeneous wireless sensor networks. In: 2007 International Conference on Wireless Communications, Networking and Mobile Computing, pp. 2469–2473. IEEE (2007)

Framework for SIP-Based VoIP System with High Availability and Failover Capabilities: A Qualitative and Quantitative Analysis

Amita Chauhan, Nitish Mahajan, Harish Kumar and Sakshi Kaushal

Abstract Real time multimedia data transmission on TCP/IP networks has evolved and expanded to take over different aspects of communication and entertainment. This paper deals with the Session Initiation Protocol (SIP) based communication encompassing audio and video data transfer. Calling configurations including one-to-one and conference type calls are taken into account to study the effect on servers. The paper describes various existing failover and load balancing configurations used to setup SIP communication environments and test setup for a highly available and scalable SIP telecommunication system is proposed. In addition to this, a qualitative and quantitative analysis of the proposed system is presented. The comprehensive study and analysis lead to better mapping of architectures and configurations corresponding to the communication needs of the service providers. This, as a whole, will lead to better Service Level Agreement (SLA) formation and better prediction of capacities of various configurations in terms of both quality and quantity.

Keywords Session initiation protocol · Failover · Load balancing · Service level agreement

A. Chauhan (✉) · N. Mahajan · H. Kumar · S. Kaushal
UIET Panjab University, Chandigarh, India
e-mail: amita2692@gmail.com

N. Mahajan
e-mail: nitish7mahajan@gmail.com

H. Kumar
e-mail: harishk@pu.ac.in

S. Kaushal
e-mail: sakshi@pu.ac.in

M. Tuba et al. (eds.), *ICT Systems and Sustainability*,
Advances in Intelligent Systems and Computing 1077,
https://doi.org/10.1007/978-981-15-0936-0_28

1 Introduction

Next Generation Networks (NGN) are believed to completely transform the communication systems as they substitute traditional circuit-based PSTN. Voice over IP (VoIP) is envisioned to become the key instrument for providing multimedia and voice communications. VoIP systems require data transfer protocols like Real-Time Transport Protocol (RTP) for transmission of voice and multimedia data, and signalling protocols for controlling and maintaining communication sessions. Session Initiation Protocol (SIP), being simple, lightweight, text-based and easy to implement, is the most favoured signalling protocol in VoIP networks. It is also opted by IP Multimedia Subsystems (IMS) for controlling call sessions for millions of users.

With the advancements in the telecommunication systems, the number of users and hence, traffic load is increasing tremendously. Although the products and standards to provide SIP services are properly developed, they still lack adequate methods and techniques to ensure the availability and reliability of IP telephony infrastructure. Due to the whopping increase in the number of SIP users and hence SIP calls, a system providing high availability and scalability is a mandate. To achieve this, efficient load balancing and failover are the two key measures to be included in the system architecture of SIP-based communication networks.

In this paper, a two-level system configuration for a SIP-based telecommunication setup is proposed that imparts high availability and scalability. Test setup for the proposed load balancing and failover configuration is presented and an analysis of the system is given with respect to quality and quantity.

The remainder of the paper is organized as follows. Section 2 provides a detailed literature survey of state of the art. Sections 3 and 4 present various failover and load balancing schemes, respectively, for SIP server configurations. Section 5 describes the proposed system architecture and test setup with its qualitative and quantitative analysis. Section 6 concludes the paper.

2 Literature Survey

This section provides the details of the various calling configurations present in state of the art to setup SIP-based communication systems.

Cheng et al. [1] considered three key issues in a clustered server architecture for SIP telephony: (1) dispatcher being the bottleneck leading to system failure, (2) supervision of SIP proxy servers residing at the back-end, and (3) efficient load sharing approach for the SIP proxy servers. Authors modelled and developed a dependable system architecture for SIP-based communication environment to achieve high-speed failover. The system adopts the redundancy model for setting up SIP proxy servers and dispatchers. The authors also proposed an effective strategy for load balancing, among various SIP proxy servers, which rests on OpenAIS (Open Application Interface Specification). The overall configuration resolves the three issues mentioned by the authors and provides carrier grade telecommunication services with high availability.

Jiang et al. [2] emphasized on balancing SIP call requests among various proxy servers in a cluster. To achieve efficient load balancing, authors proposed three novel algorithms, namely Call Join Shortest Queue (CJSQ), Transaction Join Shortest Queue (TJSQ) and Transaction Least Work Left (TLWL), that work on the basis of the knowledge about the SIP, Session-Aware Request Assignment (SARA) and approximation of the server load. Authors implemented these algorithms by including them in open source SIP load balancer called OpenSER. They compared the responses of their algorithms with those of some standard load distribution schemes like round robin, and assessed them in respect of response time, throughput and scalability. Authors concluded that two of their novel algorithms (TLWL and TJSQ) considered irregularities in SIP call lengths, the difference in SIP transaction and call, and dissimilar processing costs of various SIP transactions; and that they performed better by providing higher throughputs, better scalability and lower response times than the other strategies they evaluated.

Kundan Singh and Henning Schulzrinne [3], in their work, presented a comparison between a number of existing load balancing and failover strategies in SIP-based telecommunication environment. They described various failover and load sharing schemes. Further, they used the most advantageous schemes to setup their two-stage server architecture for failover and load balancing. Multiple proxy servers exist in the first stage, out of which one server is selected using DNS NAPTR and SRV records. The selected server then routes the call request using destination ID's hash value to one of the server clusters residing at the second stage of the system. DNS based lookup is used again to find the actual cluster member that processes the request. Authors evaluated their system with SIPstone proxy test suite having MySQL database and identified their system configuration as more reliable and scalable than other existing techniques.

Blander et al. [4] invented a mechanism to achieve load balancing and resiliency in SIP-based communication services, specifically when the servers are geographically distributed. The invention incorporates a technique for the distribution of SIP call load among multiple SIP servers by configuring a load balancer at at least two sites. The first load balancer residing at the first site receives a SIP request and determines if this request needs to be redirected to the second site. It forwards the request to the second load balancer residing at the second site if need be. The request redirection from the first site to the second one happens if the first site fails or if the second site proxy is identified to be better in serving the request according to the proximity of the SIP entity. A health monitoring component is also present in the system that checks for the availability of the SIP entities. The system configuration presented by the authors is resilient and reliable for the geographically distributed SIP load balancers and servers.

Kambourakis et al. [5] proposed load sharing and redundancy measures for SIP-oriented VoIP networks. The system presented by the authors is based on DNS SRV records. The load balancer queries the DNS and forwards the SIP request to one of the many SIP proxy servers corresponding to the SRV record. The SIP client can also communicate straightway with the DNS in case the load balancer fails. The authors tested the configuration on well-modelled test setups and found its performance to be fine with respect to the service times. They identified the system as simple to implement, efficient and stable providing high availability. The authors also validated the system under various circumstances such as massive network traffic.

Francis et al. [6] claimed a system to balance the SIP processing load among various instances of the SIP server. The authors presented a technique where the load balancer maintains a data structure to map SIP processes to an instance of the SIP server. The load balancer also receives SIP processes' state information sent by the SIP server and updates the data structure in accordance with the state data, like removing the inactive SIP processes from it. The system configuration balances the SIP load between multiple SIP server instances efficiently and manages the load well in case a server node/instance fails.

Langen et al. [7] proposed a method to achieve failover and fault tolerance in SIP-oriented telecommunication networks in which the SIP server consists of two tiers, namely a state tier and an engine tier. The engine tier has a number of engine nodes that can handle SIP messages, and can read and write information about its state provided by the state tier. The state tier is responsible for maintaining state data in various replicas. Replica failures can be reported by the engine nodes. Engine nodes while reading or writing data from or to a replica can detect failure of that replica and tell some other replica about it. Replicas can be used to detect failures in engine nodes if the replicas don't get polled by the engine node for a particular time period. The claimed system works effectively and provides sustained communication services.

Shim et al. [8] in their work presented a system that deploys a method for efficient load balancing in SIP network. Rather than using traditional scheduling schemes like round robin, the SIP messages in the proposed framework are redirected to one out of many SIP servers existing in the cluster based on the performance score of the server. The performance score is evaluated by using the performance information about the server. If the data shows that the server has undergone failure, its performance score becomes zero and the load balancer will no longer forward SIP requests to this server. The proposed system configuration is advantageous in the sense that it decreases communication latency and improves system uptime.

Aggarwal et al. [9] proposed a load balancing and clustering system based on the OpenSIPS server to provide better Quality of Service (QoS). The authors presented the design, execution and performance assessment of the proposed scheme. The system has a number of machines with different configurations and several parameters such as jitter, response time, packet loss and throughput were analyzed.

3 Failover for High Availability

Failover or high availability can be regarded as a service property enhancing the service uptime as a whole and service availability to individual users. In case of SIP communication, this means lower call drop rates and higher service availability. In an event of failure of some resource, there must be a mechanism to detect the failure of the server or resource, and proper steps must be taken to remedy the situation. The following are the various existing schemes for high availability server setups.

3.1 Client Enabled Failover

The scenario where the client takes care of the situation if the server goes down is called client enabled failover. The client knows beforehand about the multiple servers providing the same services, for, e.g. in Fig. 1, the client knows two servers S1 and S2. The users send their registeration requests to both the servers. During the normal execution, client C1 can connect to any other client by using any of the servers S1 or S2. As shown in the figure, in a normal call scenario, client 1 tries to reach client 2 via the server S1. If it is not available the client 1 sends a request to server S2 for the same services. Thus, the client itself takes care of the failover. In this configuration, the servers are acting independent of each other and need not be aware of each other. This configuration works well within a single domain but is not an efficient and agile way to implement high availability SIP services.

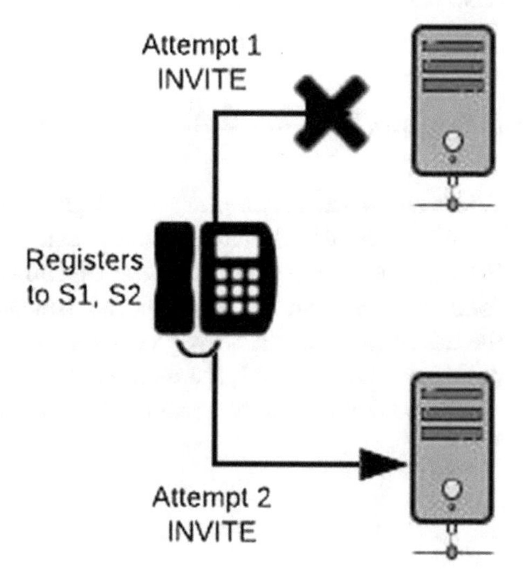

Fig. 1 Client enabled failover

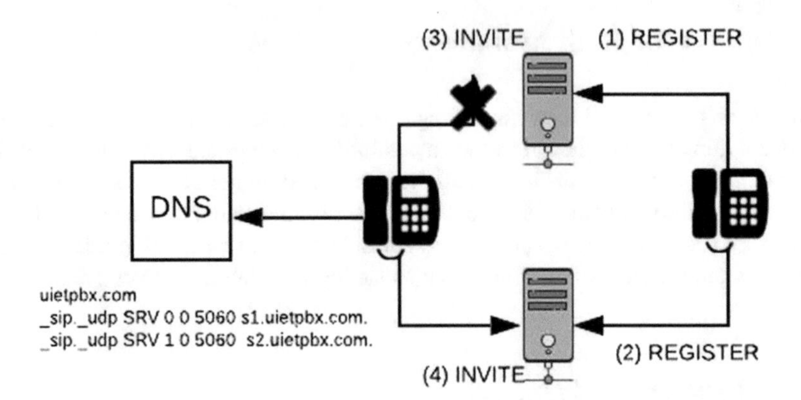

Fig. 2 DNS enabled failover

3.2 DNS Enabled Failover

DNS acts as the directional guide for clients in this approach. Figure 2 shows a DNS based failover configuration. NAPTR and SRV records are the two services provided by DNS servers which can be used for failover. SRV records provide multiple IPs for a single domain name having a priority of selecting a specific IP from the list. NAPTR records tell about the specific protocol combinations being used at the SIP server end. Similarly, another way of doing the same can be dynamic DNS. For this, S2 must monitor S1, and as soon as S1 goes down S2 must update the DNS records. In order to make this feasible, a smaller Time-To-Live value has to be assigned to DNS cache records. This technique can result in a higher delay between failovers.

3.3 Database Replication Enabled Failover

Client-based architecture for failover gives rise to a number of problems like higher failover interval and increased complexity at the client level. Besides that, all the SIP-enabled clients don't have the capability of configuring more than one server for a single account. In this case, the database D1 of the server S1 is replicated on the database D2 of the server S2. As shown in Fig. 3, client C1 sends registeration request to the server S1, the database D1 then propagates the registration to D2. If database D1 fails, server 1 can use database D2 to forward the request to the client C2.

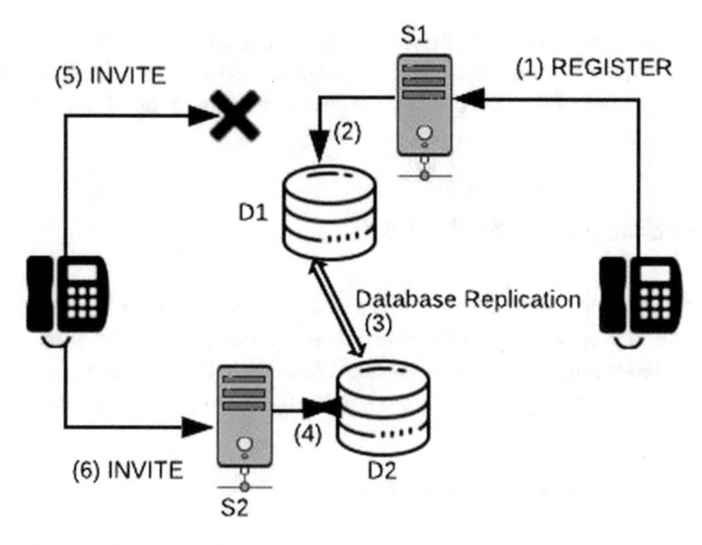

Fig. 3 Database replication enabled failover

3.4 Automated IP Assignment Enabled Failover

IP takeover can be used in instances where recovery of live calls is paramount and DNS based failover is not an option. As Fig. 4 shows, the servers S1 and S2 are configured exactly the same. The client sends a registeration request to the server S1 which is forwarded to the server S2. The database and the server can be on the same or different machines, thus binding S1 with D1 and S2 with D2 in the first configuration and separating them in the second. The registrations can be propagated from server S1 to server S2 or vice-versa. When the databases are on separate machines, the data

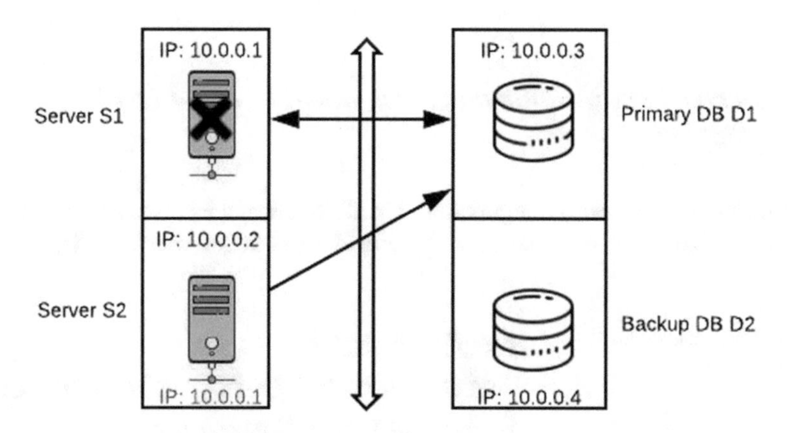

Fig. 4 Automated IP assignment enabled failover

transfer can take place between the data servers themselves. In this configuration, if S1 fails S2 takes over and can make a decision to still use D1. If only D1 fails, S1 can use the database D2. This can be true for all combinations of the available SIP and data servers.

4 Load Sharing for Scalability

In case of a failover a backup server takes over the current server and there is only a single server in play whereas in the case of load sharing all the servers are active and calls are routed through all of them. At times there is an overlap between the failover and load sharing techniques. Below are some existing load-sharing techniques.

4.1 DNS Enabled Load Sharing

The NAPTR and SRV records as discussed previously pose a power to define priority and resource weight in the records. The following example demonstrates the use of these records, Fig. 5 shows three servers that can be reached for the DNS query by the client. Each server has a priority value assigned to it in the third column. Here, the load distribution will be as 30% each to server one and two and remaining 40% to server three. There is a fourth entry for which column 2 holds value 1, which is the weight of the record. This server will take over only when none of the servers above are present.

Thus, providing failover capability combined with load distribution. This approach seems very simple but leads to a number of problems. The first problem is that each server individually is required to propagate the register requests to others present in the setup. This becomes a bottleneck for the network. Also, the DNS updating will be a hindrance in quick switch over in this case.

4.2 Server Identification and Assignment Enabled Load Sharing

This technique is a static configuration of load sharing and involves the assignment of users to particular resources, e.g. the provider has 10000 users. As depicted in

Fig. 5 DNS enabled load sharing

```
example.com
_sip._tcp 0 30 one.uietpbx.com
0 40 two.uietpbx.com
0 20 three.uietpbx.com
```

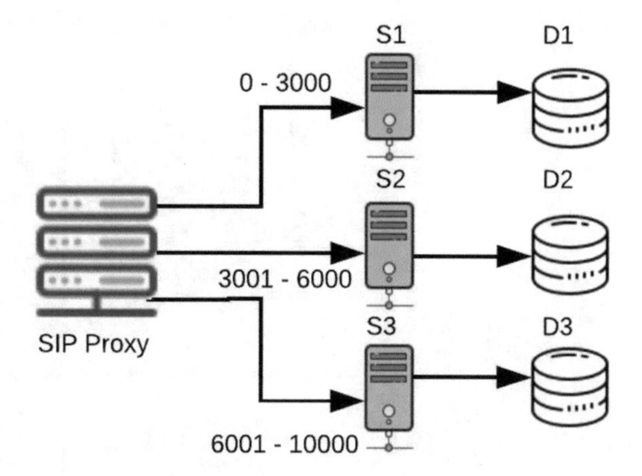

Fig. 6 Server identification and assignment enabled load sharing

Fig. 6, Server 1 is assigned to users from 0 to 3000, server 2 is assigned to users from 3001 to 6000 and server 3 is assigned to users from 6001 to 10000. The server works by using a proxy to distribute the load, as soon as a request lands at the proxy it uses a hash table to identify the destination server of the packet and forwards the packet to the respective server. The servers are independent of each other and are in no need to share the data. The configuration, however, is simple enough but poses a problem as there is no redundancy.

4.3 Single IP Enabled Load Sharing

This approach makes use of the router property of packet forwarding, multiple machines are assigned to the same IP address but a provision is made in the router to send the packets to different MAC address every time. The router can use scheduling algorithms for packet assignment. The method only works in the same subnet and is meant for stateless transactions, i.e. only UDP connections. This method is less efficient as the performance is limited by the bandwidth assigned to a subnet.

5 Architecture for Proposed High Availability and Scalability SIP System

5.1 Failover

The system presented in the paper uses a two-stage based failover for the calls. Figure 7 illustrates the setup being used for failover. The proxies act as the first contact point for a SIP client. The proxies X1 and X2 have their own IPs, IP1 and

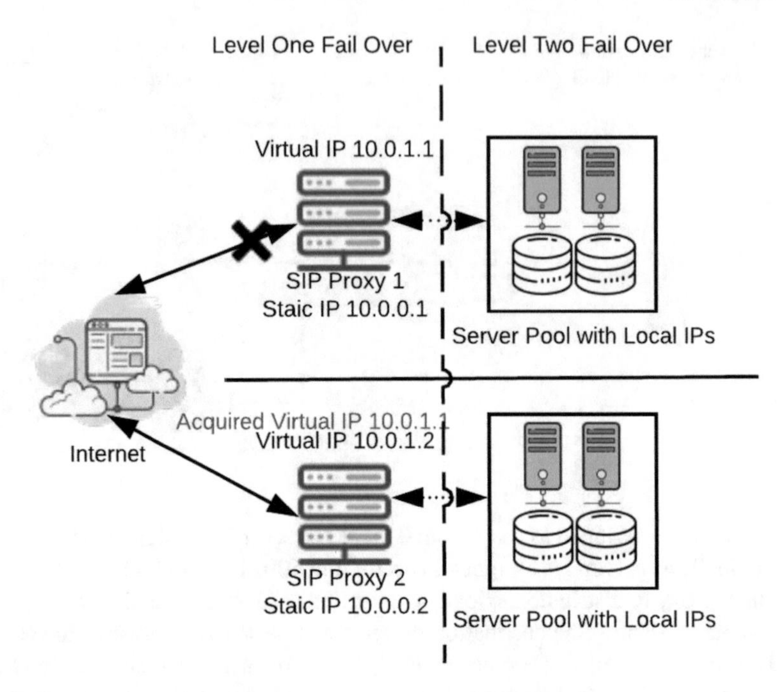

Fig. 7 Failover in proposed system

IP2, respectively. They also carry a virtual IP, VIP1 and VIP2. The proxies can act independently of each other if both of them are functioning. If proxy X1 goes down proxy X2 takes over its VIP and the users are directed to it. This acts as first level of failover.

For each proxy, there are two separate SIP servers configured. These sip servers are managed and monitored by their respective proxies. If server S1 fails, proxy detects it and sends all the requests to the server S2 providing the second level of failover. The database couples D1, D2 and D3, D4 are live replicas of each other. The backup from set 1 to set 2 takes place after an interval and registration information is shared. This delay is introduced to reduce pressure on the network.

Thus, if failure is within plain 1, it will be taken care of immediately, and the failover also can recover the state of the call, i.e. live calls can also be recovered. For an intra proxy failure, there will be a small delay for a fraction of users due to the delay in propagation of registrations, the live calls will drop but the overall system will recover in a matter of seconds.

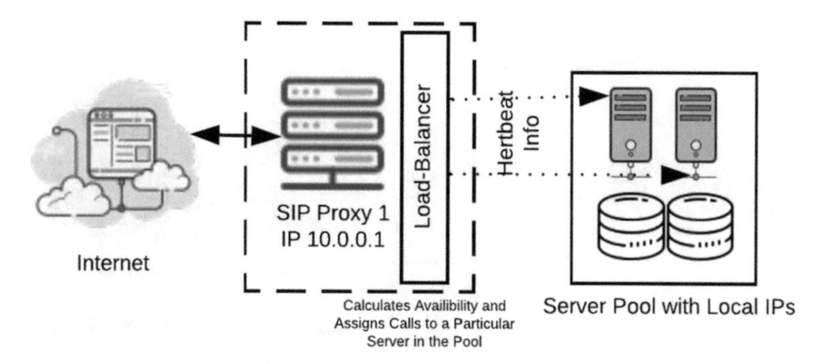

Fig. 8 Load balancing in proposed system

5.2 Load Sharing

The methods described in Sect. 4 are not sufficient in themselves to provide a reliable production-level load sharing. Also, the load sharing described in them is void of the current situation of the server, i.e. they do not consider the live situation of the servers being used. Thus, we propose an architecture combined with the failover architecture as discussed above to provide more reliable infrastructure. The proxies sitting at each point keep a track of the server variables like number of calls assigned to the server N, CPU utilization C on the server and latency L to reach the server. These factors are used to determine the availability factor for the server and then the server with the highest availability is assigned to the incoming call. The proxy and servers are configured to maintain a transaction on the same route to ensure the state fullness of the call, this only changes if the current server fails. Figure 8 depicts the load sharing architecture of the system.

In the case of REGISTER processing as soon as the packet reaches the proxy, the proxy takes care of the register packet by matching the domain names available providing a rudimentary but necessary step for authentication. After this, the proxy selects the SIP server to which the packet has to be transferred and it is done by calculating the availability value A. The availability value A is calculated using the following weighted formula.

$$A = W1 * N + W2 * C + W3 * L \tag{1}$$

Here W1, W2 and W3 weights can be calculated by applying Analytic Hierarchy Process Optimization on the historic data [10]. This optimizes the weighted importance of the variables involved and enables the proxy to select the appropriate SIP server.

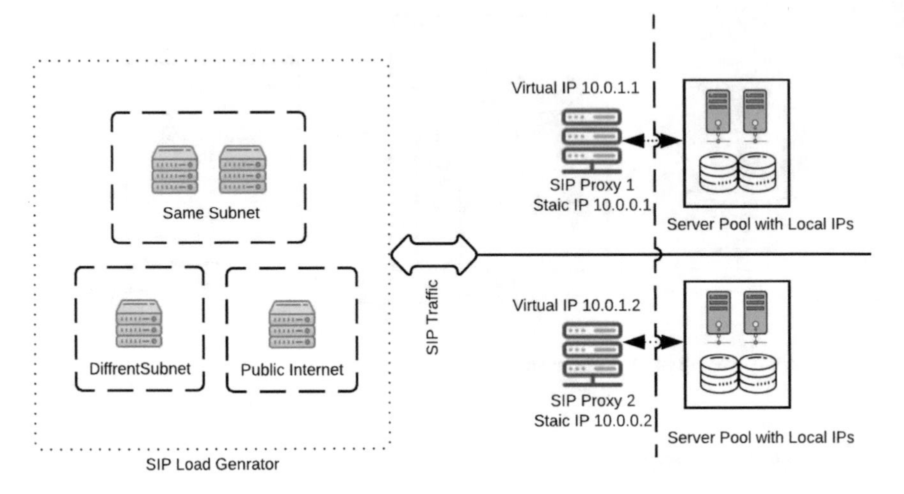

Fig. 9 Test setup

5.3 Test Setup

The server setup will be as described in Fig. 9. The load generation takes place via a traffic generator tool called Startrinity [11], a versatile tool for load generation and SIP traffic analysis. It provides a multidimensional view of the system performance and gives information about the average delay and the percentage of packet drop. The parameters that will be considered are as follows: add Startrinity parameters. As shown in the figure, we are using Startrinity for load generation and the load is generated by taking into account the load distribution among different networks. Initially, the call arrival rate follows Poisson's distribution on the public network where we have a single load generator. On the private network, the servers are situated on two locations. Two load generators are located on the same subnet and one server is on a different subnet. This is done to emulate the network configuration of big organizations which use different subnet for different services or work levels.

The load balancer sits on the proxy. The proxy uses the formula mentioned above to identify the eligible SIP server. If the identified SIP server is not responding, the proxy itself tries to switch the SIP request to the next server. Thus, in this case, the proxy is providing both load balancing and failover. The data collection takes place at two levels. The call quality data is recorded at the SIP load generator end. The number of failovers and the load in terms of CPU usage and calls assigned to a particular server can be monitored at the proxy or the SIP server.

5.4 Analysis

The analysis will be done in two perspectives, i.e. quantitative analysis and qualitative analysis. The two forms of analysis are important to evaluate the systems in terms of load bearing capacity and the quality of the calls. This analysis will result in a better estimation of load bearing capacity and better SLA formation.

The servers in the setup provide N-1 backup server architecture. The databases are implemented as clusters, and repeated exactly within the cluster for inter-cluster data backup only works for registration. The invite handling rate for the servers will be calculated in terms of calls per second. The call duration for the simulated calls will be varied between 1 and 3 min. The call arrival rate will follow Poisson distribution and the peak loads will be varied accordingly. The end results will help in the formation of empirical estimates for the proposed architecture.

The qualitative aspects deal with the packet drop ratio, latency and Mean Opinion Score (MOS) of the calls that went through. The quantitative analysis stresses the load bearing capacity in terms of calls that can be handled per second (CPS), Busy Hour Call Completion (BHCC) and Busy Hour Call Attempt (BHCA). In addition to this, a cap for the maximum load can be estimated by thresholding the CPU usage percentage of a specific SIP server. The outcomes from the above analysis will help the SIP providers to map requirements to server capacity which in turn will optimize the capital investment.

6 Conclusion

The paper focuses on SIP-oriented VoIP communication environment and describes several failover and load balancing techniques for a reliable SIP system configuration. The failover techniques include client enabled, DNS enabled, database replication enabled and automated IP assignment enabled. For scalability, various load sharing techniques explained in the paper are DNS based, server identification and assignment based and single IP based. Furthermore, the paper describes failover and load balancing configurations of the proposed SIP-based communication system which uses a two-stage configuration for failover. The first stage deploys a VRRP based technique for virtual IP switching. The second phase uses a SIP-based proxy to detect the failure and re-establish the session selecting a new server from a pool of SIP servers. The load balancing module also rides on the SIP proxy. The load balancing technique deployed here uses real-time monitoring of the servers in order to calculate their current load bearing capacity. The capacity is calculated by a formula that employs variables: CPU usage, number of calls assigned to the server and latency to the SIP server. The paper also proposes a testing setup that represents the load conditions in the real world scenario. The architecture and testing techniques discussed will lead to a better understanding of the load bearing capacity and assessment of call quality on the given setup. These parameters will enable formation of better service level agreements which will allow better user satisfaction and network optimization.

References

1. Cheng, Y.J., Wang, K., Jan, R.H., Chen, C., Huang, C.Y.: Efficient failover and load balancing for dependable SIP proxy servers. In: 2008 IEEE Symposium on Computers and Communications, pp. 1153–1158. IEEE (2008)
2. Jiang, H., Iyengar, A., Nahum, E., Segmuller, W., Tantawi, A.N., Wright, C.P.: Design, implementation, and performance of a load balancer for SIP server clusters. IEEE/ACM Trans. Netw. **20**(4), 1190–1202 (2012)
3. Singh, K., Schulzrinne, H.: Failover, load sharing and server architecture in SIP telephony. Comput. Commun. **30**(5), 927–942 (2007)
4. Blander, E., Peles, A.: Geographic resiliency and load balancing for SIP application services (2015). US Patent 9,143,558
5. Kambourakis, G., Geneiatakis, D., Gritzalis, S., Lambrinoudakis, C., Dagiuklas, T., Ehlert, S., Fiedler, J.: High availability for SIP: solutions and real-time measurement performance evaluation. Int. J. Disaster Recover. Bus. Contin. **1**(1), 11–30 (2010)
6. Francis, P.L., Collins, D.A., Dubois, G.R., Bunch, J.L., Pokala, N.R.: Load balancing for SIP services (2014). US Patent 8,775,628
7. Langen, A.R., Kramer, R., Connelly, D., Khan, R.N., Beatty, J., Cosmadopoulos, I., Cheenath, M.: SIP server architecture fault tolerance and failover (2010). US Patent 7,661,027
8. Shim, C.B., Xie, L.: System and method for load balancing a communications network (2010). US Patent 7,805,517
9. Aggarwal, S., Mahajan, N., Kaushal, S., Kumar, H.: Load balancing and clustering scheme for real-time VoIP applications. In: Advances in Computer Communication and Computational Sciences, pp. 451–461. Springer, Berlin (2019)
10. Saaty, T.: Optimization by the analytic hierarchy process p. 34 (2019)
11. Startrinity SIP tester. http://startrinity.com/VoIP/SipTester/SipTester.aspx

ICT Enabled Implementation of Rural Housing Scheme for Sustainable Rural Development

Prasant Kumar, Prashant Mittal, Ajay More and Bushra Ahmed

Abstract Digital Technology is a prominent aspect of e-governance for delivering government projects. The government schemes are now getting re-engineered to develop a modern, authentic, and informative platform for public delivery-based transparency, efficiency, and accountability. Accordingly, the focus of the Government has now shifted to the automation of internal operations of government organizations. Nowadays the development of any country can be judged by the spread of e-governance in that country as it points toward the ease of doing business and/or increased efficiency. This monograph is an overview of methodologies and progress of the various existing models of information and communication technology (ICT) use for broad-based development and economic growth specifically in the rural sector of India. The focus is chiefly on rural housing scheme which is currently running under Prime Minister's "Housing for All" vision. The rural population of India accounts for 70 percent of the total population and need crucial attention, thus several policies are embarked through digitalization. We firmly believe, greater attention is required in the rural housing sector as India has a predicament situation with an estimated shortage of around 30 million houses. Apart from those who are assured to be covered under this scheme by the year 2022. The new experience being shared is 360 degrees oriented, evidence-based and contemporary technology-centric one. The motto of this ICT solution has been "Right Benefit to the Right Beneficiary at the Right Time."

Keywords E-governance enabled rural housing · Rural infrastructure · Housing for all · Digitalizing rural development · Pradhan mantri awaas yojana-gramin

---

P. Kumar · B. Ahmed
Ministry of Rural Development, Government of India, New Delhi, India
e-mail: Prasant.kumar@gov.in

P. Mittal · A. More (✉)
National Informatics Centre, Government of India, New Delhi, India
e-mail: ajay.more@gov.in

P. Mittal
e-mail: pk.mittal@gov.in

M. Tuba et al. (eds.), *ICT Systems and Sustainability*,
Advances in Intelligent Systems and Computing 1077,
https://doi.org/10.1007/978-981-15-0936-0_29

(PMAY-G) · ICT in rural development · AwaasSoft · Direct benefit transfer scheme (DBTS) · Evidence-based monitoring

1 Introduction: de-facto Government Through Digitalization

India is the second most populous country in the world with approximately 25% of the world's poor. Identification of the right beneficiary from a population of 1.2 billion in India, providing them assistance at the right time and monitoring the utilization of the assistance on a near real-time basis is a challenge which can only be imagined.

Housing is one of the basic requirements of human being. To provide houses to the deserving rural poor, India has adopted the policy of providing them financial assistance for construction of their houses. Accordingly, rural housing schemes have been formulated since the last three decades. Till March 2016 the rural housing scheme Indira Awaas Yojana (IAY) was being implemented. Major drawbacks of this program were identified by Comptroller and Auditor General of India. The Performance Audit report included non-assessment of housing shortage, lack of transparency in the selection of beneficiaries, low quality of the house constructed and lack of technical supervision, a weak mechanism for monitoring and lack of convergence [1]. Around the same time the Government of India, resolved for "Housing for All by 2022" and aimed at construction of 29.5 million houses in the rural areas to meet this objective. Accordingly, the Government of India restructured the rural housing scheme, Indira Awaas Yojana (IAY), into Pradhan Mantri Awaas Yojana – Gramin (PMAY-G) where not only the beneficiary selection process but other drawbacks were also addressed and the processes re-engineered. Now the processes in PMAY-G are Information Technology driven that uses bottoms up approach for estimation of the target from data available in the Socio-Economic Caste Census (SECC), 2011 [2]. This approach electronically identified 40.6 million households in rural areas of India who are deprived of dignified shelter. AwaasSoft (a software application suite built by the National Informatics Centre, India for the Ministry of Rural Development, Government of India as an end to end, and workflow-based e-Governance system for implementation of PMAYG program) generated social category-wise rankings for each Gram Panchayat/village for validation of such lists. All modifications in the list, made by village's self-governance body, i.e., Gram Sabha, were captured digitally on AwaasSoft with reasons. The revised copy of the category-wise priority list of PMAY-G beneficiaries, minutes of the meeting were also uploaded on AwaasSoft for subsequent audit/scrutiny of the changes made by Gram Sabha. For transparency and citizen participation, PMAY-G data has been made available to public.

AwaasSoft has facilitated seamless convergence with other welfare schemes of various wings of the Government, such as SBM (Swachh Bharat Mission—Welfare program for Individual Household toilet of Ministry of Drinking Water and Sanitation) [3], MGNREGS for 90/95 days of unskilled wage employment (National Rural Employment Guarantee Scheme of the Ministry of Rural Development) [4] and

Ujjawala Scheme of Ministry of Petroleum and Natural Gas for providing subsidies on LPG stove and connection. AwaasSoft also helps in the quality construction of houses, inter alia, due to timely flow of funds directly into the beneficiary's account, construction of demo houses to improve quality of construction, etc.

Under PMAY-G, Government has successfully sanctioned 9.6 million houses in rural areas in 2 years since the launch of the scheme in November 2016–2017 including disbursement of funds amounting to ₹ 1.03 trillion [5]. Around 8.0 million houses have been completed [6] so far, and each house has been geotagged at multiple stages, figures for total geotagged images captured stands to be 90 million [7].

2 AwaasSoft Suite—An End to End e-Governance Solution

See Fig. 1.

2.1 Metrics for Digital Payments

Direct Benefit Transfer Scheme (DBTS), a supreme initiative of Indian government, resulted in a total saving of ₹ 900 billion till May 2018, majorly due to the removal of fake beneficiaries. The financial statistics in India marked 65% of Indian adults are considered as financially included, though only 45% of account holders report using their accounts in the last 90 days [8]. Traditionally a cash-based economy, India has physical currency in circulation estimated at over 12% of GDP. Digital India campaign has shown a fall in paper clearing specifically in banks [9]. It seeks to transform the country into a digitally empowered one which promotes rapid digitalization.

AwaasSoft is integrated with Public Financial Management System (PFMS) [10] which monitors various government programs and tracks funds dispersed, and marks itself as a platform for seamless flow of funds between different stakeholders. Under

Beneficiary Selection and Prioritization

- Downloading social category wise list from AwaasSoft at Village/Gram Panchayat level
- Segregation of minority list
- Removal of ineligibles from each list
- Prioritization and ranking
- Appellate committee approval
- Publishing permanent wait list for each Village/Gram Panchayat

→

Registration and House Sanction

- Beneficiary profile completion including capture and verification of Bank account details, Aadhaar number, unique ids for all convergent schemes
- Geo-tagging of existing house and proposed site of construction using AwaasApp
- E-Sanctioning and distribution of QR code based sanction orders to each beneficiary
- API based trigger generation for all convergent welfare schemes

→

Physical and Financial Progress

- Bank Account reverification and ordersheet generation
- Fund transfer order generation and and Digital Signing by two authorities
- Direct Benefit Transfer through public fund management system
- Physical inspection of the house construction using AwaasApp after release of each installment

Fig. 1 Core activity flow–AwaasSoft

PMAY-G, payment process to a beneficiary starts with generation of a Fund Transfer Order (FTO), an electronic cheque. The FTOs are double authenticated by two ground level authorities using their Digital Signatures. FTOs are then sent to State Nodal Bank (one DBT account in each State from which all FTOs of a State are debited) through PFMS. The FTO amount is debited from the State Nodal Bank and then sent to the beneficiary's bank for payment. The responses are consumed through PFMS and are populated on AwaasSoft. Each subsequent installment alert gets triggered when physical completion of the house to the desired level is reported using AwaasApp. PMAY-G focuses on the bank account verification and Aadhar (Unique Identification Number) verification of each beneficiary. This provides assurance of right benefit to the right beneficiary at right time. This year approximately Rs. 1.1 trillion worth funds of the Ministry of Rural Development are allocated for the rural population which accounts to be 70 percent of the total population of the country. Policymakers converted the payment methods from cheque to digital means for direct transfers to avoid inconsistency and delays. Till now under PMAY-G, funds worth ₹ 1.03 trillion have been transferred to 9.6 million beneficiaries in DBT mode.

2.2 Digitalizing Physical Landscape—A GIS Map Initiative

Under the guidance of Ministry of Electronics and Information Technology, Government of India, NIC has developed a geo-spatial mapping platform called Bharat maps. Bharat Maps depicts the core foundation data as NICMAPS, an integrated base map services using 1:50,000 scale reference data from Survey of India, ISRO(Indian Space Research Organization), FSI(Forest Survey of India), and RGI(Registrar General of India). It used ESRI Arc GIS technology service-oriented architecture which proclaims pre-cached tiled maps at 14 levels from 1:40 M to 1:4 K in seamless mosaic of IRS images in different resolutions. It encompasses 23 layers containing administrative boundaries, transport layers such as roads & railways, forest layer, settlement locations, etc., including terrain map services. The layers cover 35 states and UT's, 700 district, 6,500 blocks and sub-districts 6, 40,000 villages, 8,000 towns, 2, 50,000 panchayats. GIS-based tool has been developed on top of it for school, thermal power stations, post offices, and many others to provide landscape digitalization [11].

PMAY-G uses Bharat maps services mainly for three purposes. Firstly, for plotting the assets on map by passing geo-coordinates, this helps in visualization of assets geographically for micro monitoring, validation, and planning. Secondly, for navigation purposes in AwaasApp where a citizen wants to navigate to a location of a particular PMAY-G house. Thirdly, for GIS-based dashboards where Village/Gram Panchayat/Block/District boundaries are colored in gradient-based on their comparative performance on multiple key performance indicators. Other map services used by PMAY-G are Bhuvan map service of ISRO and Google map services to allow user the flexibility and comfort, also increasing the availability of services and their quality at the same time. Data of all images with their coordinates are made available in public in tabular format, along with the map view so that it can be utilized by other government departments and non-government organizations village level planning.

2.3 M-Governance Initiative

India has seen a remarkable shift to M-Governance as mobile subscribers' base has swollen to 616 million which is half of the population. About 330 million new mobile subscribers are expected to be added by the end of 2020 [12]. The 3G mobile network coverage is expected to be 90% of total population and 4G to be 70% by year 2020.

PMAY-G has a well-established lightweight (size less than 5 MB) M-Governance application called AwaasApp, which has approximately half a million-plus active users on Google Play store and Apple store combined. The application helps to capture each stage of physical progress starting from the existing dwelling where the beneficiary actually lived, the proposed site of construction, different phases of construction (foundation, plinth, lintel, roof cast, etc.) till completion (including toilet). The digital capture includes three good quality photographs, with time stamping and georeferencing. To address the issue of network connectivity in remote locations, AwaasApp runs in both, online and offline, modes. In remote locations the user can use AwaasApp in offline mode for data/photo capture and sync it later with servers when placed in network zone. The inspection activities can be performed on-site by village level functionaries/other ground level officials or by beneficiary himself/herself. All inspections are moderated at Block level login of AwaasSoft, thus ensuring validation and authentication.

2.4 Performance Indexes and Dashboards

To promote healthy competition among States, Districts, and Blocks, daily rankings are published on the dashboards, and these rankings are based on the pre-defined key performance indicators. Also, GIS map views are published with boundaries of the units in different colors depicting their comparative performance. Similar performance-based rankings are published on daily basis for banking institutions involved in PMAY-G e-payment transactions, based on their performance in direct benefit transfers to beneficiaries. This has induced a healthy competition among them (Figs. 2, 3 and 4).

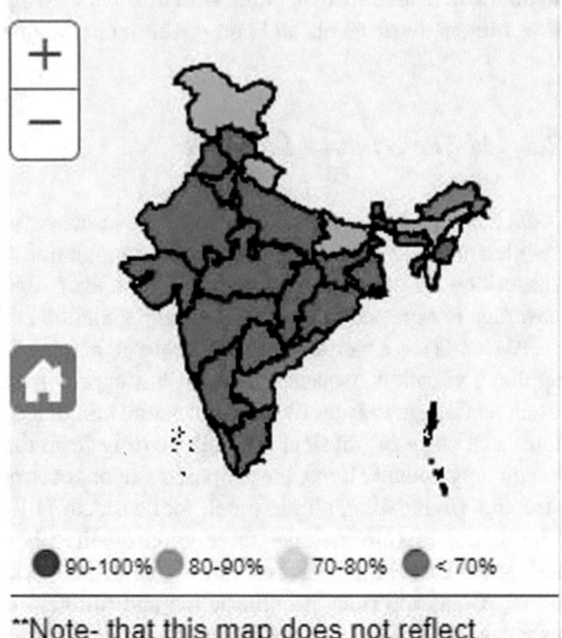

Fig. 2 A GIS map view–States are depicted in different colors based on their performance on houses sanctioned against the targets allocated by Government of India

3 Features of AwaasSoft

3.1 Open Data Paradigm

AwaasSoft puts all survey and implementation related data in public domain that is shareable, for maintaining transparency, and better governance. Geo-coordinates of images of respective houses are provided in public to locate the real-time location for all assets. The transactions are monitored and specifically every detail is provided to track FTOs along with their responses from PFMS. A dedicated platform is provided for citizens to track their welfare benefits through their FTO number which is provided at the time of its generation [13]. On AwaasSoft web portal, 40+ reports have been made available which are drillable from State up to Gram Panchayat/village level, 13 abstract reports are published for e-payment transactions which include public information of Digital Signature holding authorities, FTO summary, e-payment transaction details, pendency with banks, PFMS, etc.

Gram Samvaad [14], a citizen-centric mobile application, was developed by NIC in 2017, keeping the citizen of India at the Centre. The application is multiple local languages compliant which fetches current location with GPS tracker and allows

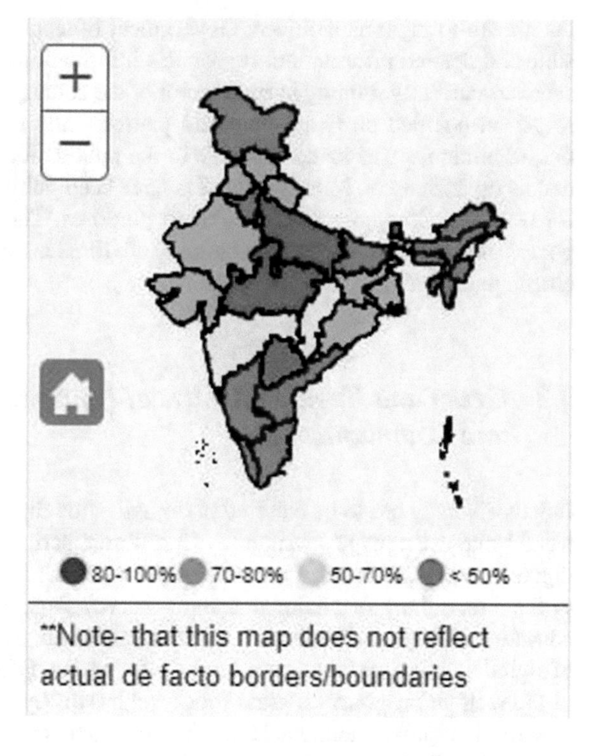

Fig. 3 A GIS map view–States are depicted in different colors based on their performance on houses completed against the targets allocated by Government of India

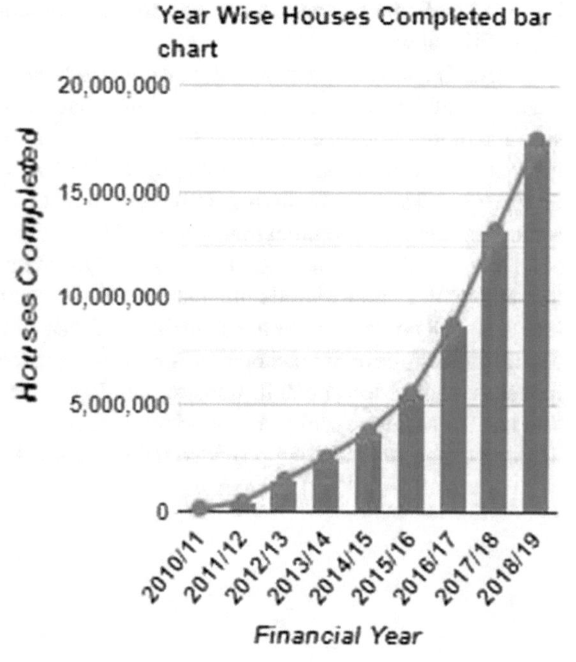

Fig. 4 A bar chart with a trend line showing year on year number of houses completed, and a sharp change in slope can be seen after revamping of IAY as PMAY-G in year 2016–17

the citizen to login as a citizen, Government officer, or as a beneficiary for accessible and desired information. It provides information of eight government welfare schemes currently running in rural sector of the country. PMAY-G is the first scheme to get on-boarded on Gram Samvaad platform and Permanent wait-list of potential beneficiaries was made available to the public along with the fund expenditure details on each asset. Many open APIs have been published for authenticated users for research and application integration purposes. The integrated digital platform, governance agility, and capacity building are three major components for integrated citizen participation and empowerment.

3.2 Green and Intelligent Ways of Information, Education, and Communication

India's diversity has to be factored in digitalization digitalizing policies; the country has 22 major languages, written in 13 different scripts, with over 720 dialects. To digitalize each platform in a communicable format is an arduous task. According to the 71st survey of National Sample Survey Organization (NSSO) of India on education 2014, just 6% of households in rural India had a computer [15], i.e., the level of digital literacy was low. Accordingly, Information, Education, and Communication of the welfare programs to grass root level has always been a challenge in the vast country like India. Internet accessibility at village level has increased many folds in last 5 years. Same has been harnessed for dissemination of information other than the traditional ways.

PMAY-G e-Gov system has online web-based training videos available in 12 languages for all 12 core modules. Also, training materials have been published on websites and mobile app for e-Learning. Concerned stakeholders, including beneficiaries, get real time notifications (SMS, Emails) at all milestones that are met under PMAY-G house construction, including daily consolidated reports on payments, progress status, mobile-based notification through FCM (Firebase Cloud Messaging,) and SMS for scheme implementation related advertisements and awareness matter. Master trainers for smooth digital implementation have been identified at state and district levels, and all contact details are available on AwaasSoft. Trainings are done through central and state-level workshops and also through video conferences. This cascading model of knowledge and skill transfer and digital ways of information dissemination has led to huge savings of public funds. Very few of the material has gone into hardcopy, majority has been provided online in digital mode.

4 How AwaasSoft Gave PMAY-G an Edge Over IAY

Indira Awaas Yojana (IAY), a flagship scheme of the Ministry of Rural Development, provided assistance to Below Poverty Line (BPL) families who were either houseless or had inadequate housing facilities, for constructing of a safe and durable shelter. It had been running under different names since 1986. In 2014, the Performance Audit of IAY brought forth many shortcomings of the program and the scheme was restructured into PMAY-G. This required re-engineering of many processes. Due to re-engineering of various modules, from completely/partially non-digitalized to digitalized modules, in PMAY-G, the results have been profound. Here is an analysis of the last 4 years of achievements of the IAY scheme where use of ICT was limited and the IT system was designed after the framing of guidelines, and initial 3+ years of achievements of PMAY-G scheme where all process are intrinsically ICT enabled through use of AwaasSoft suite and framework of implementation is designed on the plinth of ICT system (Table 1).

Impact of various digital interventions under PMAY-G has been documented in a study by National Institute of Public Finance and Policy (NIPFP), New Delhi. The study reveals that the average number of days taken for completion of a house in 2014–2015 was 314 days (House of a minimum size of 20 sq. mt.). Due to various interventions, including ICT interventions under PMAY-G, it has come down to 114 days in 2017–2018, although the PMAY-G house is larger in size, i.e., minimum of 25 sq. mt.

Monitoring under PMAY-G is evidence-based due to ICT interventions. The approach to maintain accountability has been established by eliminating the hurdles in IAY and re-engineering each module on a digital platform. In the implementation of IAY, there were many situations when the final installment got released before the completion of the house and around ₹ 1.50 billion worth of unfruitful expenditure was done on incomplete houses [16]. It was later the policy was changed and a geo-tagged photo verification got mandatorily implemented where the geocoordinates were matched at each stage completion before release of e-payments. The geotagged photos eradicated every possibility of discrepancy as each stage captures the house completion growth. The DBTS is the mandated way to transfer the installment which

Table 1 Table data source: Reports A1, B1, and B3 of www.rhreporting.nic.in

Scheme	Target	Houses completed	Percentage completion (%)
IAY + Others (2012–16)	10,337,858	5,121,939	49.55
PMAY-G + Others (2016–20)	15,085,838	11,939,499[a]	79.14

[a]This includes progress in a current financial year, count likely to be increased, the final number will be frozen on 31st March 2020

removed the cheque-based system for providing the financial aid, the main reason for delay in payments. The CAG reports once stated the approximate delay was once ranged between 14 and 1140 days [17] for the disbursements of first and second installments. In PMAY-G with the integration of AwaasSoft with PFMS in DBTS, the delay in payments is shortened to three days which usually happens due to verification of beneficiary account with the bank through PFMS before initiating payment. In IAY, ₹ 71.6 million worth of payments [18] were done as double payment to about four thousand beneficiaries in different states due to some manual glitches, In PMAY-G such cases are completely restricted through use of AwaasSoft for each payment release and assured monitoring of all payments digitally. The above analysis is a proven result for digitalizing the scheme from top to down.

5 Conclusion

PMAY-G is one of the best examples illustrating digital transformation of a welfare scheme in the public sector in India. Under PMAY-G, around 9.6 million houses have been sanctioned, 8.06 million houses completed in the last two years while target is to complete 29.5 million till 2022 [19].

With its comprehensive ICT framework, PMAY-G has accountability, accessibility, and authenticity as its underlying principles resulting in transparency in implementation life-cycle. Successfully overcoming many hurdles during its journey since 2016, rural housing has evolved over time and is contributing toward sustainable development. Successful implementation of PMAY-G in the right earnest, with transparency, DBT, evidence-based monitoring as its hallmark, has not only taken care of one of the deprivations of the rural households but also enhanced their social status, and self-esteem. Health status, especially of women has also been positively impacted with the provision of clean fuel, better and cleaner shelter with a toilet.

Though conceptualization and use of AwaasSoft suite is just a beginning toward ameliorating the concept of digitalization, ICT framework of PMAY-G is still evolving and needs work on many more aspects like centralized cross-domain convergence with other welfare schemes, capture of aspects like quality, safety, insurance, durability of the house, etc. Challenges like integration with PFMS & Banking IT systems lack maturity to process multi-level payment requests, Inconsistencies in payment credit responses from external systems, etc., also require more attention. With the proven track record of a successful project management tool, AwaasSoft has become a benchmark for executing any development program in developing countries. The foregoing discussion makes it very clear that the AwaasSoft suite has been successful in identification of the right beneficiary from a database of 180 million households, providing seamless financial assistance to each of them at the right time through Direct Benefit Transfer to their accounts, evidence-based monitoring of utilization of the assistance on a near real-time basis. Still we have to go a long way to make 1.33 billion people of this country to harness the full potential of the digitalized ecosystem. Upcoming focus could use latest technologies like big data analytics, Artificial Intelligence, and Block chain to reform every stratum of e-Governance.

References

1. Comptroller and Auditor General of India report on Indira Awaas Yojana. Report no. 37 of 2014. Ch. 6. Convergence Report. 10–11. https://cag.gov.in/content/report-no-37-2014-performance-audit-indira-awaas-yojana-union-government-ministry-rural. Retrieved 30 May 2019
2. Socio Economic Caste and Census Report. Total Households. https://secc.gov.in/statewiseTypeOfHouseholdsReport?reportType=Type%20of%20Households. Retrieved 30 May 2019
3. https://sbm.gov.in/sbmReport/home.aspx
4. National Rural Employment Guarantee Act (2005). https://nrega.nic.in/netnrega/home.aspx
5. Fund Transfer Orders Generated Since Inception of PMAY-G in the year 2016–2017. https://rhreporting.nic.in/netiay/EFMSReport/FtoTransactionSummaryReport.aspx. Retrieved 30 May 2019
6. House Completion Data for the Scheme PMAY-G https://rhreporting.nic.in/netiay/PhysicalProgressReport/YearWiseHouseCompletionReport.aspx. Report Retrieved 30th May 2019
7. Images Geo-Tagged Under the Scheme PMAY-G https://rhreporting.nic.in/netiay/GISReport/GISMobileInspectionReport.aspx.Retrieved. 30th May 2019
8. Financial Inclusion Insights Report. June-October (2015). 17-http://finclusion.org/uploads/file/reports/InterMedia%20FII%20Wave%203%202015%20India.pdf. Retrieved 30th May 2019
9. Shukla, S.: ET Bureau. Updated: Oct 03, 2017, 08.19 AM IST https://economictimes.indiatimes.com/markets/stocks/news/india-going-digital-paper-clearing-at-banks-falls/articleshow/60918886.cms. Retrieved 30th May 2019
10. Public Financial Management System for e-payment. https://pfms.nic.in/
11. Bharat Maps-Multi-layer GIS Platform. https://bharatmaps.gov.in/
12. GSMA Report on Mobile Economy India (2016). 8–9. https://www.gsmaintelligence.com/research/?file=134a1688cdaf49cfc73432e2f52b2dbe&download. Retrieved 30th May 2019
13. Fund Transfer Order Tracker. Tool for fetching data pertaining to e-payment status. https://awaassoft.nic.in/netiay/fto_transaction_details.aspx
14. Gram Samvaad Mobile Application. https://play.google.com/store/apps/details?id=com.nic.gramsamvaad&hl=en_IN
15. National Service Scheme. Education in India. Report 575(71/21.2/1). 15–16. https://www.thehinducentre.com/multimedia/archive/03188/nss_rep_575_compre_3188221a.pdf. Retrieved 30th May 2019
16. Comptroller and Auditor General of India report on Indira Awaas Yojana. Report Ch.4 4.2. https://cag.gov.in/sites/default/files/audit_report_files/Union_Performance_Indira_Awaas_Yojana%20_37_2014.pdf. Retrieved 30th May 2019
17. Comptroller and Auditor General of India report on Indira Awaas Yojana. Report no. 37 of 2014. Performance Audit. Ch.5 5.17.2 https://cag.gov.in/sites/default/files/audit_report_files/Union_Performance_Indira_Awaas_Yojana%20_37_2014.pdf. Retrieved 30th May 2019
18. Comptroller and Auditor General of India report on Indira Awaas Yojana. Report no.37 of 2014. Performance Audit. Ch.5 5.17.6 https://cag.gov.in/sites/default/files/audit_report_files/Union_Performance_Indira_Awaas_Yojana%20_37_2014.pdf. Retrieved 30th May 2019
19. Framework for implantation of PMAY-G https://pmayg.nic.in/netiay/Uploaded/English_Book_Final.pdf

Symbolic Solutions of Shortest-Path Problems and Their Applications

Mark Korenblit and Vadim E. Levit

Abstract The paper proposes a symbolic technique for shortest-path problems. This technique is based on a presentation of a shortest-path algorithm as a symbolic expression. Literals of this expression are arc tags of a graph, and they are substituted for corresponding arc weights which appear in the algorithm. The search for the most efficient algorithm is reduced to the construction of the shortest expression. The advantage of this method, compared with classical numeric algorithms, is its stability and faster reaction to data renewal. These problems are solved with reference to two kinds of n-node digraphs: Fibonacci graphs and complete source-target directed acyclic graphs. $O(n^2)$ and $O\left(2^{\lceil \log_2 n \rceil^2 - \lceil \log_2 n \rceil}\right)$ complexity algorithms, respectively, are provided in these cases.

Keywords Shortest path · DAG · Series-parallel graph · Fibonacci graph · Max-algebra · Expression

1 Introduction

Given a *digraph* (*directed graph*) G that comprises a *set of nodes* $V(G)$ and a *set of arcs* $E(G)$, a *path* starting in node v_0 and ending in node v_k in G is an alternating series of its nodes and arcs $[v_0, e_0, v_1, e_1, v_2, \ldots, v_{k-1}, e_{k-1}, v_k]$ such that e_i is an arc $(v_i, v_{i+1}) \in E(G)$ for $0 \le i \le k-1$. A *directed acyclic graph* (*dag*) has no path starting and ending at the same node. A *source-target dag* (*st-dag*) has a single source s and a single target t. A graph in which a real number (*weight*) is assigned to each arc is *weighted*. In a weighted graph, the *weight of a path* $[v_0, e_0, v_1, e_1, v_2, \ldots, v_{k-1}, e_{k-1}, v_k]$ is the sum of the weights of arcs

M. Korenblit (✉)
Holon Institute of Technology, Holon, Israel
e-mail: korenblit@hit.ac.il

V. E. Levit
Ariel University, Ariel, Israel
e-mail: levitv@ariel.ac.il

M. Tuba et al. (eds.), *ICT Systems and Sustainability*,
Advances in Intelligent Systems and Computing 1077,
https://doi.org/10.1007/978-981-15-0936-0_30

$e_0, e_1, \ldots, e_{k-1}$ which make up the path. A path of the minimum weight between nodes u and v is called a *shortest path* between u and v.

A *shortest-path problem* (*ShPP*) is dedicated to find a shortest path from one node to another. Being very essential this problem has applications in computer and telephone networks, routing systems and digital mapping services, scheduling, transportation, logic synthesis, etc. In a *single-source ShPP* (to compute a shortest path from a given node of $V(G)$ to every node $v \in V(G)$) on an n-node (*order* n) graph G with m arcs (weights of all arcs are nonnegative), Dijkstra's algorithm runs in $O\left(n^2 + m\right) = O\left(n^2\right)$ time [5]. If G is *sparse* (m is much less than n^2) modified Dijkstra's algorithms with running times $O\left((n + m)\log n\right)$ or $O\left(n \log n + m\right)$ may me applied. On an acyclic digraph, it is possible to solve this problem in $O\left(n + m\right)$ time [5]. Problems devoted to enumerating the few shortest paths in digraphs are discussed in [2, 10], and in many other works.

While the algorithms mentioned above are numeric, the technique proposed in this article is symbolic. The shortest-path algorithm on a graph is given as a symbolic expression whose formal parameters are arc tags. This expression encompasses all paths from the source to the target (*sequential paths*) in the graph. Arc tags are substituted by actual weights of corresponding arcs and the shortest-path weights are computed.

An *arc tagging* for a graph G is a transformation $E(G) \longrightarrow R$, where a set R called a *ring* includes *arc tags* as its elements and is supplied with operations $+$ (addition) and $\cdot$ (concatenation also marked by contiguity). Each sequential path in an st-dag G can be represented by a monomial consisting of all arc tags in the path. The total of arc tag monomials representing all sequential paths in an st-dag G is defined as the *customary formula* of G. A formula that is algebraically equivalent to the customary formula of an st-dag is called an *st-dag expression* and includes the tags and the ring operations $(+, \cdot)$. Given an st-dag G, its expression is symbolized by $Expr(G)$.

The amount of tags in a formula is defined as its *length*. A *shortest form of formula* F is a minimum length formula algebraically equivalent to F.

The importance of the approach proposed in this paper is the deeper understanding of the max-algebra tools [1, 4]. Max-algebra is a variation of algebra based on the pair of operations $(\oplus, \otimes)$ defined by $x \oplus y = \max(x, y)$, $x \otimes y = x + y$. Replace the operation max with the operation min [1]. Note that the commutative, the associative, and the distributive laws work for the operations min and $+$ in the same way as for the two ring operations $+$ and $\cdot$, respectively. For example, since $\min(x + y, x + z) = x + \min(y, z)$, the operations min and $+$ obey the distributive law.

For our problem, the intention is to compute the minimum sum of arc weights. Hence, denoting the operations $\oplus$ and $\otimes$ as $+$ and contiguity, respectively, allows the shortest-path problem on an st-dag (computation of the shortest path from the source to the target) to be understood as generating the st-dag expression. For this reason, the complexity of the problem depends on the number of operations $+$ and $\cdot$ in the expression. Since every two consecutive tags in the expression are separated by a ring operation, the st-dag expression length determines the complexity of the

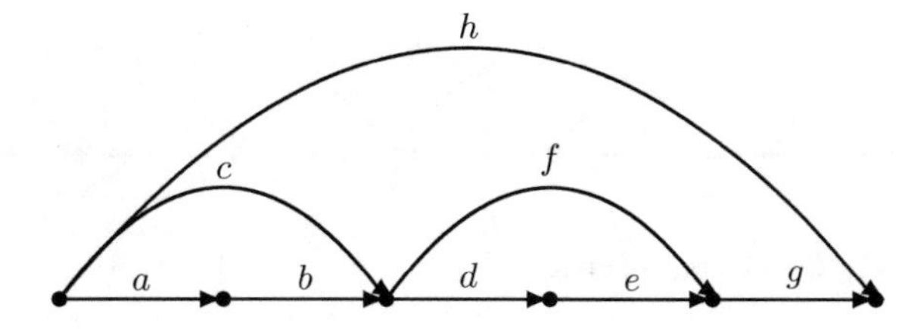

Fig. 1 A series-parallel graph

problem. Thus, for solving the shortest-path problem on an st-dag G in the shortest time, the shortest form of $Expr(G)$ should be derived.

It should be noted that while a numeric algorithm is simply a sequence of operations, the symbolic approach allows to present the algorithm as an object defined by the structure of the graph. Accordingly, direct access to separate parts (subexpressions) of this object is possible. Since different parts of the symbolic expression are independent, changes of individual arc weights do not require restarting the entire algorithm but assume the only recalculation of the corresponding subexpressions. For this reason, the benefit of this method is most manifested in real time systems for which stability and rapid response to data updates are very important indicators of their efficiency.

An st-dag called *series-parallel (SP)* is defined in the following way: an alone arc is an SP graph; given two SP graphs whose sources are s_1, s_2 and whose targets are t_1, t_2, respectively, an st-dag obtained by superposing s_1 with s_2 and t_1 with t_2 or by superposing t_1 with s_2, is SP [3]. An SP st-dag expression has a form in which each tag shows up only one time [3, 8]. This form is the shortest one for the expression of an SP graph. For instance, the customary formula of the SP st-dag depicted in Fig. 1 that is $abdeg + abfg + cdeg + cfg + h$ may be reduced to $(ab + c)\,(de + f)\,g + h$. The arcs amount in an SP graph depends linearly on the number of its nodes [8]. Therefore, an SP graph of order n provides the $O(n)$ complexity algorithm for the symbolic solution of the shortest-path problem.

This article considers a symbolic technique for the solution of the ShPP with reference to two non-series-parallel graphs: a Fibonacci graph (Sect. 2) and a complete st-dag (Sect. 3). The application of this approach to some scheduling problems is discussed in Sect. 4.

2 Symbolic Expressions for Fibonacci Graphs

A *Fibonacci graph* (FG) [7] has nodes $\{1, 2, 3, \ldots, n\}$ and arcs $\{(v, v + 1) \mid v = 1, 2, \ldots, n - 1\} \cup \{(v, v + 2) \mid v = 1, 2, \ldots, n - 2\}$. By [6], a graph is SP if it does not enclose a subgraph homeomorphic to subgraphs located between nodes $v \mid v = 1, 2, \ldots, n - 3$ and $v + 3$ of the FG in Fig. 2. Possible shortest forms of expressions of these subgraphs are $a_v\,(a_{v+1}a_{v+2} + b_{v+1}) + b_va_{v+2}$ or $(a_va_{v+1} + b_v)\,a_{v+2} + a_vb_{v+1}$. Thus for a non-SP st-dag G , $Expr(G)$ has no form in

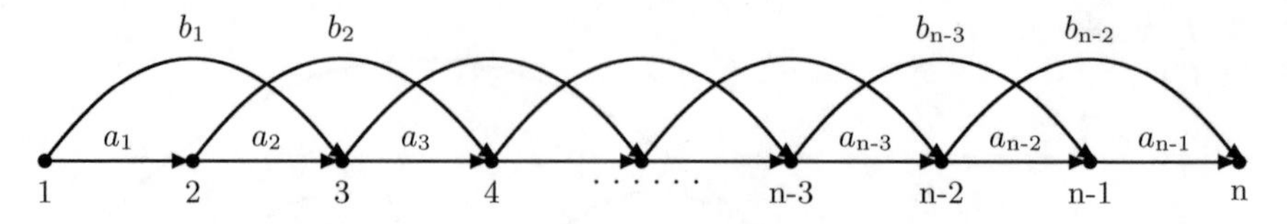

Fig. 2 A Fibonacci graph

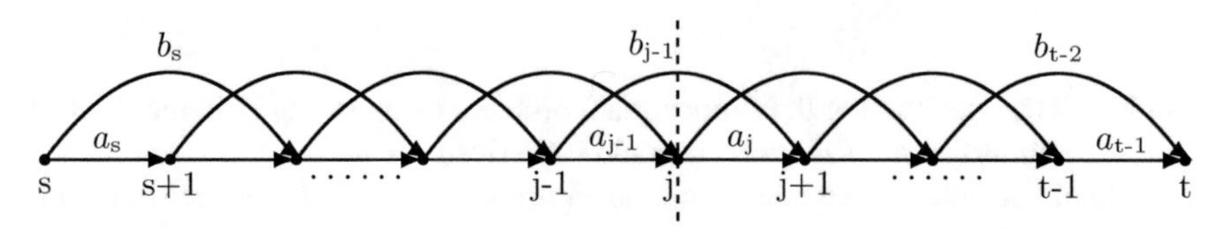

Fig. 3 A Fibonacci subgraph decomposed at node j

which each tag shows up once. Constructing the shortest form for such expressions is **NP**-complete.

For constructing an expression for an n-node FG, a *decomposition method* is used. Denote by $F(s,t)$ a subexpression of its subgraph with a source s $(1 \le s \le n)$ and a target t $(1 \le t \le n,\ t \ge s)$. If $t - s \ge 2$, then the subgraph is conditionally split at any *decomposition node* j $(s < j < t)$ (Fig. 3). Each path between nodes s and t goes via node j or through arc b_{j-1}. Therefore, every Fibonacci subgraph is divided into a tetrad of new Fibonacci subgraphs and in the general case $F(s,t)$ is generated recursively as follows:

$$F(s,t) \leftarrow F(s,j)F(j,t) + F(s,j-1)b_{j-1}F(j+1,t).$$

Subgraphs whose subexpressions are $F(s,j)$ and $F(j,t)$ comprise all paths between s and t which go via node j. Subgraphs whose subexpressions are $F(s,j-1)$ and $F(j+1,t)$ comprise all paths between s and t which go through arc b_{j-1}.

The following theorem determines the best position of the decomposition node j in an interval (s,t).

Theorem 1 *[8] The shortest form of $Expr(FG)$ generated by the decomposition method is reached off in every recursive step j equals $(t+s)/2$ for odd $t-s+1$ and $(t+s-1)/2$ or $(t+s+1)/2$ for even $t-s+1$.*

Given an FG of order $n \ge 3$, based on the *master theorem* [5], the length $L(n)$ of $Expr(FG)$ generated by the *balanced method* from Theorem 1 is

$$\begin{aligned} L(n) &= L(\lceil n/2 \rceil) + L(\lfloor n/2 \rfloor + 1) + L(\lceil n/2 \rceil - 1) + L(\lfloor n/2 \rfloor) + 1 \\ &\le 4L(\lceil n/2 \rceil) = O\left(n^2\right). \end{aligned}$$

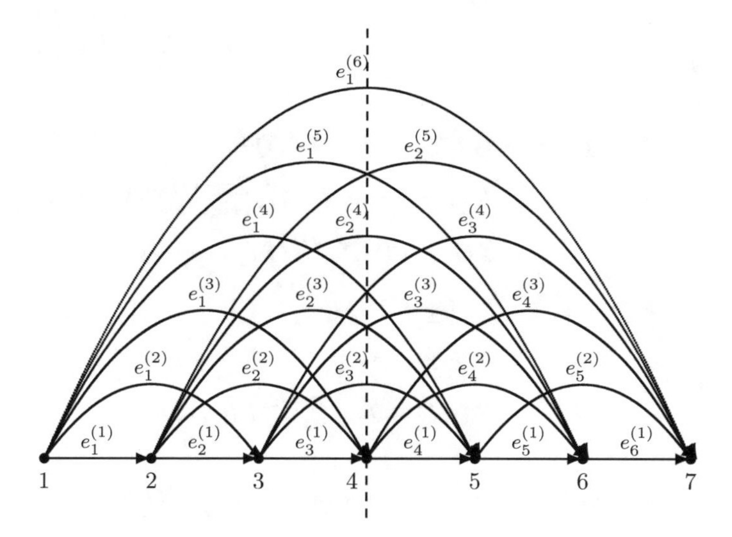

Fig. 4 A 7-vertex complete st-dag

Therefore, for n-node Fibonacci graphs there exists an $O\left(n^2\right)$ algorithm for the symbolic solution of the shortest-path problem. The balanced method is supposed to give the shortest expression possible and thus, this algorithm is the fastest one.

3 Symbolic Expressions for Complete St-Dags

A *complete st-dag* $(CST - DAG)$ has nodes $\{1, 2, 3, \ldots, n\}$ and arcs $\{(v, w) \mid v = 1, 2, \ldots, n-1 \mid w > v\}$ (see the example in Fig. 4). Arc $(v, v+l)$ in the graph is tagged by $e_v^{(l)}$ and is called an *arc of level* l. An n-node complete st-dag has $n(n-1)/2$ arcs: $n-1$ arcs of level 1, $n-2$ arcs of level 2, ..., 1 arc of level $n-1$.

Given an n-node $CST - DAG$, the decomposition method is used for constructing $Expr(CST - DAG)$. Denote by $F(s, t)$ a subexpression of its subgraph with a source s $(1 \leq s \leq n)$ and a target t $(1 \leq t \leq n, t \geq s)$. If $t - s \geq 2$, then a *decomposition node* j is chosen in the middle of a subgraph (j is $\underline{cr}$ or $\overline{cr}$, where $\underline{cr} = \lfloor (t+s)/2 \rfloor$, $\overline{cr} = \lceil (t+s)/2 \rceil$) that is split at this node (node 4 in Fig. 4). Each path between nodes s and t goes via node j or through an arc of level l $(l = 2, 3, \ldots, n-1)$ that connects revealed subgraphs (*connecting arc*). Therefore, in the general case $F(s, t)$ is generated recursively as follows:

$$\begin{aligned}F(s,t) \leftarrow & F(s,j)F(j,t) + F(s,j-1)e_{j-1}^{(2)}F(j+1,t) + \\ & F(s,j-2)e_{j-2}^{(3)}F(j+1,t) + F(s,j-1)e_{j-1}^{(3)}F(j+2,t) + \\ & F(s,j-3)e_{j-3}^{(4)}F(j+1,t) + F(s,j-2)e_{j-2}^{(4)}F(j+2,t) +\end{aligned}$$

$$F(s, j-1)e_{j-1}^{(4)}F(j+3, t)+$$
$$\dots\dots\dots\dots\dots +$$
$$F\left(s, j-\underline{cr}+s\right)e_{j-\underline{cr}+s}^{(\underline{cr}-s+1)}F(j+1, t)+$$
$$F\left(s, j-\underline{cr}+s+1\right)e_{j-\underline{cr}+s+1}^{(\underline{cr}-s+1)}F(j+2, t)+\cdots+$$
$$F(s, j-1)e_{j-1}^{(\underline{cr}-s+1)}F(j+\underline{cr}-s, t)+$$
$$F\left(s, s\right)e_{s}^{(\underline{cr}-s+2)}F(\underline{cr}+2, t)+$$
$$F\left(s, s+1\right)e_{s+1}^{(\underline{cr}-s+2)}F(\underline{cr}+3, t)+\cdots+$$
$$F\left(s, \overline{cr}-2\right)e_{\overline{cr}-2}^{(\underline{cr}-s+2)}F(t, t)+$$
$$\dots\dots\dots\dots\dots +$$
$$F(s, s)e_{s}^{(t-s-1)}F(t-1, t)+F(s, s+1)e_{s+1}^{(t-s-1)}F(t, t)+$$
$$F(s, s)e_{s}^{(t-s)}F(t, t).$$

Subgraphs whose subexpressions are $F(s, j)$ and $F(j, t)$ comprise all paths between s and t which go via node j. Subgraphs described by other subexpressions comprise all paths between s and t which go through corresponding connecting arcs.

The following theorem is proved by the iteration method.

Theorem 2 *Given an n-node $CST-DAG$, its expression has a form of length $O\left(2^{\lceil\log_2 n\rceil^2-\lceil\log_2 n\rceil}\right)$ (specifically, $O\left(n^{\log_2 n-1}\right)$ for n that is power of two).*

Therefore, the shortest-path problem can be solved by the symbolic way on a complete st-dag of order n in $O\left(2^{\lceil\log_2 n\rceil^2-\lceil\log_2 n\rceil}\right)$ time.

The following statement (so called *Monotonicity Lemma*) is proved in [9].

Lemma 1 *Given an st-dag G_1 and its subgraph G_2, the length of the shortest form of $Expr(G_2)$ does not exceed the length of the shortest form of $Expr(G_1)$.*

Since each st-dag of order n is a subgraph of a $CST-DAG$ of order n, by Lemma 1, for every n-node st-dag G and an n-node $CST-DAG\ G_c$, the length of the shortest form of $Expr(G)$ does not exceed the length of the shortest form of $Expr(G_c)$. Together with Theorem 2, it concludes with the following.

Theorem 3 *For each n-node st-dag G there exists an algorithm for the symbolic solution of the shortest-path problem with complexity bounded by $n^{O(\log n)}$.*

4 A Symbolic Technique for Robotic Line Scheduling

The problem of routing and scheduling is an important problem encountered in various applications, specifically, in robotic systems. A sequence of operations in a robotic line is very essential in product processing whose typical example is, specifically, a product assembly. Robots are responsible for jobs of a production process, for the successful transference of separate details and of whole products along the line. A special scheduler can operate a conveyor as a whole and determine the most appropriate paths of movement along the production line. Hence, the division of functions between the central scheduler and local robots takes place in the system.

A special algorithm proposed in this section handles a schedule of operations performed on several sequential machines in a production line. The problem is defined mathematically as follows. The production line is presented as a Fibonacci graph (Fig. 5). A machine is located at each node of the graph. Machine i can perform two jobs: its own job and a job of machine $i + 1$. That is, the product processed on machine i can be transferred after the end of the processing to machine $i + 1$ or to machine $i + 2$ at once. The important characteristic of such a system is its homogeneity. The conveyor consists of identical standard elements.

Arcs of the graph are tagged by weights. Each weight is generated as a complex function of the operation cost, the job performing time, the size of the queue to the corresponding machine and other parameters. Therefore, these weights are generalized costs of jobs and they are continuously renewed depending on the given situation. They can be computed in a parallel mode in sensors that are located in nodes together with machines. The intention is to ensure the minimum overall generalized cost for a product processing over the whole line. A special scheduler should dynamically determine a path for the current product (see Fig. 5). In order to provide

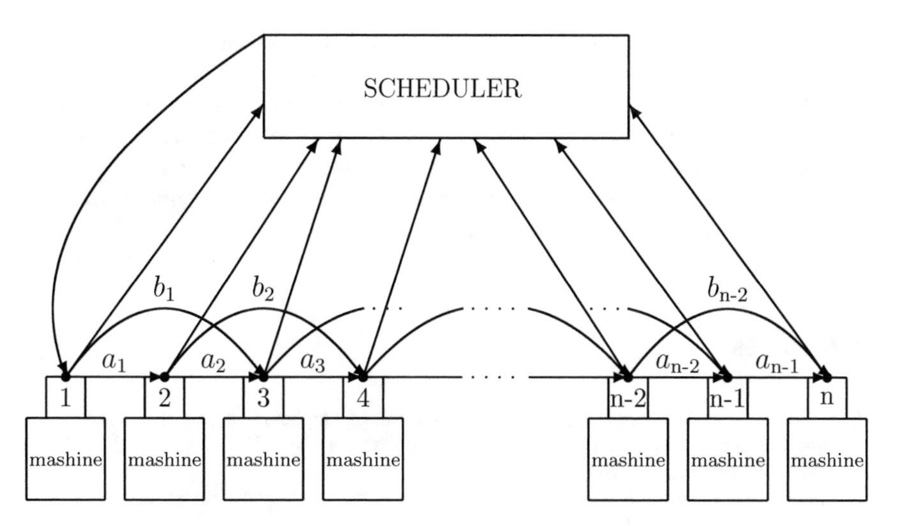

Fig. 5 The robotic line structure

the best sequencing in product processing, a scheduler must foreknow all possible appropriate sequences. The problem is that the quantity of the sequences grows with the jobs number as the number of paths in an n-node Fibonacci graph grows with n, namely, exponentially. The jobs number, in its turn, depends on the product complexity. For this reason, the scheduler should use efficient algorithms in order to generate the required sequence.

Therefore, the goal is the development of an efficient algorithm for generating the appropriate processing sequence. This problem is reduced to the shortest-path problem. The symbolic approach is proposed for its solution since stability, quicker response to the data update and the possibility of the parallel realization are very essential in such a system. Hence, the problem of optimal processing sequencing may be interpreted as an FG expression computation. The complexity of the problem depends on the length of the FG expression. The efficient algorithm of the minimum overall generalized cost calculation can be presented as the short form of corresponding algebraic expression. By the way, separate subexpressions can be computed in parallel in n processors located together with machines. In this case, the length of the FG expression determines only the upper bound of the problem complexity which is $O(n^2)$ in accordance with Sect. 2.

In more advanced production line every machine i can perform $n - i + 1$ jobs: its own job and jobs of all subsequent machines. This line can be simulated analogously by an n-node complete st-dag. According to Sect. 3, the running time of its scheduling algorithm will be bounded by $n^{O(\log n)}$.

5 Conclusion

A symbolic technique for the shortest-path problem based on the concept of a graph expression has been presented. The decomposition method applied to a Fibonacci graph that is a representative of a family of non-series-parallel graphs can provide an algorithm with polynomial complexity. The same method gives the quasi-polynomial algorithm for a complete st-dag in which there is an arc between any two nodes. It has been shown that for every st-dag of order n there exists a symbolic shortest-path algorithm with complexity bounded by $n^{O(\log n)}$. Finally, the applications of the symbolic approach have been illustrated.

References

1. Akian, M., Babat, R., Gaubert, S.: Max-plus algebra. In: Hogben, L. et al. (eds.) Handbook of Linear Algebra. Discrete Mathematics and its Applications, vol. 39, chapter 25. Chapman & Hall/CRC, Boca Raton (2007)
2. Aljazzar, H., Leue, S.: K*: a heuristic search algorithm for finding the K shortest paths. Artif. Intell. **175**, 2129–2154 (2011)

3. Bein, W.W., Kamburowski, J., Stallmann, M.F.M.: Optimal reduction of two-terminal directed acyclic graphs. SIAM J. Comput. **21**(6), 1112–1129 (1992)
4. Butkovič, P.: Max-algebra: the linear algebra of combinatorics? Linear Algebr. Its Appl. **367**, 313–335 (2003)
5. Cormen, T.H., Leiseron, C.E., Rivest, R.L.: Introduction to Algorithms, 3rd edn, The MIT Press, Cambridge, Massachusetts (2009)
6. Duffin, R.J.: Topology of series-parallel networks. J. Math. Anal. Appl. **10**, 303–318 (1965)
7. Golumbic, M.C., Perl, Y.: Generalized Fibonacci maximum path graphs. Discret. Math. **28**, 237–245 (1979)
8. Korenblit, M., Levit, V.E.: On algebraic expressions of series-parallel and Fibonacci graphs. In: Calude, C.S., Dinneen, M.J., Vajnovzki, V. (eds.) DMTCS 2003. LNCS, vol. 2731, pp. 215–224. Springer, Heidelberg (2003)
9. Korenblit, M., Levit, V.E.: Estimation of expressions' complexities for two-terminal directed acyclic graphs. Electron. Notes Discret. Math. **63**, 109–116 (2017)
10. Sedeño-Noda, A.: An efficient time and space *K* point-to-point shortest simple paths algorithm. Appl. Math. Comput. **218**, 10244–10257 (2012)

Secure Outsourced Integrity Test for Cloud Storage

Araddhana Arvind Deshmukh, Albena Mihovska, Ramjee Prasad and Naresh Gardas

Abstract Cloud-based solutions allow companies to add tools to almost any existing or future infrastructure and for client data to be stored in one place. In this way, clients can generally get to the most recent and right form of information, be that as it may, checking the respectability of the information may prompt extreme overheads, which makes re-appropriating of the honesty check to outsiders an alluring alternative. This paper proposes an AES encryption method to protect the privacy of the user data and avoid data leakage during the third party handlings. We have also shown that the proposed solution does not increase the upload time and is in fact beneficial to the system's performance

Keywords Cloud storage · Proof of retrievability · Integrity · Auditing · Privacy preservation

1 Introduction

Cloud computing has become a countless solution for providing an on-demand service, flexible network with increasing bandwidth, rapid resource elasticity, usage-based pricing, and adaptable registering framework for some applications. Cloud computing additionally gives clients critical innovation patterns, and it is now evident that it is reshaping data innovation forms and the IT commercial center. The less expensive and amazing processors with the Software as a Service (SaaS) computing

A. A. Deshmukh (✉) · A. Mihovska · R. Prasad
Aarhus University, Aarhus, Denmark
e-mail: aadeshmukh@sinhgad.edu

A. Mihovska
e-mail: amihovska@btech.au.dk

R. Prasad
e-mail: ramjee@btech.aau.dk

N. Gardas
CDAC, Pune, India
e-mail: nareshg@cdac.in

M. Tuba et al. (eds.), *ICT Systems and Sustainability*,
Advances in Intelligent Systems and Computing 1077,
https://doi.org/10.1007/978-981-15-0936-0_31

architectures are changing server farms into pools of registering administration on a colossal scale. Users can now be benefited from high-quality services using the data and software on isolated server farms.

With the cloud computing technology, users use a variety of devices, including PCs, laptops, smartphones, and PDAs to access programs, storage, and application-development platforms over the Internet, via services offered by cloud computing providers.

When implementing undertakings on isolated administrations a specific conviction level between any client is required. Trust issues happen because of numerous issues including the discovery approach of administrations and as a result of security issues. Administrations can not be trusted as their interfaces go about as secret elements with the substance variable without notice. In this manner an administration requester should make sure that what it gets to is equivalent to what was promoted by the administration. In the event that this is not the situation, at that point the VM running behind the administration would not have the capacity to comprehend the given errand perpetrating conceivable cost misfortunes because of time spent for administration determination and assignment accommodation. Security issues are likewise significant and are firmly connected to the past issue. These issues can influence both the administration requester and the specialist coop. The previous is typically influenced when the information it submits is utilized for different purposes than those chose amid arrangement (for example cloning of copyrighted information).

In spite of the fact that moving customer's information on cloud alleviates the client from the complexities included and equipment the executives errands, the cloud worldview brings numerous difficulties which impact the framework on different factors, for example, ease of use, security, unwavering quality, versatility, and generally speaking framework execution. A cloud service provider may hide data losses or discard rarely accessed data intentionally to preserve its reputation. There is need to verify that the data residing on cloud is safe and not tampered. With the large extent of outsourced data and client's resource controlled proficiencies, the problem is how a client can verify the integrity of data periodically without local replications of data. To address these issues many schemes are evolved. All schemes can be separated in two categories. First are private verification schemes in which verification can be done only by owner of the data. Another is Public verification schemes in which data integrity confirmation can be done by anyone on behalf of the data proprietor. Problem of the private confirmation schemes is that data owner gets excessively overloaded with this task. In the case of public verification schemes, user is alleviated from this task. In private verification task there is chance to crash the user's computing devices due to complex verification calculations, so there is high chance that users will accept the public verification.

All current verification schemes do not consider that cloud provides dynamic operations on the data stored on cloud. Paper [1] considers this issue also and proposes an efficient Public integrity verification scheme.

The residue of this paper is prearranged as follows. A transitory literature review is given in Sect. 2. Section 3 describes the proposed system. In Part A of Sect. 3, Disadvantages of existing systems are studied. Part B specifies the Design Goals. In

Part C, detailed Architecture and Working of the proposed system are discussed. In Sect. 4, a mathematical model is studied. In Sect. 5 experimental results are mentioned which is followed by the Conclusion. This section also highlights a scope for further research. Finally, the references are outlined.

2 Literature Survey

In the contemporary years, an extensive research is carried by eminent researchers in cloud security, fault tolerance in cloud networks, storage optimizations and other paradigms of cloud computing. Despite so many recompenses of cloud computing, it also arises that some issues related to data integrity and computation cost are still persistent. When a client stores the data on server there is a need for assurance of data security. In [1] an archetypal for "provable data possession" (PDP) is introduced. It consents the client who has stowed data on untrusted server possesses original data without retrieving it. This model reduces I/O cost by sampling random sets of blocks from server which generates probabilistic proof of possession. The issue here is that the data owner has to compute a large number of homomorphic labels for the data to be outsourced, which typically encompasses exponentiation and multiplication processes.

Scheme in [2] discussed the issues on how to frequently and securely with the efficiency verify that the client's data is safely and reliably stored on storage server. In [1], the dynamic data storage is not considered. In this paper [2] Ateniese et al. proposed a dynamic storage of its prior PDP scheme. PDP is based on symmetric key cryptography and allows dynamic data to be outsourced, i.e., it supports operations such as block modification, deletion, and append. The systems execute a priori destined in the number of queries and do not sustenance entirely dynamic data.

To ensure the user's data accuracy Wang et al. [3] proposed an elastic distributed scheme with two important features. This scheme achieves identification misbehaving server(s), i.e., integration of surety of storage precision and data error localization by utilizing homomorphic token with distributed verification of erasure-correcting coded data. However, they only measured fractional support for dynamic data operation.

In paper [4], Juels and Kaliski have defined and explored the proofs of retrievability (PORs) that enables a backend service to epitomize a proof that a user can recover desired data, i.e., the backup or archives holds and transmit sufficient data to user reliably. POR concept is designed to handle a large file. It is an important tool for online storage. To ensure privacy and integrity of retrieved file some cryptographic techniques are helpful. It's obvious that user wants to confirm that the storage do not modify or delete files before retrieval. POR aims to achieve this without users consuming to transfer the file and guarantees file retrieval within time.

Security solutions such as proof of data possession and proof of retrievability are introduced to check the data modification or deletion at storage server side, i.e., on cloud. Also the concept of Proof of Ownership (POW) evolved to relieve the cloud from storage of multiple copies of same data. This reduced the used server storage space and also the consumption of network bandwidth. In paper [5] it is showed that the two aspects (PDP and POW) can coexist within same framework. On this phenomenon they proposed proof of universe decade—"are you ready?" International Data Corporation, 2010, it is stated that only 25% of data may unique. By storing a solo replica of each data, much cloud space can be saved nevertheless of number of clients who outsourced it.

To address the different security, efficiency, and reliability issues in cloud computing different schemes and techniques are introduced. Li et al. in paper [6] proposes very efficient remote data verification scheme (audit server) which supports public verifiability and dynamic data support for PoR service simultaneously. The scheme also reduced the user's burden of computing tags of outsourced data and defines the new security model for cloud storage which is able to resist the reset attacks invoked by the cloud storage server in upload phase.

In [7], the provable data possession problem in distributed cloud stowage to sustenance the data migration and scalability of service is addressed. To solve this problem, a remote integrity checking scheme is designed with assumption of multiple cloud service providers to cooperate store and maintain the data submitted by clients.

In [8], the authors have proposed a stronger security model based on compact proof of retrievability. Under this paper, Checking scheme constraints are discussed in the security perspective. In PoR, client sends a challenge query after some time to ensure the reliability of the data stored on the cloud. An efficient PoR scheme based on BLS signatures in which client's query and response from server that are very short in length is proposed by the authors.

Wang et al. in [9] addressed the need of Third Party Auditor and designed the scheme for public integrity checking. The need of dynamic changes in the data stored on cloud by cloud users is also taken into account. Here, in this scheme, Merkle hash tree is used for full dynamic data operations. The Tag generation process is executed at client's device. However, the practical application of this scheme is significantly affected by lack of extensive performance analysis policies.

In Paper [10], the authors addressed the need of privacy preservation of the cloud data signer during public integrity verification and proposed the Privacy preserving mechanism called *Oruta,* based on ring signatures, which allows public integrity verification on shared data. Using this scheme identity of data signer at the time data upload is protected from Third Party Auditor. But the proposed scheme does not support dynamic data modification and the problem of traceability is still persistent as well as the Tag generation process is at client's side.

3 Proposed System

[A] *Disadvantages of Existing Systems*:

(1) Existing system is designed with the assumption that Third Party Public Auditor is a trustworthy agent. All the data owned by the user is exposed to the public auditor which (though not deliberately) breaches the privacy of the data.
(2) Blocks of the plain text of the files are stored on the cloud storage and here also data of the clients can be misused by the cloud storage server.

[B] *Design Goals*:

(i) An unencrypted or a plain text file can be considered as a vulnerability in the security perspective. Such data which is owned by the client is to be protected since it is exposed to the untrusted cloud audit server. Thr first design goal is to encrypt the data owned by user by using eminent Encryption technique—AES. This algorithm takes plaintext and encryption key as input and produces an encrypted file, i.e., ciphertext.
(ii) The data when demanded by the client should be unaltered or damaged, so, integrity verification of the data is to be performed. Either the client or the cloud auditer may send request the integrity verification to the cloud storage server. An Integrity Verification mechanism should be used which is same as previous approaches.
(iii) A Decryption algorithm should be used to decrypt the file after successful authentication by the verifier in response to the proof provided by the prover. The decryption algorithm takes ciphertext and decryption key as input and produces the original file to the client. The Client can now download the file.
(iv) Dyanamic Data Operations or modifications should be allowed by the system.

[C] *Architecture of Proposed System*:

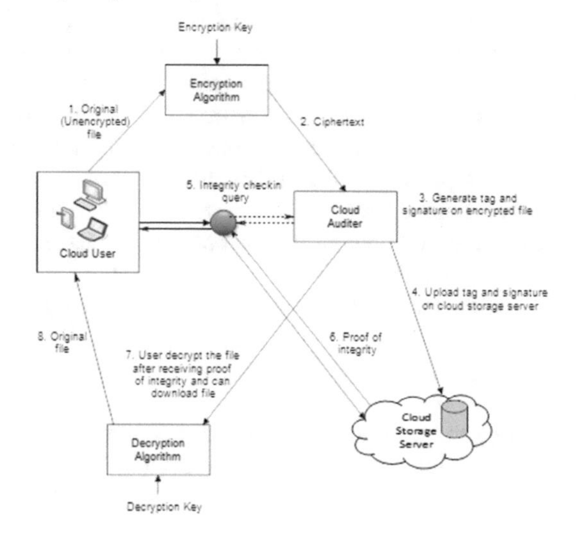

(1) *Set Up/Key Generation*:

In this step, the cloud audit server chooses a random α, u1; u2;. ..; us, and computes v ← g^a. The secret key is sk = α and the public key is pk = {v, {uj}{1 <= j <=s}}. The cloud storage server's private key is sk' = a' and public key is pk′ = v′ ← $g^{a'}$ [6]

(2) *File Upload*:

Phase 1: In this phase file is uploaded from cloud user to the cloud audit server and tag is generated for the blocks of the file. But before uploading the file, it is encrypted by using well-known AES algorithm.

To upload the file F, F is divided into n blocks M = {M1, M2, …. Mn} and each block has s sectors Mi = {Mi1, Mi2, Mi3, Mis}. Each block is encrypted with encryption algorithm(**AES**) and now M is set of the encrypted blocks. Let e : G X G → GT be a bilinear map, with three cryptographic hash functions H; h : {0,1}* → G and f:{0,1}* → Zp. Let g be the generator of G.

Cloud user uploads the M of F to the cloud audit server. Cloud audit server will then generate R root based on the Merkle hash tree, where leave nodes of the tree are an ordered set of the hashes of the file blocks H(Mi), (i = 1…..n). Then root R is signed using a private key of the cloud audit server to generate tag_t.

Phase 2 : In this step, for each block Mi, the Cloud audit server calculates signature Si. Set of the signatures is denoted by Φ = {Si} (1 <= i <= n). After this cloud auditor sends F and Φ to the cloud storage server as F*. On the reception of the F*, cloud storage server generates t' using own private key pk' on R. When audit server receives t', it deletes the original file F from local storage.

(3) *Integrity Checking*:

The client or Third Party Auditor can confirm the reliability of the outsourced data by stimulating the sloud storage server and they act as verifier. To do so, a query is generated by the challenging party to the cloud storage server which acts as the prover in this scenario. Upon delivery of the contest query, the prover generates a response and along with it an auxiliary information is also provided to the verifier. This auxiliary information contains the leaf-node information associated with the Merkle hash tree. This response is collectively called as proof.

The verifier then calculates the root of the Merkle hash tree on foundation of proof and then authenticates it. If the authentication happens to be successful then only the reliability of the outsourced data is verified.

The steps in Integrity checking are based on existing system [6] [9] enlisted as follows:

(i) *Generate Query*:

Query can be issued by the data proprietor or cloud auditor server. First step is to choose subset I (i1, i2,…., ic) of random c element from n elements, i.e., indexes of blocks to check. For each i pick an arbitrary element Vi ← f (i, t), where t is the time of the query. Let Q be the set {(i, Vi)}, which is sent to the cloud storage server. [6]

(ii) *Response from Cloud Storage Server*:

When cloud stowage server receives the Q from the audit server, it computes μj for each I

$$\mu\, j = \sum_{\{(i,Vi)\}} Vi, Mij$$

for j = 1 … s

$$\mathrm{S} = \prod_{\{(i,Vi)\}} Si^{Vi}$$

Along with this data, stowage server also provides the auxiliary data, which contains the hash of the blocks H(Mi) 1 < = i < = c and {Ωi} 1 < = i < = c.

All this computed data is sent to the audit server as a proof P.

(iii) *Actual Checking*

On the reception of the P from cloud storage server, audit server does the following checking

(1) Compute Root R using H(Mi) 1 < = i < = c and {Ωi} 1 < = i < = c and check the consistency.
(2) Check e(t, g) = e(h(R), v)
(3) Check $\mathrm{e(S, g) = e(} \prod_{\{(i,Vi)\}} H(Mi)^{Vi}. \prod_{j=1}^{s} uj^{\mu i}, v)$

if all the above checking gives true answer then data is not changed on the cloud.

(4) *Dynamic Modification*:

Data update, data insertion, data deletion is supported by this scheme. In all this three types of modification, block Mi' which needs to be modified or insert or delete is first encrypted using encryption algorithm and send to the cloud audit server. Audit server produces the monogram for Mi' and sends it to the cloud stowage server. As per the request, i.e., modify, delete or insert cloud storage server modifies, deletes or inserts Mi' block. After the modification cloud storage server generates the proof as stated in the integrity checking and sends to the cloud audit server generates the new R, authenticates new R and computes new root R' ==R, signs new root metadata by t' = Sig(sk, R') and sends it the server for the storage.

4 Possible Framework

Let I1 be the raw file.

Let P1, P2, …, P9 be the processes occurring in the system and O1,O2,…, O8 denote the output after every phase.

Before uploading the file F, it is encrypted in the following manner:

P1—I1 is divided into n equal blocks Mn and each Mi is encrypted using encryption key k of cloud user

M = Divide (F, n)

Where M is set of n equal blocks of F

For each block Mi

Mi = AESEncrypt(Mi, k)

Output will be O1

P2—Key generation
α = random (Zp)
uj = {u1,u2,u3,….us}
v $\leftarrow g^a$

Secret Key of cloud user sk = α
Public Key of cloud user pk = {v, {uj}}
Secret Key of cloud storage server sk’ = α’
Public Key of cloud user pk’ = v’

P3—O1 is uploaded to the Public audit server
R = MHT (O1)
root R is generated based on Merkle hash tree constructed using hash of file blocks

P4—Sign R using secret key sk and tag t is generated

$$\mathrm{t = Sign(R, sk)}$$

and t is sent to cloud client.

$$\mathrm{Si} \leftarrow \left(\mathrm{H(Mj)} \prod_{j=1}^{s} uj^{Mij}\right)^{a}$$

P5—is repeated for each encrypted block, $\Phi = \{\mathrm{Si}\}$ (1 <= i <= n) set of signature

P6—cloud audit server upload the File F and Φ to the cloud storage server

$$\mathrm{t' = Sign(sk', R)}$$

cloud storage server calculates the tag t′ using secret key sk′ and Root R

t’ is sent to CAS, CAS deletes F from local storage

P7—Query generation

$$\mathrm{Q = \{(i, Vi)\}}$$

where i is index of block for testing

Vi ← f(i, t) where t is time of the query

P8—Response from cloud audit server

for each i

P8—Audit server computes signature Si for each encrypted block

$$\mu j = \sum_{\{(i,Vi)\}} Vi, Mij \quad [6]$$

for each j = 1 . . . S

$$S = \prod_{\{(i,Vi)\}} Si^{Vi} \quad [6]$$

This set of ui, S, auxiliary data H(Mi) 1 < = i < = c, and {Ωi} 1 < = i < = c. is send as R response to CAS.

P9—CAS on the reception of O7 does following checks

(1) R′ is computed using H(Mi) and {Ωi} and checked the consistency with R.
(2) Check e(t, g) = e(h(R), v) [6]
(3) Check $e(S, g) = e(\prod_{\{(i,Vi)\}} H(Mi)^{Vi}. \prod_{j=1}^{s} uj^{\mu i}, v)$ [6]

If 1, 2, 3 are true then all i blocks consistent otherwise not.

O1—Set of encrypted blocks Mn
O2—Root of MHT
O3—Tag t is generated
O4—set of signature
O5—Tag t' as receipt
O6—Query Q
O7—response from CSS
O8—consistency of selected blocks in Q

5 Experimental Results

Proposed system will use 64-bit M2 high-memory quadruple extra-large Windows servers in Amazon EC2 platform as the auditing server and storage server, and a Windows machine with Intel I3 processor clocked at 2.40 GHz and 2 GB of system memory as the user.

In the existing system task of auditing and tag generation is outsourced to audit server, i.e., third party. Proposed system extends the existing system to provide privacy of the user data from audit server and storage server. To provide the privacy data

Table 1 Comparison of time essential to upload the file to cloud by proposed and existing system

Size of the file in MB	Time essential to upload file by proposed system in seconds	Time essential to upload file by ES1	Time required to upload file ES2
2	10	15	5
5	21	35	12
10	50	75	26

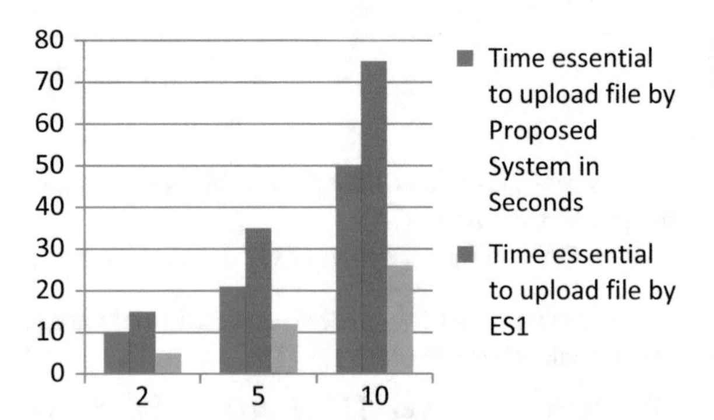

Fig. 1 Comparison of time essential to upload the file to cloud by proposed and existing system

to be uploaded is encrypted using AES encryption and then given to audit server. Proposed system will provide the privacy preservation of the data but cost of encryption of data to be uploaded and decryption of downloaded data will be on cloud user machine. Tag generation is high computation task than encryption and decryption of the data therefore proposed system will be efficient than existing systems in which tag generation is done at cloud user machine. To evaluate the effectiveness of proposed system, time for uploading same file in same setting by proposed method and existing method which does tag generation at user machine will be done. Existing system which does tag generation at user side is denoted as ES1 and Existing system which does tag generation at audit server site is denoted as ES2.

From Table 1 and graph 1, it is clear that ES1 takes highest and ES1 lowest time to upload the files. Proposed system is better than ES1 and provides the privacy preservation of the data. ES2 takes lowest time to upload the data but does not provide the privacy preservation. As privacy of the data is important to cloud user proposed system is more suitable than ES2 (Fig. 1).

6 Conclusion

Integrity checking is an issue for the user storing data on cloud storage and becomes overload for the cloud user. To mitigate this load of data integrity verification, this task is outsourced to another third party (Third party or Public verifier or Public auditor). When public auditor is not trusted there is possibility of privacy leakage of the data of the cloud user during data integrity checking by the public auditor. The proposed system is designed to overcome this problem. System alleviates the load of data integrity checking of cloud users by outsourcing the auditing task to the Third Party Auditor and also preserves the privacy of the data of the cloud user from Third Party Auditor by using Encryption technique on client's data before upload phase. For further research work, Encryption techniques can be applied on different phases to strengthen the security model and NoSQL database paradigm can be explored to enhance the performance of this scheme.

References

1. Ateniese, G., Burns, R., Curtmola, R., Herring, J., Kissner, L., Peterson, Z., Song, D.: Provable data possession at untrusted stores. In: Proceedings of the 14th ACM Conference on Computer and Communications Security, pp. 598–609 (2007)
2. Ateniese, G.L.V.M., Pietro, R.D., Tsudik, G.: Scalable and efficient provable data possession. In: Proceedings of the International Conference on Security and Privacy in Communication Netowrks, pp. 46–66 (2008)
3. Wang, C., Wang, Q., Ren, K.: Ensuring data storage security in cloud computing. In: Proceedings of the 17th International Workshop Quality Service, pp. 1–9 (2009)
4. Juels, A., Kaliski, B.S.: Pors: proofs of retrievability for large files. In: Proceedings of the 14th ACM Conference Computer and Communications Security, pp. 584–597 (2007)
5. Zheng, Q., Xu, S.: Secure and efficient proof of storage with deduplication. In: Proceedings of the ACM Conference Data Application Security Privacy, pp. 1–12 (2012)
6. Li, J., Tan, X., Chen, X., Wong, D.S. and Xhafa, F.: OPoR: Enabling proof of retrievability in cloud computing with resource-constrained devices (2015)
7. Zhu, Y., Hu, H., Ahn, G.-J., Yu, M.: Cooperative provable data possession for integrity verification in multicloud storage. IEEE Trans. Parallel Distrib. Syst. **23**(12), 2231–2244 (2012)
8. Shacham, H., Waters, B.: Compact proofs of retrievability. In: Proceedings of the 14th International Conference on the Theory and Application of Cryptology Information Security (2008)
9. Wang, Q., Wang, C., Li, J., Ren, K., Lou, W.: Enabling public verifiability and data dynamics for storage security in cloud computing. In: Proceedings of the 14th European Symposium on Research in Computer Security, pp. 355–370 (2009)
10. Li, H., Wang, B., Li, B.: Oruta: privacy-preserving public auditing for shared data in the cloud. IEEE Trans. Cloud Comput. **2**(1), 43–56 (2014)
11. Xiong, H., Zhang, X., Yao, D., Wu, X., Wen, Y.: Towards end-to-end secure content storage and delivery with public cloud. In: Proceedings of the ACM Conference on Data and Application Security and Privacy, pp. 257–266 (2012)
12. Bowers, K.D., Juels, A., Oprea, A.: Proofs of retrievability: theory and implementation. In: Proceedings of the ACM Workshop Cloud Computer Security, pp. 43–54 (2009)

13. Naor M., Rothblum, G.N.: The complexity of online memory checking. J. ACM **56**(1), 2:1–2:46 (2009)
14. Zhu, Y., Wang, H., Hu, Z., Ahn, G.-J., Hu, H., Yau, S.S.: Dynamic audit services for integrity verification of outsourced storages in clouds. In: Proceedings of the ACM Symposium on Applied Computing, pp. 1550–1557 (2011)

Sentiment Analysis for Konkani Language: Konkani Poetry, a Case Study

Annie Rajan and Ambuja Salgaonkar

Abstract Sentiment analysis is a part of NLP research. In the present work, we increased the existing corpus of Konkani senti-words by adding 75% more words and employed the Naïve Bayes classifier for the automatic senti-tagging of Konkani poems. We obtained 82% accuracy over a set of 50 poems that have been written by 22 contemporary poets, for which all the words were ensured to be the part of our corpus. We got 70% accuracy for tagging of a set of 10 randomly selected poems. The results are comparable with those of research on automatic classification of Telugu songs (Harika et al. in Proceedings of the 4th workshop on sentiment analysis where AI meets psychology, IJCAI, pp 48–52, 2016, [15]). The performance of the model could rank it third in the list of similar works reported in the literature.

Keywords Sentiment analysis · NLP research in konkani · Naïve bayes classifier

1 Introduction

Natural Language Processing (NLP) and Natural Language Understanding are one of the foremost domains where Machine Learning is employed. Sentiment Analysis (SA), the automated process of understanding an opinion about a given subject from written or spoken expression, has found a prominent place in NLP research [1]. SA of texts, poems, literatures, blogs, articles, news, tweets and messages is an interesting problem. Knowing opinions helps decision makers to resolve business problems, marketing strategies or policy decisions [2]. Sentiment classification—positive, neutral or negative, aims at determining and categorizing the sentiment of any content. Applications are in data analytics, government organizations and e-business [3]. Automatic sentiment analysis of the language that your customer speaks is an important need.

A. Rajan (✉) · A. Salgaonkar
Department of Computer Science, University of Mumbai, Mumbai, India
e-mail: ann_raj_2000@yahoo.com

A. Rajan
DCT's Dhempe College of Arts and Science, Goa, India

M. Tuba et al. (eds.), *ICT Systems and Sustainability*,
Advances in Intelligent Systems and Computing 1077,
https://doi.org/10.1007/978-981-15-0936-0_32

30% of Indians use the internet in their day-to-day business that includes calling retailers; the number is increasing; however, only 13% of Indians can barely comprehend English [4, 5]. The increase in India's digital footprint has made it necessary to have digital content in Indian languages as well. NLP for Indian languages and by extension, SA for these languages, is crucial for understanding government documents as well as informal social media data.

Konkani is the official language of the state of Goa, a state situated off the west coast of India [6]. 25 lakh people, 0.2% of the total population of India, speak the language. This is a relatively small number compared to other more widely spoken Indian languages. However, besides being one of the states with the highest GDP per capita, Goa is one of the major tourist-destinations of the country [7]. For meaningful digitization of Konkani, NLP and SA are necessary.

To the best of our knowledge this is the second attempt of sentiment analysis of Konkani text. The first one involved training an SVM for synsets in Konkani [8]. The reported accuracy is 47%. We employ a Naïve Bayes Bernoulli (NBB) classifier to identify sentiments of 50 Konkani poems by annotating 2205 words from the poems and 1934 words from the general conversations. The observed average accuracy of our model is 83% for tagging the 50 poems and is 70% for the other 10.

Section 2 provides the literature survey in sentiment analysis using NBB. Section 3 is about annotation of Konkani poems and our experiments. Section 4 presents the outcome of this research, followed in Sect. 5 by a discussion, conclusion and pointers to future work.

2 Literature Survey

Naïve Bayes classifiers, widely used for text classification, have been employed in the SA of the several Indian languages, namely, Tamil, Telugu, Kannada, Hindi, Punjabi, Nepali, Bengali and Odia. The results of a few experiments have been listed in the Table 1.

The highest accuracy is 90% for Kannada words, the second highest is 88%, for reviews of Hindi movies, and 81% for the reviews of Odia movies. The lowest performance is with Bengali tweets, only 44%, and the second lowest is with Hindi reviews of mobile apps, 47% accurate. Arguably, performance is impacted not the choice of the language but the choice of the text selected for SA. Perhaps a poem is sentiment-rich compared to prose in general. Telugu songs have shown almost 10% better result compared to Telugu text. This is the only example of SA of poetry in our literature survey, and our work presented here is the second one. Both results are in favor of our argument about poetry being sentiment-rich. However, in order to statistically establish this contention, more research is called for.

We provide details of our experiment in the next Section.

Table 1 Accuracy, precision, recall and F-score of SA with Naïve Bayes algorithm

No	Corpus	Accuracy (%)	Precision	Recall	F-score
Hindi					
1	5417 reviews on (i) Electronics (ii) Mobile apps (iii) Travel (iv) Movies [9]	50.95 46.78 56.06 87.78	48.00 59.20 20.87 56.66	45.05 54.09 31.90 63.32	46.46 56.53 25.23 59.81
2	36,465 tweets on Lok Sabha elections 2014 [10]	62.00	71.00	61.00	65.62
3	200 movie reviews [11]	80.00	80.20	80.00	80.09
4	1277 tweets of public interest for training and 467 for testing [12]	50.75	–	–	–
Tamil					
5	534 movie reviews [13]	66.17	–	–	–
Telugu					
6	1644 sentences [14]	73.85	80.00	90.00	85.00
7	100 songs [15]	70.20	–	–	–
Nepali					
8	25,345 editorial articles of daily newspapers [16]	–	77.8	70.2	73.80
9	384 reviews of movies and books [17] (i) Bag of words (BoW) (ii) BoW by removing stop-words (iii) TF-IDF (iv) TF-IDF without stop-words	65.60 63.80 60.60 58.00	– – – –	– – – –	67.20 64.20 73.40 72.00
Odia					
10	6000 movie reviews for training and 500 for testing [18]	81.00	76.00	79.00	77.47
Kannada					
11	5043 words polarity between −5 to 5 [19]	90.00	90.00	89.00	89.50
12	287 web documents on movies [20]	–	80.80	20.70	81.20
13	Product reviews for mobiles [21]	65.00	62.50	75.00	68.20
Bengali					
14	1000 tweets for training and 500 for testing [22]	44.00	–	–	–

3 Annotation of Poems in Konkani Language

This is a three step process:

3.1 Create a Senti-Tagged Corpus

The base corpus employed in this experiment has 9681 words; 5% of the words have two or three senti-tags. The details of the comprehensive corpus are given in Table 2.

The terms in the poem-words corpus (no. 2 above) and conversational-words corpus (no. 3 above) have been tagged manually. A Konkani word has been annotated with the senti-tag of its English equivalent that has been obtained by looking up an English senti-net [22].

3.2 Build a Model by Employing the Naïve Bayesian Classifier

The base corpus employed the naïve Bayesian classifier [23]. An instance is given the most probable class label, i.e., by following the maximum a posteriori (MAP) decision rule:

$$argmax p(C_k) \prod_{i=1}^{n} p(x_i|C_k) \quad where\, k \in \{1, \ldots K\} \tag{1}$$

where C_k is the class label of the k-th class, n is the cardinality of the corpus and x_i is the representation of the i-th term of the corpus in the given event.

We assume (existence-based) Bernoulli event model over (frequency-based) multinomial as each poem is a short text. Arguably, repetition of phrases in a poem adds up to the intensity of sentiment, but it does not change the sentiment. Therefore, x_i is a Boolean that takes value 1 if i-th term of the corpus occurs in the given poem, or

Table 2 Corpus for training data in Konkani language

Sr No	Corpus	Positive tags	Negative tags	Neutral tags	Total
1	Konkani synset [27]	365	407	4770	5542
2	Words from 50 contemporary poems	1138	696	371	2205
3	From day-to-day conversations	495	531	908	1934
Total		1998	1634	6049	9681

0 otherwise. We define three classes, positive, negative and neutral, with cardinalities 1998, 1634 and 6049, respectively. Therefore, k = 3 and n = 9681 in the following event model computes the likelihood of a given poem getting class label C_k.

$$p(x|C_k) = \prod_{i=1}^{n} p_{ki}^{xi}(1 - p_{ki})^{(1-x_i)} \quad (2)$$

where P_{ki} is the probability of class C_k generating the term x_i.

The data used to build and test the performance of NBB for senti-tagging Konkani poems in this paper consists of the set of 50 poems from which the poem-words-corpus (2) has been generated in Table 2. A senti-tag has been generated for each of the poems by processing it using the in-built NBB classifier provided by scikit-learn [24].

3.3 Bystander Testing to Compute the Model's Performance

Each of the 50 poems used in this experiment has been annotated as positive, neutral or negative, by three native Konkani annotators; two of them know poetry well. The Kappa statistics, a simple and popular measure of agreement between multiple annotators that essentially describes the proportion of times they would agree if they guessed on every case, has been computed (Fig. 1).

Basic formula to compute the Cohen's kappa is given below for the purpose of the novice readers,

$$\text{Cohen's kappa} = k = \frac{p_o - p_e}{1 - p_e} \quad (3)$$

$$p_o = \frac{a + d}{a + b + c + d} \quad (4)$$

$$p_e = p_{yes} + p_{no} \quad (5)$$

where

$$p_{yes} = \frac{a + b}{a + b + c + d} \quad (6)$$

Fig. 1 Confusion matrix

		Yes	No
A	Yes	a	b
	No	c	d

Table 3 Number of positive, negative and neutral poems annotated by three annotators

Sentiment of poems	Annotator 1	Annotator 2	Annotator 3
Positive	29	32	31
Negative	17	17	15
Neutral	04	01	04

Table 4 Cohen's Kappa score between annotators

Sr No	Annotators	Cohen's Kappa
1	1 and 2	0.84
2	2 and 3	0.71
3	3 and 1	0.69

$$p_{no} = \frac{c + d}{a + b + c + d} \tag{7}$$

The responses of the three annotators and the pair-wise kappa measures have been compiled in Tables 3 and, 4 respectively.

The maximum agreement is between annotator 1 and 2; maybe because both being females they share a words to sentiments mapping to a greater extent. Annotator 2 and 3 are poets. They are expected to employ similar theories in the interpretations. This could be a reason for the agreement between them being more than that of between 1 and 3. The average of the agreement between the three annotators is 0.75 which is significantly good. Events in which there are two opinions, are resolved in favor of the majority. No event was tagged with the three different tags by the three annotators. We also tested the model with unseen samples, which were 10 popular poems of Konkani [26–28]. The results of this experiment are presented in the following section.

4 Results

Table 5 is the confusion matrix [29] showing the agreement between the NBB classifier and the human annotators for senti-tagging the 50 Konkani poems.

The performance of the NBB classifier in comparison with the human annotators is shown in Table 6.

The average accuracy 82.67%, is 75.90% higher than that of reported 47% by employing SVM to identify the sentiment of words in the Konkani synset corpus (1) [30]. Performance-wise the model can be given the third rank in the list of naïve Bayes classifiers employed for the sentiment analysis of Indian language texts in Table 1. We attribute this achievement to the enhancement in the corpus that takes care of providing at least one senti-tag to each of the words in the given poems.

Table 5 Confusion matrix between NBB classifier and Human annotators

Confusion matrix	Human annotated			
	Sentiments	Positive	Negative	Neutral
Machine annotated (Predicted)	Positive	30	8	0
	Negative	1	5	0
	Neutral	0	4	2

Table 6 NBB classifier comparison with human annotators

Sentiment	% Accuracy	% Precision	% Recall	% F-score
Positive	82	78.95	96.78	86.96
Negative	74	83.33	29.41	43.48
Neutral	92	33.33	100.00	50.00
Average	82.67	71.15	69.45	60.15

Table 7 Machine classifier and human annotator outcome

Confusion matrix	Human annotated		
	Sentiments	Positive	Negative
Machine annotated (Predicted)	Positive	6	3
	Negative	0	1

We processed 10 randomly selected popular poems; the vocabulary of the poems was not explicitly considered while building the corpus. The same human experts tagged 6 with positive and 4 with negative sentiments. The confusion matrix comparing the outcome of the machine classifier with the human annotators is shown in the Table 7.

The Accuracy, Precision, Recall and F-score are 70%, 66.67%, 75% and 60%, respectively. Coincidentally and interestingly, the accuracy of automatic tagging in this case is identical with that of the 100 Telugu songs [14].

5 Discussion, Conclusion and Future Work

In this paper, we have presented an NBB classifier for the sentiment analysis of Konkani text; the corpus includes 9681 words. The model was tested on a known set of 50 poems, i.e., it was ensured that for each word in the testing sample there shall be at least one tag in the corpus. We also tested the model with 10 poems without any assurance about the tag-values of the words in the corpus. The accuracy reduces from 82.67 to 70%, though in the second experiment all the poems with positive sentiments were identified correctly. Only 29% of the poems with negative sentiment were correctly labeled in the first experiment. The number has been reduced by 13%, as 25% of the negative sentiment poems are correctly labeled the second time.

The model significantly outperformed that of the previous experiment of employing SVM. Our results are on a par with the naïve Bayesian classifier employed for tagging a set of Telugu songs. This effort has contributed an enhancement to the digital corpus of Konkani language for sentiment analysis. However, the results indicate that the corpus needs fine tuning in order to perform on a par with human annotators.

The results and the data generated through this research has been made freely available for the researchers to take the work ahead [31] and contribute to the quality and quantity of the NLP tools in Konkani, one of the important regional languages of India [4].

References

1. https://monkeylearn.com/sentiment-analysis/. Accessed 06 May 2019
2. Ferri, F., Alessia, A., Patrizia, G.: A Integrated methodology for approaching sentiment analysis in business domain. Int. Bus. Res. (2017)
3. Harshali, P., Alique, M.: Sentiment analysis for social media: a survey. In: 2nd International Conference Information Science and Security (ICISS), pp. 1–4 (2015)
4. https://www.statista.com/topics/2157/internet-usage-in-india/. Accessed 06 May 2019
5. https://www.statista.com/statistics/255135/internet-penetration-in-india/. Accessed 06 May 2019
6. https://en.wikipedia.org/wiki/Konkani_language. Accessed 06 May 2019
7. https://en.wikipedia.org/wiki/Goa. Accessed 06 May 2019
8. Ashweta, F., Jyoti, P., Ramdas, K.: Konkani sentiword—resource for sentiment analysis using supervised learning approach. In: Workshop on Indian Languages Data: Resources and Evaluation (WILDRE3), pp. 55–59 (2016)
9. Md. Shad, A., Asif, E., Pushpak, B.: Aspect based sentiment analysis category detection and sentiment classification for Hindi. In: Gelbukh, A. (Ed.) CICLing 2016, vol. 9624, pp. 246–257. LNCS, Springer International Publishing AG, part of Springer Nature (2018)
10. Parul, S., Teng-Sheng, M.: Prediction of Indian election using sentiment analysis on Hindi twitter. In: IEEE International Conference on Big Data (Big Data), pp. 1966–1971 (2016)
11. Vandana, J., Manjunath, N., Deepa, P., Venugopal, R., Patnaik, M.: HOMS: Hindi opinion mining system. In: IEEE 2nd International Conference on Recent Trends in Information Systems, pp. 366–371 (2015)
12. Kamala, S., Saikat, C.: A sentiment analysis system for Indian language tweets. In: Prasath, R., et al. (eds.) MIKE 2015, vol. 9468, pp. 694–702. LNAI (2015)
13. Shriya, S., Vinayakumar, R., Anand Kumar, M., Soman, K.P.: Predicting the sentimental reviews in Tamil movie using machine learning algorithms. Indian J. Sci. Technol. **9**(45) (2016)
14. Sandeep, M., Nurendra, C., Radika, M.: Enhanced sentiment classification of Telugu text using ML techniques. In: 4th Workshop on Sentiment Analysis Where AI Meets Psychology, 25th International Joint Conference on Artificial Intelligence, p. 29 (2016)
15. Harika, A., Eswar, A., Gangashetty, S., Radhika, M.: Multimodal sentiment analysis of Telugu songs. In: Proceedings of the 4th Workshop on Sentiment Analysis Where AI meets Psychology, IJCAI, pp. 48–52 (2016)
16. Chandan, G., Bal Krishna, B.: Detecting sentiment in Nepali texts: a bootstrap approach for sentiment analysis of texts in the Nepali languages. In: International Conference on Cognitive Computing and Information Processing (CCIP), pp. 1–4 (2015)

17. Lal Bahadur, T., Bal Krishna, B.: Classifying sentiments in Nepali subjective texts. In: International Conference on Information, Intelligence, System and Application, pp. 1–6 (2016)
18. Sanjib, S., Priyanka, B., Mohapatra, P.: Sentiment analysis for Odia language using supervised classifier: an information retrieval in Indian language initiative. CSI Trans. ICT 111–115 (2016)
19. Deepamala, N., Ramakanth, P.: Polarity detection of Kannada documents. In: IEEE International Advance Computing Conference (IACC), pp. 764–767 (2015)
20. Anil, K., Rajasimha, N., Manovikas, R., Rajanarayana, A.: Analysis of users' sentiments from Kannada web documents. In: 11th International Multi-conference on Information Processing, pp. 247–256. Proc. Comput. Sci. (2015)
21. Yashaswini, H., Padma, K.: Sentiment analysis for Kannada using mobile product reviews: a case study. In: IEEE International Advance Computing Conference (IACC), pp. 822–827 (2015)
22. Kamal, S., Mandira, B.: Sentiment polarity detection in Bengali tweets using multinomial Naïve Bayes and support vector machines. In: IEEE Calcutta Conference (CALCON), pp. 31–36 (2017)
23. https://www.cs.uic.edu/~liub/FBS/sentiment-analysis.html#lexicon. Accessed 06 May 2019
24. https://en.wikipedia.org/wiki/Naive_Bayes_classifier. Accessed 06 May 2019
25. https://scikit-learn.org/stable/modules/generated/sklearn.Naïve_bayes.BernoulliNB.html. Accessed 06 May 2019
26. http://www.konkanipoetry.com. Accessed 12 May 2019
27. http://www.kavitaa.com. Accessed 12 May 2019
28. http://p4poetry.com. Accessed 12 May 2019
29. https://en.wikipedia.org/wiki/Confusion_matrix. Accessed 11 May 2019
30. https://tdil-dc.in. Accessed 11 May 2019
31. www.annierajan.com. Accessed 11 May 2019

Meandered Slot-Line Balanced Filter with Common-Mode Noise Suppression

Ram Sidharth Nair, M. S. Kamalanadhan, Abhishek Singh, S. Abhijith and Sreedevi K. Menon

Abstract A balanced wide-band band-stop filter with high common-mode rejection at 2.45 GHz is proposed in this paper. The filter is realized using slot-line which is suitable for implementing balanced filters supporting differential TE mode signaling. An initial transmission line model for a band-stop filter employing shunt short-circuited stub is designed and differential/common-mode characteristics are examined by extracting the mixed-mode *S*-parameters. This is modified as a meander line which offers wide band-stop response, while also facilitating size reduction. Standard *S*-parameters and of the simulated slot-line structure, with and without meandered stubs are obtained. Finally, the proposed structure is implemented on FR4-Epoxy substrate and measured using a network analyzer and compared with the simulated response.

Keywords Meander · Slot-line · Balanced filter · Differential mode · Common-mode · Noise · Mixed-mode · Asymmetric coplanar strip

1 Introduction

As the scale of electronic devices continues to shrink, the distances between traces and interconnects guiding information-carrying signals shrink faster. These signals tend to be spectrum-rich and low in power, allowing them to be hijacked by noise coupled from adjacent traces or external noise sources. Maintaining signal integrity then becomes the motivation for the design of balanced networks [1]. Balanced networks are the superposition of two fundamental modes of operation: the differential mode and common mode. A differential pair comprises of two equal but opposite signals, hence, any noise coupled to one of the signals is canceled by the noise coupled to the other signal. A common-mode pair comprises of two equal and in-phase signals, hence, the immunity of this mode of operation to the effect of noise is substantially

R. S. Nair · M. S. Kamalanadhan · A. Singh · S. Abhijith · S. K. Menon (✉)
Department of Electronics and Communication Engineering, Amrita School of Engineering, Amrita Vishwa Vidyapeetham, Amritapuri, India
e-mail: sreedevikmenon@am.amrita.edu

M. Tuba et al. (eds.), *ICT Systems and Sustainability*,
Advances in Intelligent Systems and Computing 1077,
https://doi.org/10.1007/978-981-15-0936-0_33

lower. Therefore, balanced networks seek to amplify the role of the differential mode within the operating spectrum and suppress the role of the common-mode within the same band. This is accomplished in a variety of methods, some of which are delineated in [2–5].

Balanced coupled lines and quarter wave sections can be effectively utilized for noise suppression by implementing filters [2]. To achieve wide band common-mode suppression, dual mode ring resonator has been proposed [3]. Multi-sections of half wavelength resonators provide a fourth order balanced filter with limited rejection bandwidth [4]. Stepped impedance resonators along with two open stubs provide a dual band differential mode band-pass filter [5]. In all the techniques discussed, microstrip technology is utilized for noise rejection.

In the proposed work the balanced filter is realized as an Asymmetric Coplanar Strip (ACS) which provides good rejection as well as compactness for the balanced filter. ACS provides good transmission characteristics and can be used for design of filters [6] and antennas [7]. The detailed design and analysis is presented in the next sections.

2 Design Process

2.1 Band-Stop Filter Transmission Line Model

A band-stop filter network is designed using shunt short-circuited stub [8]. The stub has an electrical length of a wavelength at the frequency 2.45 GHz. The initial transmission line model is designed and simulated in ADS software. The circuit dimensions and characteristic impedance are as shown in Fig. 1.

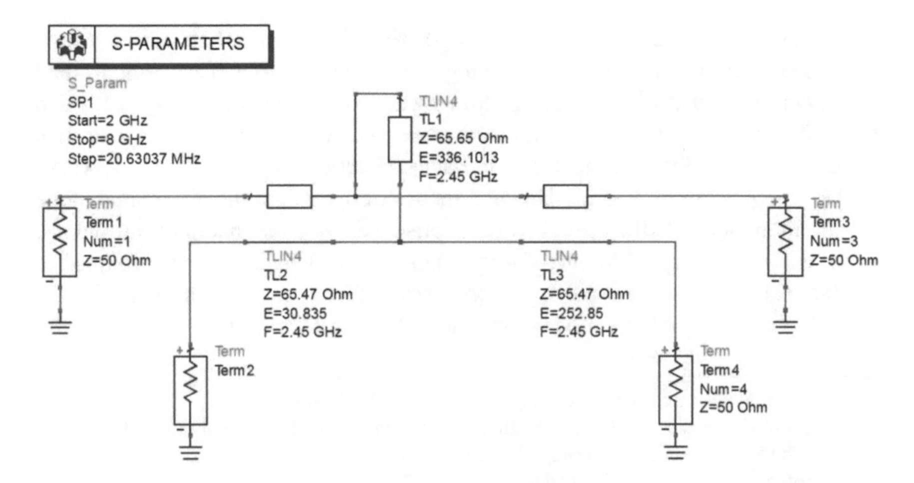

Fig. 1 Transmission line model diagram (ADS)

The network (Fig. 1) can be described as the superposition of a differential and common mode. In the differential mode, a virtual short exists along the plane of symmetry of the network, resulting in a differential mode impedance that possesses its own unique resonances. Similarly, in the common mode, a virtual open exists along the plane of symmetry of the network, resulting in a common-mode impedance and its corresponding resonances. The objective of a balanced/differential filter is to design the system in such a way as to move the common-mode resonances out of the spectral operating range of the differential mode. This ensures signals electromagnetically coupling to a system in a common-mode fashion due to external noise sources cannot affect signal integrity within the operating band, which is as plotted in Fig. 2 for the transmission line (TL) model. The broken lines in Fig. 2 show the transmission which is near to the 0 dB in the operating bandwidth which ensures signal is transmitted with less influence of noise. Also, solid lines show less reflection thus ensuring impedance matching.

Common-mode rejection ratio (CMRR) is a measure of the network's ability to suppress common-mode resonances within the operating band. It is defined as:

$$\mathrm{CMRR} = -20\log_{10}\left(S_{21,DM}/S_{21,CM}\right) \tag{1}$$

In Eq. (1), S_{21}, DM is the transmission parameter for the circuit operating in differential mode, and S_{21}, CM is the transmission parameter for the circuit operating in common mode. The CMRR metric can be extracted from the mixed-mode S-parameters [9] of the network under analysis. The mixed-mode S-parameters include descriptions of the standard scattering parameters of a network operating in more than one mode. For a two-port network, the mixed-mode scattering parameters are represented in a 4 × 4 matrix divided into 4 quadrants. Each quadrant is a standard

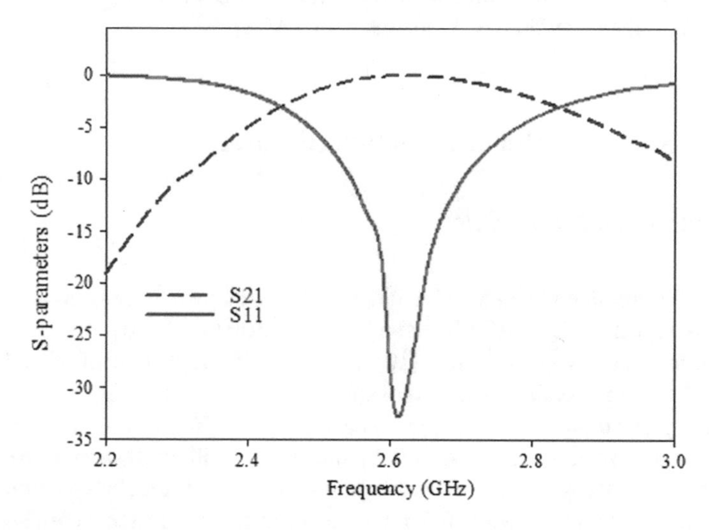

Fig. 2 Simulated two-port S-parameters transmission line model

2 × 2 scattering matrix describing circuit parameters in one of four modes, namely: purely differential mode, purely common mode, common-to-differential mode and differential-to-common mode. The last two modes describe how a common-mode stimuli can excite a differential mode resonance or vice versa. The following relations convert a two-port network excited by four single-ended terminals (resulting in a 4 × 4 standard scattering matrix) into the mixed-mode S-parameters.

$$S_{11}, DM = 0.5(S_{11} - S_{21} - S_{12} + S_{22}) \tag{2}$$

$$S_{21}, DM = 0.5(S_{13} - S_{41} - S_{32} + S_{42}) \tag{3}$$

$$S_{11}, CM = 0.5(S_{11} + S_{21} + S_{12} + S_{22}) \tag{4}$$

$$S_{21}, CM = 0.5(S_{13} + S_{41} + S_{32} + S_{42}) \tag{5}$$

The differential and common-mode reflection and transmission characteristics of the transmission line model in Fig. 1 extracted using Eqs. (2)–(5) are plotted in Fig. 3a, b.

In Fig. 3a a minimum reflection (S_{11}) with maximum transmission (S_{21}) is obtained in the operating bandwidth. So the differential mode acts as a band-pass filter at the frequency band designed.

In Fig. 3b a minimum transmission (S_{21}) with maximum reflection (S_{11}) is obtained in the operating bandwidth. So the common-mode acts as a band-stop filter at the frequency band designed. This ensures the suppression of common mode which effectively suppressing the noise. Next section discuss the planar implementation of the general transmission line model (Fig. 1) using a new approach of slot-line implemented in an Asymmetric Coplanar Strip (ACS).

3 ACS Meander Slot-Line Balanced Filter

3.1 Planar Slot-Line Filter

The transmission line model analyzed in Sect. 2 is designed as a planar slot-line on an ACS as shown in Fig. 4. ACS is designed for 50 Ω on FR4 Epoxy with central strip width 4 mm and slot width 0.35 mm [10]. This acts as a simple transmission line. For realizing this filter overall dimensions of substrate is 68 mm × 25 mm × 1.6 mm with the ground plane of 46 mm × 16 mm. The signal line is 46 mm × 4 mm. The signal line and ground plane is separated by 0.35 mm ensuring 50 for the transmission line. To ensure band-stop response on this transmission line, on the signal line of the ACS a slot of length 58 mm and width 0.5 mm is etched out. Length 58 mm corresponds to the guided wavelength at 2.45 GHz on FR4 epoxy substrate ($\varepsilon r = 4.4$, $h = 1.6$ mm).

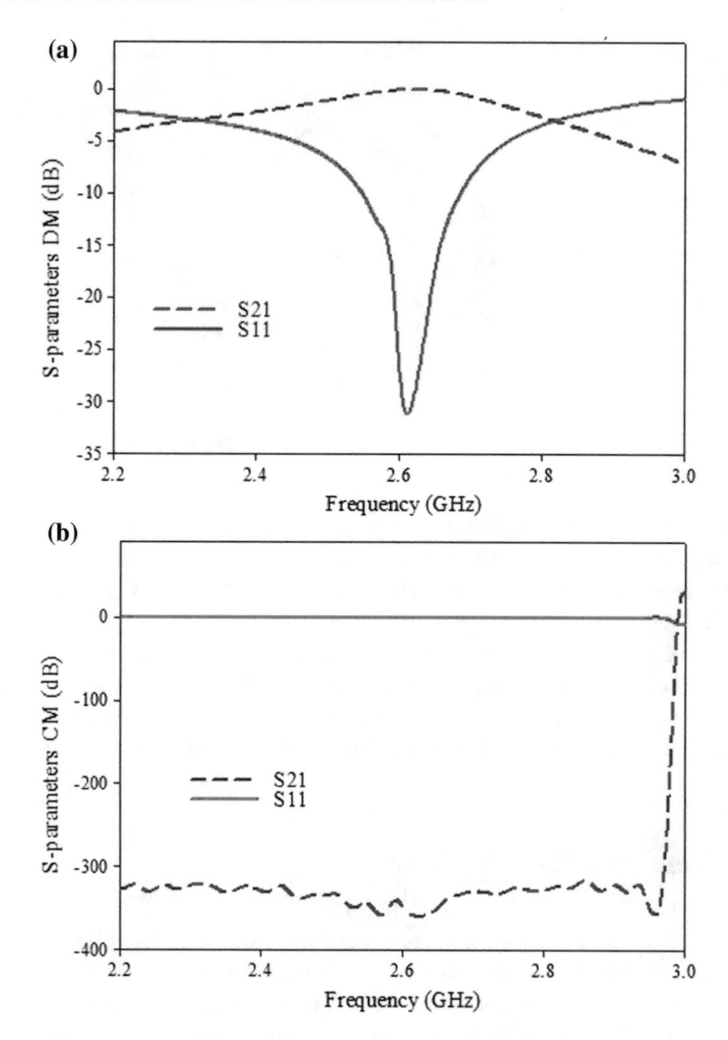

Fig. 3 **a** Differential mode *S*-parameters. **b** Common-mode *S*-parameters

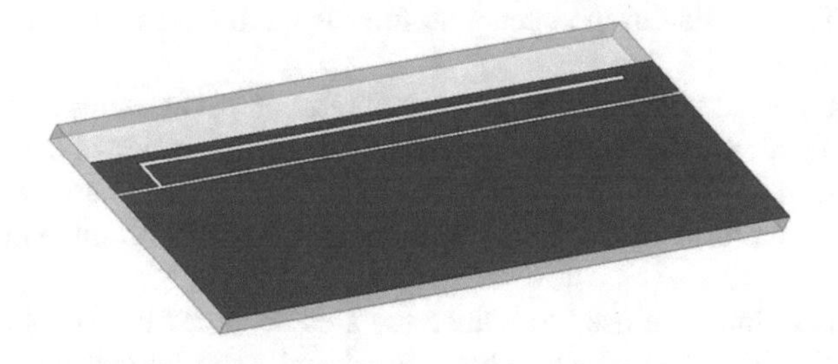

Fig. 4 ACS fed slot-line TL (black portion indicates the copper, ash is the substrate)

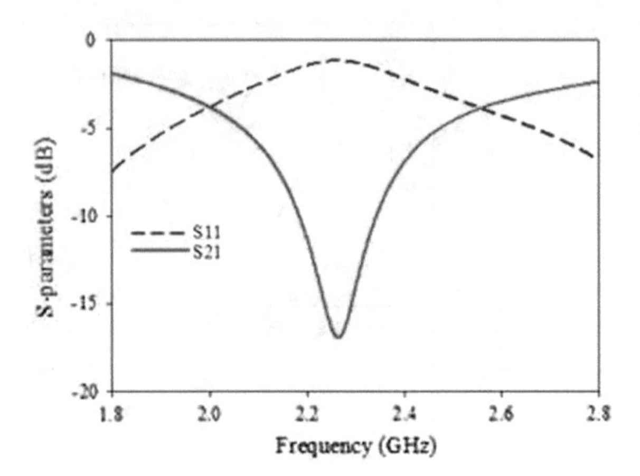

Fig. 5 *S*-parameters of ACS fed slot-line TL

Thus the etched line acts as an open circuited half wavelength line which provides a band-stop characteristics at the frequency of operation as illustrated in Fig. 5. The response ensures good reflection and transmission characteristics of a band-stop filter at 2.38 GHz.

This filter provides a 31 dB suppression in differential mode and 325 dB suppression in common-mode transmission, providing 589 dB CMRR when calculated using Eqs. (1)–(5). This open stub filter is modified as a meander line for further improvement in performance and is presented in the next section.

3.2 Meandered Slot-Line Filter

A simple open circuited stub matching network optimized for differential mode signaling by implementing the transmission line model as a slot-line etched onto a copper layer on a dielectric substrate is the initial design. The slot supports transverse electric mode of propagation parallel to the plane of the substrate, allowing for the effortless realization of short-circuits and surface-mount devices, as opposed to microstrip. The slot-line in the open stub filter is modified as a meander line as in Fig. 6.

The stub is meandered (with 6 turns) to facilitate compactness and improved wide-band characteristics in the frequency response. Length dimension of the substrate has been reduced from 68 to 46 mm which is a reduction of 33%. Increased number of meander lines can further reduce the size without serious distortion to the frequency response.

The meander lines are used to reduce the area occupied by the element. In the meander line, adjacent conductors have equal and opposite current flows, which reduce the total inductance. Mutual coupling effects are usually small if the spacing is greater than three strip widths [10]. The bandwidth and selectivity of the differential

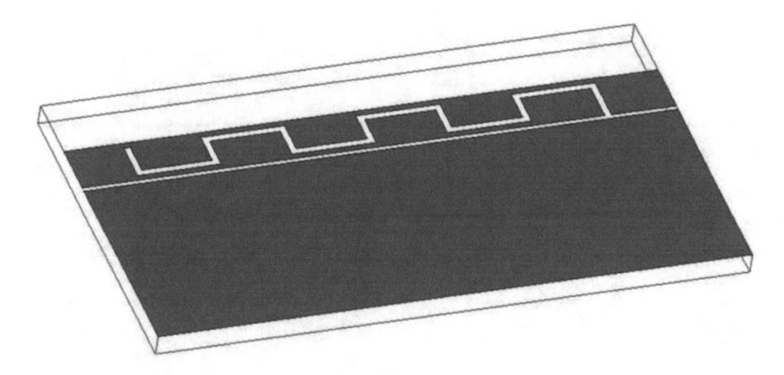

Fig. 6 Filter layout on FR4-Epoxy substrate (black portion indicates the copper, ash is the substrate)

mode is enhanced by meandering the stub slot-line, which also reduces the size of the device.

The slot-line of 68 mm is modified as meander with the different number of bends—$n = 1$–8. As the turn increases the depth of rejection increases as discontinuity increases. Beyond $n = 6$ much difference is not seen in rejection depth, optimizing $n = 6$. The design is simulated and optimized for the response. Even with 6 turns the total length of the line is maintained as 58 mm with 33% size reduction for the filter. A prototype of the optimized filter is fabricated on FR4 Epoxy and measurements are carried out using Keysight E 050A Vector Network Analyzer. The frequency response is presented in Fig. 7.

The simulated and measured results are in concurrence with each other. In comparison with the planar slot-line filter, a size reduction of 33% is achieved when slot-line

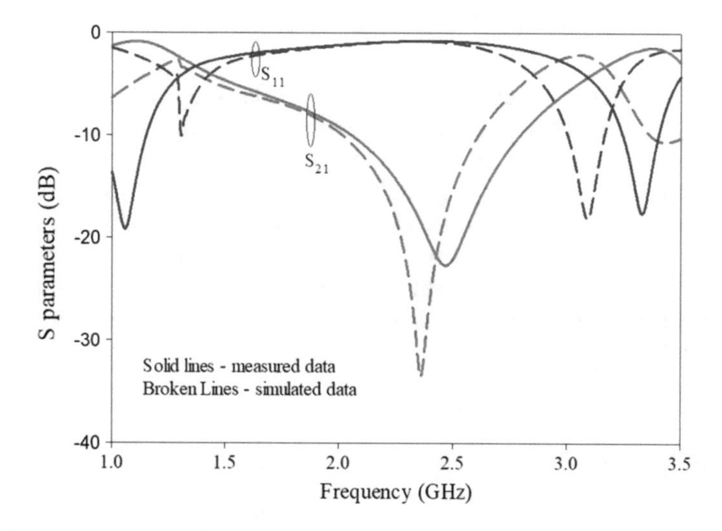

Fig. 7 Frequency response of the filter with $n = 6$, measured and simulated plots

Table 1 Performance metrics at each design level

Design level	Frequency of Max. rejection (GHz)	Max. dimension (mm)	Max. rejection (dB)
TL model in ADS	2.63	58	33
HFSS model without meandered stub	2.34	68	18
HFSS model with meandered stub	2.37	46	33
Meander slot-line balanced filter prototype	2.46	46	23

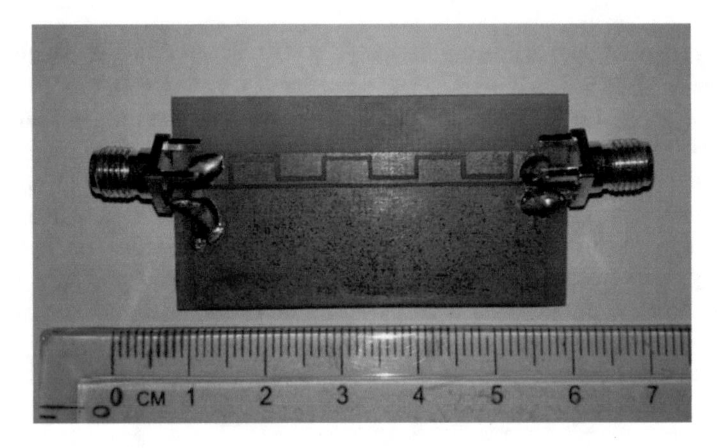

Fig. 8 Photo of the fabricated prototype

is meandered with 6 turns (effective length of meandered signal line is 37 mm). Performance of each filter from the initial transmission line model is presented in Table 1 and the fabricated prototype is shown in Fig. 8.

Compared to the structures proposed in the referred studies, the proposed structure involves straight-forward transmission line single-stub matching analysis (as opposed to coupled resonator theory, which is complex), is intrinsically balanced due to its asymmetric coplanar nature and is easier to fabricate. Even though the rejection is small compared to the other filters this can be increased by cascading multiple slot-lines.

4 Conclusion

A shunted short-circuited stub transmission line network with appropriate lengths and characteristic impedance is designed and analyzed for band-stop response at 2.45 GHz and high common-mode rejection ratio (CMRR) at the same frequency in ADS. Subsequently, the TL model is implemented as a planar slot-line structure in HFSS. In order to make the device more compact and wide-band at the desired frequency, the short-circuited slot-line stub is meandered to 6 turns to reduce the structure size by 33% along one dimension. The new structure is simulated once again in HFSS before being fabricated and tested using Keysight Vector Analyzer. The proposed device can find applications as low-noise filters that suppress interference due to electromagnetic coupling from adjacent signal paths or noise sources.

References

1. Feng, W., Che, W., Xue, Q.: The proper balance: Overview of microstrip wideband balance circuits with wideband common mode suppression. IEEE Microw. Mag. **16**(5), 55–68 (2015)
2. Wu, C.H., Wang, C.H., Chen, C.H.: Novel balanced coupled-line bandpass filters with common-mode noise suppression. IEEE Trans. Microw. Theory Tech. **55**(2), 287–295 (2007)
3. Feng, W., Che, W., Xue, Q.: Balanced filters with wideband common mode suppression using dual-mode ring resonators. IEEE Trans. Circuits Syst. I Regul. Pap. **62**(6), 1499–1507 (2015)
4. Wu, C.H., Wang, C.H., Chen, C.H.: Balanced coupled-resonator bandpass filters using multisection resonators for common-mode suppression and stopband extension. IEEE Trans. Microw. Theory Tech. **55**(8), 1756–1763 (2007)
5. Shi, J., Xue, Q.: Novel balanced dual-band bandpass filter using coupled stepped-impedance resonators. IEEE Microw. Wirel. Compon. Lett. **20**(1), 19–21 (2010)
6. Menon, S.K., Donelli, M.: Compact CPW filter with modified signal line. In: 2017 IEEE Applied Electromagnetics Conference (AEMC). IEEE (2017)
7. Meenu, L., Aiswarya, S., Menon, S.K.: Compact monopole antenna with metamaterial ground plane. In: 2017 Progress in Electromagnetics Research Symposium—Fall (PIERS—FALL), Singapore, pp. 747–750 (2017)
8. Pozar, D.M.: Microwave Engineering. Wiley (2009)
9. Bockelman, D.E., Eisenstadt, W.R.: Combined differential and common-mode scattering parameters: theory and simulation. IEEE Trans. Microw. Theory Tech. **43**(7), 1530–1539 (1995)
10. Garg, R., Gupta, K.C.: Expressions for Wavelength and Impedance of a Slotline (Letters). IEEE Trans. Microw. Theory Tech. **24**(8), 532–532 (1976)

A Novel Approach for Patient-Centric and Health Informatics Using Mobile Application

Bhagiya Kiran and Amit Kumar

Abstract Nowadays Indian Health Governance focuses on patient treatment as its key success factors with Hospital administrative and measures the quality of services of a hospital by adopting IT Technology. Health management has brought many enhancements through IT Technology for patient-centric that have a positive impact on the patient experience in hospitals as well as taking administrative decisions based on health indicators. Maharashtra Public Health Department through CDAC Noida developed and run its own e-Sushrut Hospital Management Information System. CDAC's e-Sushrut HMIS has integrated computerized clinical information system for improved hospital administration management system and patient health management. This paper is introduced adapting Mobile Technology, to design Android-based e-Sushrut Hospital Management information system easy to use and user-friendly. We here propose to develop an application where a patient can book an appointment with the authenticated secure payment process, track health records, download, and view investigation reports as well as Health Indicators which helps hospital administration to take decisions and facilitate the patient in a most efficient manner. The e-Sushrut mobile application is implemented by Java, providing the REST API and Android studio. Server uses JavaEE three-tier architecture.

Keywords e-Sushrut mobile application · Health indicators HMIS web application · Secure payment process

---

B. Kiran (✉) · A. Kumar
Centre for Development of Advanced Computing, Anusandhan Bhawan, C-56/1, Sector 62, Noida, India
e-mail: bhagiyakr@cdac.in

A. Kumar
e-mail: amit_kumar@cdac.in

M. Tuba et al. (eds.), *ICT Systems and Sustainability*,
Advances in Intelligent Systems and Computing 1077,
https://doi.org/10.1007/978-981-15-0936-0_34

1 Introduction

Healthcare domain is growing fast with the help of IT Technology. Nowadays the government health sector has implemented Hospital Management Information System(HMIS) at government hospitals in order to improve patient-care quality and make its decision supportive [1]. HMIS is a web-based application that has the facility to coordinate all patient & hospital information on single platform to enable hospital administration to do their work efficiently. HMIS has an Electronic Medical Records System (EMR) designed to manage all aspects of a hospital clinical operation [2] of patient care at hospital. This customizable HMIS solution includes Out Patient Management and In Patient Management, Inventory Management, Ward Management, Laboratory, Billing of Patient, Accounting of Hospital, HR, Alert system, HL7, and DICOMS standard [3].

HMIS has brought many improvements in patient expectations from hospital and have a positive impact on the patient experience to avail facilities in hospitals. HMIS is computerized clinical, nonclinical information system for improved hospital administration system and patient care [2] effectively. It also provides and maintain electronically stored medical record of the patients [2, 4]. Medical data of such records can be utilized for statistical requirements and research purpose of health care. The HMIS gives better monitoring and controlling the functionality of hospital. HMIS enables monitoring predefined health indicators by health department and provided exception reporting which facilitates decision making to state-level administrators for strategic decisions and policy.

Several factors contribute to delay in hospitals to provide better care to patients, like wait time at the registration counter, billing counter, waiting time for Investigation reports and waiting hours for treatment. In India many patients are not aware about the financial obligations of hospital charges. Healthcare systems need to design for patient's better care, it is very crucial to reduce patient waiting time at the hospital, convenient, transparent and reliable payment methods. However, the recent rapid progress in the mobile industry with increased mobile access [5], patients want more from their healthcare providers [6]. Some facilities are using mobile technology originally designed for patient-centric and health-centric is improve waiting time at the hospital, and improve treatment facility of care, transparency of hospital bill, and provide effective and responsive hospital facilities [7, 8]. The Health industry continues to upgrade patient-centric model of care as well as other facilities of hospital like approach of billing.

According to the above all expectations it seems necessary to set up a proper Mobile-based HMIS to fulfill the strategic goal of achieving patient-care quality, effective clinical & administrative processes of hospital. Now Mobile-based HMIS in our pocket, easy to use and user-friendly. The e-Sushrut Mobile-based HMIS has been developed to not only help the hospital administrators to have better monitoring and controlling the functionalities of hospitals over the state using pre-defined decision support indicators but also to assist the doctors and medical staff to improve health services. With basic demographical details patient, patient can book an appointment with different secure payment option, track demographic and medical records using QR code, download reports, etc. In this paper, we shall briefly discuss our e-Sushrut HMIS application and then goto the main theme of this paper is highlight the e-Sushrut Mobile HMIS application designed on Android platform.

In our proposed e-Sushrut application we have concentrated on web-based HMIS and Mobile application for Public Health Department of Maharashtra state in Sect. 2. The paper is organized as per Sect. 3 describes the proposed e-Sushrut mobile HMIS and functionalities. Section 4 finally covers the simulation setup, results obtained and analysis of the data results on a mobile screen.

2 e-Sushrut HMIS Background

In India, the majority of government hospitals are facing day to day operational challenges in administration and ineffective of handling patient-care services which result in waste of time, manpower, nontransparent and unaccountable working environment. Major Government hospitals follow paper base manual processes for patient registration, patient billing, investigations, OPD, and IPD patient treatment, and hospital administration. There is no any standardized administration of health-oriented data, improper inventory management, accounting system, and no such good records to generate various health indicators for the state-level officials, due to lack of decision making information. The hospital administrators have face difficult to get the accurate clinical information from each department on time, resource management, patient's demographic details, and clinical data online. Forgetting the same records take a lot of time and manpower was wasted unnecessary. Day to day such challenges in hospital administration and monitoring activities reflects the crucial need to adopt standardized automated platform which will manage the huge amount of information of government hospitals and meet the demands of modern computerized healthcare delivery.

Fulfilling this need, The Public Health Department of Maharashtra launched CDAC's Hospital Management Information System at two hospitals on a pilot basis in 2018, with the aim of other Hospitals keeping in mind. PHDM state has categorized the hospitals into different categories based on bed count at hospitals, CDAC has developed HMIS application for the different type of hospital categories like District Hospital, Sub District Hospital, Mental Hospital, Women Hospital, Regional Referral Hospital, General Hospital, Rural Hospital, and PHC.

e-Sushrut HMIS incorporates an integrated computerized clinical and nonclinical information system for improved patient health care and hospital administration. It also provides an accurate, Electronic medical record (EMR) of the patients. HMIS has facilitates to monitoring of pre-defined health indicators by generating reports on a dashboard for the hospital administration and state-level officials of Maharashtra. Due to queue management in application patient has to spend less time in queue at registration Counter, OPD, laboratory, dispensary, admission counters. Due to the digitization of data leads to transparency and accuracy in billing. Hospital Administration can view daily/weekly/monthly/yearly status on a dashboard. Medical officers can view and analysis patient treatment records, patient hospital visit summery on EMR Desk and laboratory investigations reports online at any place, which saves patients treatment time. Auto-generated emailing facility of daily stats like OPD & emergency registration, admission of the day, operated OT, a stock of blood bag, death report, Test performed, etc., to hospital administration as well as state-level officials.

The HMIS application has 20 modules covering patient, clinical, and administrative services. These 20 modules cover the entire scope of hospital activities from patient OPD registration, Bed Management, OPD Management, IPD Management, Operation Theatre management, Investigation, Inventory management, Diet Kitchen, Centralized sterilization System, Accounting, Transportation, Equipment management system, Alert System and overall monitoring functions. It is built in compliance with international and national standards like ICD-10, International codification for diseases and SNOMED, etc., HL7, DICOM: Imaging standard, Messaging standard, it is integrated with barcode technology for blood bank, drug management, and investigation. It has extensive MIS based reporting facility. The HMIS has been integrated with different applications like e-aushadhi, e-raktkosh, and MeraAsptal which are running at Maharashtra.

3 Proposed Mobile e-Sushrut HMIS

In the government hospitals manual process was installed decades ago, hence the hospital staff and patients are used to of these manual processes and are being unable to easily shift and accept the new system. The government hospitals have large volume of patients that makes the process of migrating to automated computerized processes very difficult. Patients do not have the patience to wait for online registration and data entry. Patients who are wanting quickly have better outcomes than those who have to wait a long time for completed online registration. Although it's a tough task, there are certain steps hospitals can take to more improve patient waiting time of registration process. Patients want more transparency on bill paid to hospital and hospital have to change the approach of billing for more accurate accounting of the hospital.

Most processes to collect healthcare payments today are paper-intensive. HMIS has to find ways to meet patients demand for transparent, convenient and secure reliable payment methods. It is quite difficult for a medical officer to use HMIS and view EMR of the patient online during the OPD hours. Some Patients are coming from rural areas are not able to collect investigation report physically. Pharmacy officers too faced some problems to manage online inventory in HMIS and handle patients at OPD Hours. It was difficult for nursing staff too, to handle application with patient care. All the above challenges and expectation of patient, hospital staff, and state-level officials from HMIS is increased.

Overcoming the above challenges CDAC has enhanced HMIS application and developed Mobile-based HMIS for Public Health Department of Maharashtra. Awareness and use of mobile phone become very common in villages also, everyone wants to get all the things very quickly and easily. A simpler and easy to use version of the HMIS needs to be developed instead of all-encompassing HMIS software. Such a step can introduce both hospital administration and citizens to the new HMIS system and prepare them over time for a much more detailed Mobile-based HMIS system. Mobile HMIS is easy to use and user-friendly.

The system is integrated with the web application in order to patients can schedule appointments of doctor easily, and provide an interactive, simplified version of the pay patient's bill. The Mobile HMIS gives the multiple simple and secure options to pay what they owe from mobile devices using a secure digital payment option. The patient can register with basic demographical details using HMIS Mobile application with digital payment options for a hospital visit. The system has a QR Code facility to download investigation report. The patient can view and download investigation report any time anywhere using the mobile-based HMIS system by registered mobile no also. The patient can view last visited hospital details with multiple options so do not need to carry bulky medical files and doctor's prescription for the next visit. Citizens can inquire about department wise OPD doctor visit time, charges of the hospital for different hospital services, available test at the hospital with charges, get IPD patients details, etc. Healthcare staff and doctors increased efficiency due to easy access of electronic medical records, templates for treatment recording cycle with ICD10, SNOMED and International codification for diseases. Hospital staff and State Administrators can view the hospital's day-to-day functioning, management information system, accurate patient information & can monitoring of pre-defined health indicators in Mobile device at any time easily. Mobile-based HMIS helps state-level officials to take efficient decisions with the help of pre-defined Health indicators. Hospital Staff and state-level officer can log in Mobile HMIS using same credentials of web-based HMIS.

3.1 Technological Features

Mobile-based HMIS is built on the Android OS using technology Java with the Postgres server as the backend. The Mobile HMIS solution using three-tier distributed architecture accessible over the Internet. The three-tier architecture enables the easy replacement of any tier without affecting the other two tiers. Mobile HMIS has used payment gateway facilitates the transfer information between a payment portal such as a mobile phone and the Front End Processor or acquiring a bank. Payment gateways encrypt sensitive information, such as credit card numbers, debit card details to ensure that information passes securely between the customer and the merchant. Implemented Smart Library to an application and pass relevant information required. All complex algorithm and encryption are handled by a framework on behalf of the application. SQL reporting and Crystal reports services are used for the reporting purpose on dashboard. A CDAC's central data server is connected to each hospital.

4 Simulation and Result

Mobile e-Sushrut HMIS has been implemented on Android OS and for this APK file has to be transferred and installed into a mobile phone. The user can download APK file from Google play store. This application has two facilities one for citizens and second for hospital administrators and State-Level officers.

4.1 For Hospital Staff and State-Level Officers

The interfaces have been designed and implemented to facilitate user as shown in Fig. 1 for the hospital administrator.

The application has Health Indicators which helps to hospital administration to facilitate the patient in the most efficient manner and help to make a decision for health care. The user can select any health indicator shown in Fig. 1 on click of an icon. This application has designed and implemented nine different health indicators.

Figure 2 is shown a graphical view of month wise OPD and month-wise admission for a selected hospital. Based on ADT management design few health indicators like month wise admission, month wise death, Bed Occupancy Ratio, Average Length of Stay of a hospital is shown in Fig. 2.

Fig. 1 Health indicators of e-sushrut mobile

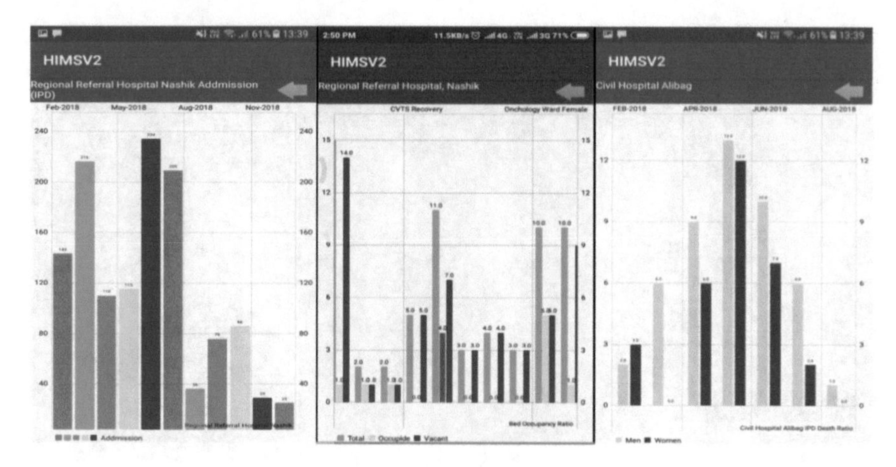

Fig. 2 Month wise admission, bed occupancy ratio & month wise death

Stock position of a drug in dispensary and Blood component-wise stock can view by a user in mobile HMIS dashboard as shown in Fig. 3. This all the health indicators can view on mobile by a single click, which is very useful to a hospital administrator.

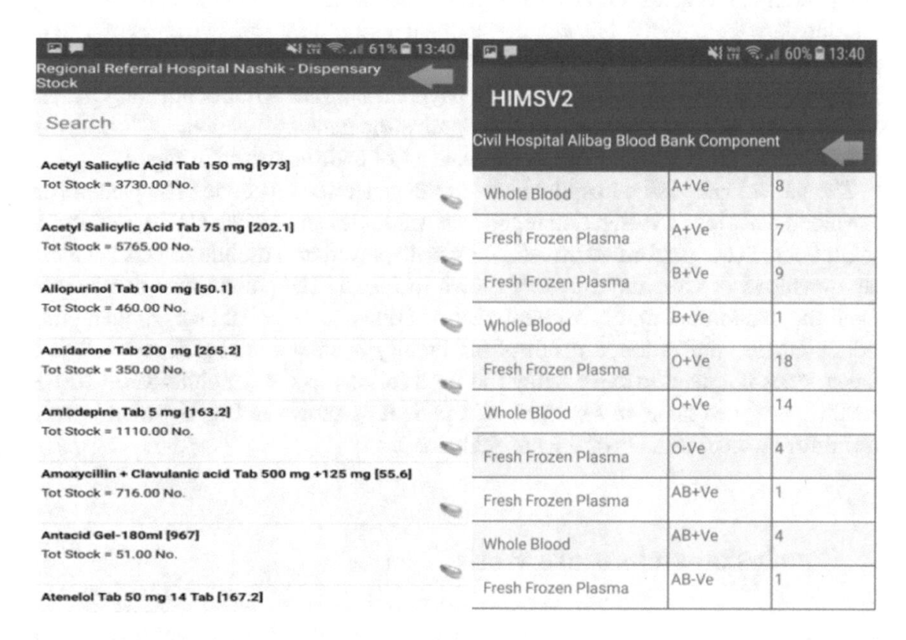

Fig. 3 Stock position of drug & blood component

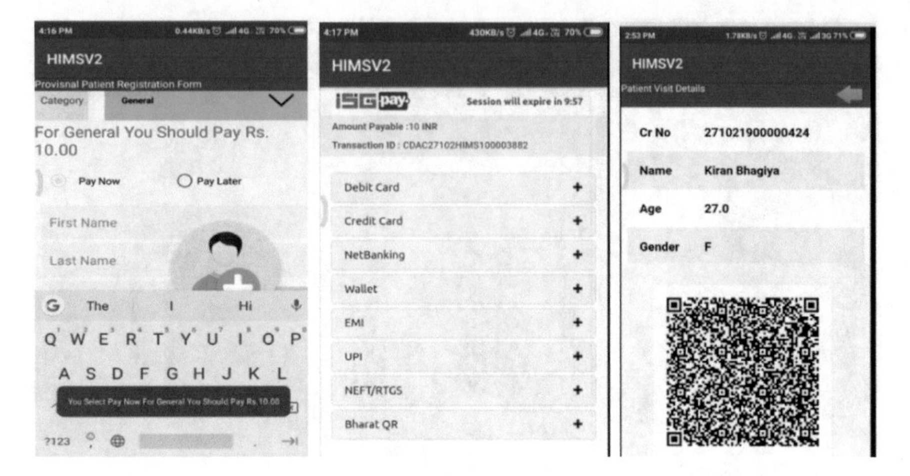

Fig. 4 Patient registration with payment & QR code for patient registration

4.2 *For Citizens*

The patient can register OPD visit of a hospital using this mobile application by entering demographical detail with the payment option. There are two types of patient registration with payment mode and without payment option. Figure 4 is shown patient registration with different secure payment options. After confirming patient registration details and payment option application generates patient's CR no. with basic detail of OPD visit as well as generate a QR code as shown in Fig. 4.

The patient can visit a hospital base on the generated QR code. The patient can download and view investigation report with multiple options like Cr.No, mobile no or QR code, list of performed investigation is displayed on a mobile screen, a patient can download or view the report as shown in Fig. 5. The patient also can inquire about the department unit consultant visit of OPD and Hospital charges using this mobile HMIS application, a result of the inquiry is shown in Fig. 5. As well as a patient does not need to carry bulky file for a follow-up visit, mobile-based HMIS has all the visited hospital summary of a patient as shown in Fig. 6 which is very useful for medical officers of the hospital also.

5 Conclusion and Future Work

An e-sushrut HMIS is enhancing and replacement of manual processes into computerized system of government hospitals of Maharashtra. We have enhanced web-based HMIS to Mobile-based application for more convenient daily use for patients and hospital administration. Mobile-based HMIS solution reduced patient waiting time for registration, a long queue of registration counter, billing process and effective

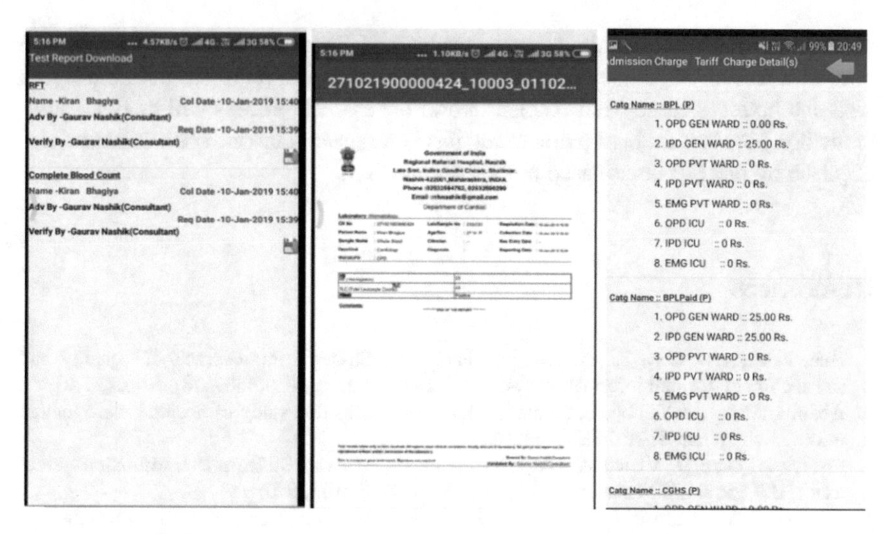

Fig. 5 Investigation report view & hospital charge

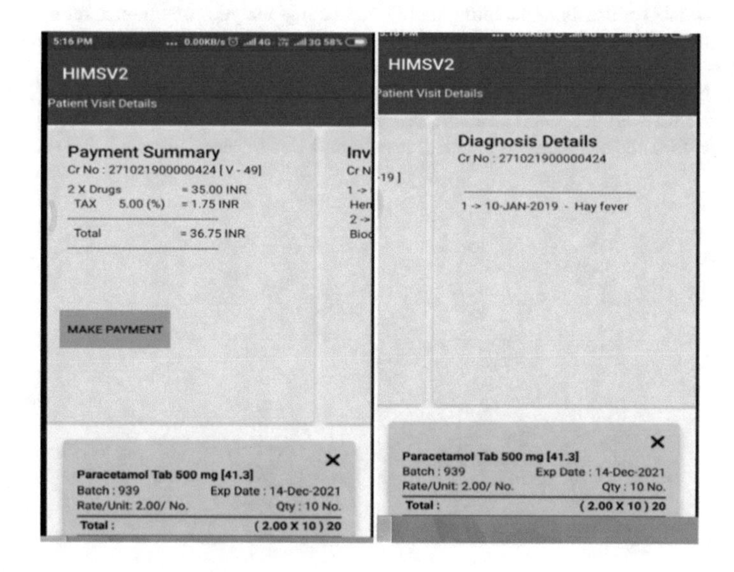

Fig. 6 Patient pre-visited details

health services at hospitals due to a digitized history of records at EMR Desk. The State Administrators can monitor hospitals day to day functioning, management information system and pre-defined health indicators on a mobile phone screen on single click. In future work focuses on enhancing application for more indicators, nonclinical HMIS services, machine interface with lab equipment and bar-coding system for a Blood bank. In the future, the proposed Mobile Patient Registration system can be extended to include the ability to develop the system towards further notification of appointments using messages and Email. In addition, System will possess a feature of providing automatic system calls as a reminder before hospital visit and better accuracy and efficient workflow as each process will be done automatically. Last but not least patients can just be registered by one click of button and check-in by just QR code-based mobile application.

References

1. Kuo, K.M., Liu, C.F., Talley, P.C. and Pan, S.Y.: Strategic improvement for quality and satisfaction of hospital information systems. J. Healthc. Eng. **2018**(3689618), 14 p, (2018)
2. Abdulla, M.N., Al-Mejibli, I., Ahmed, S.K.: An investigation study of hospital management information system. IJARCCE **6**(1) (2017)
3. Cordos, A., Orza, B., Vlaicu, A., Meza, S., Avram, C., Petrovan, B.: Hospital information system using HL7 and DICOM standards. ISSN: 1790–0832, **7**(10) (2010)
4. Adebisi, O.A., Oladosu, D.A., Busari, O.A., Oyewola, Y.V.: Design and implementation of hospital management system.|| Int. J. Eng. Innov. Technol. (IJEIT), **5**(1) 2015
5. Anyl Bosomworth (2015 Jan 15). Mobile Marketing Statistics 2015. Retrieved from http://www.smartinsights.com/mobile-marketing/mobile-marketing-analytics/mobile-marketing-statistics/
6. Nizar, S.S., Khadamkar, P., Kumar, M., Maramwar, V.: Healthcare services using android devices. Int. J. Eng. Sci. (IJES) **3**(4), 41–45 (2014)
7. Zarka, N., Mansour, M.M., Saleh, A.: Mobile Healthcare System. ICYRIME **1712** (2016)
8. Talwar, Y.K., Karthikeyan, S., Bindra, N., Medhi, B.: Smartphone'-A user-friendly device to deliver affordable healthcare–a practical paradigm. J Health Med Informat 7, **7**(3) (2016)

Policies to Mitigate Select Consequence of Social Media: Insights from India

Naganna Chetty and Sreejith Alathur

Abstract Apart from the benefits to humanity, often, social media possess consequences such as online hate content, fake news, online abuse, cyber-bullying and other demeaning expressions. As the increased hate content causes several health issues, it is necessary to mitigate it. Hate content mitigation may attain sustainable development goals of United Nations. Therefore, the objective of the article is to identify the possible policies to mitigate social media consequence-online hate content. In this regard, online and offline opinions from the Indian respondents are gathered through the questionnaire designed for the purpose. The software which is developed in R programming language is used to analyse opinions. The analysis results reveal the role of government and non-government authorities for digital hate content reduction. The non-governmental communities-civil societies, private sectors, and intermediaries are more important to reduce digital hate content.

Keywords Social media · Online hate content · India · Mitigation · Policy

1 Introduction

Advances in Internet Technologies (ITs), online social media, network connectivity and reduced cost of communication devices are attracting citizen towards online forums. Irrespective of the location and background, social media communities support interaction among the users conveniently [1, 2]. Often, the other side of the social media results in antisocial activities [1], depression [3] and generation of problematic content. Some of the problematic content is harassment, pornography, hate content [4] and fake news [5]. Expression of hatred through social media is prevalent and a serious issue, because it is difficult to provide barriers [6] and social media supports rapid amplification [7] to hate content.

N. Chetty (✉) · S. Alathur
National Institute of Technology Karnataka, Surathkal, Mangalore, India
e-mail: nsc.chetty@gmail.com

S. Alathur
e-mail: sreejith.nitk@gmail.com

M. Tuba et al. (eds.), *ICT Systems and Sustainability*,
Advances in Intelligent Systems and Computing 1077,
https://doi.org/10.1007/978-981-15-0936-0_35

The hate content is bias motived with hostile and malicious nature and exhibited against a group or an individual. This content is expressed by considering the innate characteristics of the victim such as gender, religion, race, disability, etc. [8, 9]. The content which is expressed over the Internet and its associated social media based on any of the protected characteristics is referred to online hate content.

Regulations are the legal landscape [10], applied to ensure the appropriate use of available information or content. The digital content regulations are measures to ensure the appropriate use of digital content. An increase in the access to online content, particularly the social networking sites by individuals and corporate bodies leads to draw a legal framework and some of the key issues that affect users of social media. Regulation is a major subset of governance and tries to control events flow and behavior. Governance is meant to "provide, distribute and regulate" [11]. As the traditional technologies of digital content communication are already regulated separately, it is essential to regulate digital content, in particular, digital hate content.

United Nation (UN) is putting more efforts to attain sustainable development goals agenda by 2030. As the increased hate content [12] causes health issues, it is required to take corrective actions to control it. Hate content control may attain sustainable development goals of UN such as "gender equality by suppressing sexist language and behavior" (UN, 2015: Goal 5), "reduce inequality within and among countries by raising voice against discrimination" (UN, 2015: Goal, 10) and "promoting peaceful and inclusive societies by electing right persons, addressing homicide, violence and human trafficking" (UN, 2015: Goal, 16). Often, the hate content control may reduce health issues and support health improvement (UN, 2015: Goal 1).

With this introduction, the rest of the article is organized as follows. Section 2 provides the background of the research as a review. In Sect. 3, the research model and methodology are developed. Results of the analysis are discussed in Sect. 4. Section 5 draws some hate content reduction policies based on the results. Finally, in Sect. 6, the article concludes with the outcome and future scope of the work.

2 Background

Nowadays, hatred and harassment can be expressed easily online because anonymity and mobility well possible in Internet communities. These features increased the generation of online hate content and demanded the stricter laws to combat. The hate content and its impact can be reduced by strengthening legislations along with technological innovations [13, 14]. Depending on the context and target of a hate incident, the penalties can be increased to control it [15].

Racism is a prevalent form of hate content and is defined as "hatred behavior, exhibited in written, verbal or physical form against the ethnicity or physical appearance of a group or an individual" [16, p. 1]. In 1976, a campaign was organized by the rock against racism (RAR) organization in Briton with a title "love music, hate racism" to fight against racism. The organizers of the civil society campaign attained the popularity to some extent [17]. In 2012, though, the politicians and some citizens were talking about the war between countries, the messages of love exchanged over social media by civilians were powerful in controlling hatred against each other [18]. To reduce occurrence and impact of online hate content users are required to be careful during their online presence. They can avoid unnecessary and unknown connections online for individual's betterment [14].

Anonymity and different legal jurisdiction in online world made difficult to trace generator of the digital hate content. In India, the possible strategies for controlling hate content are helplines, Internet disconnection and counter-speech [19]. With this background on influencers for hate content reduction, the following two hypotheses are designed, tested and validated.

Hypothesis H1: Governmental policy attributes reduce online hate content.
Hypothesis H2: Non-governmental policy attributes reduce online hate content.

3 Research Model and Methodology

3.1 Data Collection and Preparation

Both online and offline modes are used for data collection. The people with Internet awareness are identified as target respondents to answer the questionnaire. In the offline mode of data collection, the students and faculty of institutes are approached. By personally visiting them, the questionnaires are get filled from both students and faculty. For convenience, related terms of the questionnaire are briefed to the respondents. Google document is created with all the questions to collect data online. Online data is collected through the platforms—E-mail, Twitter, and Facebook by posting a request message and the link. A total of 716 samples are collected through both online and offline modes.

3.2 Research Model

With the help of a literature review, different dependent and independent variables related to hate content regulations have been identified. These variables are used in model design.

Dependent variable: Hate content reduction (HCR) is identified as the dependent variable and can be measurable by governmental and non-governmental policy attributes.
Independent variables: governmental and non-governmental policy attributes are identified as independent variables and can be measured through the questionnaire

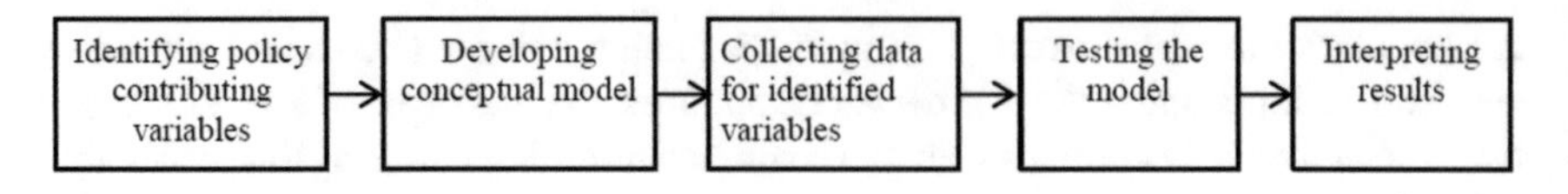

Fig. 1 Methodology

surveys. Strengthening legislation, increasing penalties and campaigns by the government are the contributing factors to governmental policy attribute [20, 21]. Campaigns by private sectors, campaigns by civil societies and efforts by social media are the contributing factors to non-governmental attribute [22, 23].

3.3 Methodology

A conceptual model is designed using identified attributes. To ensure the strength and contribution of policy attributes to reduce hate content, the model is tested with the help of software developed using R programming language. The overall steps of a process are shown in Fig. 1.

Structural equation modeling (SEM) takes into account the relationship between the more than one variable and the measurement errors simultaneously [24]. The variance based SEM, i.e., partial least square SEM (PLS-SEM) is used to develop the model. The model is developed in R software using partial least square-path modeling (PLS-PM) package. The model involves two parts such as structural/inner model and measurement/outer model.

4 Results and Discussion

4.1 Measurement Models Assessment

These models are evaluated for unidimensionality and contribution of the measuring indicators. Cronbach's alpha and principal component analysis approaches are used to assess unidimensionality. Unidimensionality evaluation values are shown in

Table 1 Factors and unidimensionality

Factor	Mode	Type	Size	C. alpha	Eigen value 1st	Eigen value 2nd
GPA	Formative	Exogenous	3	1.0	3	0.000224
NPA	Formative	Exogenous	3	1.0	3	0.000260
HCR	Formative	Endogenous	2	1.0	2	0.000347

Table 1. As Cronbach's alpha values are greater than the specified value 0.7, the first and second Eigenvalues are larger and smaller than 1 respectively, the measurement model is unidimensional.

Further, the measurement models are assessed by measuring the contribution of individual indicators of a latent construct. The measurement models are shown in Fig. 2. Each path of a model possesses magnitude and direction. Magnitude indicates the weight of the predictor whereas direction indicates the effect of the predictor on the response variable.

The construct governmental policy attribute is measured by strengthening legislation, increasing penalties and campaigns by the government. The contribution from strengthening legislation, increasing penalties and campaigns by the government is 34%, 26%, and 40% respectively. The results indicate that strengthening legislation, increasing penalties and campaigns by the government all plays an important role in forming governmental policies to reduce hate content.

The indicators campaigns by private sectors, campaigns by civil societies and efforts by social media are the contributing factors to non-governmental policy attribute with an individual contributions of 27%, 31%, and 42% respectively.

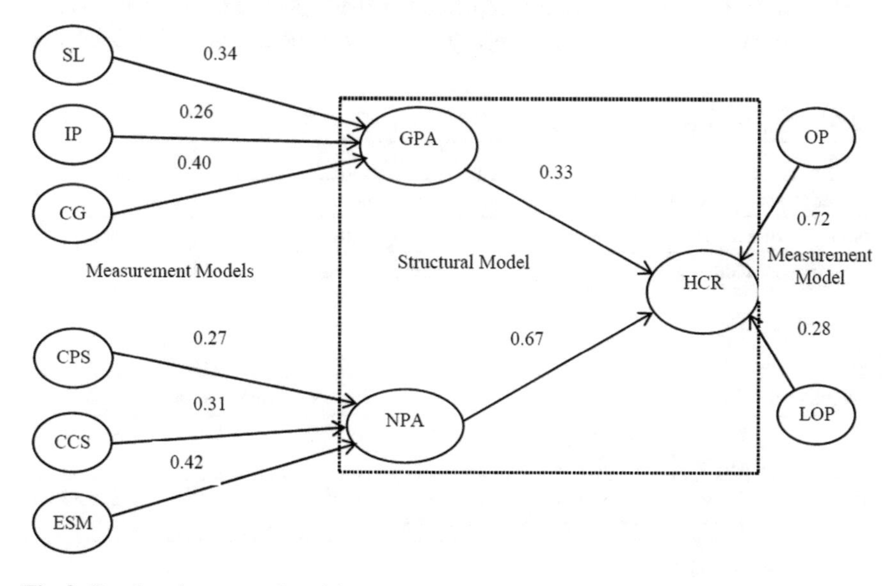

Fig. 2 Developed conceptual model

This indicates that all the indicators play an important role in constructing non-governmental policies. The contribution of online precaution to hate content reduction is 72%, and the limited online presence of an individual is 28%. The contribution of online precaution reveals that by being careful while accessing online media the hate content and its impact can be reduced largely.

4.2 Structural Model Assessment

The structural model is shown in Fig. 2 within a dotted rectangle. Coefficients of determination/R^2, the fitness of good and the regression analysis are used to evaluate structural models. R^2 represents the percentage of variance in the dependent variable with the changes in independent variables. The exhibited R^2 value of developed model is 1.0 and is shown in Table 2. Referring to R^2 value, the model is good as it is above 0.60 [25] and is substantial/high as it exceeds 0.26 [26]. The R^2 value 1 also indicates that all necessary covariates are considered for model development [27, 28].

The goodness of fit (GoF) is another criterion to measure the overall performance of the model. Usually, GoF requires no threshold for its statistical significance, but some authors suggested the possible range of values. The GoF value of the developed model is 0.99 which is much more than the cut-off 0.36 specified for substantial/large models [29]. The path analysis results of the structural model are shown in Table 3.

A path GPA → HCR with coefficient 33.5 at t-value = 7.56 and p-value < 0.001 represents that the predictor GPA is significant and hence by rejecting the null hypothesis the alternate hypothesis H1 is accepted. Hypothesis H2 tested by a path NPA → HCR. The path values reveal that it is significant with coefficient 66.5 at t-value =

Table 2 Coefficient of determination

Block	Type	R^2
Governmental policy attribute	Exogenous	0.00
Non-governmental policy attribute	Exogenous	0.00
Hate content reduction	Endogenous	1.0

Table 3 Path analysis results

Hypothesis	Path	Path coefficient	t-value	Result
H1	Governmental policy attribute → Hate content reduction (GPA → HCR)	33.5[a]	7.56	Supported
H2	Non-governmental policy attribute → Hate content reduction (NPA → HCR)	66.5[a]	15.0	Supported

[a]Significance at the $P < 0.001$ level

15.0 and p-value < 0.001. Therefore, the null hypothesis is rejected and the alternate hypothesis H2 is accepted.

5 Policy Guidelines

Based on analysis results, the following policy guidelines are recommended to reduce online hate content in the Indian context.

Governmental policies

- Strengthen legislations against hate content expression. The possible means are sections 153(A), and 295(A) of Indian penal code (IPC) can be improved and more sections under IPC to punish the perpetrator of hate content against each protected characteristics can be included.
- Increase penalties on hate content perpetrators. In this policy, a fine to be paid by the perpetrator can be revised regularly. The fine amount may be paid to the victims so that the impact of hatred can be reduced.
- Organize campaigns to bring awareness on hate content and its impacts. This policy can be implemented by conducting campaigns at educational institutes, organizations with more workers, at religious events and through social media sites.

Non-governmental policies

- Organize informational campaigns by private sectors on hate content and its impacts. This policy can be implemented by conducting the campaigns for employees of the organization. Further, if the sectors are non-profit organizations such as educational institutes and charitable trusts, the campaigns can be extended to more citizens.
- Organize informational campaigns by civil societies. The civil societies such as NGOs can organize the campaigns on hate content impacts for the citizen to implement this policy.
- Active involvement of social media companies with a sufficient amount of resources such as man power, technologies, and infrastructure is essential to mitigate hate content. By adapting appropriate techniques and required man power, social media may detect the hate content at its first occurrence. Then by identifying the source of the content, the hate content and the perpetrator social media account can be made inactive.

Online usage related policies

- Encouraging people to be careful during online presence as a personal responsibility may reduce the impacts of hate content.
- Convenience people for limited presence with online platforms to reduce the impact of hate content.

6 Conclusion and Future Work

The conceptual model is developed and tested with R programming against the hypotheses and validated the model. The results reveal that the model is fit and valid for the purpose. Both governmental and non-governmental policy attributes play an important role in hate content reduction. The variables strengthening legislation, increasing penalties and campaigns by the government are the influencers to governmental policy attributes. Similarly, the non-governmental policy attribute is influenced by the variables campaigns by private sectors, campaigns by civil societies and efforts by social media. Apart from these governmental and non-governmental policy attributes, two standalone variables- online precaution and the limited online presence of an individual also influences hate content reduction.

As the policies are suggested based on the respondents views, implementing them may reduce online hate content. The reduction of online hate content may lead to attain UN's sustainable development goals 3, 5, 10 and 16. To conclude, non-governmental policy attributes such as campaigns by private sectors, civil societies, and efforts by social media are vital to reducing hate content than the governmental policy attributes. In future, to gain broader control on hate content, variables size can be increased and respondents' online presence can be considered for model development.

References

1. Amedie, J.: The Impact of Social Media on Society. Advanced Writing: Pop Culture Intersections 2. http://scholarcommons.scu.edu/engl_176/2 (2015)
2. Kapoor, K.K., Tamilmani, K., Rana, N.P., Patil, P., Dwivedi, Y.K., Nerur, S.: Advances in social media research: past, present and future. Inf. Syst. Front. **20**(3), 531–558 (2018)
3. Shensa, A., Escobar-Viera, C.G., Sidani, J.E., Bowman, N.D., Marshal, M.P., Primack, B.A.: Problematic social media use and depressive symptoms among US young adults: a nationally-representative study. Soc. Sci. Med. **182**, 150–157 (2017)
4. Gillespie, T.: Custodians of the Internet: Platforms, Content Moderation, and the Hidden Decisions that Shape Social Media. Yale University Press (2018)
5. Caplan, R., Hanson, L., Donovan, J.: Dead reckoning navigating content moderation after "Fake News". Data Soc. https://datasociety.net/output/dead-reckoning (2018)
6. Mondal, M., Silva, L.A., Correa, D., Benevenuto, F.: Characterizing usage of explicit hate expressions in social media. New Rev. Hypermed. Multimed. **24**(2), 1–21 (2018)
7. Jubany, O., Roiha, M. (2016). Backgrounds, experiences and responses to online hate speech: a comparative cross-country analysis. Universitat de Barcelona. abs/1504.02305
8. Chetty, N., Alathur, S.: Hate speech review in the context of online social networks. Aggress. Violent. Beh. **40**(3), 108–118 (2018)
9. Seglow, J.: Hate speech, dignity and self-respect. Ethical Theory Moral Pract. **19**(5), 1103–1116 (2016)
10. Cohen, R.: Regulating hate speech: nothing customary about it. Chi. J. Int'l L. **15**, 229 (2014)
11. Braithwaite, J., Coglianese, C., Levi-Faur, D.: Can regulation and governance make a difference? Regul. Gov. **1**(1), 1–7 (2007)

12. Dutton, M.A., Green, B.L., Kaltman, S.I., Roesch, D.M., Zeffiro, T.A., Krause, E.D.: Intimate partner violence, PTSD, and adverse health outcomes. J. Interpers. Violence **21**(7), 955–968 (2006)
13. Banks, J.: Regulating hate speech online. Int. Rev. Law Comput. Technol. **24**(3), 233–239 (2010)
14. Penney, J.W.: Internet surveillance, regulation, and chilling effects online: a comparative case study. Internet Policy Rev. **6**(2) (2017). https://doi.org/10.14763/2017.2.692
15. Jenness, V.: The emergence, content, and institutionalization of hate crime law: How a diverse policy community produced a modern legal fact. Annu. Rev. Law Soc. Sci. **3**, 141–160 (2007)
16. Chetty, N., Alathur, S.: Racism and social media: a study in Indian context. Int. J. Web Based Commun. **15**(1), 44–61 (2019)
17. Dawson, A.: Love Music, Hate Racism: the cultural politics of the rock against racism campaigns. Postmod. Cult. **16**(1) (2005). https://muse.jhu.edu/article/192260
18. Kuntsman, A., Raji, S.: Israelis and Iranians: Get A Room!: Love, Hate, and transnational politics from the "Israel Loves Iran" and "Iran Loves Israel" Facebook campaigns. JMEWS: J. Middle East Women's Stud. **8**(3), 143–154 (2012)
19. Arun, C., Nayak, N.: Preliminary Findings on Online Hate Speech and the Law in India. Berkman Klein Center Research Publication No. 2016–19 (2016). https://papers.ssrn.com/sol3/papers.cfm?abstract_id=2882238. Accessed 10 Jan 2019
20. Breen, D., Nel, J.A.: South Africa-A home for all: the need for hate crime legislation. South Afr. Crime Q. **38**, 33–43 (2011)
21. Dharmapala, D., Garoupa, N.: Penalty enhancement for hate crimes: an economic analysis. Am. Law Econ. Rev. **6**(1), 185–207 (2004)
22. Henry, J.S.: Beyond free speech: novel approaches to hate on the Internet in the United States. Inf. Commun. Technol. Law **18**(2), 235–251 (2009)
23. Gagliardone, I., Gal, D., Alves, T., Martinez, G.: Countering Online Hate Speech. Unesco Publishing, Paris (2015)
24. Schumacker, R.E., Lomax, R.G.: A Beginner's Guide to Structural Equation Modeling, 3rd edn. Routledge, New York (2012)
25. Sanchez, G.: PLS path modeling with R. Berkeley: Trowchez Editions **383** (2013)
26. Cohen, J.: Statistical Power Analysis for the Behavioral Sciences, 2nd edn. Lawrence Erlbaum Associates, New York (1988)
27. Miaou, S.P., Lu, A., Lum, H.S.: Pitfalls of using R2 to evaluate goodness of fit of accident prediction models. Transp. Res. Rec. **1542**(1), 6–13 (1996)
28. Shtatland, E.S., Kleinman, K., Cain, E.M.: One more time about R2 measures of fit in logistic regression. NESUG 15 Proc. **15**, 222–226 (2002)
29. Wetzels, M., Odekerken-Schröder, G., Van Oppen, C.: Using PLS path modeling for assessing hierarchical construct models: guidelines and empirical illustration. MIS Q. **33**(1), 177–195 (2009)

Exploring Demographics and Personality Traits in Recommendation System to Address Cold Start Problem

Vivek Tiwari, Ankita Ashpilaya, Pragya Vedita, Ujjwala Daripa and Punya Prasnna Paltani

Abstract Several different approaches in recommender system have been suggested based on underlying rating history, but the majority of them suffer from the cold start problem (i.e., an inability to draw inferences to recommend items to new users. In this paper, a hybrid method has been proposed that combines personality traits using myPersonality application created by Facebook and the demographic characteristics into the traditional rating-based similarity computation. Majorly, personality information has been categorized into five personality traits: Openness, Conscientiousness, Extraversion, Agreeableness, and Neuroticism. Aforesaid personality characteristics can efficiently address the new user problem as it will accurately predict the similarity between users.

Keywords Personality trait · Cold start problem · Recommender system · Collaborative filtering · Demographic attribute · Cosine similarity

1 Introduction

Nowadays, data is available in huge amount and continuously growing minute basis, especially through e-business and socially sites. This is possible due to cheap and efficient data and pattern storage system such as data warehouse, pattern warehouse, Hadoop, etc. [1, 2]. Business industries can utilize such data through data mining techniques [3, 4]. In this view, we are living in a world, where recommendations are so frequent and use in a variety of purpose [5].

Majority of the recommender framework deals with two sorts of information [5, 6]: (i) data related to user's buying and rating pattern (user-item interactions) and (ii) data related to users and items such as textual profiles, underlying keywords, etc. Methods that use the users' explicit ratings on items or their purchase history are referred to as collaborative filtering methods. Collaborative Filtering builds the system using the past behavior of users [7, 8]. The methods that use the items attributes

V. Tiwari (✉) · A. Ashpilaya · P. Vedita · U. Daripa · P. P. Paltani
IIIT Naya Raipur, Raipur, CG, India
e-mail: vivek@iiitnr.edu.in

M. Tuba et al. (eds.), *ICT Systems and Sustainability*,
Advances in Intelligent Systems and Computing 1077,
https://doi.org/10.1007/978-981-15-0936-0_37

or specific keywords are known as content-based recommender framework. Content-based recommendations framework is based on the content of items rather than on other user's opinions, interactions, etc. [5, 6]. It suggests an item which the user has already enjoyed in the past and are looking for the same features in the present day. Another approach which combines the two methods is a hybrid approach based recommendation. This uses a linear combination of recommendations and presents them together. Another approach, which utilizes user demographic data to recommend items. Demographic-based recommender [9, 10] uses users' demographic information stored in profiles such as age, gender, location, etc., it assumes that users with correlated demographic characteristics might rate items similarly. The major problem in recommender system is the cold start problem.

2 Brief Literature Review

Buettner [11] has discussed personality prediction using online social media dataset and tried to get underlying behavior and trends. They incorporated three well know machine learning algorithm (Naïve Bayes classifier, a Support Vector Machine, and a Multilayer Perceptron neural network.) for prediction. Lima and De Castro [9] introduced a multi-label approach for user behavior analysis using social media data but followed a different approach. They consider a group of text rather than single and have given less weight on profile. Braunhofer et al. [12] have taken a little bit different problem and so tried to predict about a new user (there no data except his profile). They developed Context-Aware Recommender using user demographic details. Pappas et al. [13] carried out their investigation on online purchase behavior and intention. They studied how Price sensitivity and promotion sensitivity are important to an individual.

According to "Xavier Amatriain (former Research/Engineering Director at Netflix)" recommendation system has emerged as one of the major assets of Netflix [14] to personalize the services. Zhao et al. proposed a user-based collaborative filtering recommendation methodology on Hadoop. It employed a collaborative filtering technique [8] to partition users into groups. But the Cold start problem and Stability vs. plasticity issue could not be solved. Netto et al. [10] discussed briefly five personality characteristics and incorporated in the recombination system. They suggested how recommendations can be improved by adding personality traits. Hu and Pu [15] have proposed a system that uses personality information in collaborative filtering for new users. It combines human personality characteristics into the traditional rating-based similarity. Delic et al. [16] carried out work for the observational user to study group recommender systems in the tourism domain and suggested the importance of group recommender systems. Wu et al. [17] proposed a generalized, dynamic personality-based greedy re-ranking approach to generate the recommendation list.

3 Methodology

Researchers use demographic filtering [9, 10] to take care of cold start problem if users don't have rating history the comparability among them will be computed in light of their demographic properties. This implies if user "A" is comparative with user "B" as indicated by age, gender, and other demographic information, this suggests user "B" can prescribe to user "A" and the other way around. The problem happens when two users in the comparable age level, gender, and with a similar occupation have distinctive identity characteristics. We propose a combined technique in which personality trait is used to determine the Big Five personality score Openness, Conscientiousness, Extraversion, Agreeableness, and Neuroticism whose data is collected from myPersonality.org [18] along with demographic filtering. The personality traits dataset was extracted from myPersonality.org, which is provided by Facebook, and the demographic attributes data was collected from MovieLens dataset [19].

The data from these sources are more accurate than that acquired by taking a personality test because users are more active on social networking [18, 20] sites. An analysis has been carried out on the user's data based on status, followers, a number of friends, Facebook likes, and other closely connected attributes. There are mainly five types of social sites users: listeners (who follow a large number of users), popular (many people follow them), highly-read (they are found in the reading list of others), and two types of influential's (how many times they are clicked). Extrovert and emotionally stable (low in Neuroticism) comes under the category of popular users, and popular users are also high in Openness (Imaginative in nature). Influential's people consider being high in Conscientiousness (organized in nature). Openness is the easiest to predict, while difficult one is Extraversion (outgoing). Agreeableness is generally peacemaker. Similarity function uses users demographics and personality data to obtain several similar users having similar demography and personality traits. Finally, find out the items which are commonly liked by neighborhood users and then suggest to the target user. It assumes that users with similar demographic attributes and personality traits will rate items similarly. This is a technique which computes the relationship between two users in light of their rating that they provide for comparable things and consider the time the user appraised the thing as factors in deciding the closeness. This implies the rating a user gives for a thing now might be changed through time, and hence, this implies that rating isn't a constant factor. This strategy is superior to the customary Pearson connection coefficient to handle the cold start problem.

In Fig. 1 depicts the flowchart of the proposed recommender framework where demographic and personality data of the new user is stored in the database. It goes through data transformation and data cleaning steps. Then we create similarity matrices between user-user based on personality traits and demographic characteristics. Next, these matrices need to be combined and try to recommend top 20 movies to new users by finding k nearest neighbors who are similar to them.

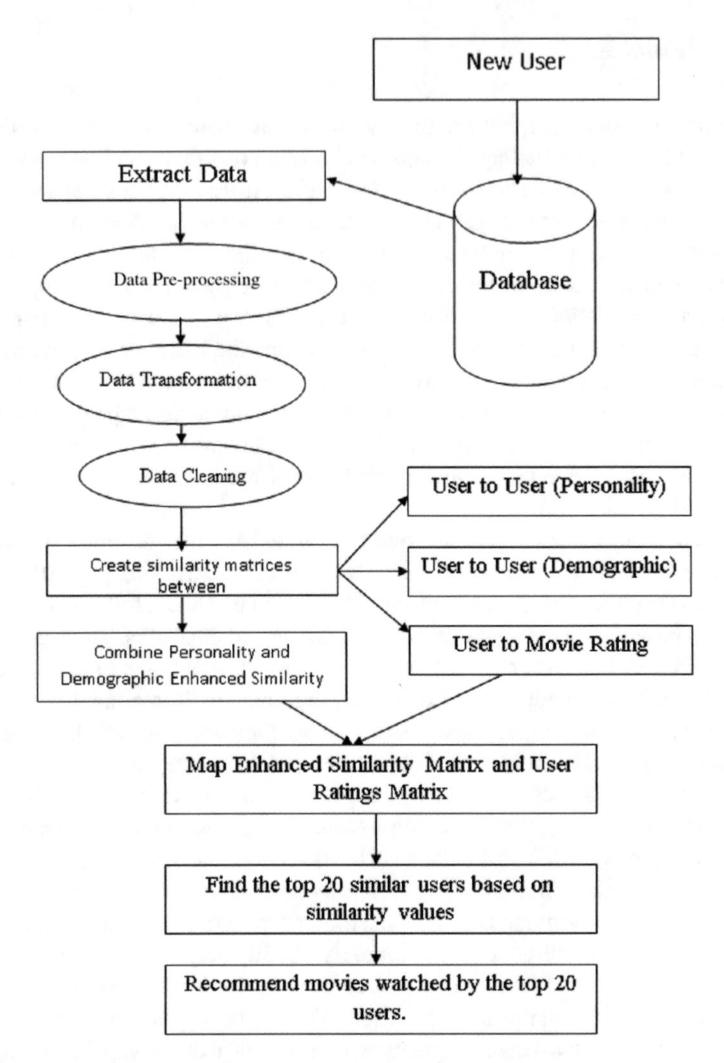

Fig. 1 Flowchart of the proposed recommender system

1. *Similarity calculation of two users' x and y* [8, 11, 20]

$$(x, y) = \frac{\sum_{j=1}^{m}(R_{x,j} - \overline{R_x})(R_{y,j} - \overline{R_y})}{\sqrt{\sum_{j=1}^{m}(R_{x,j} - \overline{R_x})^2}\sqrt{\sum_{j=1}^{m}(R_{y,j} - \overline{R_y})^2}}$$

where, $R_{x,j}$ and $R_{y,j}$ are the rating of user x and y for item j, and, $_x$ and $_y$ are their respective average rating and "m" is the number of items.

Table 1 User demographic table

UserId	Age	Sex	Occupation	Zipcode
1	24	F	Technician	85,711
2	53	F	Writer	94,043
3	23	M	Lawyer	32,067
4	24	M	Student	43,537
5	33	M	Administrator	18,213

2. *Similarity Based on users Demographic attributes*: this method considers only the demographic (Table 1) attributes of users, and most researchers have used this method to alleviate the cold start problem [21].

$$Dem\_sim(x, y) = \cos\_\mathrm{sim}(\vec{x}, \vec{y}) \text{ or } \mathrm{vect\_sim}(\vec{x}, \vec{y}) = \frac{\vec{x} \cdot \vec{y}}{\|\vec{x}\| * \|\vec{y}\|}$$

$$= \frac{\sum_{j=1}^{m} x_j y_j}{\sqrt{\sum_{j=1}^{m} xj^2}\sqrt{\sum_{j=1}^{m} yj^2}}$$

Dem_sim is the demographic similarity, cos_sim is the cosine similarity, vect_sim is the vector similarity. Where x_j and y_j are components of vector u and v, respectively.

3. *Similarity-Based on user personality*: The personality similarity between two users' x and y has been computed (Table 2). Personality model [10] categorizes human personality traits into five dimensions:

   Openness to Experience: appreciation for art, sentiment, adventure, new ideas, interest, and differences of experience.
   Conscientiousness: self-discipline, act dutifully, aim or objective, plan rather than spontaneous behavior.
   Extroversion: liveliness, positive emotions, importance, and the tendency to seek incentive in others influence.
   Agreeableness: concerned and cooperative rather than suspicious or hostile and antagonistic toward others.

Table 2 User personality table

UserId	E	A	C	O	N
1	4.56	5.26	5.47	5.43	4.85
2	4.43	5.21	5.45	5.53	4.90
3	4.12	5.14	5.11	5.07	4.62
4	4.38	5.37	5.57	5.53	5.14
5	4.07	5.21	4.34	5.43	4.89

Neuroticism: a tendency to experience unpleasant emotions easily, such as irritation, restlessness, depression, or vulnerability.

$$simp(x, y) = \frac{\sum_k (P_x - \overline{P_x})(P_y - \overline{P_y})}{\sqrt{\sum_k (P_x - \overline{P_x})^2} \sqrt{\sum_k (P_y - \overline{P_y})^2}}$$

4. *Enhanced Demographic similarity*:

It consists of both demographic similarity and personality similarity.

$$Edem\_sim(x, y) = Dem\_sim(x, y) + simp(x, y)$$

$$Edem\_sim(x, y) = \frac{\sum_{j=1}^{m} x_j y_j}{\sqrt{\sum_{j=1}^{m} xj^2} \sqrt{\sum_{j=1}^{m} yj^2}} + \frac{\sum_k (P_x - \overline{P_x})(P_y - \overline{P_y})}{\sqrt{\sum_k (P_x - \overline{P_x})^2} \sqrt{\sum_k (P_y - \overline{P_y})^2}}$$

4 Implementation

The research shows that as the similar attributes of two users increase their correlation also increases. In this view, demographic attributes have been mixed with their personality traits. Demographic characteristics alone cannot be a complete solution because users with the same demographic characteristics may have a different personality or behavior.

Most of the real-world data is raw and cannot fit for data mining since data is inaccurate, inconsistent, has lots of missing values, contains noisy data, and outliers. Hence, data preprocessing is required to make data consistent and accurate for the data mining processes. Data preprocessing consists of data transformation which included data discretization in which a numerical value replaces demographic attributes, i.e., occupation, 1 for an engineer, 2 for an educator, etc., and the same way, gender is changed to 1 for male and 0 for female.

Similarity based on item evaluates the similarities between users based on the ratings given to a particular movie. Although rating given to a particular movie may change depending on the mood of users and their priority keep on changing as time evolve. This means that the rating is not a constant value. But cold start problem still exists in this traditional method. The main goal of this paper is to overcome the cold start problem by combining a user's demographic attributes with their personality traits. We calculated three types of similarity. The similarity between users, based on rating history (sim_r), the similarity between users based on demographic attributes (sim_dem) [21] and similarity between users, based on their personality traits (sim_per) [8, 11, 20]. Our next step is to add the similarity on demographic attributes and personality traits called the enhanced demographic similarity (Enh_dem), the total similarity is calculated as:

$$Enh\_sim = \alpha * (sim\_r) + (1 - \alpha) * (Enh\_dem)$$

where α is a weight parameter which controls the percentage, cosine similarity is popular and efficient as it considers only non-zero dimension to find similarity. Cosine similarity values range from 0 to 1.

Figure 2 depicts the similarity between users based on ratings given to movies by them, which is calculated by using cosine similarity. Figure 3 depicts similarity between users based on big five personality traits, which is calculated by using cosine similarity.

Figure 4 depicts similarity between users based on their demographic traits like age, gender, occupation. Figure 5 depicts enhanced similarity between users by combining the similarity calculated based on ratings, demographic attributes, and personality traits.

In our work, similarities between new users and their neighbors have been found according to their demographic information. Demographic similarities search users with the almost same age, occupation, and gender. However, user with similar demographic feature might have differences with respect to personality traits based on Big Five [10] personality score (Openness, Conscientiousness, Extraversion, Agreeableness, Neuroticism). The proposed recommender system utilizes both demographic

	0	1	2	3	4	5
0	0.000000	0.166931	0.047460	0.064358	0.378475	0.430239
1	0.166931	0.000000	0.110591	0.178121	0.072979	0.245843
2	0.047460	0.110591	0.000000	0.344151	0.021245	0.072415
3	0.064358	0.178121	0.344151	0.000000	0.031804	0.068044
4	0.378475	0.072979	0.021245	0.031804	0.000000	0.237286
5	0.430239	0.245843	0.072415	0.068044	0.237286	0.000000

Fig. 2 Similarity matrix between users based on rating matrix

	0	1	2	3	4	5
0	0.000000	0.967444	0.995470	0.967444	0.970861	0.979928
1	0.967444	0.000000	0.960038	1.000000	0.968324	0.984882
2	0.995470	0.960038	0.000000	0.960038	0.961198	0.965800
3	0.967444	1.000000	0.960038	0.000000	0.968324	0.984882
4	0.970861	0.968324	0.961198	0.968324	0.000000	0.989692
5	0.979928	0.984882	0.965800	0.984882	0.989692	0.000000

Fig. 3 Similarity matrix between users based on personality traits

	0	1	2	3	4	5
0	0.000000	0.998642	0.998865	0.998642	0.998412	0.998628
1	0.998642	0.000000	0.998568	0.998853	0.998482	0.998758
2	0.998865	0.998568	0.000000	0.998568	0.998255	0.998492
3	0.998642	0.998853	0.998568	0.000000	0.998482	0.998758
4	0.998412	0.998482	0.998255	0.998482	0.000000	0.998754
5	0.998628	0.998758	0.998492	0.998758	0.998754	0.000000

Fig. 4 Similarity matrix between users based on demographic traits

	0	1	2	3	4	5
0	0.000000	1.581297	1.602244	1.560782	1.629073	1.653933
1	1.581297	0.000000	1.558179	1.635624	1.563914	1.624980
2	1.602244	1.558179	0.000000	1.604891	1.542165	1.559763
3	1.560782	1.635624	1.604891	0.000000	1.555680	1.589420
4	1.629073	1.563914	1.542165	1.555680	0.000000	1.630964
5	1.653933	1.624980	1.559763	1.589420	1.630964	0.000000

Fig. 5 Enhanced similarity matrix

and personality feature to find the top 20 similar neighbors for a given user based on calculated similarity values. The movies rated by these similar users would be recommended to the new target user.

5 Conclusion

The proposed methodology has been implemented by using significant personality traits along with suggestions based on neighborhood users. The result generated through the proposed approach would help to handle the problem of Cold start and generate a better and usable prediction. The implementation results are relevant as it considers distinguishing personality features of users for more customized recommendations. In this work, a unique filtering approach has been discussed that draws ideas from the existing state of the art algorithms and consolidates them with demographic and personality traits data from social networking sites to enhance user-based collaborative filtering to recommend items to the users.

References

1. Tiwari, V., Thakur, R.S.: Pattern warehouse: context based modeling and quality issues. Proc. Natl. Acad. Sci., India Sect. A: Phys. Sci., Springer **86**(3), 417–431 (2016)
2. Tiwari, V., Thakur, R.S., Tiwari, B., Gupta, S.: Handbook of Research on Pattern Engineering System Development for Big Data Analytics. IGI-Global (2018)
3. Tiwari, V., Tiwari, V., Gupta, S., Tiwari, R.: Association rule mining: a graph based approach for mining frequent itemsets. In: International Conference on Networking and Information Technology (ICNIT), pp. 309–313. IEEE (2010)
4. Kunal, S., Saha, A., Varma, A., Tiwari, V.: Textual dissection of live twitter reviews using Naive Bayes. In: International Conference on Computational Intelligence and Data Science (ICCIDS), pp. 307–313. Elsevier (2018)
5. Ricci, F., Rokach, L., Shapira, B.: Recommender systems: introduction and challenges. In: Recommender Systems Handbook, pp. 1–34. Springer (2015)
6. Kunaver, M., Požrl, T.: Diversity in recommender systems—a survey. Knowl.-Based Syst. **123**, 154–162 (2017)
7. Elahi, M., Ricci, F., Rubens, N.: A survey of active learning in collaborative filtering recommender systems. Comput. Sci. Rev. **20**, 29–50 (2016)
8. Zhao, Z.-D., Shang, M.-S.: User-based collaborative-filtering recommendation algorithms on Hadoop. In: Third International Conference on Knowledge Discovery and Data Mining (WKDD '10), pp. 478–481. IEEE (2010)
9. Lima, A.C.E., De Castro, L.N.: A multi-label, semi-supervised classification approach applied to personality prediction in social media. Neural Netw. **58**, 122–130 (2014)
10. Netto, M.A.S., Cerri, S.A., Blanc, N.: Improving recommendations by using personality traits in user profiles. In: International Conferences on Knowledge Management and New Media Technology, pp. 92–100 (2008)
11. Buettner, R.: Predicting user behavior in electronic markets based on personality-mining in large online social networks. Electron. Mark. **27**(3), 247–265 (2017)
12. Braunhofer, M., Elahi, M., Ricci, F.: User personality and the new user problem in a context-aware point of interest recommender system. In: Information and Communication Technologies in Tourism. Springer, pp. 537–549 (2015)
13. Pappas, I.O., Kourouthanassis, P.E., Giannakos, M. N., Lekakos, G.: The interplay of online shopping motivations and experiential factors on personalized e-commerce. A complexity theory approach. Telemat. Inform. **34**(5), 730–742 (2017)
14. Amatriain, X., Basilico, J.: Netflix recommendations: beyond the 5 stars (part 1). Netflix Tech Blog 6 (2012)
15. Hu, R., Pu, P.: Using personality information in collaborative filtering for new users. Recomm. Syst. Soc. Web **17** (2010)
16. Delic, A., Neidhardt, J., Nguyen, T.N., Ricci, F.: An observational user study for group recommender systems in the tourism domain. Inf. Technol. Tour. **19**(4), 87–116 (2018)
17. Wu, W., Chen, Li, Zhao, Yu.: Personalizing recommendation diversity based on user personality. User Model. User-Adapt. Interact. **28**(3), 237–276 (2018)
18. Harper, F.M., Konstan, J.A.: The movielens datasets: history and context. ACM Trans. Interact. Intell. Syst. (TIIS) **5**(4) (2016)
19. Datasets: myPersonality data (myPersonality.org). Accessed Oct 2017
20. Quercia, D., Michal, K., David, S., Jon, C.: Our twitter profiles, our selves: predicting personality with twitter. In: 2011 IEEE Third International Conference on Social Computing (SocialCom), pp. 180–185. IEEE (2011)
21. Jyoti, G., Jayant, G.: A framework for a recommendation system based on collaborative filtering and demographics. In: International Conference on Circuits, Systems, Communication and Information Technology Applications (CSCITA), pp. 300–304. IEEE (2014)

Automatically Designing Convolutional Neural Network Architecture with Artificial Flora Algorithm

Timea Bezdan, Eva Tuba, Ivana Strumberger, Nebojsa Bacanin and Milan Tuba

Abstract Convolutional neural network has demonstrated high performance in many real-world problems in recent years. However, the results and accuracy of a CNN that are applied for a specific problem are highly controlled by the architecture and its hyperparameters. The process of finding the right set of hyperparameters for the network's architecture is a very time-consuming process and requires expertise. To address this problem, we present a powerful method that does automatic hyperparameter search for designing CNN architecture. The hyperparameter optimization is performed by the artificial flora optimization algorithm. The proposed framework has the ability to explore different architectures and optimize the values of hyperparameters for a given task. In this research, the proposed framework is validated on MNIST image classification task and it can be concluded from the experimental results that the suggested search method accomplishes promising achievement in this domain.

Keywords Convolutional neural network architecture · Swarm intelligence · Metaheuristics · Artificial flora Optimization

T. Bezdan · E. Tuba · I. Strumberger · N. Bacanin · M. Tuba (✉)
Singidunum University, Danijelova 32, 11000 Belgrade, Serbia
e-mail: tuba@ieee.org

T. Bezdan
e-mail: timea.bezdan.17@singimail.rs

E. Tuba
e-mail: etuba@ieee.org

I. Strumberger
e-mail: istrumberger@singidunum.ac.rs

N. Bacanin
e-mail: nbacanin@singidunum.ac.rs

M. Tuba et al. (eds.), *ICT Systems and Sustainability*,
Advances in Intelligent Systems and Computing 1077,
https://doi.org/10.1007/978-981-15-0936-0_39

1 Introduction

Convolutional neural network (CNN) is a particular type of neural network that is broadly used in computer vision algorithms. CNNs are mainly intended for image classification, image segmentation [1], object detection [2], and face recognition [3].

The classic CNN [4–6] architecture consists of different types of layers, such as a convolutional layer, pooling layer, and dense layer (also called fully connected or FC layer) at the end of the architecture. More modern architectures, such as GoogleNet [7], ResNet [8], DensNet [9], and SENet [10] are different from the typical CNN architectures, and they are much deeper and use different operations to avoid vanishing and exploding gradient descent. The convolutional layer is composed of a group of learnable kernels and by sliding the filter on the input volume it performs convolution operation and computes the scalar products between the filter and the input image at any position. The activation function is applied after the convolution operation that results with features (activation maps). After the convolutional layer, for dimension reduction (subsampling or downsampling), pooling layer is used. After several convolution and pooling layers, the architecture ends with several dense layers, the final dense layer in the architecture represents the classifier that has equal output unit numbers as the different categories to be recognized.

Hyperparameter optimization for CNN architecture design optimizes the number of different hyperparameters, such as convolutional layers, filters, selects the right filter size, stride, padding, number of dense layers, along with the number of its neurons, determines the learning rate, epoch number, and batch size. Besides the aforementioned, choosing the appropriate activation function, regularization technique to prevent overfitting, and optimization algorithm for training the network is important for obtaining high classification accuracy.

CNN architectures are handcrafted by using different techniques, such as grid search, random search [11], and brute-force methods that all require a lot of effort, domain expertise, and time. Since metaheuristics found to be effective in tackling NP-hard problems like this [12, 13], hyperparameter optimization problem could be solved by using some of the available metaheuristics. According to the literature survey, just a few researchers employed nature-inspired metaheuristic [14, 15] approach for optimizing the set of hyperparameters for the network.

We propose an automatic approach for the selection of the appropriate hyperparameters by using a swarm intelligence approach, the artificial flora optimization algorithm (AF) [16]. Swarm intelligence belongs to the category of population and iterative-based techniques and it is very efficient in solving different NP-hard optimization tasks [17–21].

The software package that is proposed in this paper is named AF-CNN framework. Our framework optimizes certain hyperparameters, it defines how many convolutional layers, filters, and dense layers are needed for architecture as well as the filter size and the number of neurons within the dense layers.

We have structured the remainder of this paper in the following way: artificial flora swarm intelligence metaheuristics are described in Sect. 2. Section 3 presents experimental results and discussion. The conclusion is in Sect. 4.

2 Artificial Flora Optimization Metaheuristics

The AF [16] is a recent swarm intelligence optimization algorithm proposed by Cheng et al. [16] and the algorithm is based on the flora migration and reproduction process. AF was tested on six benchmark functions and it showed strong exploration ability and fast convergence speed. In the AF, original plants are created initially and they scatter seeds, so-called offspring plants, within a certain distance. The algorithm of AF described as follows.

In the first step, the initial population is generated randomly with N original plants.

$$X_{i,j} = r \times D \times 2 - D \tag{1}$$

where the original plants' location is denoted by $X_{i,j}$, i and j are dimension and number of plants, respectively, and r is uniformly distributed random number in the interval (0,1).

In the next step, the propagation distance D_j is calculated for each plant. The calculation of the spreading distance of a plant depends on the spreading distance of the previous two generations'.

$$D_j = D_{1j} \times r \times c_1 + D_{2j} \times r \times c_2 \tag{2}$$

where the spreading distances of the parent plant and grandparent plant are D_{2j} and D_{1j}, respectively, c_1 and c_2 denote the learning coefficients, and r is the uniformly distributed random number in the interval (0,1).

The new grandparent plant's spreading distance is

$$D'_{1j} = D_{2j} \tag{3}$$

The new parent propagation distance is calculated as

$$D'_{2j} = \sqrt{\frac{\sum_{i=1}^{N}(X_{i,j} - X'_{i,j})^2}{N}} \tag{4}$$

The offspring plants' location around the original plant is calculated by the following formula:

$$X'_{i,j\times m} = R_{i,j\times m} + X_{i,j} \tag{5}$$

where $X'_{i,j}$ is the location of the offspring plant, m denotes the number of offsprings that is generated by one plant, and $R_{i,j\times m}$ is a normally distributed random number with mean 0 and variance D_j.

Some of the generated offsprings will survive and some of them will not, whether an offspring is alive or not, determined by the probability of surviving that is calculated by using the proportionate-based selection.

$$p = \left| \sqrt{\frac{F(X'_{i,j\times m})}{F_{\max}}} \right| \times Q^{(j\times m-1)} \tag{6}$$

where the selective probability is denoted by $Q^{(j\times m-1)}$ and its value is between 0 and 1, $F_{\max}$ denotes the maximum fitness of all offsprings, and the fitness of each individual offspring plant is $F(X'_{i,j\times m})$.

The offspring will survive if the probability of surviving p is greater than r, where r is a uniformly distributed random number in the interval (0,1).

The N offspring plants are selected of the survived offsprings as new original plants for the next iteration. The AF algorithm is repeated until it meets the termination criteria. In the end, the best solution is chosen as an optimal solution.

Algorithm 1 Pseudocode of the AF algorithm

```
Initialization: Generate N original plants randomly using Eq. (1); evaluate fitness of each individuals; choose the best solution
while t < MaxIter do
  for i from 1 to N * M do
    Calculate the spreading distance using Eqs. (2), (3), (4)
    Generate offspring plants using Eq. (5)
    if p > r then
      Offspring plant live
    else
      Offspring plant not live
    end if
  end for
  Evaluate new solutions
  Randomly choose N new original plants
  The new solution replaces the old one if its value is better
end while
return the best solution
```

3 Experimental Results and Discussion

Java SE 10, version 18.3 was used for the AF metaheuristic development, and a programming library called Deeplearning4j was used for the implementation of AF-CNN framework. To improve the speed of the execution of the conducted experiments, the framework is executed on NVIDIA's CUDA™ GPU.

Table 1 Control parameters of AF

Parameter	Value
N: original plant number	10
M: offspring plant number	100
$c1$: grandparent learning coefficient	0.75
$c2$: parent learning coefficient	1.25
$MaxIter$: number of iterations	20

For validation of our framework, as benchmark dataset, the MNIST dataset is selected for the image classification task in the simulation. More information about MNIST dataset can be found in [22].

The control parameters of AF that is used in the proposed AF-CNN framework are represented in Table 1.

In this AF-CNN, the optimization algorithm for CNN architecture is specified by the parameter h:

$$h = \{\{C_0, \ldots, C_{n-1}\}, \{S_0, \ldots, S_{m-1}\}\} \tag{7}$$

where the number of convolutional layer is defined by C_n and S_m denotes the number of dense layers. The *ith* layer of the convolutional layer contains two hyperparameters:

$$C_i = (k_{\text{count}}, k_{\text{size}}) \tag{8}$$

where the kernel number is denoted by k_{count} and its size is denoted by k_{size}. The set of all possible CNN architectures is denoted by $\mathcal{H}$ and the aim is to find the right set of hyperparameters for designing an architecture $h \in \mathcal{H}$ that minimizes the classification error rate and maximizes the classification accuracy.

Since the hyperparameters' search space is large and computationally expensive, it is reduced by specifying the range for every hyperparameter: the range of the convolutional layers is between 0 and 6 and their initial number is 1 and 2, fully connected layers ranges from 1 to 4 and their initial number is 1 and 2, the kernel has a size between 1 and 8, the filter number per conv layer is set to be between 1 and 128, and the number of neurons in the FC layer is between 16 and 2048.

In the model, all convolution uses ReLU (rectified linear unit) activation function [23]. As pooling layer, max pooling is used with the size of 2×2. As an optimizer, Adam optimizer is selected with a learning rate of $\theta = 0.0001$, and to measure the loss, MSE function is applied on the convolutional neural network, and the training of the network is conducted in one epoch with the batch size of 54. The initial parameter configuration of the model is similar like in [15].

The obtained experimental results are compared to the experimental results in the paper [15] that has very similar initial hyperparameter setup and the experiments were

Table 2 Error rates (%) on the test set

Method	Error rates (%)
LeNet-5 [24]	0.95
Deeply Supervised Net [25]	0.39
Shallow CNN [26]	0.37
Recurrent CNN [27]	0.31
Gated Pooling CNN [28]	0.29
IEA-CNN [15]	0.34
CEA-CNN (where k=1) [15]	0.26
CEA-CNN (where k=2) [15]	0.24
CEA-CNN (where k=3) [15]	0.28
AF-CNN	**0.23**

also performed on the same benchmark dataset. The configuration of the AF-CNN architecture is less complex than in [15].

The solutions of the final population resulted in three convolutional layers and one FC layer. The obtained optimal filter number is between 60 and 132, 2×2 and 4×4 are the optimal sizes of the filters, and the optimal number of the neurons in the FC layers is between 58 and 142. In [15], the CEA-CNN method has notably much larger filter sizes, the sizes are between $4x4$ and 3×3 and between 8×8 and 7×7.

An architecture, generated by AF-CNN, can be described as follows:

$$h = \{\{(70, 3), (140, 3), (140, 3)\}, \{128\}\}\} \tag{9}$$

The best performing network of AF-CNN was created only after the first 18–20 iterations.

The IEA-CNN and the CEA-CNN methods in [15] resulted with networks, where the best architecture resulted with 0.24% test error rate, while the architecture configuration generated by our AF-CNN framework has a test error rate of 0.23%. It should be also noted that our method, the AF-CNN, ran in only one epoch, while the number of epoch in [15] is much larger; consequently, the computation is less expensive of AF-CNN.

A comparison is made between our obtained experimental results of the AF-CNN framework and other methods that used metaheuristics and it is presented in Table 2. The same dataset was used in all compared methods. AF-CNN has a lower error rate than other methods that are compared to, besides that our method is computationally less expensive.

4 Conclusion

The method presented in this paper is proposed for automatically designing a ConvNet architecture by using a swarm intelligence artificial flora optimization algorithm (in short, named AF-CNN), which has the ability to explore the best architecture for a specific task. One of the most challenging tasks in the CNN design is to find the right set of hyperparameters for a particular task.

The aim of this research is to determine the number of certain hyperparameters for an architecture that will produce higher training and test accuracy for a given classification task. The proposed method optimizes certain hyperparameters, it defines how many convolutional layers, filters, and dense layers are needed for architecture as well as the filter size and the number of neurons within the dense layers.

The AF-CNN is applied to image classification tasks with images of handwritten digits from the well-known benchmark dataset MNIST. The experimental results show that AF-CNN shows high performance in exploring the best CNN architecture and achieves higher accuracy than other methods tested on the same dataset and using metaheuristics.

Acknowledgements Ministry of Education and Science of Republic of Serbia, Grant No. III-44006 supported this research.

References

1. Farabet, C., Couprie, C., Najman, L., LeCun, Y.: Learning hierarchical features for scene labeling. IEEE Trans. Pattern Anal. Mach. Intell. **35**, 1915–1929 (2013)
2. Ren, S., He, K., Girshick, R., Sun, J.: Faster R-CNN: towards real-time object detection with region proposal networks. In: Advances in neural information processing systems, pp. 91–99 (2015)
3. Taigman, Y., Yang, M., Ranzato, M., Wolf, L.: Deepface: closing the gap to human-level performance in face verification. In: IEEE Conference on Computer Vision and Pattern Recognition, pp. 1701–1708 (2014)
4. Lecun, Y., Bottou, L., Bengio, Y., Haffner, P.: Gradient-based learning applied to document recognition. Proc. IEEE **86**, 2278–2324 (1998)
5. Krizhevsky, A., Sutskever, I., Hinton, G.E.: Imagenet classification with deep convolutional neural networks. In: Pereira, F., Burges, C.J.C., Bottou, L., Weinberger, K.Q. (eds.) Advances in Neural Information Processing Systems 25, pp. 1097–1105, Curran Associates, Inc., (2012)
6. Simonyan, K., Zisserman, A.: Very deep convolutional networks for large-scale image recognition. In: 3rd International Conference on Learning Representations (ICLR) (2015)
7. Szegedy, C., Liu, W., Jia, Y., Sermanet, P., Reed, S., Anguelov, D., Erhan, D., Vanhoucke, V., Rabinovich, A.: Going deeper with convolutions. In: IEEE Conference on Computer Vision and Pattern Recognition (CVPR), pp. 1–9 (2015)
8. He, K., Zhang, X., Ren, S., Sun, J.: Deep residual learning for image recognition. In: Proceedings of the IEEE Conference on Computer Vision and Pattern Recognition, pp. 770–778 (2016)
9. Huang, G., Liu, Z., Weinberger, K.Q.: Densely connected convolutional networks. CoRR, vol. abs/1608.06993 (2016)

10. Hu, J., Shen, L., Sun, G.: Squeeze-and-excitation networks. In: Proceedings of the IEEE Conference on Computer Vision and Pattern Recognition, pp. 7132–7141 (2018)
11. Bergstra, J., Bengio, Y.: Random search for hyper-parameter optimization. J. Mach. Learn. Res. **13**(1), 281–305 (2012)
12. Dolicanin, E., Fetahovic, I., Tuba, E., Capor-Hrosik, R., Tuba, M.: Unmanned combat aerial vehicle path planning by brain storm optimization algorithm. Stud. Inform. Control **27**(1), 15–24 (2018)
13. Strumberger, I., Tuba, E., Bacanin, N., Beko, M., Tuba, M.: Wireless sensor network localization problem by hybridized moth search algorithm. In: 2018 14th International Wireless Communications Mobile Computing Conference (IWCMC), pp. 316–321 (2018)
14. Young, S.R., Rose, D.C., Karnowski, T.P., Lim, S.-H., Patton, R.M.: Optimizing deep learning hyper-parameters through an evolutionary algorithm. In: Proceedings of the Workshop on Machine Learning in High-Performance Computing Environments, pp. 4:1–4:5, ACM (2015)
15. Bochinski, E., Senst, T., Sikora, T.: Hyper-parameter optimization for convolutional neural network committees based on evolutionary algorithms. In: 2017 IEEE International Conference on Image Processing, pp. 3924–3928 (2017)
16. Cheng, L., Wu, X.-h., Wang, Y.: Artificial flora (AF) optimization algorithm. Appl. Sci. **8**, 329:1–22 (2018)
17. Tuba, E., Tuba, M., Dolicanin, E.: Adjusted fireworks algorithm applied to retinal image registration. Stud. Inform. Control **26**(1), 33–42 (2017)
18. Tuba, E., Strumberger, I., Bacanin, N., Tuba, M.: Bare bones fireworks algorithm for capacitated p-median problem. In: LNCS: Advances in Swarm Intelligence, (Cham), pp. 283–291, Springer, Berlin (2018)
19. Alihodzic, A., Tuba, E., Simian, D., Tuba, V., Tuba, M.: Extreme learning machines for data classification tuning by improved bat algorithm. In: International Joint Conference on Neural Networks (IJCNN), pp. 1–8, IEEE (2018)
20. Tuba, E., Strumberger, I., Zivkovic, D., Bacanin, N., Tuba, M.: Mobile robot path planning by improved brain storm optimization algorithm. In: 2018 IEEE Congress on Evolutionary Computation (CEC), pp. 1–8 (2018)
21. Strumberger, I., Tuba, E., Bacanin, N., Beko, M., Tuba, M.: Modified monarch butterfly optimization algorithm for RFID network planning. In: 6th International Conference on Multimedia Computing and Systems (ICMCS), pp. 1–6 (2018)
22. LeCun, Y., Cortes, C., Burges, C.: MNIST handwritten digit database. AT&T Labs. http://yann.lecun.com/exdb/mnist, vol. 2, p. 18 (2010)
23. Nair, V., Hinton, G.E.: Rectified linear units improve restricted boltzmann machines. In: Proceedings of the 27th International Conference on International Conference on Machine Learning, ICML'10, (USA), pp. 807–814. Omnipress (2010)
24. Lecun, Y., Bottou, L., Bengio, Y., Haffner, P.: Gradient-based learning applied to document recognition. Proc. IEEE **86**(11), 2278–2324 (1998)
25. Lee, C.-Y., Xie, S., Gallagher, P., Zhang, Z., Tu, Z.: Deeply-supervised nets. In: Proceedings of the Eighteenth International Conference on Artificial Intelligence and Statistics **38**, pp. 562–570 (2015)
26. Mcdonnell, M., Vladusich, T.: Enhanced image classification with a fast-learning shallow convolutional neural network. In: 2015 International Joint Conference on Neural Networks (IJCNN), pp. 1–7 (2015)
27. Liang, M., Hu, X.: Recurrent convolutional neural network for object recognition. In: Proceedings of the IEEE Conference on Computer Vision and Pattern Recognition, pp. 3367–3375 (2015)
28. Lee, C.-Y., Gallagher, P.W., Tu, Z.: Generalizing pooling functions in convolutional neural networks: mixed, gated, and tree. In: Artificial Intelligence and Statistics, pp. 464–472 (2015)

Electro-Thermo-Mechano-Magneto Characteristics of Shape Memory Spring Actuator/Sensor in Assistive Devices

T. Devashena and K. Dhanalakshmi

Abstract In recent years, shape memory alloys (SMAs) have been used widely as actuators in modern systems but it can also be used as sensor since it offers unique self-sensing property. Shape memory alloys exhibit parameter variations that include electrical properties such as resistance in case of SMA wire and impedance in case of SMA spring, as a nonlinear hysteretic behavior within a transformation cycle. The actuation and sensing characteristics of SMA can be obtained from the material's electromechanical impedance. This work presents about the electro-thermo-mechano-magneto effects exhibited in commercially available NiTiNOL spring with 90 °C phase transition temperature during phase transformation. The phase-dependent electrical resistance, inductance, impedance, and magnetic field of the SMA spring is measured using impedance analyzer and milli gauss meter respectively. The results opens the door to a new dimension of application of SMA spring, i.e., the investigation explores a possibility for the design of electro-thermo-mechano-magneto characteristics based assistive systems and in robotics.

Keywords NiTi spring · Joule heating · Temperature · Displacement · Force · Magnetic field · Electrical resistance · Series inductance · Duty cycle

1 Introduction

Evolution in technology is matched by evolution in its component parts. Smart material based systems have revolutionized technologies by being able to be functional to many different domains. Functional developments of these materials over conventional ones are, these material exhibits unique characteristics of both shape memory

T. Devashena · K. Dhanalakshmi (✉)
Department of Instrumentation and Control Engineering, National Institute of Technology, Tiruchirappalli, Tiruchirappalli 620015, Tamil Nadu, India
e-mail: dhanlak@nitt.edu

T. Devashena
e-mail: tdevashena@gmail.com

M. Tuba et al. (eds.), *ICT Systems and Sustainability*,
Advances in Intelligent Systems and Computing 1077,
https://doi.org/10.1007/978-981-15-0936-0_40

effect and super-elasticity. Shape memory effect in SMA arises from the existence of two temperatures dependent crystal structure which exhibits solid to solid phase transformation as a result two phases occur namely martensite (low temperature) and austenite (high temperature). Their most smart ability is that they return to a predetermined shape when it is heated. As such, shape memory alloy actuators have both large force-to-weight ratio and force-to-length ratio; hence, they are used in applications where size and weight are considered as boundaries. To date very scanty literature is found with regards to presenting about the electro-magnetic features of SMA. A particular citation is on the sensorless displacement estimation using inductance of SMA spring actuators [1]. The inductance is determined from the time constant of the SMA spring along with the resistance under actuation; since inductance is a geometrical property, the exact position is estimated from its inductance measurement. The multiphysics behavior, i.e., electro-thermo-mechano-magneto characteristics exhibited by SMA spring during its phase transformation offers a feasibility in the design and function of soft robotics and assistive devices.

1.1 Electro-Thermo-Mechano-Magneto Characteristics of SMA Spring

The most commonly used thereby available forms of SMA element are the helical/spring and wire. There are practical differences between them in their displacement, force generated and electrical resistance and inductance variation during activation. SMA springs have higher strain/displacement (up to 200%) than SMA wires (4%) but with reduction in resistance variation. The intention of the work is to determine and study the electro-thermo-mechano-magneto features revealed by nitinol SMA spring.

The shape memory alloy spring under test is a product of the Dynalloy with 90 °C phase transition temperature; its coil diameter is 2.54 mm, the diameter of the wire is 0.381 mm; 2 springs of 32 and 26 turns each are used in the study. SMA spring actuator offers low force and high travel requirements. An experimental setup is built to enable the measurement of current, force, displacement, temperature, resistance, inductance, impedance and magnetic field by activating the SMA spring through joule heating. The schematic and photograph of the experimental setup are shown in Fig. 1. The specifications of SMA are listed in Table 1.

Experimentation. SMA spring is connected to the impedance analyzer (Keysight E4990A) with DC bias unit via a shielded two wire cable and configured with the steel bias spring that recovers force during cooling so that the SMA returns to the

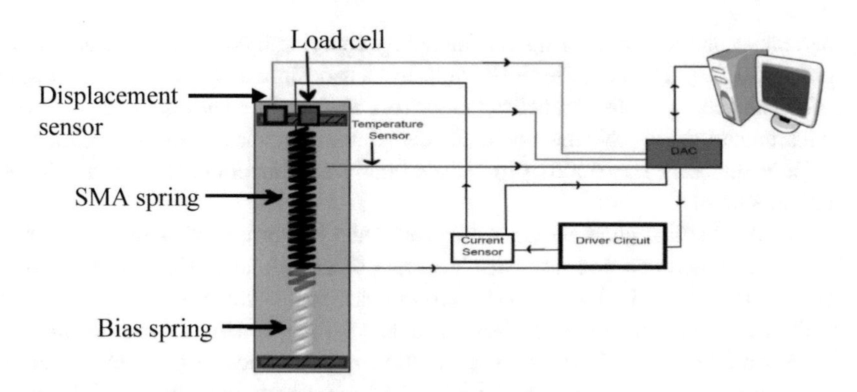

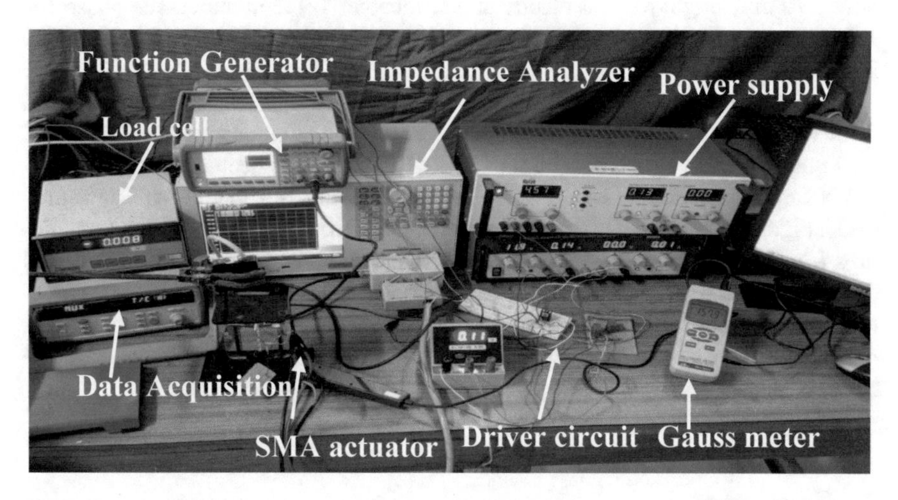

Fig. 1 Experimental setup: schematic diagram and photograph

Table 1 Specifications of SMA spring

Specifications of SMA spring	Value
Wire diameter (mm)	0.381
Spring diameter (mm)	2.54
No. of turns	32, 26
Displacement per coil (mm)	1.1
Transition temperature (°C)	90
Approximate safe heating current (A) for 2 s	1.9
Composition	NiTi

martensite phase. SMA spring is actuated by applying a constant safe heating current in 10, 16, 21 and 25% duty cycle by self-designed amplifier driver circuit. SMA exhibits two phase transformation from martensite to austenite and austenite to martensite, during this the electrical characteristics namely resistance, inductance and impedance are measured by impedance analyzer; temperature by means of J-type thermocouple.

The mechanical characteristics like force and displacement exhibited are measured using load cell and laser displacement sensor (Acuity AR200) respectively, whereas the magnetic field revealed during phase transformation is measured using Milli gauss meter (Lutron GU-3001). The data is logged using data acquisition unit (Keysight 34972A) which is interfaced with computer and sensors. Shape memory alloy spring of 0.381 mm wire diameter, 2.54 mm spring diameter and 32 turns of length 70 mm length possess electrical resistance of 1.9 Ω in martensite phase and 1.7 Ω in austenite phase. During phase transformation the SMA spring displaces by 28 mm.

Experimental results. Figure 2 shows the temperature profile and displacement, force relationship with temperature for the entire actuation current range of 1–1.8 A. Temperature reaches a maximum value of 125 °C, displacement as 28 mm and 2.2 N force is exerted at 1.8 A, when the SMA spring is actuated with 25% duty cycle. Figure 3 depicts the magnetic field exhibited during phase transformation. Figures 4 and 5 represents the electromechanical characteristics of SMA spring exhibited during phase transformation; electrical characteristics namely resistances, inductances are measured using impedance analyser for various duty cycle as mentioned, result depicts the hysteresis is wider for higher duty cycle in displacement-resistance relation.

2 Conclusion

The phase-dependent electro-thermo-mechano-magneto effects exhibited during phase transformation in the NiTi spring is experimentally measured. The measurements herein show that the SMA spring exhibit minimal, resistance variation (0.2 Ω) and inductance variation (0.015 μH) whereas the displacement (28 mm) and force (2.2 N) revealed is higher. Likewise, the magnetic field varied between 310 and 355 milli gauss for a current activation between 0.5 and 1.8 A. The resistance variation follows the geometrical relationship unlike the inductance variation. The measurements and the nature of variation of the electrical and mechanical characteristics inform that this NiTi spring is strong in actuation but weak in sensing with the number of turns considered. Suitable selection of the number of turns and wire diameter will exhibit better electrical resistance to make it suitable for self-sensing with resistance,

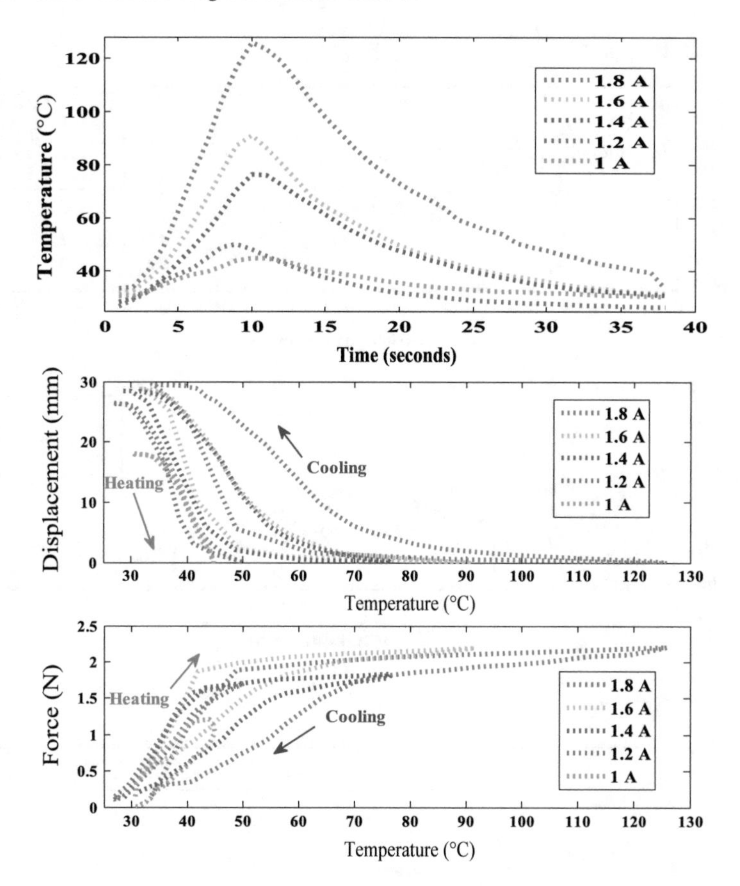

Fig. 2 Displacement–force–temperature relations obtained for 32 turns SMA spring

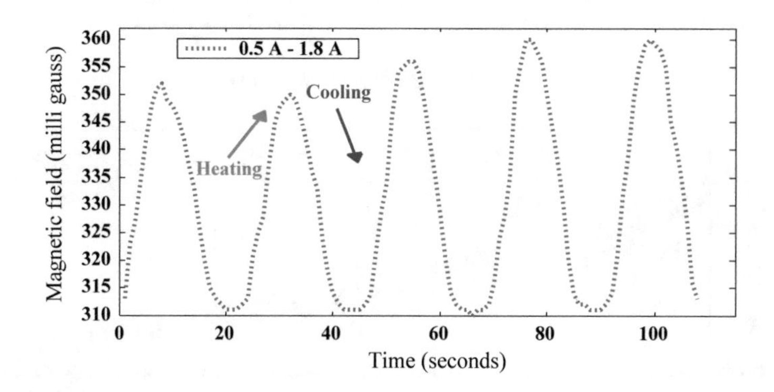

Fig. 3 Magnetic fields revealed during actuation

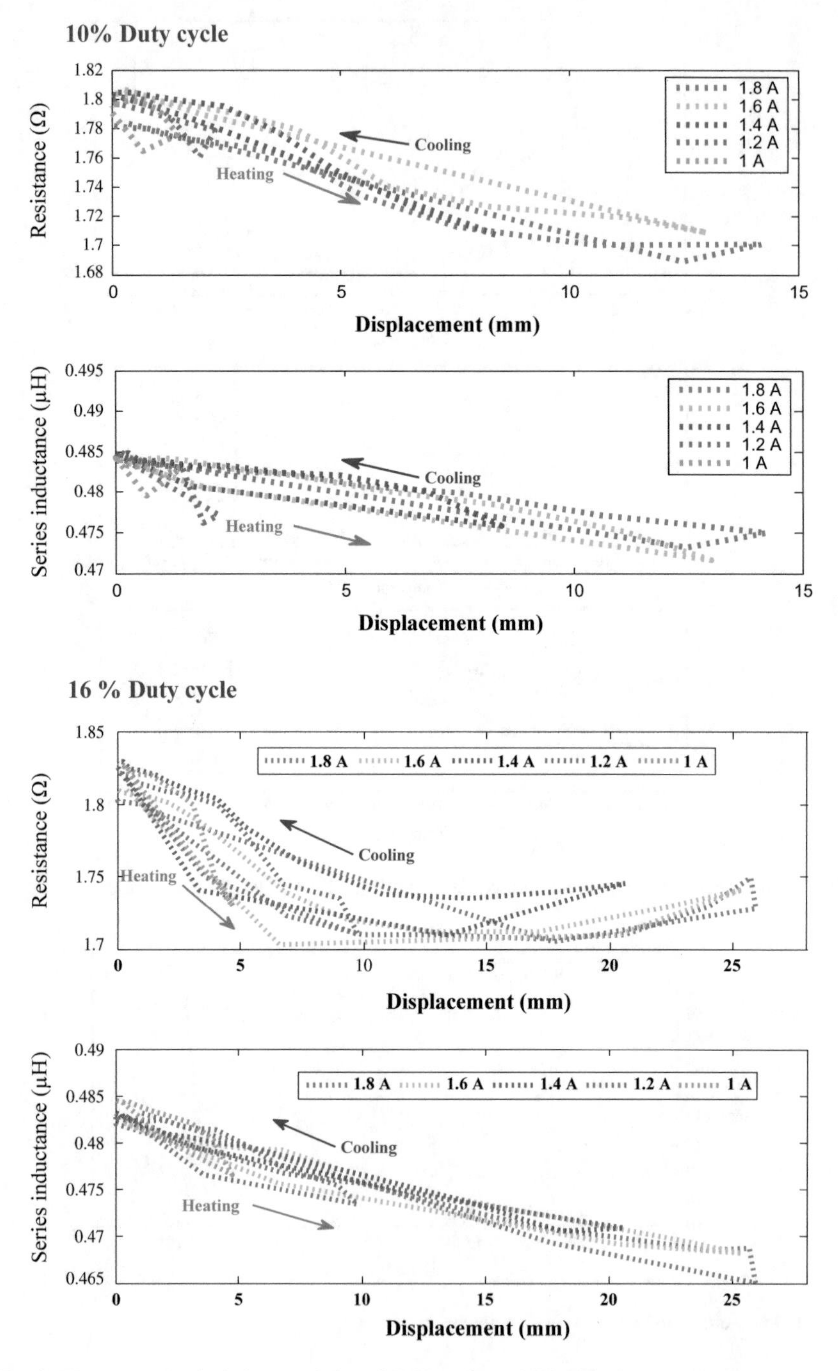

Fig. 4 Electro-mechanical characteristics of SMA spring exhibited during actuation

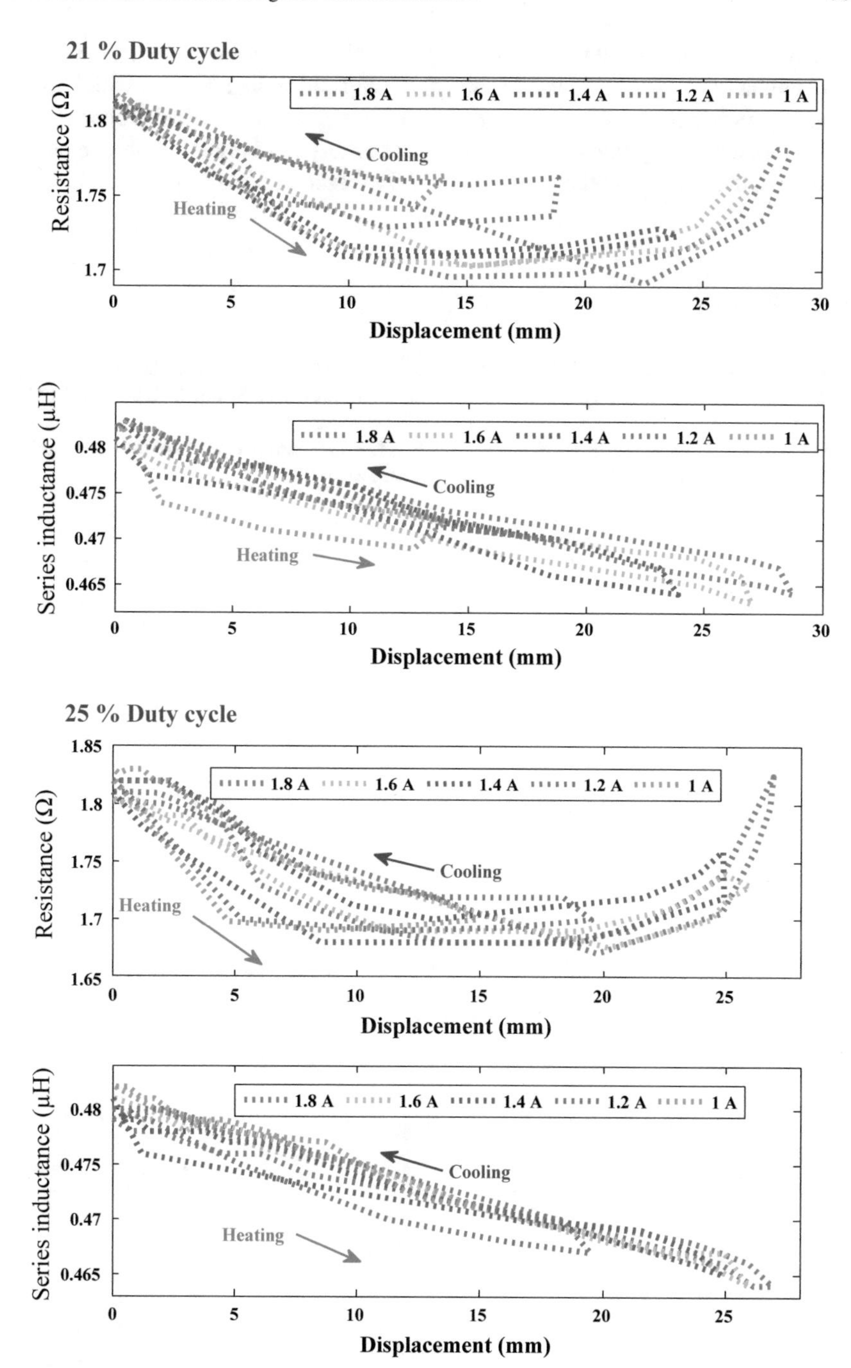

Fig. 5 Electro-mechanical characteristics of SMA spring exhibited during actuation

thereby offer resistance feedback for control. Notable electrical characteristics can be obtained by proper selection of material composition like NiTiCu which offers relatively higher electrical conductivity in addition to the geometrical properties like wire diameter, spring diameter, and number of turns [2]. Likewise, usable magnetic characteristics can be obtained by using NiTiFe-based SMA spring.

References

1. Kim, H., Han, Y., Lee, D.-Y., Ha, J.-I., Cho, K.-J.: Sensorless displacement estimation of a shape memory alloy coil spring actuator using inductance. Smart Mater. Struct. **22**(2), 025001 (8 pp) (2013)
2. Devashena, T., Dhanalakshmi, K.: Electromagnetic characteristics of shape memory spring. Materials Science Forum (ICPCM 2018) (8 pp) (2019)

Task-Aware Energy-Efficient Framework for Mobile Cloud Computing

Priyanka Sharma, Raj Kumari and Inderdeep Kaur Aulakh

Abstract Mobile devices, including smartphones, are becoming an important part of our daily lives. These devices have powerful features and useful applications that help us to accomplish multiple tasks in minimum time, especially in cloud services. Mobile devices, on the other hand, have limitations like battery life, computing power, and storage capacity. Mobile Cloud Computing (MCC) is a rising technology that helps mobile users to keep away from these limitations, primarily to save energy. In this article, we have discussed energy efficiency in MCC because it is the most important design requirement for mobile devices. Modified Best Fit Decreasing (MBFD) algorithm is used to sort the users as per their task. To minimize energy consumption and completion time required for completing the tasks optimization algorithm named as Artificial Bee Colony (ABC) with supervised learning technique Support Vector Machine (SVM) is used. The performance of the proposed MCC model is analyzed on the basis of energy consumption and completion time. It is analyzed that energy consumption and completion time are reduced by 36.12% and 8.12% respectively.

Keywords MCC · Task allocation · Energy consumption · Completion time · MBFD · ABC · SVM

1 Introduction

Cloud computing is a part of Network-based computing service in which there is a sharing of information on the basis of computing resources among the different user with the help of Internet. The user's access or share data on the basis of pay per use model. The cloud consists of different computing resources that are responsible to provide services to their users. The resources are stored in a database of cloud named as Infrastructure as a service (IaaS) cloud. The service providers uses three types of services such as: Infrastructure as a Service (IaaS), Software as a Service, and Platform as a Service [1–3]. The cloud computing gives various advantages to their users like as flexibility of accessing services, cost minimization, sharing of location

P. Sharma (✉) · R. Kumari · I. K. Aulakh
Department of Information Technology, UIET, Panjab University, Chandigarh, India

M. Tuba et al. (eds.), *ICT Systems and Sustainability*,
Advances in Intelligent Systems and Computing 1077,
https://doi.org/10.1007/978-981-15-0936-0_41

and hardware independence. The applications of clouds are in general deployed in isolated Data Centers in which storage systems and high capacity servers are established. The quick growth of demand for varied data centers consumes more electrical power amount. There is a necessity of an energy effective model for the reduction of functional expenses with the maintenance of Quality of Service (QoS) [4]. The optimized energy could be attained with the amalgamation of resources according to effective virtual network topologies, current utilization with the computing nodes and hardware thermal condition. On the contrary, the main aim of cloud computing is associated with the resources flexibility. A QoS-constrained resource allocation problem is termed with this, where service demanders intend to solve sophisticated 254 G. Wei et al. have resolved the problem by accessing cloud resources with minimum cost [5–8]. A number of research has been conducted to determine the performance of cloud computing. The researchers have a research such as how to assign tasks to ensure the quality of service provided by these dynamic mobile servers. This is a challenging task. To solve this problem, an artificial intelligence scheme is used to allocate a task to an appropriate user with high accuracy. The aim of this research is to develop an energy efficient mechanism in the cloud network and the best solution has been delineated. Modified Best Fit Decreasing (MBFD) algorithms are considered for sorting the Virtual Machines (VMs) in decreasing manner with CPU utilization and for allocating every VM to the host that provides the less amplification in power consumption because of allocation [9–11]. It controls the heterogeneity of the resources by selecting the most appropriate power consumption initially. ABC optimization algorithm is utilized for lessening the energy consumption for the allocation of mobile users. The important use of the ABC algorithm is to determine the issue of false allocation of tasks. SVM classification algorithm has a role to train the system from the optimized solution discovered from the ABC algorithm. SVM helps in classifying the false and true allocation. QoS parameters, such as, Task Allocation, Energy Consumption and Completion Time are computed to depict the proposed work performance. Comparative analysis has been performed that has helped in enlightening the fruitfulness of proposed work.

The organization of this paper is such that Sect. 2 delineates the existing work related to the energy consumption in cloud computing and has drawn a problem according to which proposed methodology has been developed as depicted in Sect. 3. The result and analysis are shown in Sect. 4 that is obtained after the evaluation of the proposed work. The conclusion has been drawn in Sect. 5 following the references.

2 Related Work

Yang et al. [7] have optimized the overall performance of cloud by using the concept of dynamically offloading along with the Android portion of code that is executed on the smartphone. Satyanarayanan et al. [12] presented a framework by utilizing the VM conception that was operated on a trusted and sophisticated computer, which uses cloudlet computers. Zhang et al. [9] have segregated a flexible application into a number of elements known as Weblets. Yang et al. [10] have researchers on multi-user computing schemes in such a way so that the performance can be optimized. Yang et al. [11] have worked to make the best use of MCC and hence provide the best services to the MCC users. Kaewpuang et al. [13] have designed a framework for resource allocation in mobile applications environment, fund management and collaboration among different service providers. However, the above-mentioned studies have not to pay attention to the problem of minimizing energy efficiency, the impact of task dependence as well as task completion rate have not considered. Kosta et al. [14] offered an offloading scheme for allocating the resources on dynamic basis. Another source planning strategy for cloud computing was proposed by Singh and Chana [15] Cloud computing issues have been discussed by Durao et al. [16]. While cloud servers are used remotely, communication time and power are turning into problems when communicating with remote cloud servers. Cloudlet is a well-liked approach for low-power. The concept of offloading mobile data has been used by Satyanarayanan et al. [12]. Also, the effect of data offloading according to the mobility of human has been discussed by Lima et al. [17]. Protection of energy services for MMC has been discussed by Chunlin and Layuan [18]. For elastic mobile applications, the distribution of work has been discussed by Shiraz et al. [19].

Currents studies have concluded that energy consumption linearly scaled by resource utilization. This promising factor has enlightened the considerable involvement of tasks consolidation for energy consumption reduction. Though, task consolidation might also result in releasing the resource that could sit inactively but simultaneously, consumes power as well. Some efforts have been noticed for the reduction of power by placing the resources in power saving or sleep mode. This article has dealt with reducing energy consumption by introducing a novel mechanism using MBFD, ABC and SVM as well.

3 Proposed Modified Best Fit, Artificial Bee Colony Algorithm and SVM Algorithm

Initially, a request is sent by mobile devices to the cloud server for performing or executing a task. The cloud server takes some time to perform the task and mobile users losses its connection to the cloud server. Now, the smartphone user has reconnected with the cloud network and the resultant task is allocated to the user by performing two operations such as (i) sorting of users and (ii) Optimize and classification of tasks in case of false allocation Tziritas et al. [20].

3.1 Sorting of Tasks

In this research, the MBFD algorithm is used to sort the users as per their task. Let us considered that there is n number of mobile users and should be allocated with m number of the server in MCC, with a search space of mn. The main aim of this research is to determine the best energy efficient solution. The flow of MBFD algorithm is defined as below:

Algorithm 1: MBFD

Input: server_list, mobile user_ list
Output: Sorted task
1. Sort Task in decreasing order of utilization ( )
2. For each mobile user in mobileuser_ list do
3. minpower← max
4. allocated server← $null$
5. for each server in the server, list do
6. if the server has enough resource for the mobile user then
7. $power \leftarrow estimate\ power\ (server, mobile\ user)$
8. If power<mean power then
9. Allocate server← $server$
10. Mini power← $power$
11. if Allocated server≠ $null\ then$
12. Allocate mobile user to the allocated server
13. Return Sorted task

3.2 Allocation Algorithm Design and Implementation of ABC

We have implemented an ABC algorithm to minimize or save energy, which is being consumed by mobile user during the allocation of tasks. ABC algorithm is used to solve the problem of false allocation of task towards a mobile user.

ABC is a new meta-heuristic approach developed in 2005. Smart nutrition behaviour of honey bees is motivated and simulated by minimalist nutrition honey bee model in the search process for real, consistent, and inaccurate optimization problems. This algorithm is composed of three types of artificial bees, including employed, onlooker and scout bees [21].

Employed bee: It is responsible to detect a novel food source by searching the food source in the neighbourhood of its present position.

Onlooker bees: After seeing the dance of the employed bees, onlooker bees select one of the food sources on the basis of their dance and then proceed towards the selected source.

Scout bees: The number of scout bees is not predetermined in the colony but is generated as per the situation such as of food source, where these bees are taken into account or not.

Algorithm 2: ABC Algorithm

Input: Sorted task with mobile users properties, Servers and Fitness Function
Output: Allocated Servers and Overloaded Servers

1. Define: N – Number of Servers
2. Fit Fun – Fitness Function:
3. $Fit\ Fun = \begin{cases} True; & if\ \text{Servers}_P < Threshold_P \\ False; & Otherwise \end{cases}$

Where, Servers_P : is the properties of Servers and $Threshold_P$ is the threshold properties of Servers

4. Calculate Length of all sorted task in terms of R Length
5. Initialize, Allocated Servers = []
6. for i = 1→R Length
7. E_{Bee} = Servers (i) = $Servers_P$
8. O_{Bee} = $Threshold_P$
9. $Fit\ Fun = Fit\ Fun\ (E_{Bee}, O_{Bee})$
10. S_{Bee} = ABC (Fit Fun, Sorted Task)
11. end
12. index = Find the index of S_{Bee} in the sorted task list
13. if index = Normal
14. Allocated Servers = Users (index)
15. else
16. Mark as Overloaded Servers
17. end
18. Returns: Allocated Servers & Overloaded Servers
19. end

3.3 Supporting Vector Machine (SVM)

The optimized solution obtained from the ABC algorithm is used to train the system by using a classification algorithm named as SVM. The standard SVM takes a set of input data from the ABC algorithm and predicts which of the two possible categories for each given input is formed such that the SVM becomes a binary linear classifier Zhong et al. [22].

Algorithm 3: Support Vector Machine Algorithm

Input: Properties of Overloaded Servers, Types of Servers (Cat), Kernel Function
Output: Allocation of Overloaded Server

```
1. Training : Initialize the SVM training data T is the total property list of servers with RBF as Kernel
function

2. for I = 1→ All Overloaded Servers
3. if Property of Overloaded Servers (I) == Real
4.      Defined the Cat as a category of training data
5.      Cat (1) = Overloaded Servers (I)
6.   else
7.       Cat (2) = Overloaded Servers (I)
8.   end
9.  end
10. Train_Structure=SVMTRAIN (T, Cat, Kernel function)
11. Testing :
Current Overloaded Servers = Properties of currently overloaded servers in a network
12. Allocation = simulate (Net, Current Vehicles)
13. if Allocation is valid
14. Allocation of Overloaded Server = Right
15. else
16. Allocation of Overloaded Server = Wrong
17. end
18.Return;Allocation of Overloaded Server as Output
19end
```

Initially, n number of users with m number of servers is generated. MBFD algorithm is applied to allocate servers to mobile users. According to server different properties of mobile users such as CPU utilization, RAM, the cost function is verified. Apply the ABC algorithm and generate a random solution. If the threshold value of the ABC algorithm is less than the server load then apply the SVM algorithm otherwise mobile user allocation is done. SVM is used to classify false and true allocation. The host is allocated to the user which is nearby the server. This is done by using Euclidian distance formula as written below.

$$\mathrm{d} = \sqrt{(x_2 - x_1)^2 - (y_2 - y_1)^2}$$

At last, calculate performance parameters described in the subsequent section.

4 Result Analysis

In this section, the experimental results that are executed in MATLAB simulator are listed in the table as well as display in the form of a graph. The experiment has been performed by using a different number of user's ranges from 10 to 300. The parameters such as task allocation, energy consumption and completion time are evaluated and discussed.

Figure 1a represents the graph of task allocated by different user's ranges from 10 to 300. From the above graph, it is clear that when swarm intelligence technique along

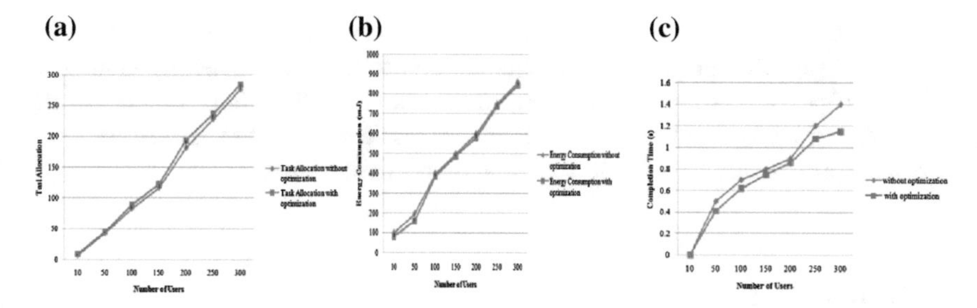

Fig. 1 **a** Task allocation, **b** energy consumption, and **c** completion time (s)

with supervised learning algorithm is used the task allocation has been increased. The average number of task allocated by the server without and with optimization techniques is 86.2 and 91.8 respectively. Thus there is an increase of 6.5% while the optimization technique is employed.

The energy consumption for the entire MCC system is shown in Fig. 1b. From Fig. 1 it is analysed that with the increase in the number of users the energy consumption also increases. The average energy measured with and without optimization is 487.14 mJ and 477.14 mJ respectively. Thus there is a reduction of 2.05% while utilizing ABC along with the SVM algorithm.

From Fig. 1c, it is clear that as the numbers of users increase the tasks requires more time for the completion of a task. The average completion time measured with optimization and without optimization algorithm are 1 s and 0.82 s respectively. Thus, there is a reduction of 18% in the completion time while utilizing an optimization algorithm with a novel fitness function for scheduling tasks in MCC. With the increase in the number of users the time taken to complete tasks is also increases which represent the dependency of completion time with number of users.

Also, to show the enhancement of proposed work comparison between energy consumption and completion time is illustrated in the Fig. 2a, b.

Figure 2a, represents the graphical representation of the average energy obtained by Guo et al. [6] and the presented work using (ABC and SVM) algorithm. From the above graph, there is a reduction of 36.12% in energy consumption has been determined compared to the existing work.

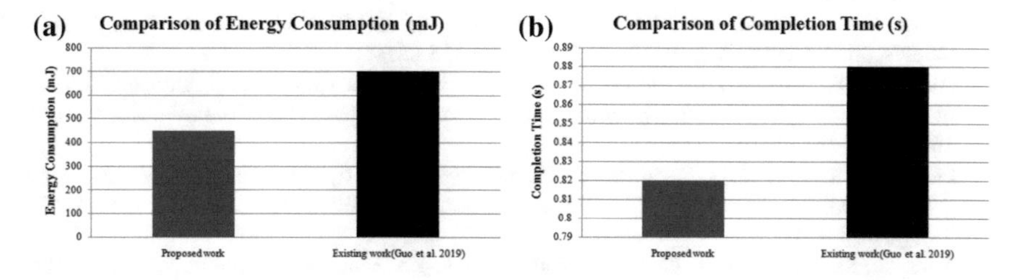

Fig. 2 **a** Comparison of energy consumption (mJ) and **b** comparison of completion time

The comparison of average completion time is depicted in Fig. 2b. The black colour and grey colour graph represents the completion time taken by Guo et al. [6] and proposed work respectively. From the above graph, it is observed that the completion time required by proposed work is less compared to the existing work, which is approximately 6.82% less compared to existing work.

5 Conclusion

We have studied the issue of online task assignment in MCC to maximize the cloud provider's services. The main challenges include the NP-hard nature of the task assigned to the cloudlet and the online nature of the user's task arrival. In MCC, the applications with large computing tasks are assigned to the cloud and as per the user's request, MCC send results back to the mobile device. This process consumes a large amount of energy and hence task completion time also increases. Thus to reduce this problem, MBFD algorithm is used which sorts the tasks in decreasing order that is the tasks that consume highest energy is computed first. The sorted tasks are assigned to ABC algorithm, which provides optimized tasks on the basis of fitness function according to task properties and at last supervised learning technique is used to distinguishing between faulty and normal tasks. In this way, energy consumption and completion time are reduced to 36.12% and 6.82% respectively.

In future, the work can be extended by considering input data size instead of users and other optimization technique can be used in hybridization with ABC. Also, multiclass classifier usability can enhance the performance of the proposed work.

References

1. Ren, J., Zhang, Y., Zhang, K., Shen, X.: Exploiting mobile crowdsourcing for pervasive cloud services: challenges and solutions. IEEE Commun. Mag. **53**(3), 98–105 (2015)
2. Gong, Y., Zhang, C., Fang, Y., Sun, J.: Protecting location privacy for task allocation in ad hoc mobile cloud computing. IEEE Trans. Emerg. Top. Comput. **6**(1), 110–121 (2018)
3. Tang, J., Quek, T.Q.: The role of cloud computing in content-centric mobile networking. IEEE Commun. Mag. **54**(8), 52–59 (2016)
4. Othman, M., Madani, S.A., Khan, S.U.: A survey of mobile cloud computing application models. IEEE Commun. Surv. Tutor. **16**(1), 393–413 (2014)
5. Abolfazli, S., Sanaei, Z., Ahmed, E., Gani, A., Buyya, R.: Cloud-based augmentation for mobile devices: motivation, taxonomies, and open challenges. IEEE Commun. Surv. Tutor. **16**(1), 337–368 (2014)
6. Guo, S., Liu, J., Yang, Y., Xiao, B., Li, Z.: Energy-efficient dynamic computation offloading and cooperative task scheduling in mobile cloud computing. IEEE Trans. Mob. Comput. **18**(2), 319–333 (2019)
7. Yang, S., Kwon, D., Yi, H., Cho, Y., Kwon, Y., Paek, Y.: Techniques to minimize state transfer costs for dynamic execution offloading in mobile cloud computing. IEEE Trans. Mob. Comput. **13**(11), 2648–2660 (2014)

8. Satyanarayanan, M., Bahl, P., Caceres, R., Davies, N.: The case for vm-based cloudlets in mobile computing. IEEE Pervasive Comput. (4), 14–23
9. Zhang, X., Jeong, S., Kunjithapatham, A., Gibbs, S.: Towards an elastic application model for augmenting computing capabilities of mobile platforms. In: International Conference on Mobile Wireless Middleware, Operating Systems, and Applications, June, pp. 161–174. Springer, Berlin, Heidelberg (2010)
10. Yang, L., Cao, J., Yuan, Y., Li, T., Han, A., Chan, A.: A framework for partitioning and execution of data stream applications in mobile cloud computing. ACM SIGMETRICS Perform. Eval. Rev. **40**(4), 23–32 (2013)
11. Yang, L., Cao, J., Cheng, H., Ji, Y.: Multi-user computation partitioning for latency sensitive mobile cloud applications. IEEE Trans. Comput. **64**(8), 2253–2266 (2015)
12. Satyanarayanan, M., Bahl, P., Caceres, R., Davies, N.: The case for vm-based cloudlets in mobile computing. IEEE Pervasive Comput. **4**, 14–23 (2009)
13. Kaewpuang, R., Niyato, D., Wang, P., Hossain, E.: A framework for cooperative resource management in mobile cloud computing. IEEE J. Sel. Areas Commun. **31**(12), 2685–2700 (2013)
14. Kosta, S., Aucinas, A., Hui, P., Mortier, R., Zhang, X.: Thinkair: dynamic resource allocation and parallel execution in the cloud for mobile code offloading. In: 2012 Proceedings IEEE Sethi Infocom, March, pp. 945–953. IEEE (2012)
15. Singh, S., Chana, I.: QRSF: QoS-aware resource scheduling framework in cloud computing. J. Supercomput. **71**(1), 241–292 (2015)
16. Durao, F., Carvalho, J.F.S., Fonseka, A., Garcia, V.C.: A systematic review on cloud computing. J. Supercomput. **68**(3), 1321–1346 (2014)
17. Lima, E., Aguiar, A., Carvalho, P., Viana, A.C.: Impacts of human mobility in mobile data offloading. In: ACM CHANTS, October, vol. 4 (2018)
18. Chunlin, L., Layuan, L.: Cost and energy aware service provisioning for mobile client in cloud computing environment. J. Supercomput. **71**(4), 1196–1223 (2015)
19. Shiraz, M., Ahmed, E., Gani, A., Han, Q.: Investigation on runtime partitioning of elastic mobile applications for mobile cloud computing. J. Supercomput. **67**(1), 84–103 (2014)
20. Tziritas, N., Loukopoulos, T., Khan, S., Xu, C.Z., Zomaya, A.: A communication-aware energy-efficient graph-coloring algorithm for VM placement in clouds. In: 2018 IEEE SmartWorld, Ubiquitous Intelligence & Computing, Advanced & Trusted Computing, Scalable Computing & Communications, Cloud & Big Data Computing, Internet of People and Smart City Innovation (SmartWorld/SCALCOM/UIC/ATC/CBDCom/IOP/SCI), October, pp. 1684–1691. IEEE (2018)
21. Sethi, N., Singh, S., Singh, G.: Multiobjective Artificial Bee Colony based Job Scheduling for Cloud Computing Environment (2018)
22. Zhong, W., Zhuang, Y., Sun, J., Gu, J.: A load prediction model for cloud computing using PSO-based weighted wavelet support vector machine. Appl. Intell. **48**(11), 4072–4083 (2018)

Fake Review Prevention Using Classification and Authentication Techniques

P. Prakash, N. Shashank, M. Arjun, P. S. Sandeep Yadav, S. M. Shreyamsa and N. R. Prazwal

Abstract Day by day people shopping via E-commerce sites is burgeoning. Decision of placing orders relies on product and service reviews provided by the customers. The importance of the reviews has increased tremendously because they provide information about the quality of product and service. This stimulates sellers to exploit these reviews to increase their sales by deceiving the customers with false information. Thus, detection and prevention of fake reviews becomes pivotal. This paper focuses on detecting and preventing fake reviews using classification and authentication techniques. Review content and Reviewer behavior-based features were used to train different machine learning algorithms such as SVM, Random forest, and Decision tree; among these, Random forest classification algorithm had the highest accuracy of fake review detection with 73.33%. Prevention of fake reviews is achieved by making sure the right person gets to write the review by sending the review writing link to the registered email-id. This research is also concerned over preventing bots to write review by examining the keyboard and mouse activities of the machine. Captcha authentication method has been adopted to prevent bots from writing reviews.

Keywords Fake reviews · Detection · Prevention · Reviewer · Spammer · Classification · Bot · E-commerce

1 Introduction

Internet has influenced people in more ways than it can be imagined. The concept of E-commerce started as more people started using Internet. E-commerce means buying or selling a good or a service for which monetary value has been paid. As technology improved, humans have become lazier and they need everything at their hands reach. People now prefer to buy online rather than physically visiting the store. This is where E-commerce has tremendous scope. The retail E-commerce revenues

---

P. Prakash (✉) · N. Shashank · M. Arjun · P. S. S. Yadav · S. M. Shreyamsa · N. R. Prazwal
Ramaiah University of Applied Sciences, Bengaluru, India
e-mail: prakashp.cs.et@msruas.ac.in

M. Tuba et al. (eds.), *ICT Systems and Sustainability*,
Advances in Intelligent Systems and Computing 1077,
https://doi.org/10.1007/978-981-15-0936-0_42

are projected to grow up to 4.88 trillion US dollars by 2021 [1]. So, one can clearly see that E-commerce holds lots of promises and opportunities. Articles have been published about fake reviews written either by bots or paid persons [2, 3].

Reviews play an important role in any consumer's buying choices [4]. It influences the consumers buying thoughts by forming an opinion about the product [5]. Humans generally tend to find out how a product is, by asking people about it and testing it themselves. With online shopping the physical aspect of the buying is missing, so one tends to rely on reviews written on the E-commerce sites. So, one can fabricate reviews to increase average rating or decrease competitors' average ratings. So these fake reviews damage the buying experience of consumers, as sometimes they may not get the standard of product they expected. This affects the E-commerce sites a lot, since consumers start losing their trust. So they tend to take many measures to detect and prevent these fake reviews.

As there is a famous saying "Prevention is better than cure," it is better to stop a problem from happening than to fix the problem once it has happened. This paper proposes the prevention of the fake reviews by detecting spammers and fake reviews, combined with various authentication techniques. In this paper, three new features are introduced for detecting fake reviews.

2 Related Work

The paper [6] discussed about the supervised learning [7, 8] for fake review detection using Amazon Mechanical Turk (AMT) that [9] generated fake reviews and real-life fake reviews. SVM algorithm is used for model building and the results have shown that the real-life data is much harder to classify with an accuracy of only 67.8%. Yelp filtering accuracy is more of random guessing of 50%. This paper concluded that models trained using AMT generated fake reviews perform poorly in detecting real fake reviews in Yelp, which indicates that the AMT fake reviews are probably not representative of the real-life fake reviews.

The paper [10] proposed review density, semantic- and emotion-related models and features to identify fake reviews from professional review website. This paper briefly discussed about Review Ratio, Average Review Density, and Rating Variance. They collected the review dataset from www.dianping.com which is the most famous review website in China. They have used different types of classifiers to detect the fake reviews based on selected features and used fivefold cross-validation to train and test. The accuracy of every classifier was more than 0.9. Naïve Bayes performed weak because of its simplicity. SVM and Decision tree performed quite well with accuracy of 0.92 and 0.93, respectively.

The paper [11] considers four different datasets from TripAdvisor–Gold, TripAdvisor–Heuristic, Yelp, and deceptive essays collected using Amazon Mechanical Turk. This paper used SVM classifier with fivefold cross-validation to build the model. The features driven from Context-Free Grammar (CFG) parse trees consistently improve the detection performance over several baselines that are based only on shallow lexico-syntactic features.

The paper [12] focused on detecting fake reviews from a set of product reviews by simulating spam reviews. In these product reviews, data is extracted using import.io. This paper proposes two new features: user review frequency on the same product and term frequency. And, they have used two classification methods such as Naïve Bayes and Random forest. The classification is done by the measure of accuracy, positive predictive value or precision, negative predictive value, recall or sensitivity and specificity all of which are calculated from the confusion matrix. They observed 98% accuracy in Naïve Bayes and 99% accuracy in Random forest. They have concluded that Random forest performs better than Naïve Bayes and can be used to detect the genuine as well as the fake reviews.

3 Proposed Model

The proposed model aims at preventing fake reviews, which involves authentication and real time detection of spammers and fake reviews by classification. The features considered to classify the reviews are of two types: Reviewer behavior based and Review Content based. Following are the features' definition:
Reviewer behavior-based features:

- **Number of reviews on same category products (F1)**: Fake reviewers sometimes post reviews on many products of same category to make more money. This is not the usual behavior of a normal user [10].

$$n_{cat}(x) = |\{rvw\ on\ p\ by\ x : p \in cat\}|$$

  where x—reviewer, cat—category, rvw—review, p—product
- **Average review density (F2)**: This is the average number of reviews that a reviewer is posting per day. Fake reviewers post large number of reviews compared to a normal user [10].

$$density(x) = \frac{no\ of\ reviews\ posted\ by\ x}{no\ of\ days\ x\ was\ active}$$

  where '*active*' denotes reviewer has posted at least one review on that day.
- **Average break b/w reviews on same category products (F3)**: There will be some days break between the purchases done by customers for writing a review. This gap will be very less for the reviews posted by fake reviewers.

$$N_{cat}(x) = \frac{no\ of\ days\ b/w\ 1\text{st}\ \&\ last\ review}{n_{cat} - 1}$$

- **Standard Deviation (SD) of ratings (F4)**: This gives the variation of the ratings given by a reviewer [10].

$$\sigma_r(x) = \sqrt{\frac{1}{n}\sum_{i=1}^{n}(r_i - \bar{r})^2}$$

where r—rating given out of 5.
- **Average helpful votes (F5)**: People read the reviews which are most helpful [10].

$$\frac{Total\ no\ of\ helpful\ votes\ x\ got}{total\ no\ of\ reviews\ by\ x}$$

- **SD of number of reviews written per day (F6)**: Fake reviewers maintain the consistency in the number of reviews written per day.

$$\sigma_d(x) = \sqrt{\frac{1}{n}\sum_{d=1}^{n}(n_d - density(x))^2}$$

where σ—standard deviation, n—no of days x was active, n_d—no of reviews posted on day d.

Review Content-based features:

- **Ratings (F7)**: Rating of the review given to the product out of 5 [10].

$$r = rating/5$$

- **Length of the review (F8)**: Number of words in the review text. Fake reviews tend to be very long since it contains many unnecessary details [6] n_w.
- **Similarity with other reviews (F9)**: Fake reviewers create multiple user accounts and post same/similar reviews from all those accounts to a product.

$$Similarity = \text{cosine}\big(wordVec_i, wordVec_j\big)$$

The similarity is found for the reviews to a product with similar ratings and maximum of the similarities is considered [10].
- **Deviation from average rating (F10)**: Fake reviews try to affect the overall rating of the product by deviating significantly from the average rating of the product before.

$$r - \overline{r_{before}}$$

where r—rating of the review, $\overline{r_{before}}$—the average rating of the product before posting the review.

- **Diversity in emotion (F11)**: Fake reviews are not likely to have balanced emotion because their main aim is to leave either positive or negative impression about the product in the reader's mind which cannot be achieved by balanced statement [10].

$$diversity = \frac{1}{3}\left(\left|\frac{n_{+ve}}{n_w} - \frac{1}{3}\right| + \left|\frac{n_{-ve}}{n_w} - \frac{1}{3}\right| + \left|\frac{n_{ne}}{n_w} - \frac{1}{3}\right|\right)$$

where n_{+ve}—no of positive words, n_{-ve}—no of negative words, n_{ne}—no of neutral words.

- **Use of Numerical and Capital words (F12)**: Fake reviews try to grab the reader's attention by the usage of capital words and numerical data in their content [6].

In Fig. 1 of the model, the user is classified as a spammer or normal user using the Reviewer behavior-based features (F1–F6). Users are directly allowed to write a review if they are not detected as spammers and are in the same device where purchase of the product was done, or else a link to write a review is sent to their registered email. The model also detects bots—since bots are also being used to post large number of reviews—by tracking various behaviors like time taken to write review, mouse/touch and keyboard inputs; and prevents them from posting reviews by showing captcha.

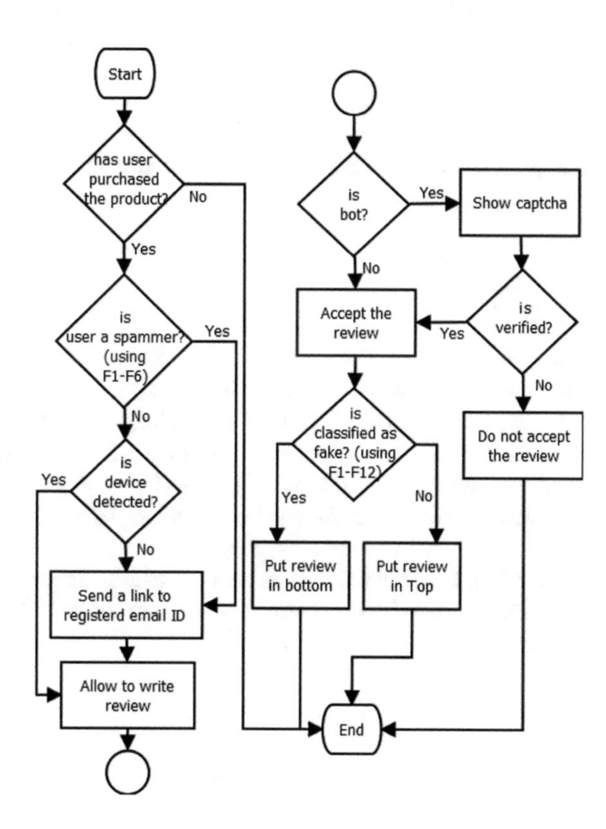

Fig. 1 Flowchart of proposed model

After posting the review, it is classified as fake or genuine using both reviewer behavior- and content-based features (F1–F12). The reviews which are classified as fake are put into the bottom of review list rather than completely discarding them because prediction of the classification cannot be completely accurate.

4 Experiment and Results

4.1 Data Preprocessing

For this paper, the dataset used was that of Amazon Reviews [13]. The raw dataset contained details about the reviewers like their respective names, IDs, reviews written by them, details of number of people who found the review helpful, short summary of the review, and many other information. All the data in this raw dataset is not required for building this model. So, the raw dataset was preprocessed to extract the required data for building this model, which included reviewerID, reviewerName, review text, summary, ratings, time of the review written, productID, and productName of products of 18 categories which were mainly electronics and accessories. The reviews were also condensed by a small extent by removing stop words. The removal of such words (the, an, a, is) would not change the meaning or context of the reviews. Data reduction is also one of the major tasks in data preprocessing which helps in obtaining a reduced representation of the dataset that is much smaller in volume yet produces the same (or almost the same) analytical results. A python code was used to extract and preprocess the required data from the raw set of Amazon Reviews.

4.2 Detection

Some reviews were labeled as 'True,' 'Fake,' or 'Not a review' manually by reading the review and looking at few features like rating, review density, similarity, break b/w reviews, etc. The labeled reviews along with the computed 12 features were divided into training and testing data in the ratio of 7:3, respectively. This label was verified twice in order to avoid any errors in training and testing data. These training data was used to train different classifiers such as SVM, Random forest, and Decision tree to detect the fake reviews (Table 1).

Random forest classifier has better detection accuracy compared to other classifiers. Hence, this classifier can be used for detecting fake reviews and spammers for prevention.

Table 1 Accuracy of Detection with different classifiers

Classifier	Features	
	F1–F6 (detects reviewer) (%)	F1–F12 (detects fake review) (%)
SVM	60	62.67
Decision tree	58.67	64
Random forest	66.67	73.33

4.3 Prevention

A simple online shopping website with the extracted data was built using python flask for implementing the prevention model (Fig. 2).

Users who have not purchased the product are not allowed to write a review for that product. If the user is detected as spammer or has purchased the product in a different device, he/she is not directly allowed to write review. User is detected as spammer or genuine, using the classifier with reviewer behavior-based features (F1–F6). As shown in Fig. 3, clicking "Write Review" button sends the link to the user's email through which he/she can write the review.

User can write and post the review through the interface shown in Fig. 4. If user has purchased the product in the same device and is not detected as spammer, then

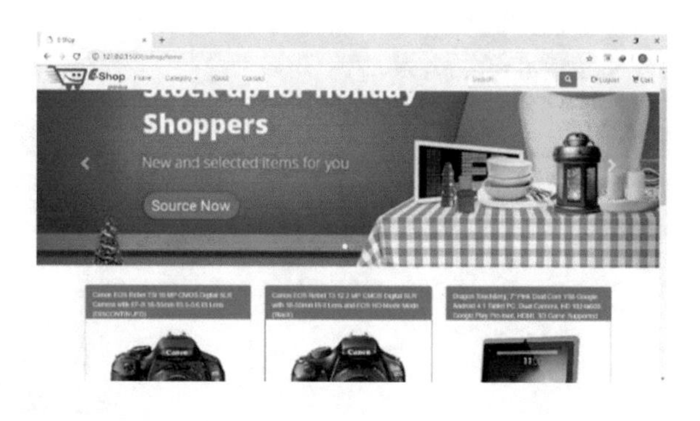

Fig. 2 An online shopping website created to implement the fake review prevention algorithm

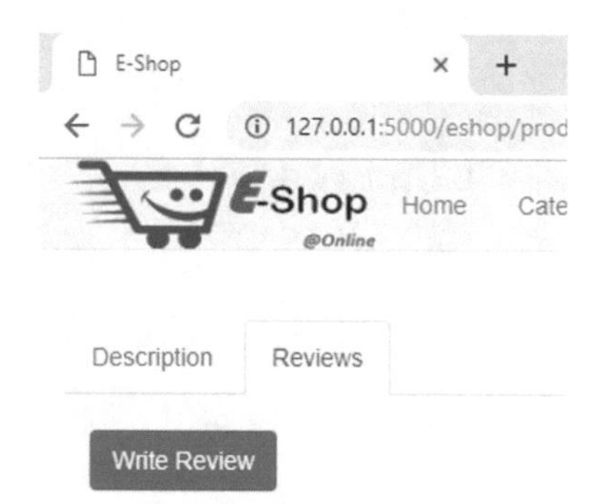

Fig. 3 Not allowing to write review

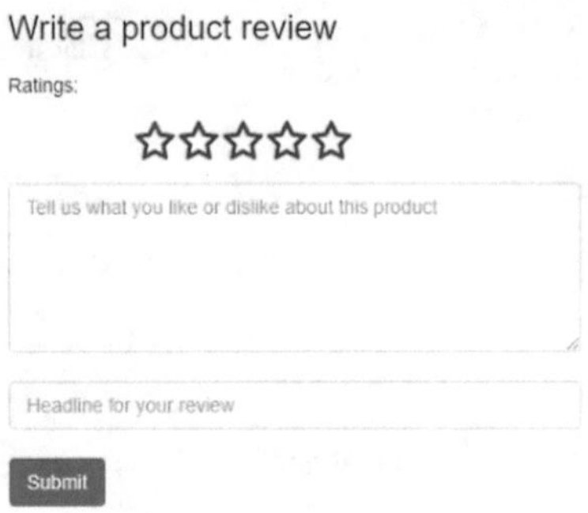

Fig. 4 Interface for writing review

he is allowed to write the review. Device is uniquely identified using session cookies, device name, platform, and browser used. Users who come through the link sent to their email will also be allowed to write the review.

The bot is detected by tracking time taken to write the review, monitoring mouse/touch input, and keys pressed in keyboard. If bot is detected, captcha is shown to stop the bot from continuing further. Review is accepted only if captcha is verified. This prevents bots from posting reviews (Fig. 5).

After accepting the reviews, they are classified as fake or not by the classifier with both Reviewer behavior-based and Review content based-features (F1–F12). The reviews which are classified as true are given "Verified" tag and put in the top

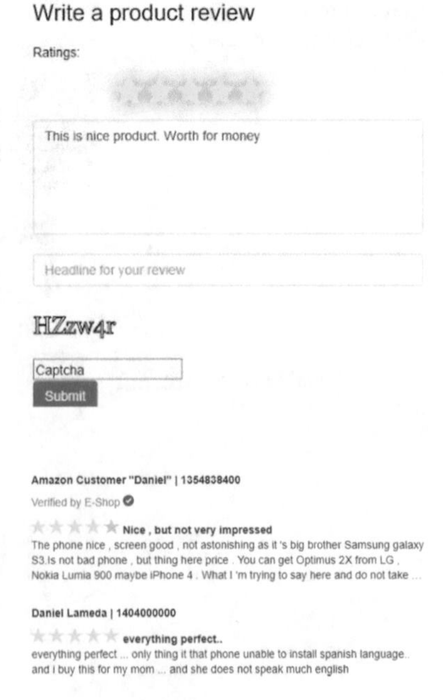

Fig. 5 Scenario of captcha being shown when bot is detected

Fig. 6 Displaying reviews

of the review list as shown in Fig. 6. The reviews which are detected as fake are not given any tag and put in the bottom where readers are unlikely to make till there.

5 Conclusion and Future Work

This paper proposes three new features to classify real-world Amazon product reviews as fake such as average break between reviews on same category products (F3), standard deviation of number of reviews written per day (F6), and deviation from average rating (F10). Moreover, this paper also proposes new feature of preventing the fake reviews using device detection, bot detection, and sending review writing link to the registered email-id. Experiment shows that the proposed model accurately classified the reviews as fake and genuine, and perfectly prevented the spammers and fake reviews. Experimental results concluded that the Random forest performs better than SVM and Decision tree to detect fake and genuine reviews.

References

1. Statista—Global retail e-commerce market size 2014–2021. https://www.statista.com/statistics/379046/worldwide-retail-e-commerce-sales/. Last accessed 6 May 2019
2. Marketing Land—Sterling, G., Sterling, G.: Study finds 61 percent of electronics reviews on Amazon are 'fake'. https://marketingland.com/study-finds-61-percent-of-electronics-reviews-on-amazon-are-fake-254055. Last accessed 11 May 2019
3. Dwoskin, E., Timberg, C.: How merchants use Facebook to flood Amazon with fake reviews. The Washington Post. https://www.washingtonpost.com/business/economy/how-merchants-secretly-use-facebook-to-flood-amazon-with-fake-reviews/2018/04/23/5dad1e30–4392-11e8-8569-26fda6b404c7_story.html. Last accessed 10 May 2019
4. The Drum—Fullerton, L.: Online reviews impact purchasing decisions for over 93% of consumers, report suggests. https://www.thedrum.com/news/2017/03/27/online-reviews-impact-purchasing-decisions-over-93-consumers-report-suggests. Last accessed 11 May 2019
5. von Helversen, B., Abramczuk, K., Kopeć, W., Nielek, R.: Influence of consumer reviews on online purchasing decisions in older and younger adults. Decis. Support Syst. **113**, 1–10 (2018)
6. Mukherjee, A., Venkataraman, V., Liu, B., Glance, N.: Fake review detection: classification and analysis of real and pseudo reviews. Technical report. UIC-C S-03–2013 (2013)
7. Ott, M., Choi, Y., Cardie, C., Hancock, J.T.: Finding deceptive opinion spam by any stretch of the imagination. In: ACL, pp. 309–319 (2011)
8. Mukherjee, A., Kumar, A., Liu, B., et al.: Spotting opinion spammers using behavioral footprints. In: ACM SIGKDD International Conference on Knowledge Discovery and Data Mining, pp. 632–640 (2013)
9. Paolacci, G., Chandler, J., Ipeirotis, P.G.: Running experiments on amazon mechanical turk. Judgm. Decis. Mak. **5**(5), 411–419 (2010)
10. Li, Y., Feng, X., Zhang, S.: Detecting fake reviews utilizing semantic and emotion model. In: 3rd International Conference on Information Science and Control Engineering (ICISCE), pp. 317–320. IEEE (2016)
11. Feng, S., Banerjee, R., Choi, Y.: Syntactic stylometry for deception detection. In: Proceedings of the 50th Annual Meeting of the Association for Computational Linguistics: Short Papers-Volume 2, pp. 171–175. Association for Computational Linguistics (2012)

12. Chowdhary, N.S., Pandit, A.A.: Fake review detection using classification. Int. J. Comput. Appl. **975**, 8887
13. McAuley, J.: Amazon review data. http://jmcauley.ucsd.edu/data/amazon/links.html. Last accessed 17 April 2019

A Study of Cataract Patient Data Using C5.0

Mamta Santosh Nair and Umesh Kumar Pandey

Abstract Cataract is one of the common problems among the humans related to eye. Clouding of lens termed as cataract leads to blindness. Various causes of cataract are identified by the ophthalmologist. Data mining has become popular in the past few years because of information extracted from the dataset using algorithm and computational capability. In this paper, cataract patients' historical data is used to build the predictive model. C5.0 algorithm is one of the decision tree algorithms used for predictive modeling. Present study uses C5.0 method to predict cataract status on various parameters. Data used in this research paper is the primary data collected from the Raigad Maharashtra, India.

Keywords Classification · C5.0 algorithm · Cataract

1 Introduction

One of the major causes of avoidable blindness in the world is cataract. Cataract-affected blindness can be prevented if treated in time. Many agencies across globe are working toward awareness and treatment of cataract. The International Agency for Prevention of Blindness (IAPB) mentions that 48% reason behind blindness is cataract. The World Health Organization [1] defines cataract as a disease of the eye that happens because of clouding on the lenses of the eye, making lenses opaque and reducing visibility. The World Health Organization [1] considers cataract as aging process, but there are certain other factors because of which occasionally children are born with the cataract conditions. The World Health Organization [1] highlights some other reasons for cataract, such as eye injuries, inflammation, and some other eye diseases. Statistics from several literatures and survey of different institutions are very huge and the amount of work required is gigantic. Cataract patients' data needs to be

M. S. Nair (✉) · U. K. Pandey
MATS School of Information Technology, MATS University, Raipur, India
e-mail: mamtanair@yahoo.com

U. K. Pandey
e-mail: umesh6326@gmail.com

M. Tuba et al. (eds.), *ICT Systems and Sustainability*,
Advances in Intelligent Systems and Computing 1077,
https://doi.org/10.1007/978-981-15-0936-0_43

studied and analyzed to understand the patterns underlying in it and create awareness in people about it. Hidden patterns of cataract data can be uncovered by data mining algorithms. Data mining includes decision tree algorithms which standout among the most broadly used for assigning the class. One very strong algorithm used for classification is C5.0. In this paper, we collected data of patients with eye problems, some of whom have tested positive for cataract and some are negative. So, with the different parameters about their medical history, food habits, and environment factors, we try to assess their possibility of getting a cataract. Data mining is the tool helping in all types of analysis.

2 Literature Review

Gajpal and Pandey's [2] research work was conducted on student dropout classification using c5.0. The classification error reported in this research work was 34.9%. Decision maker will utilize the rule to take necessary action and to classify the likely students going to be dropouts.

Kesavaraj and Sukumaran [3] did a research on the classification methods. They concluded in their research work that every decision tree algorithm has its own pros and cons. Correct algorithm selection for the problem is necessary before using any decision tree algorithm.

The phrase "read between the lines" applies to any text; similarly, data mining does the job of uncovering what exists but is not visible. It finds patterns or trends, which are interesting and useful too. It helps to see beyond the obvious which is the insight of data. And finally, it helps to take decisions and make predictions while the pearls of wisdom are retrieved from the oceans of data.

Pandya and Pandya's [4] research work studies ID3, C4.5 and C5.0 algorithm. They highlighted that C5.0 has advantages over other studied algorithms like high accuracy, low memory usage, fewer rules, etc.

Satteler et al. [5] address the issue of database technology used in the data mining. The research work demonstrates the implementation of primitive and performance benefits. In this study, they identified various primitive attributes. They concluded that for scalable data mining, data mining and database system need to be coupled tightly for better performance of the algorithm.

Sharma and Kumar's [6] research work focuses on majorly four decision tree algorithms comparison. In their research work, they highlighted the advantage and the comparison of the decision tree algorithms.

Anyanwu and Shiva [7] did experimental research for evaluating the performance of studied decision tree algorithms, i.e., ID3, C4.5, CART, SLIQ, and SPRINT. In their research work they highlighted that there are two ways to implement decision tree, i.e., serial or parallel. Their research work focuses on serial implementation and reported C4.5 and CART have better accuracy for classification.

Kaur and Grewal [8] did study on classification techniques for lung cancer detection. Classification uses attributes to classify the objects to the class.

Pang and Gong [9] applied C5.0 decision tree algorithm for classification. They used attributes to evaluate individual's credit value and build the model. The model classifies each individual into good and bad.

Classification algorithms like SimpleLogistic, IBK, Naïve Bayes, SGD, LMT, and SMO were used for medical data, and LMT was found to be the most appropriate for medical data decision making [10].

See5/C5.0 has been designed to analyze big database with large records containing numerical, categorical, and ordinal data. See5/C5.0 is simple to use and anyone who does not have knowledge about statistics or machine learning can also use it [6].

C5.0 model splits the sample on the attributes giving highest information gain. The sample subset obtained from the previous split will be split more. The splitting process will go until further splitting is not possible for the sample subset. Finally, the lowest level split contains those which do not have significant contribution to the model and not included in the model. C5.0 is capable to handle multi-valued attribute. It also handles the missing value from the attribute [11].

C5.0 decision tree algorithm is developed to overcome the limitation of C4.5 and ID3. C5.0 works with multivariate attribute and missing attribute also. C5.0 is based on entropy and information gain. These entropy and information gains are used to build the tree [2].

3 Data Collection and Research Instrument

The research work used cataract patients' primary data for the study. Questionnaire was designed in consultation with Ophthalmologists. After consultation with ophthalmologist, a total of 41 different parameters were selected for the data collection. These parameters included personal details, food habits, medical, birth history, addictions, etc. The target location of the data collection is Raigad District of Maharashtra, India. Questionnaire was prepared in English and Marathi language. This questionnaire was distributed among 700 cataract patients approximately. Because of low education, most of the respondents are not familiar with the questionnaire system, thus assistance was provided for the form filling. The data includes people of both genders of different age groups. The data also had good mix of rural, including tribal, as well as urban population. A total of approximately 500 forms were received and filled at the camps and outpatient department (OPD) of doctors. In most of the columns, respondents did not provide value or very few respondents gave answer. Thus, those columns were removed from the dataset. In all, only 297 forms were found complete and were selected for analysis. Response sheet had many factors like drugs, hereditary disease, metabolic disease, congenital disease, prematurity, gestation methods, dysthyroidism, birth weight, ventilation/incubator, etc., which were either blank or had very low amount of data in respective attributes compared to 297 records. Thus, those attributes were removed from the dataset and only 17 attributes were considered for the study. For the data study R software is used. R software has package named "C50" for using C5.0 for analysis of the data.

Table 1 Attribute name, type of data, possible value and symbolic name

Attribute symbol	Attribute name	Type of data	Possible values
a	Age	Continuous	
b	Gender	Categorical	F—1; M—2
c	Occupation	Categorical	Housewife—1 Farming—2 Office—3 Teacher—4 Driver—5 Watchman—6 Laborer—7, Business—8 Retired—9
d	Height	Continuous	
e	Weight	Continuous	
f	Diet	Categorical	Veg—1 Nonveg—2
g	Addiction	Categorical	No addiction—1 Any addiction—2
h	Hypertension duration	Continuous	
i	Diabetes duration	Continuous	
j	Cholesterol duration	Continuous	
k	Surgical history	Categorical	No—1 Yes—2
l	Type of surgery	Categorical	No—1 Any surgery (other than cataract)—2 Cataract—3
m	Occupation history in years	Continuous	
n	Sun exposure in hours	Continuous	
o	History of trauma	Categorical	No—1 Yes—2
p	Spectacle use duration	Continuous	
q	Cataract	Categorical	No—1 Yes—2

Attributes used for C5.0 are shown in Table 1. Symbolic name mentioned before the attribute name was used only for programming easiness. Type of Attribute name column is considered for the prediction of the cataract. Type of data is the nature of value received in the column. This column may store either continuous integer or categorical value. Possible value column represents the possible category value with number through which it is used in the dataset.

4 Analysis and Discussions

Total numbers of records are 297, out of which 70% records, which constitutes 207 records, have been used as training set and for rule generation, whereas 30% records which constitutes 97 records are used for testing and prediction accuracy of the C5.0. Developed code in R runs at various set.seed values ranging from 1 to 200. Highest accuracy received at set.seed value is 118. Table 2 shows only those set.seed values

Table 2 set.seed value and prediction accuracy

Set.seed value	Prediction accuracy
61	0.7
118	0.76666
137	0.72222
148	0.74444
173	0.71111
191	0.71111
192	0.71111

where prediction accuracy is more than 0.7.

C5.0 function in R software generates the following rules to using train set records. These rules are shown in Table 3. Rules shown in Table 3 are generated on 207 records of training dataset. Total size of the tree is 27 with error rate of 13.5%. Table 4 shows attribute usage in generation of rule. Order of attribute and usage percentage indicates the importance of attribute in prediction of cataract.

Rules established by C5.0 algorithm show that a person with longer occupational history is affected by the cataract. If occupational history is less than 42 years, then age becomes the prominent feature for cataract. After age, the type of surgery is a prominent factor. Similarly, other factors are used at various levels for prediction of cataract status.

Out of 16 factors only 14 factors are used as listed in Table 4. Attributes' weight and surgical history have no role in prediction of cataract presence.

Table 5 shows model evaluation metrics. Model evaluation metrics indicate that accuracy of the model is 76.66% with misclassification rate of 0.23333. Precision, F1 score, and G score values are 0.75, 0.758620, and 0.132064, respectively, which indicate that model prediction is good.

5 Conclusion

This research discusses cataract patients' data for the prediction of cataract presence. Data used in this research work is primary data collected form Raigad district Maharashtra and processed using C5.0 algorithm in R Software available in "C50" package. Algorithm finds a set of rules as shown in Table 3 using attributes shown in Table 4 with usage percentage. It is concluded from the study that occupational history duration is more responsible for cataract. Next factor is age. It is true that longer occupational history indicates more age but the occupational history is considered separately because it could be possible that a person has come in occupation at an early age. Compete hierarchy represented in the Table 3 will help decision makers to predict the cataract without going into any medical checkup. Model evaluation metrics shown in Table 5 estimates the model validity.

Table 3 Rules identified using C5.0

```
m <= 42:
:...a <= 54:
    :...l = 2: 2 (9/2)
    :   l = 3: 1 (2)
    :   l = 1:
    :   :...o = 2:
    :       :...m <= 2: 1 (4/1)
    :       :   m > 2: 2 (4)
    :       o = 1:
    :       :...h <= 0: 1 (43/6)
    :           h > 0:
    :           :...j > 0: 1 (2)
    :               j <= 0:
    :               :...p <= 12: 2 (6)
    :                   p > 12: 1 (4)
    a > 54:
    :...l = 3: 2 (5)
        l in {1,2}:
        :...c in {4,5,9}: 1 (11/2)
            c in {6,8}: 2 (5/1)
            c = 2:
            :...d <= 5.3: 2 (5)
            :   d > 5.3: 1 (4/1)
            c = 7:
            :...f = 1: 1 (2)
            :   f = 2: 2 (3)
            c = 3:
            :...b = 1: 1 (2)
            :   b = 2:
            :   :...n <= 0: 1 (2)
            :       n > 0:
            :       :...h <= 3: 2 (16/2)
            :           h > 3: 1 (5/1)
            c = 1:
            :...i > 4: 1 (4)
                i <= 4:
                :...g = 2: 2 (4)
                    g = 1:
                    :...f = 1: 2 (21/5)
                        f = 2:
                        :...i > 0: 2 (2)
                            i <= 0:
                            :...p <= 1: 1 (8/1)
                                p > 1:
                                :...d <= 4.9: 1 (5/1)
                                    d > 4.9: 2 (16/4)
```

Table 4 Attribute usage in generation of rule

Usage (%)	Attribute name
100.00	m (Occupation history in years)
93.72	a (Age)
93.72	l (Type of surgery)
55.56	c (Occupation)
36.71	h (Hypertension duration)
30.43	o (History of trauma)
28.99	i (Diabetes duration)
27.54	f (Veg/Nonveg)
27.05	g (Addiction)
18.84	p (Spectacle use duration)
14.49	d (Height)
12.08	b (Gender)
11.11	n (Sun exposure in hours)
5.80	j (Cholesterol duration)

Table 5 Model evaluation metrics

Accuracy of classifier	0.766666
Misclassification rate	0.233333
True positive rate (recall)	0.767444
False positive rate	0.234042
True negative rate	0.765957
False negative rate	0.232558
Precision	0.75
Prevalence	0.477777
F1 score	0.758620
G score	0.132064

6 Future Work

This research work discusses cataract patients' data for prediction of cataract presence. Data used in this research work is primary data collected form Raigad district Maharashtra and processed using C5.0 algorithm. In this research work out of 41 only 17 parameters were used. Reason behind this is that respondents did not provide value. So, there is further space to conduct larger area to find the values missing in 41 variables. This will increase the effectiveness of the study.

References

1. WHO homepage. https://www.who.int/blindness/causes/priority/en/index1.html. Last accessed 25 May 2019
2. Gajpal, A.L., Pandey, U.K.: Identifying dropout factor order using C5.0 decision tree. Int. J. Adv. Res. Sci. Eng. **6**(4) (2017). ISSN(o) 2319–8354. ISSN(P) 2319-8346
3. Kesavaraj, G., Sukumaran, S.: A study on classification techniques in data mining. IEEE-31661, July 4–6 (2013)
4. Pandya, R., Pandya, J.: C5.0 algorithm to improved decision tree with feature selection and reduced error pruning. Int. J. Comput. Appl. **117**(16), 18–21 (2015)
5. Sattler, K.U., Dunemann, S.O.: SQL database primitives for decision tree classifiers. CIKM '01 atlanta, ACM, GA USA (2001)
6. Sharma, H., Kumar, S.: A survey on decision tree algorithms of classification in data mining. Int. J. Sci. Res. (IJSR) **5** (2016)
7. Anyanwu, M.N., Shiva, S.G.: Comparative analysis of serial decision tree classification algorithms. Int. J. Comput. Sci. Secur. (IJCSS) **3**(3)
8. Kaur, S., Grewal, A.K.: A review paper on data mining classification techniques for detection of lung cancer. Int. Res. J. Eng. Technol. (IRJET) **3**(11) (2016). e-ISSN: 2395-0056, p-ISSN: 2395-0072
9. Pang, S.-L., Gong, J.-Z.: C5.0 classification algorithm and application on individual credit evaluation of banks. Syst. Eng.-Theory Pract. **29**(12), 94–104
10. Alaoui, S.S., Labsiv, Y., Aksasse, B.: Classification algorithms in data mining. Int. J. Tomogr. Simul. **31**, 34–44 (2018)
11. Kumar, S.V.K., Kiruthika, P.: An overview of classification algorithm in data mining. Int. J. Adv. Res. Comput. Commun. Eng. **4**(12) (2015). ISSN (Online) 2278–1021 ISSN (Print) 2319-5940

Electrochromic Properties of Vanadium Pentoxide Thin Film Prepared by Sol–Gel Process

R. Varshini, Ancy Albert and C. O. Sreekala

Abstract Prepared by chemical sol–gel process, thin films of ~500 nm thickness from the Vanadium pentoxide sol by spin coating over indium tin oxide (ITO) glass plates. These are annealed at different temperatures to identify the best sample. The field emission scanning electron microscopic (FESEM) analysis is used to view the morphology of the prepared film. The optical band gap of the film is evaluated from absorbance spectrum. For the prepared films, the transmittance study is also carried out. From the studies, it is inferred that the sample thin film annealed at 300 °C (S300) shows the best crystallinity and has lesser optical band gap of 2.16 eV. The electrochromic behaviour of the film is studied from cyclic voltammetric analysis. The reversible multichromism of yellow, green and blue is exhibited by the film annealed at 300 °C.

Keywords Vanadium pentoxide · Electrochromic · Chemical sol–gel process · FESEM · Cyclic voltammetry

1 Introduction

Vanadium is a transition element with various oxidation states whose compounds have numerous uses in chemical, biological, optical fields, etc. [1, 2]. Vandadium pentoxide is the significantly used inorganic compound of vanadium as it has high coefficient of thermal resistance and exhibit electrochromism which has applications in making detectors, bio-sensors, optoelectronic devices and in thin films [3, 4]. Based on the applications, microporous sol–gel thin films or any other form may be applicable. Vanadium pentoxide thin films are prepared from gels of corresponding sols using either spin coating [5–8] or dip coating [9–11]. The reversible switching from semiconductor to metal phase of thin films made by a sol–gel method deposited on Indium Tin oxide coated glass (ITO) annealed at various temperatures is explained

R. Varshini · A. Albert · C. O. Sreekala (✉)
Department of Physics, Amrita School of Arts and Sciences, Amritapuri,
Amrita Vishwa Vidyapeetham, Kollam 690525, Kerala, India
e-mail: sreekalaco@am.amrita.edu

M. Tuba et al. (eds.), *ICT Systems and Sustainability*,
Advances in Intelligent Systems and Computing 1077,
https://doi.org/10.1007/978-981-15-0936-0_45

in this work. The morphological, structural, electrochromical, and optical behaviour of the prepared V_2O_5 thin film is also presented in the paper.

2 Experimental Studies

2.1 Preparation of Vanadium Pentoxide Sol

5 g of vanadium pentoxide (V_2O_5) is heated at 800 °C for 30 min in a crucible. The molten V_2O_5 is transferred directly to 50 ml of distilled water and stirred well. It is then filtered and allowed to settle for sol formation for one week.

2.2 Thin Film Fabrication

Indium Tin Oxide (ITO) glass slides are cleaned with distilled water after washing them in water. The glass slides are then immersed in a mild soap solution and sonicated for 15 min. After rinsing with distilled water it is again sonicated for 5 min in distilled water. An adequate amount of isopropanol is taken in a beaker in which the ITO's are immersed and sonicated for 10 min. Later the ITO's are kept in an oven overnight for 70 to 80 °C. The prepared V_2O_5 sol is spin coated over the cleaned ITO at 3000 rpm for 5 min and allowed to dry at room temperature for 10 min. The samples are annealed at 100 °C, 200 °C, 300 °C and 400 °C, respectively, in a hot air oven for 1 hour. Thus, five samples were made (i) film as deposited indicated hereafter as S0 (ii) film heated at 100 °C as S100, (iii) film heated at 200 °C as S200, (iv) film heated at 300 °C as S300 and (v) film heated at 400 °C as S400. The complete experimental process is indicated in Fig. 1.

3 Result and Discussion

3.1 FESEM Analysis

FESEM images of two of the fabricated films are shown in Fig. 2. The film morphology is very clear from the images. The film S0 which is as deposited and S100 annealed at 100 °C shows amorphous surfaces while S300 shows crystalline structure. This indicates that the film shows uniform crystal morphology at 300 °C, which indicates a change in film structure from amorphous to crystalline at 300 °C.

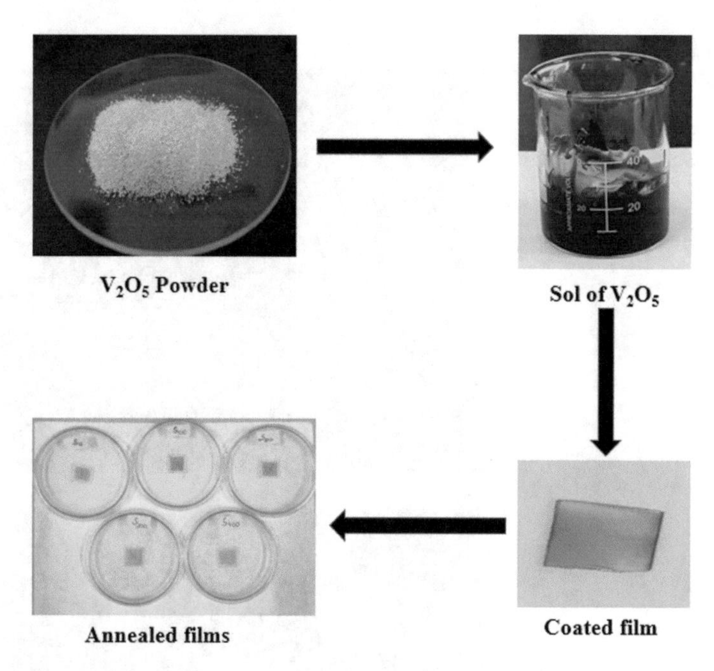

Fig. 1 Diagrammatic representation of V_2O_5 thin film preparation from V_2O_5 Powder

3.2 *Optical Analysis of the Fabricated Thin Films*

Absorbance spectra: Absorbance spectra of the fabricated thin films are shown in Fig. 3.

From the absorbance spectral analysis, the energy gap diagram is drawn. The film thickness is about 500 nm and is measured using ellipsometer. V_2O_5 is a direct bandgap semiconductor. The energy gap of the deposited film S0 is found to be 2.28 eV, S100 as 2.24 eV, S200 as 2.2 eV, S300 as 2.16 eV and for S400 it is found to be 2.18 eV. The lowest band gap is found for the sample S300. This may be due to the high crystallinity of the fabricated device annealed at 300 °C.

Transmittance Spectra: The transmittance spectral analysis is done for the fabricated thin films and is shown in Fig. 4. All the films show very good transmittance from 550 nm onwards. As the annealing temperature increases, the transmittance intensity decreases. From the graph, it is obvious that S300 shows stable transmittance, because of the best morphology of the film at 300 °C.

Electrochromic Properties: To study the electrochromic properties, an electrochemical cell is made with a saturated calomel electrode (SCE) as reference electrode and a platinum wire as counter electrode of that cell. The spin-coated vanadium pentoxide thin films annealed at different temperatures are used as other electrodes for electrochromic measurements. Electrolyte solution consists of 5.2 g of lithium per chlorate ($LiClO_4$) added to 50 ml of propylene carbonate. Using an electrochemical

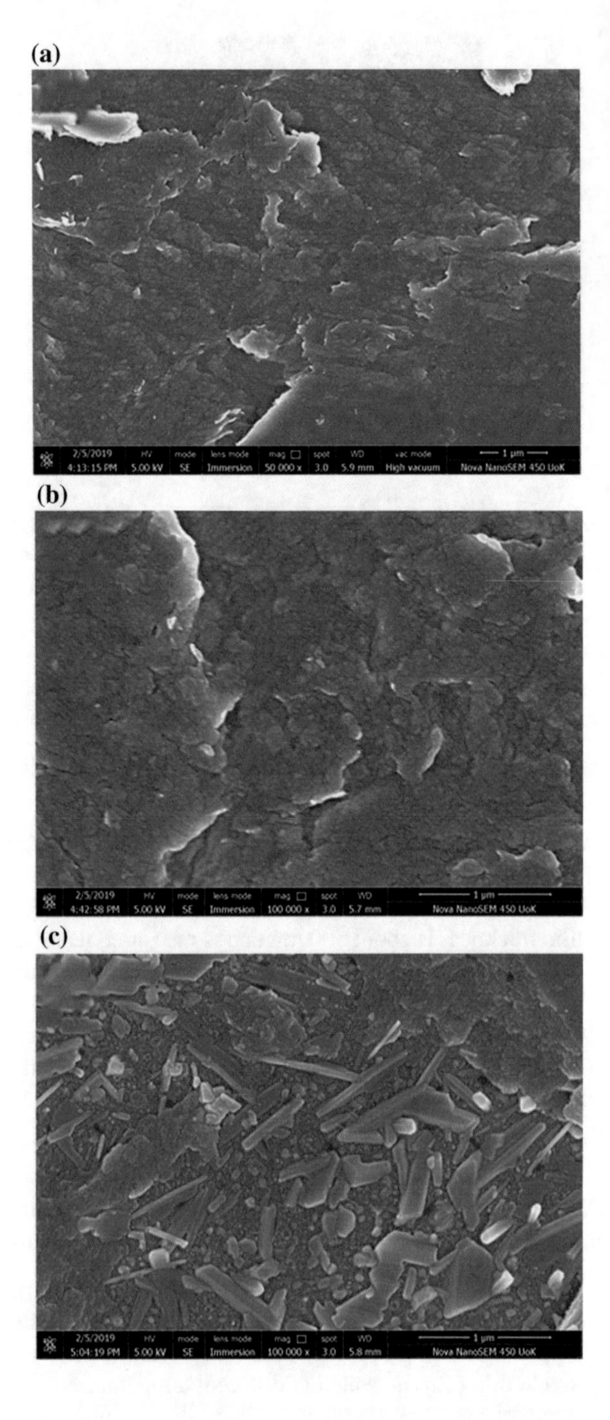

Fig. 2 FESEM Images of **a** film as deposited S0, **b** film annealed at 100 °C S100 and **c** film annealed at 300 °C S300

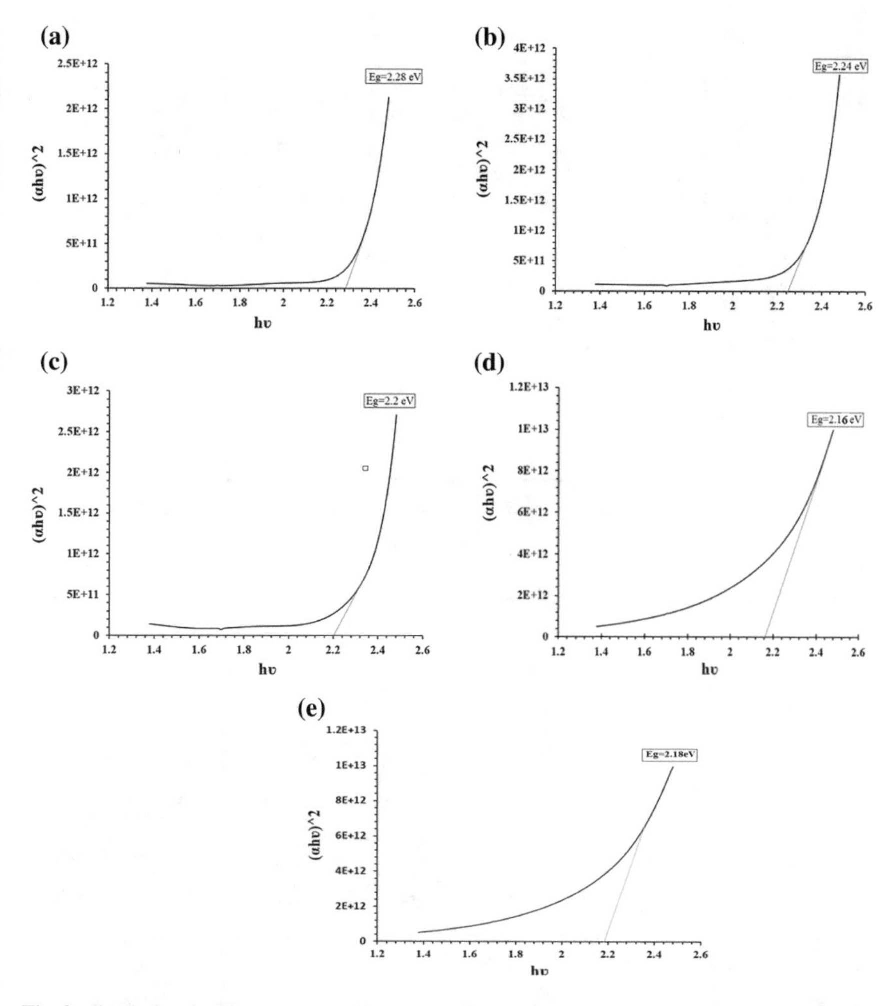

Fig. 3 Graph showing the energy gap diagram of fabricated thin films **a** S0, **b** S100, **c** S200, **d** S300, **e** S400

analyser Cyclic voltammograms are measured. Figure 5 shows the linear-sweep IV at 100 mV/s from −1 to +1.4 V (vs. SCE) potential range. Among the samples, S300 shows the best performance. The film shows brilliant reversible Cyclic voltammogram with electrochromism of yellow to green to blue. The oxidation/reduction peaks are visible in Fig. 5. As reported in the literature, only a fraction of the V5+ ions is reduced to V4+ before the reduction peak leading to a (V4+, V5+) mixture. But in the reduction peak it is observed that the remaining ions of V5+ are reduced to V4+. This explanation can also be used for the oxidation peak, i.e. the V4+ ions undergo incomplete oxidation leaving behind a mixture of V4+ and V5+ before the

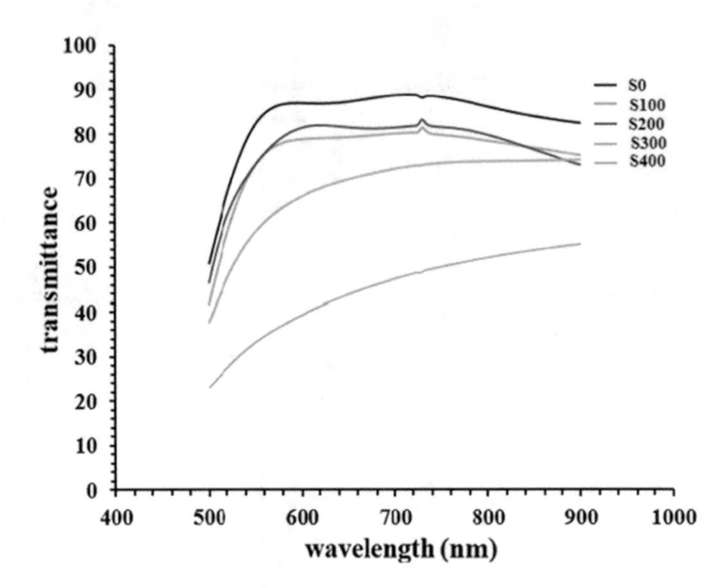

Fig. 4 Transmittance spectra of the fabricated thin films

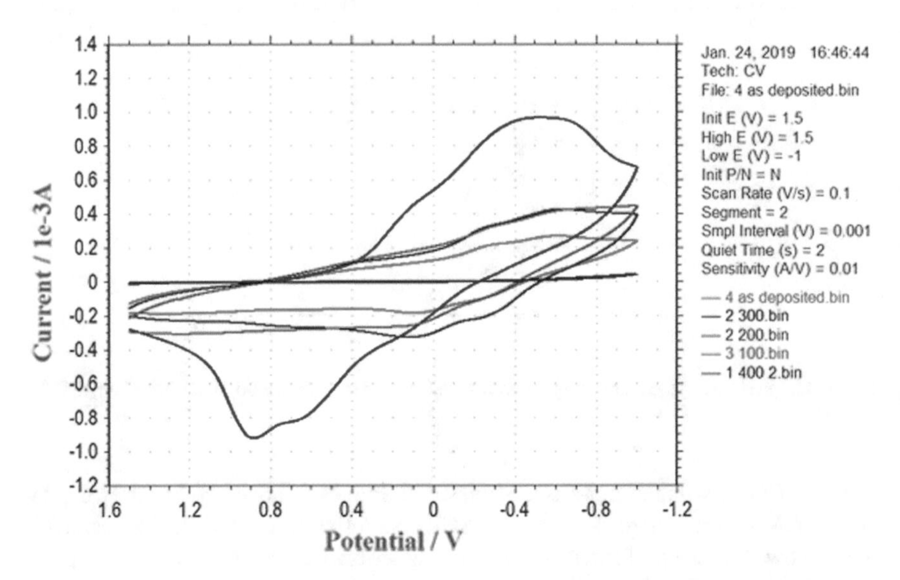

Fig. 5 Cyclic voltammograms of fabricated thin film samples annealed at different temperatures

oxidation state and the oxidation of the rest of the V4+ ions leads to V5+ ions [12]. This shows the crystalline behaviour of the fabricated film S300.

In the cathodic scan, at potentials below +0.9 V and down to −0.4 V the current density grow significantly. Consistently, the film gradually loses its original yellow colour, turn out to be green, succeeding blue when the falling potential touches −0.4 V. On the reverse scan, −0.5 V showed the peaks of oxidation, the green colour disappears and the film recovers its unique yellow colour. This transition of colour occurs due to the removal of electrons from the oxide framework which is under anodic potential.

4 Conclusion

In this work, V_2O_5 thin films are fabricated by spin coating the sol prepared by chemical sol–gel process. The thickness of the films are more or less same and is found to be 500 nm. Five samples annealed at different temperatures are fabricated. FESEM analysis of the sample shows that the amorphous nature of the V_2O_5 sol is changed to crystalline at 300 °C. From the absorbance and transmittance spectral analysis the optical properties are studied. The films have sufficient optical properties along with perfect low-absorption state and high transparency. The bandgap of S300 film is found to be 2.16 eV. The film annealed at 300 °C shows exceptional reversible electrochromism showing a transition from yellow to green and then blue. This change in colour is due to the progressive pentavalent/tetravalent Vanadium ions reduction/oxidation by the injection/abstraction of electrons. Only moderate time is needed for the colouration and decolouration. The fabricated V_2O_5 thin film are most applicable for fast switching electrochromic coating.

References

1. Yang, Z., Ko, C., Ramanathan, S.: Oxide electronics utilizing ultrafast metal-insulator transitions. Annu. Rev. Mater. Res. **41**, 337–367 (2011)
2. Gao, Y., Luo, H., Zhang, Z., Kang, L., Chen, Z., Jing, D., Kanehira, M., Cao, C.: Nanoceramic VO_2 thermochromic smart glass: a review on progress in solution processing. Nano Energy **1**(2), 221–246 (2012)
3. Wang, Z., Chen, J., Hu, X.: Thin Solid Films **375**, 238 (2000)
4. Granqvist, C.G.: Handbook of Inorganic Electrochromic Materials, p. 295. Elsevier, Amsterdam (1995)
5. Özer, N., Lampert, C.M.: Thin Solid Films **349**, 205 (1999)
6. Özer, N.: Thin Solid Films **305**, 80 (1997)
7. Shimizu, Y., Nagase, K., Miura, N., Yamazoe, N.: Jap. J. Appl. Phys. **29**, L1708 (1990)
8. Nagase, K., Shimizu, Y., Miura, N., Yamazoe, N.: Appl. Phys. Lett. **60**, 802 (1992)
9. Partlow, D.P., Gurkovich, S.R., Radford, K.C., Denes, L.J.: J. Appl. Phys. **70**, 443 (1991)
10. Hirashima, H., Sudoh, K.: J. Non-Cryst. Solids **145**, 51 (1992)
11. El Mandouh, Z.S., Selim, M.S.: Thin Solid Films **371**, 259 (2000)
12. Benmouss, M., Outzourhit, A., Jourdani, R., Bennouna, A., Ameziane, E.L.: Structural, optical and electrochromic properties of sol–gel V_2O_5 thin films. Act. Passive Electron. Compon. **26**(4), 245–256 (2003)

Mathematical Word Problem Solving Using Natural Language Processing

Shounaak Ughade and Satish Kumbhar

Abstract Natural language processing (NLP) is generally done on large data. Due to limited data word problem solving is challenging using NLP. There are some approaches proposed which could solve basic arithmetic problems like addition/subtraction. Knowledge representation is the main task to be done by NLP. Each kind of problem has its own approach. In this paper three types of mathematical word problems have been solved. Two of them are aptitude problems while the other two are reasoning problems. The spacy library has been used for effective use of Named Entity Recognition (NER) and word vectors. Stepwise solution has been generated instead of just answers which helps in improving understanding. The quite generic rather intuitive approach can be extended to solve some other kind of aptitude problems.

Keywords Knowledge representation · Natural language processing · Sentiment analysis · Named entity recognition

1 Introduction

In today's rapidly developing digital world, people tend to use technology to the maximum possible extent to lessen dependency on manual work. We identified one such activity that could be automated to make life easier for certain students. Our project appeals to the students preparing for competitive examinations. The proposed system of "Aptitude and Reasoning Problem Solver" intends to simplify the barrier between learning and technology by providing a system that can efficiently solve certain Word problem categories that are asked frequently in almost all the competitive examinations such as finding day on given date, analogy, time workmen, and direction sense test problems.

S. Ughade (✉) · S. Kumbhar
Department of Computer Engineering and IT, College of Engineering, Pune, India
e-mail: ughadess17.comp@ceop.ac.in

S. Kumbhar
e-mail: ssk.comp@oep.ac.in

M. Tuba et al. (eds.), *ICT Systems and Sustainability*,
Advances in Intelligent Systems and Computing 1077,
https://doi.org/10.1007/978-981-15-0936-0_46

The approach to solve mathematical word problems can be seen a paradigm shift. Systems work range from working on well-formed input to template generation. It is important how one represents knowledge in the question. One innovative way can be found in which a new representation language is created.

In this paper, three types of problems have been classified, generated stepwise solution. The paper is structured as follows. Section 2 presents literature review. The methodology is described in Sect. 3, Results in Sect. 4, Conclusion in Sect. 5. The future work is discussed in Sect. 6.

2 Literature Review

Paper name	Results	Issues
Natural language processing for solving simple word problems [1]	It solves addition/subtraction word problems within quite accurate manner	The system is not able to solve problems such as Shyam has 12 Basketballs. He lost 4 of them. How many Basketballs does he have?
Neural math word problem solver with reinforcement learning [2]	Reinforcement learning is used for optimization and improving solution accuracy	Mathematical concept and commonsense knowledge, has not yet incorporated in the proposed system
Automatically solving number word problems by semantic parsing and reasoning [3]	A high precision of 95.4% and a reasonable recall 60.2% on test set of over 1,500 problems is obtained	Sometimes it fails to derive math expressions in the reasoning stage This is often because the language of the problem is modified
Machine-guided solution to mathematical word problems (MWP) [4]	Text is classified in multiple stages to solve arithmetic problems of elementary level The accuracy is quite improved, also the performance gains of the solver is more compared with the State-of-the-art MWP solvers such as Wolfram Alpha	However, only elementary grade questions are focused and complex problems has not been taken into consideration
A novel framework for math problem solving [5]	Four types of word problems, which are investment, distance, projectile, and percent are solved	Classification of the problem is done using basic classifier. The accuracy is not mentioned

(continued)

(continued)

Paper name	Results	Issues
Learning to automatically solve algebra word problems [6]	Use of different features. The overall performance is improved using all features and that the pair features, followed by the document and solution features	The system cannot recognize the word like "twice", "thrice", "more/less", etc.
Word problem solver system [7]	Experimental results showed that that the system was quite helpful in clarifying doubts as well as helps in improving conceptual clarity	The authors did not claim any specific problem set used. It mainly focused on solving simultaneous linear equations
Constructing the representation model of arithmetic word problems for intelligent tutoring system [8]	Different representative models are proposed and worked on	
Two-phase classifier for automatic answer generation for math word problems [9]	System's performance is increased as time in training phase has significantly reduced	Problems involving complex grammar cannot be solved
Unit conflict resolution for automatic math word problem soling [10]	100% accuracy to deal with unit conflicts	

3 Methodology

3.1 Determine Type of Question

The main task of classifier as the name suggests is to classify the given problem. Each kind of problem has some unique words. The key words help to identify the type of problem and hence the first step to solve.

In the project, time workmen problem has key words like "men, women, girls, boys" which are worker type; "time, days, hours" which are time quantum.

In Direction Sense test problems, "east, west, north, south" are directions and "left, right" are turns, are key words.

In finding day problems, months i.e. January, February etc. are key words.

Here, the probability of important key words is calculated. The words which might occur in two or more type of problems are not considered as unique. The classified output as type of problem is the one which has higher probability of unique words/key words.

```
In [10]: 1 classifyProblem("A man walks 5 km toward south and then turns to the east. After walking 3 km he turns to the left \
         2 and walks 5 km.Now in which direction is he from the starting place?")

         Probability score is :0.6666666666666666
         Direction Test Problem
```

Fig. 1 Direction sense test problem is determined

Example: "A man walks 5 km toward south and then turns to east. After walking 3 km he turns to the left and walks 5 km. Now in which direction is he from starting place?"

In this question, key words are east, south, left.

The overall probability of the question that it is Direction sense question is 0.66 (Fig. 1).

3.2 Finding Day on Given Date

To find day on given date the generic algorithm is as follow:
Concept.:
Odd days.: Given number %7 = odd day count
Example: 18%7 = 4
Consider we have to find a day on 6 November 2016
Year = 2016:
x = (2016%100) − 1 = 15
If (Year %100) = 0, Total odd days in year =0
Else, Total odd days in year = ((int(x/4) * 2) +(x − int(x/4))) %7…(I)
Century = 2000
Odd days array for century = [5, 3, 1, 0]
Value = (2000/100) %4 = 0
The value gives index position of Odd day array.
So, number of odd days in century = odd days array [value]…(II)
Here, odd days for century = odd days array [0] = 5
Odd days for moths = [3, 0, 3, 2, 3, 2, 3, 3, 2, 3, 2, 3]
This array is obtained as (number of days in a month %7).
Odd days of each month are computed and stored in the Odd days for month array.
Till the given month, the values are added.
In our example: November month.
So, till November all odd days are added as
Month odd = 3 + 0 + 3 + 2 + 3 + 2 + 3 + 3 + 2 + 3 = 24
If the year is leap year, then add 1 to month odd.
Here, Month odd = 25. … (III)
Days = {0:'Sun', 1:'Mon', 2:'Tue', 3:'Wed', 4:'Thu', 5:'Fri', 6:'Sat'}
Now add (I), (II), (III).
Result = (I) + (II) + (III)
The (Result %7) gives "Key" in dictionary i.e. required day.
Here, Result = 0 Therefore answer is Sunday.

3.3 Time Work Men

Time work men problems are quantitative aptitude problems. In this category of problem based on capacity of workers, the time required to finish the work is computed.

The generic method to solve this type of problem is:

Find solution string

Solution string is basically all important words and numbers required for computation. It does not include other irrelevant data.

For example, 2 men and 3 boys can do a piece of work in 10 days while 3 men and 2 boys can do the same work in 8 days. In how many days can 2 men and 1 boy do the work?

Solution string is: [2, 'man', 3, 'boy', 10, 'days', 3, 'man', 2, 'boy', 'same', 8, 'days', 'many', 2, 'man', 1, 'boy']

Compute equation table.

In EqTable we write list as [work (partial/complete), days required, worker type1, worker type2].

In the above question EqTable obtained is: [[1, 10, 2, 3], [1, 8, 3, 2], [1, 0, 2, 1]]

Find capacity of each kind of worker.

Capacity is measured after comparing worker type * days.

Here, 1 man-work = 3.5 boy-work

Based on question asked find required days or number of workers.

To find a solution string, important words in a way are filtered. In spacy NER (Named Entity Recognition) is used to extract information from given text and classify it in predefined categories. In Spacy, word vectors are used. Based on word occurrence and their proximity with other words it is classified in one of the following categories.

In Python, displacy.render() function is used to get NER (Fig. 2).

In our example, by default words like "men, women, children, boys, girls, complete, same" are neither recognized in any given category and hence no question of categorizing. However, words like "men, women, children, boys, girls" can be classified under PERSON category. Also, words like "complete, same" can be classified under EVENT category.

To do this, we need to update existing NER model. For this, PhraseMatcher class is used. It let one match terminology list. It accepts patterns in the form of Doc object [11].

```
matcher = PhraseMatcher(nlp.vocab)
phrase_list = ["Men","men","Women","women","Children","children",
"boys","girls","boy","girl","man","woman"]
phrase_patterns = [nlp(text) for text in phrase_list]
matcher.add("updated",None,*phrase_patterns)
matches = matcher(doc) #returns(hash,start,end)
```

TYPE	DESCRIPTION
'PERSON'	People, including fictional.
'NORP'	Nationalities or religious or political groups.
'FAC'	Buildings, airports, highways, bridges, etc.
'ORG'	Companies, agencies, institutions, etc.
'GPE'	Countries, cities, states.
'LOC'	Non-GPE locations, mountain ranges, bodies of water.
'PRODUCT'	Objects, vehicles, foods, etc. (Not services.)
'EVENT'	Named hurricanes, battles, wars, sports events, etc.
'WORK_OF_ART'	Titles of books, songs, etc.
'LAW'	Named documents made into laws.
'LANGUAGE'	Any named language.
'DATE'	Absolute or relative dates or periods.
'TIME'	Times smaller than a day.
'PERCENT'	Percentage, including "%".
'MONEY'	Monetary values, including unit.
'QUANTITY'	Measurements, as of weight or distance.
'ORDINAL'	"first", "second", etc.
'CARDINAL'	Numerals that do not fall under another type.

Fig. 2 Predetermined set of entities identified by NER

Token and Doc are types of containers in Spacy. Token could be a word, punctuation symbol, whitespace, etc. A Doc is sequence of Token objects [12].

In PhraseMatcher class, PhraseMatcher add method is used to add a rule to the matcher. It has ID key, list of patterns and callback function. The callback function will receive the arguments matcher, doc, I, matches. If pattern exists for the given ID the pattern will be extended. And on match callback will be overwritten.

However, by default, only a few person names are recognized by Spacy. This is one drawback which is yet to be solved.

Thus, after updating NER, key words are obtained and put in Token list.

In the next step, based on token and its relevant context, i.e., if token is "men" preceded by "12", the value 12 is appended in worker type column in EqTable.

Then the worker capacity is measured. There is powerful concept called man-days. It is product of number of workers and days. For example, if 10 men can do a work in 5 days. Here, $10 * 5 = 50$, i.e., 50 man-days are required to finish a work.

Based on capacity, number of workers or days as asked in question is computed.

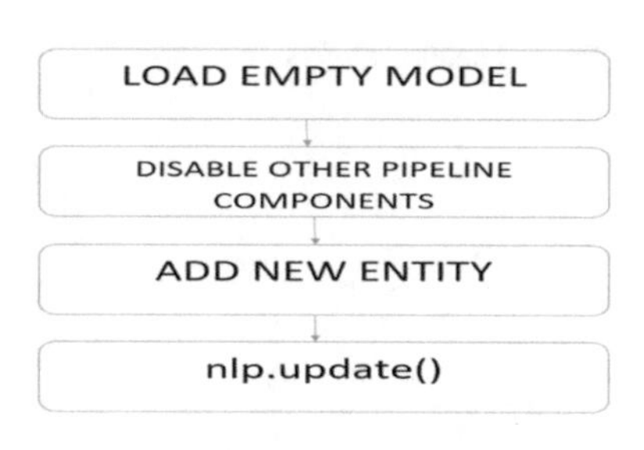

Fig. 3 Flowchart for creating new/updated NER model

3.4 Direction Sense Test

Direction sense test is a verbal reasoning problem. In this movement of person in different directions is given and final displacement or direction one is facing is asked.

The reasoning problem generally requires knowledge base. Here, information about directions, their difference in angle, etc. is required. In analogy detection word2vec tool is used to find similarity. However, it is interesting fact that word2vec does not understand English syntax or semantics.

NER has been used to counter this issue. To solve in generic approach, first we need to find solution string. However, in Direction sense test the words like "east, west, north, south, left, right, twice" cannot be categorized in any given category. Therefore, a new model is created which can categorize words "east, west, north, south" as DIRECTION category; words like "left, right" as TURNS.

Following algorithm is used to create a new model (Fig. 3).

Load the model:
An empty language model is created. Empty language class of English language is created as spacy.load('en')
It returns language object with the loaded model.
As we have loaded blank model, it is necessary to disable other pipeline components such as tokenizer, tagger, and parser. The command used is nlp.disable_pipes()
Add entity recognizer in the pipeline.
Thus only entity recognizer will be trained.
Model is trained over training data.
For example: "He went 5 km to East.", {"entities": [(16, 20, LABEL1), (8, 13, LABEL2)]}, Here labels are number/direction/turns. Thus, the data in output will be labeled.
Training data is shuffled and loop over again and again using minibatch, compounding.
minibatch: Iterate over batches of items. Size may be an iterator so that batch-size can vary on each step.
compounding: Yield an infinite series of compounding values. Each time the generator is called, a value is produced by multiplying the previous value by the compound rate.

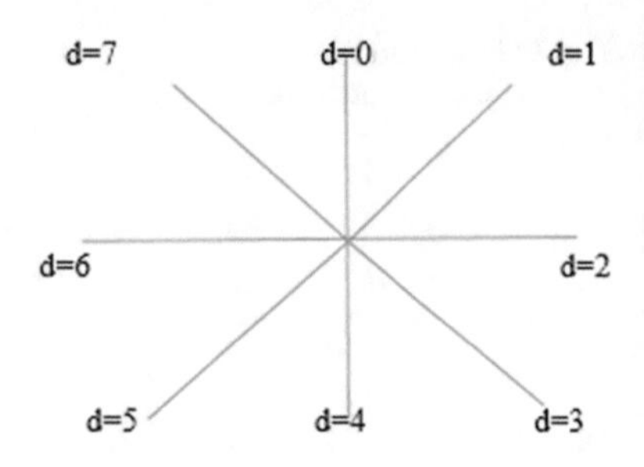

Fig. 4 Labelled directions

For each example in training data, model is updated using nlp.update()
At each word, it makes prediction. It consults annotations to verify the correctness. If it is wrong it adjusts the weights so that correct action will score higher next time [13].
nlp.update(texts, annotations, sgd = optimizer, drop = 0.35, losses = losses)
texts.: text in training data
annotations: obtained from training data
sgd: optimizer obtained after nlp.begin_training() which holds states between updates.
drop: dropout rate in neural network which is used to avoid over-fitting. Hidden/ visible layer neurons are hidden. Sometimes, model does not recognize important key words, then dropout rate is changed.
Once the new model is trained, we get solution string.
Example: A man walks 5 km toward south and then turns to east. After walking 3 km he turns to left and walks 5 km. Now in which direction is he from the starting place? (Fig. 4).
Solution string is: [5, 'km', 'south', 'east', 3, 'km', 'left', 5, 'km', 'which']
Here, d = 0: North, d = 1: Northeast, d = 2: East, d = 3: Southeast, d = 4: South, d = 5: Southwest, d = 6: West, d = 7: Northwest
In our question as person moves, coordinates are updated as:
solution list [5, 'south']
Here, d = 4 The coordinates are: 0, −5.0
solution list [5, 'south', 'east', 3]
here, d = 4 − 2 = 2 The coordinates are: 3.0, −5.0
solution list [5, 'south', 'east', 3, 'left', 5]
The coordinates are: 3.0, 0.0
solution list in questions [5, 'south', 'east', 3, 'left', 5, 'which']
'Endpoint is in East to starting point.: -> 3.0, 0.0; result'].

4 Results and Analysis

Results should be verified at four stages:

1. Is problem classified correctly?
2. Is solution string generated is correct?
3. Is stepwise solution generated?
4. Is final answer is correct?

Following are the output screenshots for all four categorized problems (Figs. 5, 6 and 7).

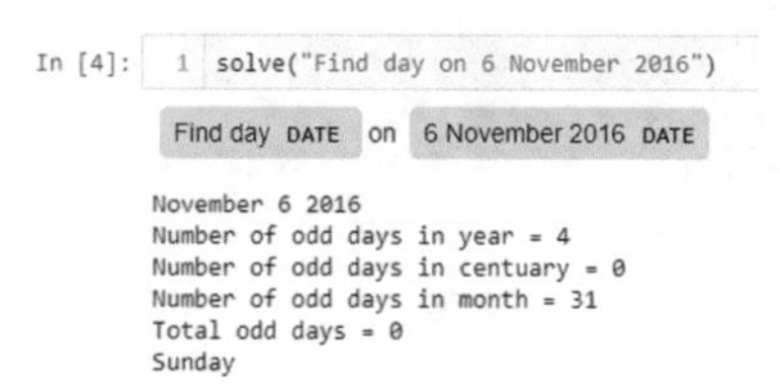

Fig. 5 Output screen of finding day problem

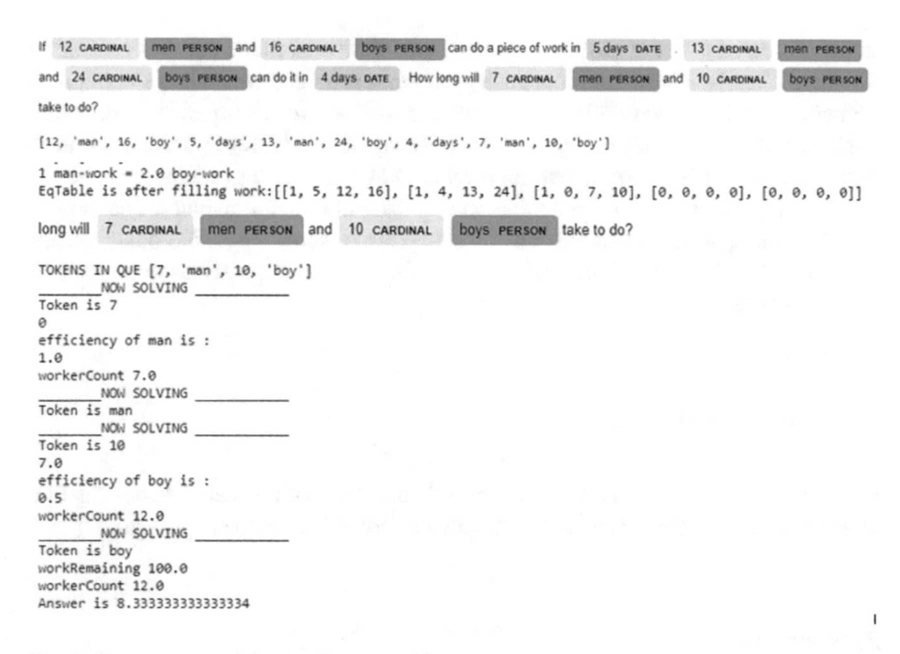

Fig. 6 Output screen of time work men problem

a man walks 5 km DISTANCE toward south DIRECTION and then turns to the east DIRECTION . after walking 3 km DISTANCE he turns to the left LEFT and walks 5 km DISTANCE . now in DIRECTION which direction is he from the starting place?

```
[5, ['km'], 'south', 'east', 3, ['km'], 'left', 5, ['km'], 'in']
[5, 'km', 'south', 'east', 3, 'km', 'left', 5, 'km', 'in', 'which']
5
solution list in int[5]
km
south
solution list in dirctions[5, 'south']
solution list in distance & turns [5, 'south']
The cor-ordinates are : 0,-5.0
east
solution list in dirctions[5, 'south', 'east']
3
solution list in int[5, 'south', 'east', 3]
solution list in distance & turns [5, 'south', 'east', 3]
The cor-ordinates are : 3.0,-5.0
km
left
towards left
solution list in turns[5, 'south', 'east', 3, 'left']
5
solution list in int[5, 'south', 'east', 3, 'left', 5]
solution list in distance & turns [5, 'south', 'east', 3, 'left', 5]
The cor-ordinates are : 3.0,0.0
km
in
which
solution list in questions [5, 'south', 'east', 3, 'left', 5, 'which']
Steps in ques :[]
Steps after 2nd:['Endpoint is in  East to starting point.:-> 3.0 , 0.0;result']
Out[6]: ' in East direction'
```

Fig. 7 Output screen of direction sense test

5 Conclusion

In this project, 3 types of problems 2 are of quantitative aptitude and 1 of reasoning problems. Step-by-step solution of each problem is provided which helps in improving understanding. The problem is classified based on probability of important key words occurrence. Spacy's Named Entity Recognition (NER) is used as default, modified and updated form to extract important key words. It is useful for students to solve the concerned word problems in a quick and efficient manner. The system can solve most of the questions and has exception in some cases. No other system has been developed yet which can solve multiple dynamic questions and generate stepwise solution.

6 Future Work

The system can be extended to solve several categories of problems. Problems like simple interest, set theory questions can also be solved using the approach.

References

1. Sundaram, S.S., Khemani, D.: Natural language processing for solving simple word problems (2015)
2. Huang, D., Liu, J., Lin, C.-Y., Yin, J.: Neural math word problem solver with reinforcement learning (2018)

3. Shi, S., Wang, Y., Lin, C.-Y., Liu, X., Rui, Y.: Automatically solving number word problems by semantic parsing and reasoning (2015)
4. Amnueypornsakul, B., Bhat, S.: Machine-guided solution to mathematical word problems (MWP) (2014)
5. A novel framework for math word problem solving. Int. J. Inf. Educ. Technol. **3**(1) (2013)
6. Kushman, N., Artzi, Y., Zettlemoyer, L., Barzilay, R.: Learning to automatically solve algebra word problems. Computer Science and Artificial Intelligence Laboratory, Massachusetts Institute of Technology and Computer Science & Engineering, University of Washington
7. Miyani, M., Doshi, S., Jain, J.: Word problem solver system using artificial intelligence. Proc. Comput. Sci. (ICACTA) **45**, 800–807 (2015)
8. Ma, Y., Tan, K., Shao, L., Shang, X.: Constructing the representation model of arithmetic word problems for intelligent tutoring system (2011)
9. Hevapathige, A., Wellappili, D., Kankanamge, G.U., Dewappriya, N., Ranathunga, S.: Two-phase classifier for automatic answer generation for math word problems (2018)
10. Dewappriya, N., Kankanamge, G.U., Wellappili, D., Hevapathige, A., Ranathunga, S.: Unit conflict resolution for automatic math word problem solving (2018)
11. https://spacy.io/api/phrasematcher
12. https://spacy.io/api/doc
13. https://spacy.io/usage/training
14. spacy web: https://spacy.io/models/en

Computer Vision Based Position and Speed Estimation for Accident Avoidance in Driverless Cars

Hrishikesh M. Thakurdesai and Jagannath V. Aghav

Abstract In the field of driverless cars, safety is the main concern. The safety systems for these cars are mainly dependent on the inputs of cameras and sensors like Light Detection and Ranging (LIDARs). Lidar is an essential component in driverless cars which creates 3D map of the surrounding and assists the car for driving. Lidar based safety systems are effective but are very expensive as the lidars are costly. Hence, there is a need to design a system which can assist the car effectively on the road, without using the lidar. This paper presents a proposed design of accident avoidance system for driverless cars, which uses computer vision techniques and works on the input of single video camera. It analyzes the video frames and shows accident warnings while on turn or lane change. The system is cost effective and can work with IoT devices having low computational power. To verify the performance of the system, we have performed experiments on various scenarios including cars at different positions moving with different speed and we obtained satisfactory results.

Keywords Machine learning · Driverless cars · Computer vision · Accident avoidance

1 Introduction

Driverless cars are expected to reduce the number of accidents happening due to human errors, and hence should be equipped with effective safety systems. These systems work on the inputs from cameras and sensors [1–3]. Major sensors are Light Detection and Ranging (LIDAR), Radio Detection and Ranging (RADAR), and Sound Detection and Ranging (SONAR). Radar is used to identify the objects in fog or during rainy season using radio waves. Sonar is used for identifying objects mainly below the water as it uses sound waves. Lidar is an essential component

H. M. Thakurdesai (✉) · J. V. Aghav
College of Engineering, Wellesley Rd, Shivajinagar, Pune 411005, Maharashtra, India
e-mail: thakurdesaihm17.is@coep.ac.in

J. V. Aghav
e-mail: jva.comp@coep.ac.in

M. Tuba et al. (eds.), *ICT Systems and Sustainability*,
Advances in Intelligent Systems and Computing 1077,
https://doi.org/10.1007/978-981-15-0936-0_47

which is considered as an eye of the driverless car. It uses pulsed laser waves to create three-dimensional map of the surrounding and assists the vehicle by estimating positions and speed of the surrounding objects. Existing safety systems mainly use lidars but the cost of it is very high, which increases the overall cost of the car [4]. Uber and Tesla have tried reducing the lidars from five to one (mounted on the top of the car) but faced some accident issues because of it [5, 6]. Lot of research is going on for making the cost-effective accident avoidance systems which can assist the car like human brain [3, 7–10]. Existing systems use various techniques for car detection which includes histogram of gradient decent (HOG) and You Look Only Once (YOLO)-based identification [11–14]. In this paper, we have made a sincere attempt to implement the accident avoidance system for driverless car. This system takes input from video camera and analyzes the video frames using trained TensorFlow (TF) model. It estimates the positions of the vehicles and provides warnings for accidents. We have used TF-object detection model for detection of cars and computer vision techniques for position & speed estimation. We have compared the results of hog and yolo but the accuracy was not satisfactory. We obtained more accurate results by training the TensorFlow model on the dataset, and hence we used the same for implementation.

1.1 Frameworks and Libraries Used

In the proposed system, we have mainly used two frameworks; TensorFlow and Open Source Computer Vision (OpenCV). Tensorflow framework provides various pretrained models on various datasets which can be retrained for specific application. OpenCV library contains implementation of computer vision algorithms which can be used out of the box. In this system, the initial task is to detect the cars in the frame. It is done by using faster Region-Convolutional Neural Network (R-CNN) model. Post detection, next task is to calculate the distance and speed of the car. To do so, we have used functionalities in openCV library.

2 Dataset Selection

We have used "Open Images Dataset-V4" for the proposed system. It is a dataset of approximately 9 million images that have been annotated with image-level labels, object bounding boxes, and visual relationships. The training set of V4 contains 14.6 million bounding boxes for 600 object classes on 1.74 million images, which makes it the largest existing dataset with object location annotations [15]. Out of 600 classes, we have used images of only "Car" class. These images can be downloaded with bounding box coordinates using a python script. Figure 1 shows the filtered dataset for cars' images. We have used 20,000 training images and 2000 test images for training the model.

Fig. 1 Open images dataset filtered for the "Car" class

The dataset includes images and a csv file which contains bounding box coordinates for each image file. The csv file contains name of the file along with top-left and bottom-right coordinates, respectively. To make the system more accurate, we can include other vehicle class images like "Truck" or "Bus", but we have only considered a "Car" class for the implementation. Accuracy of the system also depends on the number of training images and the number of steps of training.

3 Architecture

Following Fig. 2 shows the architecture of the proposed system. Object detection module identifies the cars in the frame. It provides the coordinates of the bounding box of the detected car. These coordinates are given to the position and speed calculator for distance and speed calculation. The time is calculated using these parameters and warnings are shown for accident avoidance while on turn or lane change.

The sections below explain the implementation of modules used in the system.

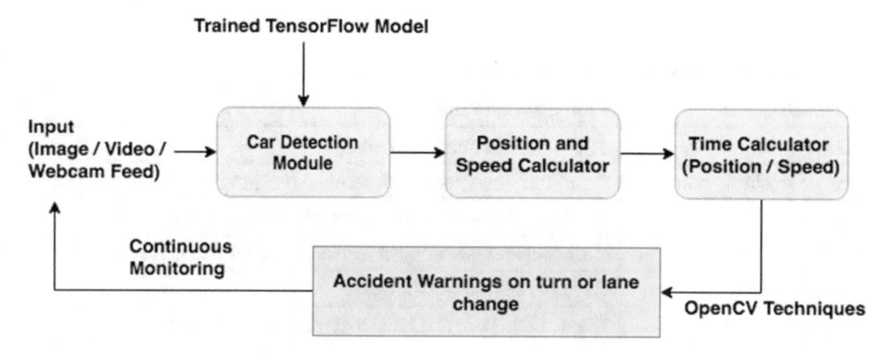

Fig. 2 Architecture of the system

4 Implementation of Modules

In the following sections, we have explained the implementation of car detection module and position estimation module. The first module includes training of the TF model for detection of car in the frame. Second module contains computer vision functions used for distance and speed estimation. For displaying warnings on the screen, we have used methods in opencv library.

4.1 Car Detection Module

The input to car detection module is a video or an image taken by the camera. If the camera video is getting used as an input, then the identification process works similar to image. Only the additional part is dividing the video into frames and each frame is given as an input for the detection of cars. We have trained faster R-CNN model for 200,000 steps by giving 20,000 training images and 2000 test images of open images dataset. Following Table 1 shows the parameters in the configuration file of the model that we have modified before starting the training.

In the training process, the loss is recorded at each step. Initially the loss value is high and it reduces as the training progresses. We have stopped the training after 20,000 steps and the loss value was approximately below 0.3. Following graph shows the classification of loss obtained during the training. Various loss graphs are shown on the TensorBoard interface (Fig. 3).

The trained model can be converted into "Lite" version using Tensorflow Lite utility and it can be used In the IoT devices like mobiles which do not have high computational power. The below sections explain the implementation of position and speed calculation module.

Table 1 Configuration parameters for training

Parameter name	Configuration value
num_classes	1(Depends on number of class)
fine_tune_checkpoint	".../faster_rcnn_inception_v2/model.ckpt"
input_path	".../models/research/object_detection/train.record"
label_map_path	"../models/research/object_detection/training/labelmap.pbtxt"
num_examples	(Number of training images)
eval_input_reader	(Same input path and label map path is configured)

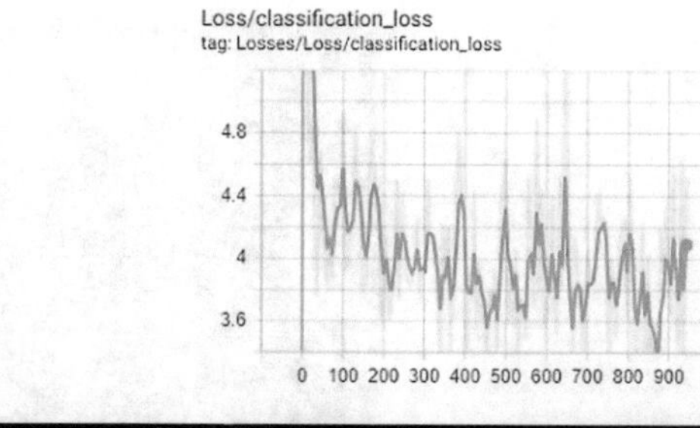


```
INFO:tensorflow:global step 200000: loss = 0.2795 (0.781 sec/step)
INFO:tensorflow:Stopping Training.
INFO:tensorflow:Stopping Training.
INFO:tensorflow:Finished training! Saving model to disk.
INFO:tensorflow:Finished training! Saving model to disk.
```

Fig. 3 Classification loss graph and training completion state

4.2 Distance and Speed Calculator Module

Distance Calculation:
Distance of the car in the image can be found using the formula below. It uses focal length of the camera which is camera specific.

$$\text{Distance} = \frac{\text{Focal Length} * \text{Actual width of the object}}{\text{Width of the object in the image}}$$

For distance calculation, the car detection module provides coordinates of the bounding box. These coordinates are used to get the width of the box, which is same as width of the car in the image. Actual width of the car is the average width of the car which we have considered as 1 m for simplicity in calculations. Focal length of the camera can be found using calibration process. The camera which we have used for the testing was having focal length of 750 pixels. Following Table 2 shows some of the results obtained using this method.

From the above table, we have calculated the accuracy by considering the error in sample measurements and we obtained it **94% accurate**. Figure 4 shows the car with the width of 250 pixels.

Velocity Calculation:
Velocity calculation involves detection of the same car in the next frame. This detection is done by calculating the movement of the center of the bounding box in two

Table 2 Sample results for distance calculation

Actual position (In meter)	Width of the bounding box (In pixel)	Calculated position (In meter)
3	250	3.15
2	374	2.005

Fig. 4 Car present three meters away from the camera with width = 250 pixel

consecutive frames. If the movement is less than a threshold value, we conclude that same car is present and we take the difference of the distance in old and new frame. This difference gives the distance traveled by the car in two frames. The time between the two frames is calculated by taking the reciprocal of the frames per second (fps) value of the input video. The fps value can be found using the built-in function in opencv library. As shown in Fig. 4, the car is detected at a distance of 17.84 m with a speed of 85.02 km/h. We have used a python dictionary to store the distances and coordinates. Figure 5 shows the dictionary values for an instance. It stores the following parameters, respectively (Table 3).

As shown in Fig. 5, the car is moved from 9.32 to 9.47 m in one frame. Hence, the speed is 13.83 km/h (Fig. 6).

```
Old dictionary=  [[1072, 9.323899773991137, 396, 1174.0, 456.0]]
New Dictionary [[1066, 9.47761112999336, 398, 1171.5, 456.5]]
13.834022040200047 (1066, 338)
```

Fig. 5 Dictionary output

Table 3 Parameters stored in the dictionary

Sr. No	Parameter name	Description
1	x_min	X coordinate of the top left vertex of the bounding box
2	distance	Distance of the car from the camera
3	y_min	Y coordinate of the top left vertex of the bounding box
4	mid_x	X coordinate of the midpoint of the bounding box
5	mid_y	Y coordinate of the midpoint of the bounding box

Fig. 6 Distance and velocity calculation output

5 Algorithm for Accident Avoidance System

This algorithm is used to design an accident avoidance system for driverless cars. It continuously monitors the video frames and shows the warnings for accidents.

Steps of algorithm:

Step 1: Detect the cars in the frame using the TF model.
Step 2: Calculate the distance of the car using the dimensions of the bounding box.
Step 3: Calculate the relative speed of vehicle using difference in distance and frames per second (fps).
Step 4: Depending on the car positions, identify if the car is in warning area or collision area. Give appropriate warnings on the screen. Calculate the time that the vehicle will take to cross. If the time is less than certain threshold value, give warning for taking turn or changing lane.
Step 5: If "Safe Move" indication is present on the screen, it is safe to take turn.
Step 6: Continue to follow the steps 1 to 5 for each frame.

---

6 Results

We have verified the proposed system in various scenarios including cars at different distance and lane. We obtained satisfactory results for inputs on images and videos. Figure 7 shows the results in the image where three cars are identified at a different distance.

7 Conclusion

In this paper, we have used trained TF model and computer vision techniques for estimation of distance and speed. These parameters are used to show the warnings for accident avoidance. If the car is present at very less distance (we have taken

Fig. 7 Results of the accident avoidance system

threshold as 15 m), then the warning is shown. The system worked with the input of single camera in the form of image or video. We have verified the proposed system in various scenarios including cars at different positions and obtained satisfactory results. Hence, the driverless cars can be equipped with this system for accident avoidance.

References

1. Yu, J., Petng, L.: Space-based collision avoidance framework for autonomous vehicles. Proc. Comput. Sci. **140**, 37–45 (2018)
2. Liu, G., Zhou, M., Wang, L., Wang, H., Guo, X.: A blind spot detection and warning system based on millimeter wave radar for driver assistance. Optik **135**, 353–365 (2017)
3. Park, M.W., Jang, K.H., Jung, S.K.: Panoramic vision system to eliminate driver's blind spots using a laser sensor and cameras. Int. J. ITS Res. **10** (2012)
4. Stoffel, M.: Biggest challenges in driverless cars. https://9clouds.com/blog/what-are-thebiggest-driverless-car-problems. Accessed 29 Jan 2019
5. Stewart, J.: Tesla's autopilot was involved in another deadly car crash. Wired (2018)
6. Naughton, K.: Ubers fatal crash revealed a self-driving blind spot. Night Vision, Bloomberg (2018)
7. Wu, B.F., Huang, H.Y., Chen, C.J., Chen, Y.H., Chang, C.W., Chen, Y.L.: A vision based blind spot warning system for daytime and nighttime driver assistance. Comput. Electr. Eng. **39**(3), 846–862 (2013)
8. Chen, Y.L., Wu, B.F., Huang, H.Y., Fan, C.J.: A real-time vision system for night time vehicle detection and traffic surveillance IEEE Trans. Ind. Electron. **58**(5), 2030–2044 (2011)
9. Kuo, Y.C., Pai, N.S., Li, Y.F.: Vision-based vehicle detection for a driver assistance system. Comput. Math. **61**, 2096–2100 (2011)
10. Chen, C.T., Chen, Y.S.: Real-time approaching vehicle detection in blind-spot area. In: Proceedings IEEE International Conference Intelligence Transport System, pp. 1–6 (2009)
11. Ajsmilutin CarND-Vehicle Detection. https://github.com/ajsmilutin/CarND-Vehicle-Detection. Accessed 29 Mar 2017
12. Tensorflow Object Detection API Tutorial. http://Github.com/EdgeElectronics. Accessed 10 Apr 2019
13. Rosebrock, A.: Find distance from camera to object using python and opencv. https://www.pyimagesearch.com/2015/01/19/find-distance-camera-objectmarker-using-python-ovencv. Accessed 19 Jan 2015
14. Hendry, Chen, R.C.: Automatic license plate recognition via sliding-window darknet-YOLO deep learning. Image Vis. Comput. **87**, 47–56 (2019)

15. Kuznetsova, A., Rom, H., Alldrin, N., Uijlings, J., Krasin, I., Pont-Tuset, J., Kamali, S., Popov, S., Malloci, M., Duerig, T., Ferrari, V.: The open images dataset V4: unified image classification, object detection, and visual relationship detection at scale (2018). arXiv:1811.00982

Analysis and Prediction of Customers' Reviews with Amazon Dataset on Products

Shitanshu Jain, S. C. Jain and Santosh Vishwakarma

Abstract The main objective of this paper is to get a deeper knowledge of the text classification methods used in text mining. This paper describes different methods and algorithms used in text mining. Various text preprocessing steps have been performed like tokenization, case folding, stemming, stopword removal, etc. The customer reviews posted in the amazon website have been used as the training set and used with various classifiers like Naive Bayes, KNN, random forest and decision tree. The performance parameter of each method is determined with standard evaluation parameters such as precision, recall, and kappa measures. The results show that K-nearest neighbor method gives the optimal performance with the same dataset.

Keywords Text mining methods and techniques · Naive Bayes · KNN · Decision tree · Performance parameter

1 Introduction

1.1 Text Mining

Text mining is one of the most significant methods of data mining used for getting information from the dataset. It automatically extracts the information from multiple sources. The prime aim of the text mining is to find out the unidentified information from multiple sources; also, as well as the main obstacle encountered is that the result is irrelevant to user needs [1].

S. Jain (✉) · S. C. Jain
Amity University, Madhya Pradesh, India
e-mail: shitanshujain00@gmail.com

S. C. Jain
e-mail: scjain@gwa.amity.edu

S. Vishwakarma
Gyan Ganga Institute of Technology and Sciences, Jabalpur, India
e-mail: santoshscholar@gmail.com

M. Tuba et al. (eds.), *ICT Systems and Sustainability*,
Advances in Intelligent Systems and Computing 1077,
https://doi.org/10.1007/978-981-15-0936-0_48

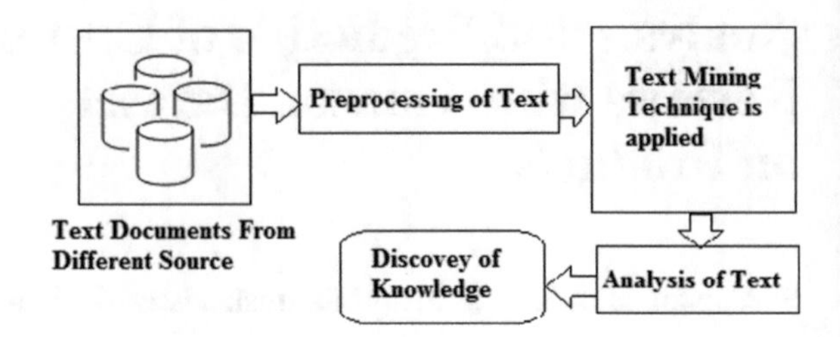

Fig. 1 Text processing steps

1.2 *Text Processing*

Text processing focuses on data mining, information retrieval, and web search. Text processing basically avoids the problems of high dimensionality and sparsity.

Text mining process involves the following steps [2]:

1. Multiple text files from various sources that are in semi-structured or unstructured format are collected to perform text mining.
2. Preprocessing involves cleaning the data that is collected.
3. Many techniques and methods are used in text mining which are discussed later and then applied to extract meaningful information.
4. The data obtained is analyzed to extract knowledge and meaning out of it.
5. Finally, the required knowledge is obtained and can then be used for further analysis (Fig. 1).

1.3 *Text Classification Techniques*

Mining the information from large amount of data has become a difficult task. To troubleshoot this problem some important text classifications techniques are introduced to split the text document into predefined classes [3].

Nearest- Neighbor Classifier: KNN is a method based on instance learning and its algorithm is based on closest sample dataset which needs two parameters to deal with loud dataset.

Bayesian Classifier: It is a technique of the probabilistic classifier used for the classification of text document. Naive Bayes Model is highly responsive to attribute selection which handles only low dimensions and it is fast and easy to implement.

Support Vector Machine: The SVM is a greatly precise machine learning scheme used for text classification. SVM tries to find a best possible hyperplane within the input space so as to exactly categorize the multi-class problem.

Association-Based Classification: Association-based classification integrates association rule mining which generates class association rule and classifies extra correct then decision tree and C-4.5.

Centroid-Based Classification: Centroid-based classification is mostly used to form centroid for each class of the document. It is easy to implement and provide high flexibility for text data.

Decision Tree Induction: A well-accepted decision tree classification algorithm is ID-3, C-4.5. A decision tree is represented as a structure of tree. Decision tree is understandable when dealing with loud data.

Classification Using Neural Network: Neural network is one of the advanced mechanisms of text classification. It performs excellently while given assumptions are fulfilled as per the given set of documents.

2 Review of Literature

In the year 2015, Eduard Alexandru Stoica and EsraKahyaÖzyirmidokuz proposed a system named "Mining Customer Feedback Documents" [4]. By analyzing the customer complaint data they extract some useful information that is hidden in the database. They apply text mining on the document which contains the feedback of customer, and further they perform classification and clustering.

In the year 2015, Cai-Nicolas Ziegler, Michal Skubacz, and Maximilian Viermetz did an extensive work on "Mining and Exploring Unstructured Customer Feedback Data Using Language Models and Tree map Visualizations" [5]. They proposed a system to exploring the corpora of textual customer feedback in a guided fashion, bringing order to massive amounts of unstructured information.

In the year 2015, Murali Krishna Pagolu worked on "Analysis of Unstructured Data: Applications of Text Analytics and Sentiment Mining" [6], which gave the quick look on how to organize and analyze textual data for extracting some useful information about the customer response from large collection of documents, and then use these information for betterment and good performance of business

In the year 2016, Marcelo Drudi Miranda and Renato José Sassi worked on sentimental analysis named "Using Sentiment Analysis to Assess Customer Satisfaction in an Online Job Search Company" [7]. This work proposes a tool for aiding the evaluation of customer satisfaction in a Brazilian Online Job Search Company through the use of Sentiment Analysis. After performing Sentiment Analysis for evaluating the customer satisfaction in a Brazilian online job searches company.

3 Proposed Methodology

3.1 Methodology

1. The first step performed is the creation of the corpus. Corpus is the collection of the documents (here: Customer feedbacks) taken from Amazon which is a popular e-commerce online shopping website. These documents contain customer feedbacks from verified customers of Amazon.
2. Data mining is the extraction of useful information from huge dataset. Once the customer feedbacks are taken from Amazon, the reviews are merged into one and the dataset is integrated into one unit. Then the textual data is taken from these documents that contain customer feedbacks.
3. Preprocessing—This is a mandatory step of text mining where the noise and inconsistencies are removed. There are four different steps in preprocessing using RapidMiner: Tokenization, Transform case, Filter stop, and Stemming. Once the dataset passes from all the four phases, we get the preprocessed data which is free from noise or any inconsistencies.
4. Training and testing dataset are created for providing training to the Naive Bayes and KNN classifiers, respectively. The system is trained against 2000 customer feedbacks. Once the training is done, the testing dataset is applied where predictions are made for new and unknown customer feedbacks (Fig. 2).

3.2 Classification

Classification is a very popular data mining technique based on machine learning [8]. Machine learning gives the learning ability to the computers where they can learn so that the system can be trained using past decisions, and predictions can be made for the future. These predictions are made such as the objects belong to which category or class. Here, we train the system by giving some examples and known outputs to the system for the objects. Later the system automatically makes predictions on the learning it acquired.

In this work, we have used different classification algorithms to implement our dissertation work and accuracy of each classifier is determined. In this work, Naive Bayes classifier (which is based on Bayes theory) and KNN is used to classify customer feedbacks.

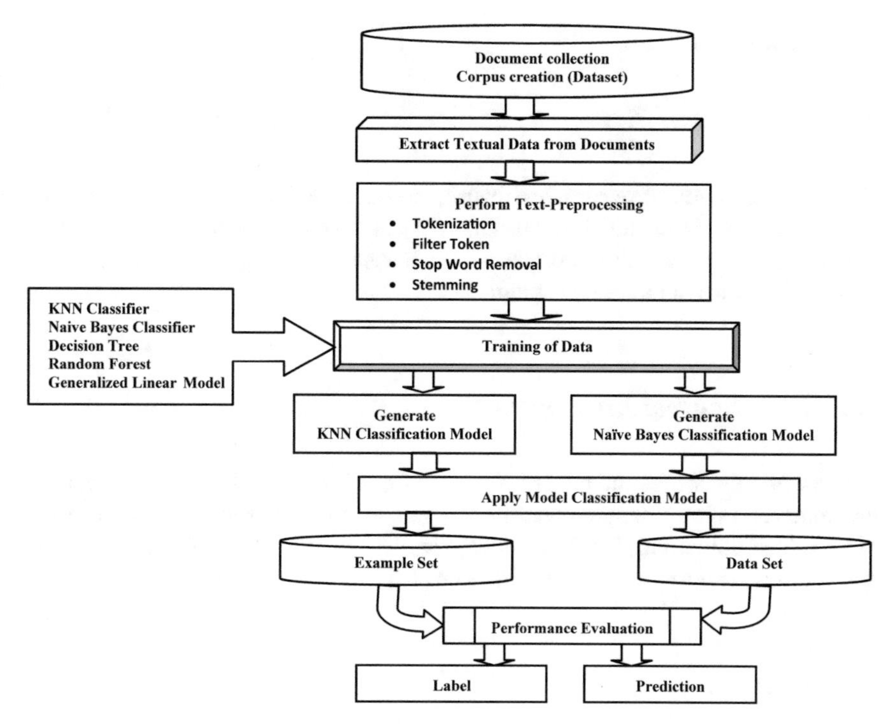

Fig. 2 Methodology work flow

3.3 *Naive Bayes Classifier*

Naive Bayes classifier is a classification algorithm/technique which is based on supervised learning approach [9].

Naive Bayes takes the objects as inputs and predicts the target class for the object, on the basis of certain probabilities. Naive Bayes classifier is based on Bayes' theorem.

3.4 *K-Nearest Neighbor Classifier*

KNN is one of the earliest known techniques and simplest classification algorithms used in data mining. Value of "k" is arbitrary and is always given by the user. This algorithm is based on neighbors and the deciding feature for the target class [10].

The factor for predictions is the distance of the object among its neighbors. The measure used for determining neighbors is done using distance functions such as Euclidean, Manhattan, Hamming distance, etc.

4 Implementation

4.1 RapidMiner Tool

For implementation of our research work, we have used a very popular data mining software platform called RapidMiner [11]. It is an open source platform which empowers and allows all the organizations/businesses to apply data analytics for making decisions in their organizations.

4.2 Data Source: Amazon Reviews

In this work, feedbacks and reviews are taken from Amazon, which is a very popular e-commerce, online shopping website. Nowadays, millions and thousands of users buy products online and give their feedback which helps thousands of other customers and is also useful for research and analysis [12].

4.3 Training Dataset

The system is given training, where it is able to make predictions as to which class a particular review/feedback belongs. The system learns and trains itself to make predictions for the new feedback where the target class can be predicted for this unknown review on the basis of its past learning which is given through training

To implement this work, we have taken 2000 customer reviews for training our system. These reviews are taken from verified purchases done by verified customers from Amazon. These feedbacks are stored in an excel sheet. Certain keywords are used for training as if there are keywords such as “Excellent,” “Great,” “Good,” it goes to the GOOD class/category.

4.4 Text Preprocessing

Pre processing is removing the inconsistencies and noise from the data. We have used “Process Documents from File” operator in RapidMiner. This operator is widely used to perform Preprocessing which is a mandatory step to remove errors from our dataset (Fig. 3).

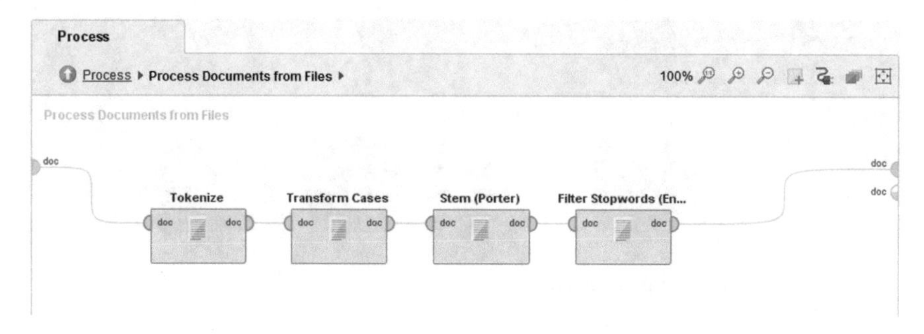

Fig. 3 Text preprocessing

4.5 *Main Process for Training Dataset*

There are two processes to be defined in RapidMiner for execution. One is the main process where we train the system and check its accuracy against various performance parameters. The classifier is defined in this process, and different operators are used for determining the accuracy of the system. Different operators are used for execution (Fig. 4).

Training of the system is performed using Naive Bayes operator, KNN operator, and Decision tree operator. We have used these classifiers for classifying customer reviews. The system learns from previous examples, the known outputs. Once the learning is complete, we apply this learnt model to the testing subprocess (Fig. 5).

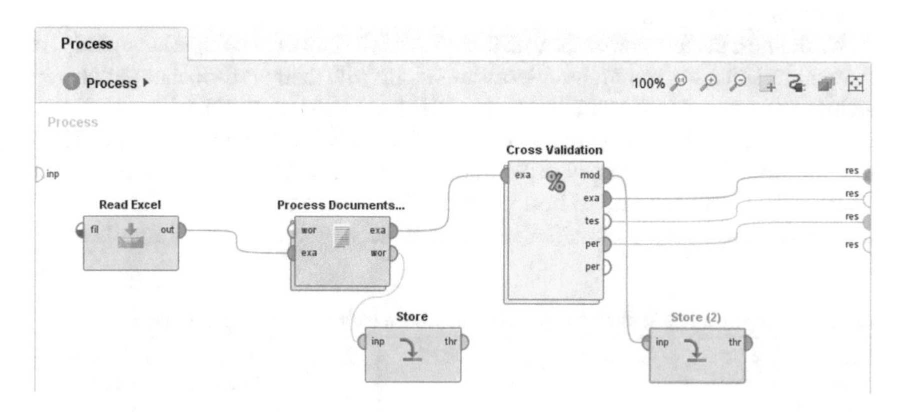

Fig. 4 Main process for training dataset

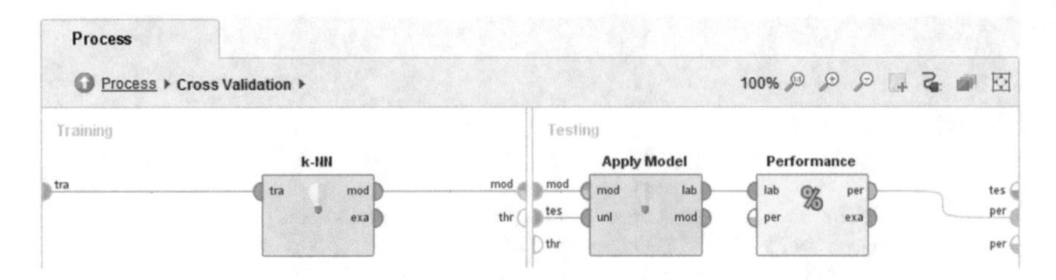

Fig. 5 Cross validation for training dataset

5 Result and Discussion

In this work, we perform a comparative study where we take into account several classification algorithms along with different performance parameters.

We began our implementation using Naive Bayes classifier and used KNN alongside with it; but we are not limited to only these two classification algorithms in data mining. There are several other popular classification techniques, decision tree, random forest, and generalized linear model which can be used for classifying dataset and analyzing the output of all of them.

Once we run the RapidMiner main process, we get results for different performance parameters such as accuracy, classification error rate, Kappa, precision, and recall for Naive Bayes, KNN, and decision tree classifier which is used to classify Amazon reviews. We analyze the performance of the system using these performance parameters. These performance parameters are analyzed against 2000 records taken for training dataset.

We define a Performance table, in Table 1, which shows the comparative study for various classification techniques/algorithms along with their performance parameters such as accuracy, classification error rate, Kappa, precision, and recall.

The Performance table gives us a comparative study for different classifiers. KNN has achieved highest accuracy of 96.38% and very low classification error rate of 3.62%. KNN has been proven best in terms of performance in comparison to other classifiers. Naive Bayes classifier which is widely used for text mining has also

Table 1 Comparison table of classification method with their performance parameter

Classification method/Performance parameter	Accuracy	Classification error	Kappa	Weighted mean recall	Weighted mean precision
Naïve Bayes	90.09	9.91	0.816	92.65	79.97
KNN	96.38	3.62	0.929	95.59	95.86
Decision tree	77.90	22.10	0.499	77.01	91.01
Random forest	68.04	31.96	0.201	43.06	71.99
Generalized linear model	93.01	6.99	0.856	89.81	96.14

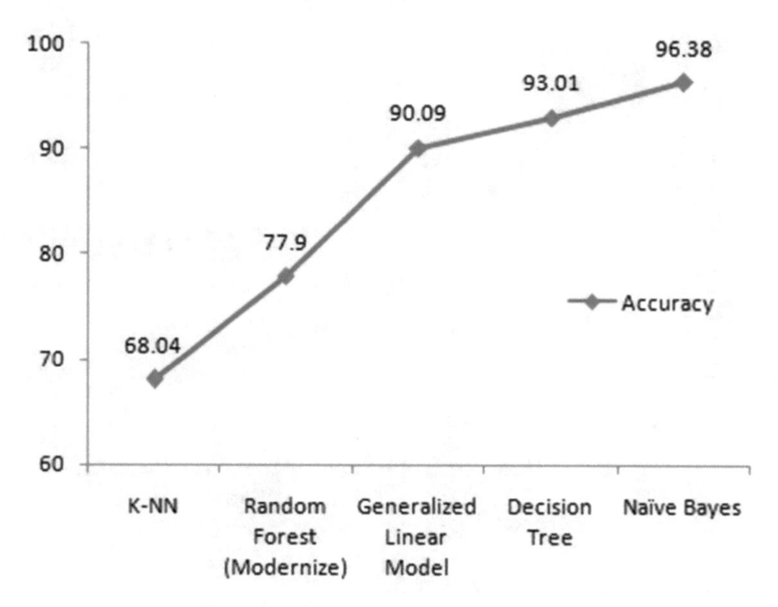

Fig. 6 Accuracy of Different Classification Methods

achieved high accuracy of 90.09%. Decision tree which makes classification on the basis of decisions of if-else conditions has achieved accuracy of 77.90%. Random forest which uses combination of decision trees with random function has given low accuracy and little high error rate as it requires more features for classifying data. This may not be appropriate for classifying Amazon reviews where there are not a lot of features but the focus and emphasis is on feedbacks. Generalized linear model has also achieved high accuracy of 93.01%. The classifiers which can effectively handle polynomial data have given high accuracy results. KNN has achieved overall best performance in comparison to other classifiers as it has achieved high Kappa values and high recall and precision values. This is further depicted in the graph section next.

6 Conclusion and Future Work

In this work, we surveyed the feedbacks of verified customers and built a system where the reviews can be given to the system, and it can automatically give the analysis which can instantly help the consumers in deciding whether they should buy a particular product or not.

We trained the system using 2000 records to correctly predict the class for reviews. We also made a comparative study for various different classification algorithms such as decision tree, random forest, and generalized linear model, along with their

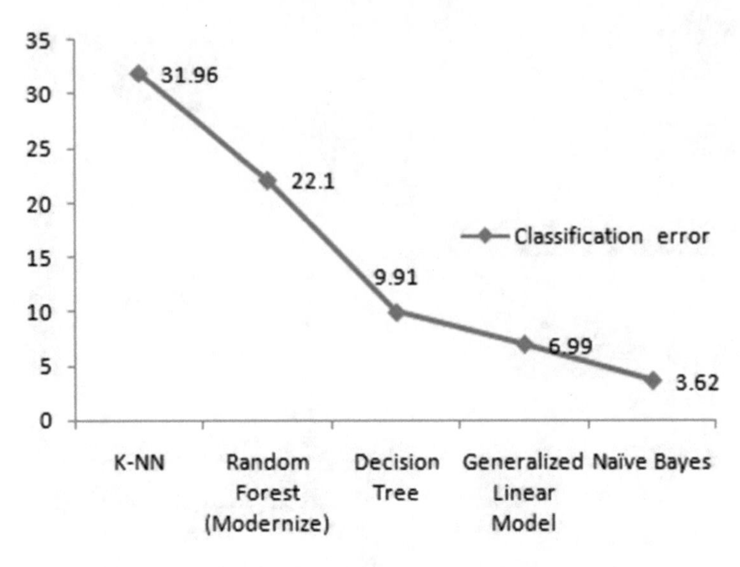

Fig. 7 Error of Different Classification Methods

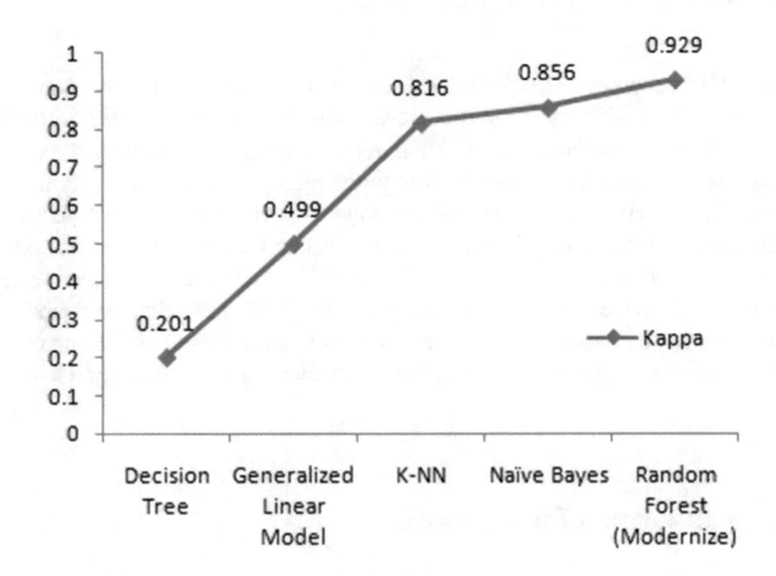

Fig. 8 Kappa Measure of Different Classification Methods

performance parameters such as accuracy, classification error rate, Kappa, precision, and recall which were used.

As we focused on 4 categories in this dissertation work, in future we can extend our research work covering more different categories from Amazon such as Cameras, Gadgets, Grocery items, etc. We can also extend this work by taking Customer

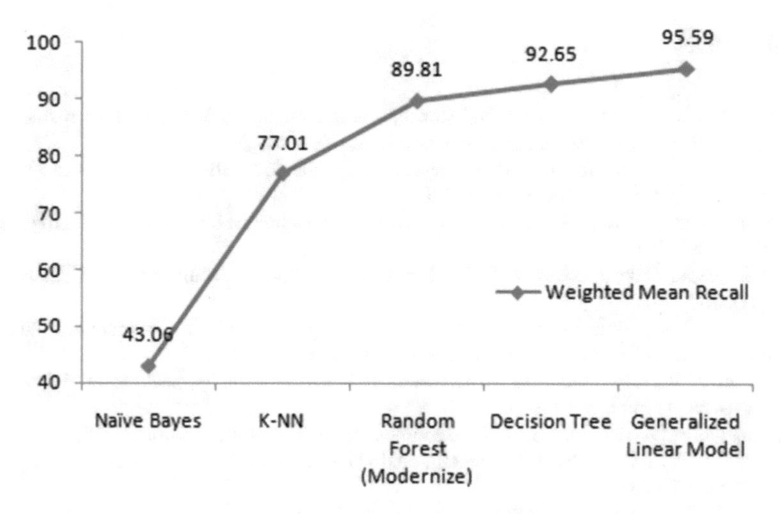

Fig. 9 Weighted Mean Recall of Different Classification Methods

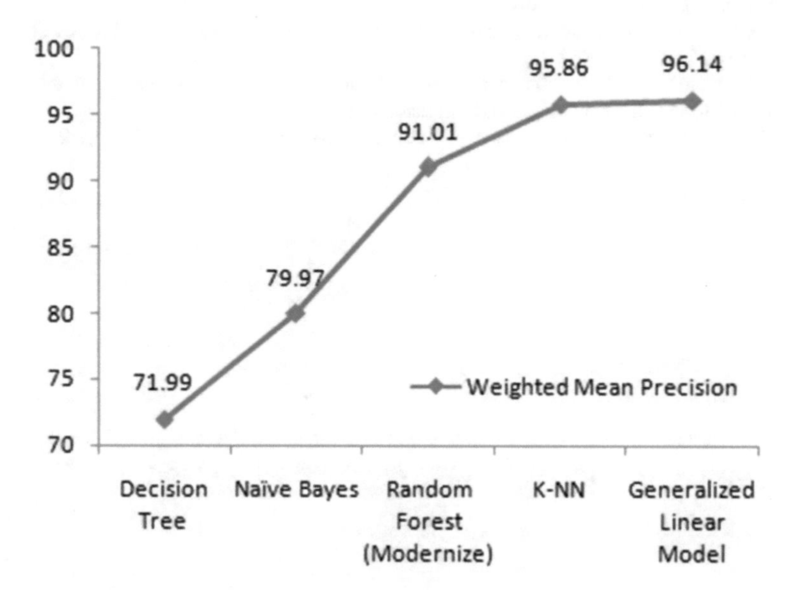

Fig. 10 Weighted Mean Precision of Different Classification Methods

Feedbacks from other websites such as Flipkart, Shopclues, etc. The research work can also be extended for offline markets. In this work, we have focused on KNN and Naive Bayes classifiers but there are other advanced classification algorithms such as ID3 and multiple support vector machine, which can also be studied and used in future for performing data analytics.

References

1. Han, J., Kamber, M.: Data Mining: Concepts and Techniques, 3rd edn. Morgan Kaufmann, Boston, MA, USA. Elsevier, San Francisco, CA, USA (2006)
2. Vijayarani, S., Ilamathi, J., Nithya: Preprocessing techniques for text mining—an overview. Int. J. Comput. Sci. Commun. Netw. **5**(1), 7–16
3. Nidhi, Gupta, V.: Recent trends in text classification techniques. Int. J. Comput. Appl. **35**(6), 0975–8887 (2011)
4. Stoica, E.A., Özyirmidokuz, E.K.: Mining customer feedback documents. Int. J. Knowl. Eng. **1**(1) (2015)
5. Ziegler, C.-N., Skubacz, M. Viermetz, M.: Mining and exploring unstructured customer feedback data using language models and tree map visualizations
6. Pagolu, M.K., Chakraborty, G.: Analysis of unstructured data: applications of text analytics and sentiment mining
7. Miranda, M.D., Renato, J.S.: Using sentiment analysis to assess customer satisfaction in an online job search company. In: International Conference on Business Information Systems. Springer, Cham (2014)
8. Voznika, F., Leonardo, V.: Data Mining Classification (2007)
9. Leung, K.M.: Naive Bayesian classifier. Department of Computer Science/Finance and Risk Engineering, Polytechnic University (2007)
10. Dey, L., et al.: Sentiment analysis of review datasets using Naive Bayes and K-NN classifier (2016). arXiv:1610.09982
11. Jungermann, F.: Information extraction with rapid miner. In: Proceedings of the GSCL Symposium Sprachtechnologie und eHumanities (2009)
12. Dataset Amazon: https://www.dataworld.com/amazonreviews/. Accessed 05 June 2019

Challenges in Implementation of Safety Practices for Building Implosion Technique in India

Asifali Khan, Herrick Prem and Siddesh Pai

Abstract In the controlled demolition industry, building implosion is the strategic placing of explosive material and timing of its detonation so that a structure collapses on itself in a matter of seconds, minimizing the physical damage to its immediate surroundings. Despite its terminology, building implosion also includes the controlled demolition of other structures, such as bridges, smokestacks, towers, and tunnels. Every structure/building after its useful life needs to be taken care of as it poses a threat to the adjacent structures, and as the purpose of the building is fulfilled the structure should be demolished to construct a new structure. The techniques of selective demolition aim to maximize the recovery of reusable and recyclable building materials; however, a being labor-intensive and dilatory process it is mostly adopted in the Indian industry, as the building demolition activity is very complicated and involves high risk in terms of safety and cost. A minute deviation with respect to the timing of successive detonators, placement of detonators, and miscommunication can cause the demolition process to go for a toss. This paper deals with the various challenges faced by the major demolition techniques in India. The study also includes the precautionary measures regarding machinery or equipment, scaffolding, public safety, and worker safety. Various strategies for the demolition of waste have been reported in literature for implementing good practices for demolition of buildings.

Keywords Controlled demolition · Building implosion · Safety issues

A. Khan (✉) · H. Prem · S. Pai
National Institute of Construction Management and Research (NICMAR), Ponda, India
e-mail: asifali.acm12goa@nicmar.ac.in

H. Prem
e-mail: herrik.acm12goa@nicmar.ac.in

S. Pai
e-mail: siddeshp@nicmar.ac.in

M. Tuba et al. (eds.), *ICT Systems and Sustainability*,
Advances in Intelligent Systems and Computing 1077,
https://doi.org/10.1007/978-981-15-0936-0_49

1 Introduction

The construction industry in India has been growing at rapid rate in the past few decades; all the cities are being flooded with the high-rise buildings which are being used for both residential and commercial purposes. These buildings have to be demolished in the upcoming future for different reasons like end of life of the building, construction of more advanced and space-efficient buildings, etc. Demolition is being done throughout the world but India is relatively new to this field. In India, demolition is generally done manually or by using machines; the use of explosives is generally less when compared to many developed countries. In the upcoming future, the use of explosives will be increased as it is less time consuming even compared to the other types of demolition. Demolition is considered to be one of the most risky fields of construction as the chance for accidents is very high when compared with other fields of construction. The contractor has to follow strict norms in the field of demolition in order to prevent any major tragedy from ever happening. The demolition safety is not only about the safety of laborers in the site but also it deals with the safety of the people of the surrounding areas and also the safe disposal of debris. Accidents could be easily prevented if proper preplanning is done and safety rules are properly followed.

2 Methodology

2.1 Primary Data Collection

This research commenced by reviewing the relevant literature on construction safety published by the Health and Safety Executive as well as academic journals. The operatives were chosen as the subjects of this study because their positions are more directly related to the construction environment and their positions interact with the organizational policies and practice (Fig. 1).

2.2 Sampling Frame

The data collection steps include setting boundaries for the research, gathering information by unstructured or semi-structured observations and interviews, documents, and visual materials (Table 1 and Fig. 2).

For the primary data collection, 50 respondents namely safety engineers, consultants, and project managers were interviewed personally to understand the challenges in implosion technique implemented in India; the effective sample size is 50. Company archives and documents were also gathered and reviewed.

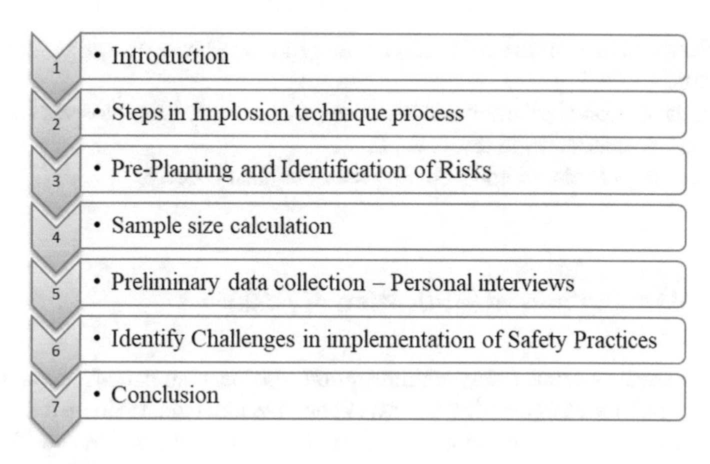

Fig. 1 Methodology flowchart

Table 1 Total sample interviewed

Designations	Number
Safety engineers	17
Safety consultants	21
Project managers	12
Total sample	50

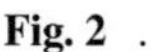

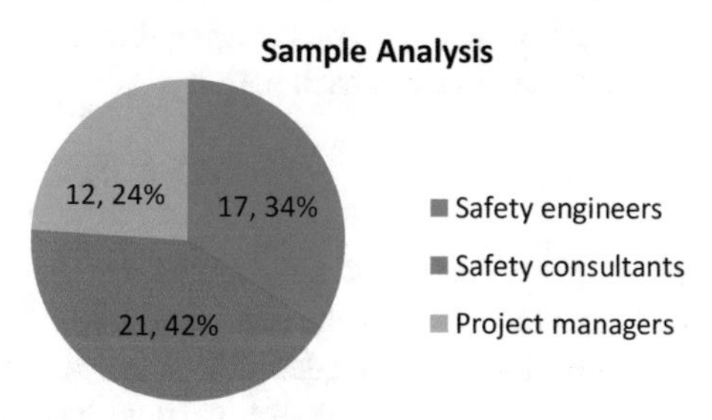

Fig. 2 .

3 Implosion Technique Procedure

In controlled demolition, a structure is blasted with the help of explosive material, i.e., by placing the explosion in strategic location of the structure, and timing the detonation to collapse the structure within minutes, however without physical damage to the adjacent structure.

It involves the following steps:

1. Planning and analysing the structure as to understand the condition of the structure.

2. Drilling of Holes in the columns according to the plan approved by demolition experts.
3. Placing of explosives after understanding the quantity of explosives to be used to displace concrete and direct the fall.
4. Connecting the circuit and initiating the final blast process.

4 Preplanning and Identification of Risks

OSHA (Occupational and safety administration, 2007) recommends for conducting a proper and elaborate engineering survey. A competent person has to visit the site, do a proper investigative survey, and create a detailed report. This report should contain the full evaluation of the project, the possible hazards that could occur for the project, labors, and environment; and proper safety precautions that should be taken and plans for reuse, recycle, and disposal of waste materials from the structure. It should also contain information like what kind of temporary structures are preferred and a proper place for the disposal of the waste material. The main steps for producing safety plan are

1. Establishing the tasks to be undertaken,
2. Classifying the existing risk and issues,
3. Defining how each risk and issues will be eliminated.

The ideal planning to attain zero accident targets is highly impossible, so the construction manager tries to bring down the risk to an acceptable level by estimating the risk factor within each activity.

$$\text{Total risk (TR)} = \text{P} \times \text{H} \times \text{T} \times \text{E}$$

where P—process risk, H—human resource risk, T—technology risk, E—physical risk. The total risk for the blast demolition is very high compared to the other types of demolitions. So blast demolition is used only in the case of absolute necessity. All the procedures have to be followed with complete care.

5 Material Handling Issues

In the demolition process, a lot of waste materials are formed which can be properly recycled or reused. There are also materials that are hazardous when not handled properly or left unattended. OSHA also deals with the handling of these materials with a very expansive section (Occupational and safety administration, 2007). There are guidelines from the Central Pollution Control Board and the Ministry of Environment, Forest and Climate Change regarding the construction and demolition (C&D) waste, 2016, for the different materials that are encountered in the site.

Some of the hazardous materials include

Lead: Lead is well known for its use in the manufacture of plumbing and painting material, and also significant amount of lead is used in different lead alloys that are used throughout in the process of construction. Exposure to lead even in a very small quantity in the form of inhaling or ingestion will present serious damages to humans. It should be made sure that the lead exposure of the laborers is within a certain level that is prescribed within the OSHA. Proper monitoring, PPE, and hygiene practices could be very helpful in preventing lead poisoning. Lead is also an easily recyclable material, so it should be properly recycled rather than ending up poisoning a land fill.

Asbestos: Asbestos is a widely used material in the field of construction. This makes it one of the most important hazards that has to be considered during the process of demolition. Since asbestos is airborne it could easily spread up to a great length and become a huge threat to the laborers and the people in the nearby area. The construction manager should be very knowledgeable about all the regulation, the planning that is required, protective practices, and a proper method for the disposal of asbestos.

6 Challenges in Implementation of Safety Practices in Implosion Technique

6.1 Hazard Awareness

The construction manager has to be aware of the possible hazards of the site that could occur in the process of demolition. All the workers working in the site should be given proper safety drills and be aware of the hazards in the site, the protective measures, and what to do in case of any emergency that could occur in the site.

6.2 Personal Protective Equipment (PPE)

The PPE is not just for the demolition of the site; it is compulsory in each and every type of construction site irrespective of whatever construction work is going on. PPE generally includes the equipments like helmet, gloves, safety harness, reflective jacket, and safety shoes. These should be made compulsory for all the workers as these could help to reduce the accidents up to a great extent in the construction site.

6.3 Safety Components

In addition to the PPE which are, generally, to be followed by the laborers themselves, there are also other components that are to be provided by the construction company like safety nets, hand rails for the stairs, and barricades for any open areas. Using safety nets should not be neglected since the safety nets not only prevent the death of humans by falling but also the falling of objects on a passerby.

6.4 Safe Blasting Procedures

Even though the explosion is only a small portion of the total demolition, it is considered to be one of the most risky of all the demolition procedures. It generally involves different risks like process, human resource, technology, and physical environment.

6.5 Safety for Confined Spaces

Confined spaces are those areas where the access is limited, natural movement is restricted, and air supply is limited. These include places like tanks, silos, tunnels, vessels, etc. These places hold toxic gases, flammable gases, corrosive gases, and poor working environment. It has to be made sure that all the toxic and flammable gases in these places are within the permissible limits. The excess amount of oxygen or flammable gases could cause the sparks while cutting to explode. Less amount of oxygen could be serious for the laborers.

6.6 Prestressed and Post-tensioned Concrete Safety

Prestressed and Post-tension are widely used nowadays. These structures have steel reinforcement in them which are in very high tension all the times. So, if these structures are demolished without proper precaution there is a possibility for the release of the steel reinforcement with a violent or explosive force. So the construction managers have to take necessary planning and engineering advice before the demolition of these structures.

6.7 Competent Person

OSHA requires a competent person for the inspection of the process. The person should be able to predict the hazards that could occur in the site. A well-competent person should be well prepared for all the risks and issues that could occur in the site, and should be able to stop the work if any unsafe condition is detected. The ultimate target of a competent person is to achieve a zero accident target.

6.8 Public Health Hazards

Demolition does not only has an impact on the people working in the site but also has huge impact on the people who are living near the site. Lots of the demolition activities are generally done in the urban setting where lots of people live in close proximity to the building under demolition. The people who are living nearby are most likely to be affected by the dust and other harmful materials that are formed as a result of the demolition process. The demolition process releases both airborne and waterborne contaminants into the area where it is done. Demolition also produces a large amount of debris which has to be handled and disposed properly or else there is a high chance of it falling on some person. Some of the public health hazards due to demolition and reconstruction include.

6.9 Vibration

Demolition work will cause vibration to the adjacent buildings or structures to various extents, depending on the adopted method of demolition. The most serious vibration is caused by implosion. The effect of vibration caused by implosion is categorized as follows:

a. Permanent ground distortion produced by blast-induced pressures,
b. Vibratory settlement of foundation materials;
c. Projectile impact (blast fly rock),
d. Vibratory cracking from ground vibration or air blast.

6.10 Dust Exposure

When a building is being demolished, it produces a lot of dust which could easily spread throughout the area where the demolition takes place. Demolition also causes diseases like asthma and other lung diseases. To further reduce the dust, the columns should be covered with geotextile fabric.

6.11 *Water Droplet Dust Control*

The most common method for solving this problem is by wetting the area. Presently, most of the big cities are facing water scarcity, which could be a problem. But demolition could expose a huge population to dust, which could affect the elderly and/or individuals with weak immune system. Hence, greater care should be taken for it.

6.12 *Debris Disposal*

Demolition produces a lot of construction waste which mostly is bricks, concrete, and reinforcement. These three are easily recyclable or reusable. Broken bricks and concrete can be used as fillers in the construction. Reinforcement can be recycled for the production of new rods or any other goods. There are also small quantities of different materials which could be hazardous to the environment; some of the most common materials that go to the land fill include

- Thermostats—Mercury
- Batteries—Lead, Mercury and Cadmium
- Flashing and Pipes—Lead
- Smoke Detectors—Radioactive Material
- Treated Wood—Arsenic

These materials are generally thrown in the landfills and end up contaminating the environment.

The demolition experts aim at the critical portion of the structure so that as soon as these sections are imploded using explosion the remaining section would fall on its own footprint by the law of gravity, as after the explosion of the critical section the structure is incapable of giving stability due to its self-weight.

7 Conclusion

Demolition is considered to be an art of dismantling and destroying the existing old structures without disturbing the structural ability of the structures surrounding it. Government has to make proper norms for the process of demolition and it has to be followed during the demolition. The demolition experts aim at the critical portion of the structure so that as soon as these sections are imploded using explosion the remaining section would fall on its own footprint by gravity, as after the explosion of the critical section the structure is incapable of giving stability due to its self-weight. The advantages of implosion technology method over conventional methods are as follows: It is comparatively a less expensive and quick method when used for huge structures like multistoried structures/high piers, etc. The disadvantages of this method over conventional methods are: Large pieces of debris might project toward spectators during the explosion; a small miscalculation or a mistake will lead to a huge irreversible damage; need of expert hands. The advantages of this method make this method more acceptable over the other demolition methods.

References

1. Arie, G.: Health and Safety in Refurbishment Involving Demolition and Structural Instability (2004)
2. Buildings Department: Code of practice for demolition of buildings. Government Logistics Department (2004)
3. Anumba, C.J., Abdullah, A., Ruikar, K.: An integrated system for demolition techniques selection. Arch. Eng. Des. Manag. 130–148
4. Recommended management practices for the removal of hazardous materials from buildings prior to demolition. Department of Environmental Engineering Sciences, Grainesville, Florida (2004)
5. Diven, R.J., Shaurette, M.: Demolition: Practices, Technology, and Management
6. Dorevitch, S., Demirtas, H., Perksy, V.W., Erdal, S., Conroy, L., Schoonover, T., et al.: Demolition of high-rise public housing increases particulate matter air pollution in communities of high-risk asthmatics. J. Air Waste Manag. Assoc. **56**(7), 1022–1032 (2006)
7. Fauzey, I.H., Nateghi, F., Mohammadi, F., Ismail, F.: Emergent occupational safety & health and environmental issues of demolition work: towards public environment. Proc. Soc. Behav. Sci. **168**, 41–51 (2015)
8. Hecker, S., Gambatese, J., Weinstein, M.: Designing for worker safety. Prof. Saf. (n.d.)
9. Hon, K.H.: Relationships between safety climate and safety performance of repair, maintenance, minor alteration and addition (RMAA) works. Ph.D. dissertation, The Hong Kong Polytechnic University (2012)
10. Patel, R.: Demolination method & techniques. Int. J. Res. Sci. Eng. Manag. (n.d.)
11. Pranav, P., Pitroda, J., Bhavsar, J.J.: Demolition: methods and comparision. Engineering: Issues, Opportunities and Challenges for Development. Umrakh, Bardoli (n.d.)
12. Rathi, S.O., Khandve, P.: Demolition of buildings—an overview. Int. J. Adv. Eng. Res. Dev. (IJAERD) **1**(6) (2014)
13. Shaurette, M.: Safety and health training for demolition and reconstruction activities. In: Proceedings of the CIB W99 International Conference, Gainesville, Florida, USA
14. Sobotka, A., Radziejowska, A., Czaja, J.: Tasks and problems in the buildings demolition works: a case study. Arch. Civ. Eng. 3–18 (2015)
15. Ulang, N.M., Kadar, N.A.: Study on health and safety aspects of demolition projects in Penang. Jurnal Teknologi **75**(5)
16. Yi, K.-J., Langford, D.: Scheduling-based risk estimation and safety planning for construction projects. J. Constr. Eng. Manag. **132**(6), 626–635 (2006)
17. Zhen, C., Abdullah, A.B., Anumba, C.J., Li., H.: ANP experiment for demolition plan evaluation. J. Constr. Eng. Manag. **140**(2)

Large-Scale Data Clustering Using Improved Artificial Bee Colony Algorithm

M. R. Gaikwad, A. J. Umbarkar and S. S. Bamane

Abstract Clustering is grouping the similar data points in the clusters. Large-scale data grouping has discovered wide applications in many fields, particularly in big data analytics. Traditional clustering algorithms do not explore and exploit all feasible solutions of clustering. Artificial Bee Colony algorithm (ABC) is a metaheuristic algorithm applied for clustering. ABC suffers from slow convergence. Hence, Improved ABC (IABC) is used for experimentation. UCI datasets—wine and seed are modified to large scale and used for experimentation. Experimental results show that IABC give the quality clusters than ABC and K-mean for large-scale dataset. ABC gives the better clusters than K-mean.

Keywords Artificial bee colony algorithm · ABC · Data clustering · Large-scale data · Evolutionary algorithm · EA · Swarm intelligence · SI

1 Introduction

Data clustering is collection of the information with the goal that comparable things are set together. Clustering algorithms have been connected to a wide scope of fields, including exploratory information investigation, information mining, artificial intelligence, picture recovery, and numerical programming. There are two conventional techniques: partitional and hierarchal grouping. The limitations of the traditional algorithms like K-means, C-means, etc, are affecting the underlying centroids. The change in positions of centroids produces distinctive answers. The algorithm is likely to stick to the local minima and declare it as the optimal solution.

Evolutionary algorithms are used to overcome these drawbacks. Evolutionary algorithms are soft computing methods used to provide optimal solutions. Optimization is meant by selecting the best possible solution to a problem from some set of available alternatives. Karaboga introduced a metaheuristic Artificial Bee Colony

M. R. Gaikwad (✉) · A. J. Umbarkar · S. S. Bamane
Department of Information Technology, Walchand College of Engineering, Shivaji University, Sangli, India
e-mail: madhuragaikwad56@yahoo.com

M. Tuba et al. (eds.), *ICT Systems and Sustainability*,
Advances in Intelligent Systems and Computing 1077,
https://doi.org/10.1007/978-981-15-0936-0_50

(ABC) that has been used in several optimization problems [1–3] to discover an ideal solution [4]. This algorithm is influenced when searching for a good food source by the conduct of honey bees. It is basic in and easy to enforce because it needs only one attribute in its algorithm called a "limit" [5] to control the discarded answer. Like other metaheuristic algorithms, ABC's convergence velocity relies on a successful trade-off between search strategies of exploration and exploitation. Exploration implies an algorithm's ability to find the best possible solution in the unknown fields of search space, whereas exploitation is the method of discovering the ideal solution by taking benefit of prior excellent alternatives in its area. Another important factor in the algorithm is the ABC search operator.

Improvement in ABC presents a new search operator. This new operator improves the exploitation approach of ABC, which takes more data from the feasible solutions of the present population by choosing a random combination of them and the present best source of food.

Banharnsakun et al. proposed parallel processing by dividing the data into multiple small groups. Each subgroup then executes a local scan simultaneously on every processing node. Local best answers are compared between processing nodes and optimal solutions are given [6]. Xing described artificial bee colony algorithm and various optimization algorithms. It provides the basic implementation of ABC in the paper [7]. Shi et al. presented a new way of ABC. A better dimension selecting plan for employed bees to get the global search and local adapting capabilities is given in the work [1]. Banharnsakun experimented the large-scale data clustering on map-reduce framework using ABC algorithm. The optimal clusters of the large-scale data set are determined in this work [2]. Cai et al. evaluated, ant colony algorithm by improving for the searching time and convergence [5]. Ghambari introduced improvements in the original ABC algorithm which was used to solve optimization problems in the paper [8].

The following is the organization of this paper. A short overview of the ABC Algorithm and IABC Algorithm is provided in Sect. 2. Section 3 describes the tests, summarizes the outcomes and discusses the algorithm's efficiency. Section 4 concludes.

2 Artificial Bee Colony Algorithm (ABC) for Data Clustering

The fundamental idea of ABC is to imitate the searching behavior, of honey bees. ABC algorithm comprises of two clusters of artificial bees: utilized artificial bees and unemployed artificial bees (for example searching of a food origin to utilization). The unemployed artificial bees will be divided in two group's onlooker bees and scout bees which will be looking for solutions in exchange of abandoned solutions.

Step 1: The employed bees are randomly sent to search the food sources. When an employed bee discovers a better possible solution, it stores in the memory. Every single employed bee completes search procedure; the solution will be given to the onlooker bees.

$$v_{ij} = x_{ij} + \Phi\left(x_{ij} - x_{kj}\right) \tag{1}$$

where

x_{ij} = old data point, x_{kj} = neighbor centroid data point chosen randomly, v_{ij} = new data point for x_{ij}, Φ is a random number between 0, 1.

Step 2: An onlooker bee examines all the solutions from employed bees and chooses a better solution in terms of objective function.

The closest data point to the centroid is chosen among all the data points updated by the previous step.

Step 3: Stopping the utilization procedure of the solutions searched by the bees.

The value for objective function for every cluster is calculated. And, the cluster with maximum value of objective function is discarded.

$$MinD(w, c) = \sum_{i=1}^{N} w_{ij} \sum_{j=1}^{K} \sum_{d=1}^{D} o_{id} - c_{jd} \tag{2}$$

where, w_{ij} is the association weight of the objects o_i, o_{ij} is the object i of cluster j, c_j is the centroid of cluster j.

Step 4: Send the scouts for locating the new food sources, randomly.

New values of data points are searched and made to follow the same procedure.

Step 5: Store the best solution up till now.

Step 6: Repeat until requirements meet. The stopping criteria can be number of iterations for the procedure.

2.1 *Improved ABC for Data Clustering [8]*

There are two phases in the algorithm, exploration and exploitation phase. Classic ABC algorithm creates another food source by shifting the present food source to the randomly chosen food source from the population. This solution gives a condition in which ABC gives a decent capacity to investigate the search space. The new search operator is given in Eq. (3) [8].

$$v_{ij} = x_{ij} + sf * x_{best,j} - x_{ij} + x_{r1,j} - x_{r2,j} \tag{3}$$

where x_{ij} = old data point, $x_{r1,j}$ and $xr_{2,j}$, randomly selected data points, v_{ij} = new solution for x_{ij}., sf is random number between (0, 1), $x_{best,j}$ is a data point with the best fitness value.

Figures 1 and 2, explain the flow of data processing that is creating clusters of large-scale data using ABC and IABC respectively.

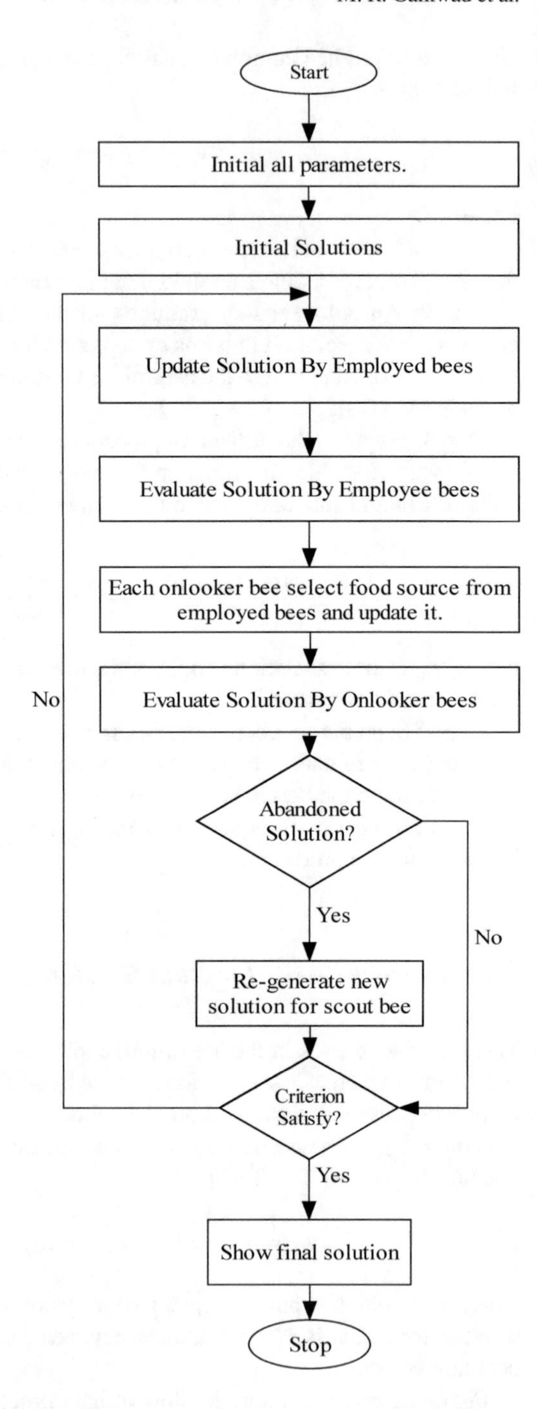

Fig. 1 ABC algorithm

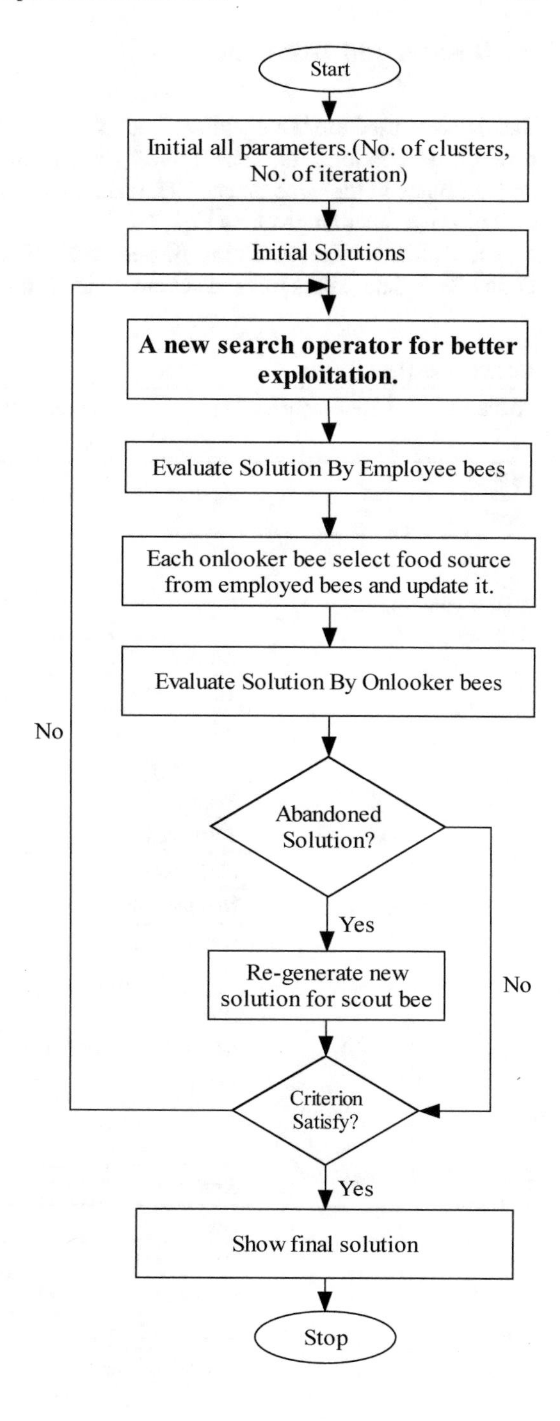

Fig. 2 IABC algorithm

3 Results and Discussion

The datasets used are wine and seed datasets from the UCI repository [9]. Table 1 gives the description of datasets. Table 2 gives the information of various dimensions and attributes in the wine dataset. The data points are divided in the clusters using the objective function given by Eq. (2) (Table 3).

The results for the clustering of Seed and Wine datasets are shown in Fig. 3. Figure 3a is data points of seed set are divided in to three clusters using ABC and

Table 1 Used UCI datasets

Datasets	Number of attributes	Number of attributed in each cluster		
		Cluster 1	Cluster 2	Cluster 3
Wine	13	59	71	48
Seed	7	70	70	70

Table 2 Wine dataset attributes [9]

Attribute	Values
Alcohol	14.23
Malic acid	1.71
Ash	2.43
Alkalinityof ash	15.6
Magnesium	127
Total phenols	2.8
Flavanoids	3.06
Non flavanoid phenols	0.28
Proanthocyanins	2.29
Color intensity	5.64
Hue	1.04
OD280/OD315 of diluted wines	3.92
Proline	1065

Table 3 Seed dataset attributes

Attribute	Values
Area: A	15.26
Perimeter: P	14.84
Compactness: C	0.87
Length of Kernal	5.76
Width of Kernal	3.31
Asymmetry coefficient	2.22
Length of kernel groove	5.22

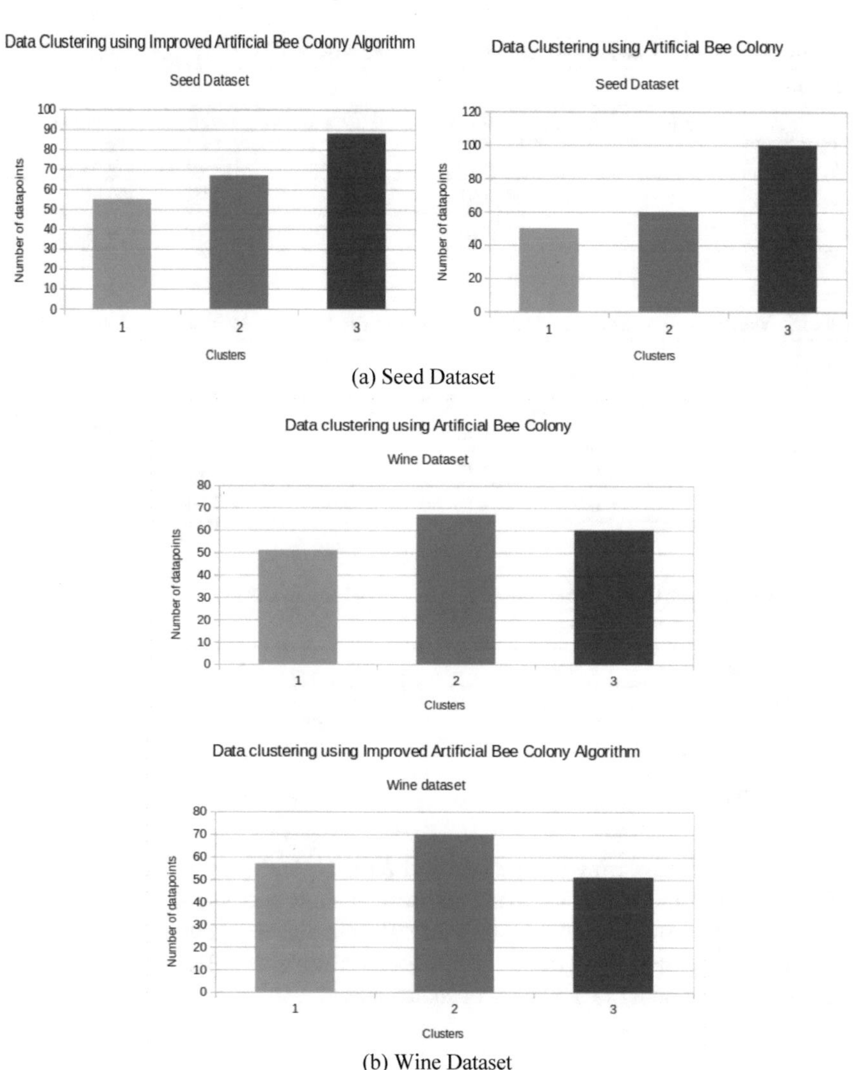

(a) Seed Dataset

(b) Wine Dataset

Fig. 3 Number of data points in the clusters for **a** seed dataset and **b** wine dataset [9]

IABC. Figure 3b is data points of wine set are divided in to three clusters using ABC and IABC. Clustering quality F-Measure is given in Tables 4 and 5 with precision concept and recall for both datasets.

Table 6 shows the comparison of F-measure both the datasets. It is observed that, the IABC performs better than ABC and K-mean clustering algorithms.

Table 4 Values for F-measure on clusters formed using ABC and IABC for seed dataset

Cluster	ABC			IABC		
	Precision	Recall	F-measure	Precision	Recall	F-measure
1	1	0.71	0.83	1	0.78	0.87
2	1	0.85	0.91	1	0.95	0.97
3	0.7	1	0.82	0.79	0.79	0.78

Table 5 Values for F-measure on clusters formed using ABC and IABC for wine dataset

Cluster	ABC			IABC		
	Precision	Recall	F-measure	Precision	Recall	F-measure
1	1	0.86	0.92	1	0.96	0.94
2	1	0.94	0.94	1	0.94	0.98
3	0.8	0.8	0.8	0.94	0.94	0.94

Table 6 Comparison of F-measure values of ABC, IABC and K-means

	ABC	IABC	K-mean
Seeds dataset	0.91	0.97	0.85
Wine dataset	0.94	0.98	0.85

4 Conclusion

Large-scale data clustering using Improved ABC algorithm is experimented using UCI datasets in the paper. Experimental results show that IABC performs better than ABC and K-mean in terms F-measure. Results further show that ABC outperforms over traditional K-means. The IABC outperform over ABC in large-scale data clustering due to better exploitation of search space.

References

1. Singh, S.J., Tiwari, R., Sharma, H.: Hybrid artificial bee colony algorithm with differential evolution. Appl. Soft Comput. **58**, 11–24 (2017)
2. Maulik, U., Bandyopadhyay, S.: Genetic algorithm-based clustering technique: Pattern Recogn. **33**, 1455–1465 (2000)
3. Hruschka, E.R., Campello, B.: A survey of evolutionary algorithms for clustering. IEEE Syst. Man Cybern. 379–390 (2007)
4. Shi, Y., Pun, C., Hu, H., Gao, H.: An improved artificial bee colony and its applications. Knowl. Based Syst. **16**, 0950–7051 (2016)
5. Silva, J., Hruschka, E.R., Gama, J.: An evolutionary algorithm for clustering data streams with a variable number of clusters. Expert Syst. Appl. **67**, 228–238 (2017)

6. Banharnsakun, A.: MapReduce-based artificial bee colony for large-scale data clustering. Pattern Recogn. Lett. **65**, 125–165 (2016)
7. Banharnsakun, A., Achalakul, T., Sirinaovakul, B.: Artificial bee colony algorithm on distributed environments. Nature and Biologically Inspired Computing, vol. 12, pp. 4244–7376 (2010)
8. Ghambari, S., Rahati, A.: An improved artificial bee colony algorithm and its application to reliability optimization problems. Appl. Soft Comput. **17**, 1568–4946 (2017)
9. UCI Repository of Machine Learning Databases. http://archive.ics.uci.edu/m1/index.php. Accessed 21 June 2019
10. Xing, B., Gao, W.: Innovative Computational Intelligence: A Rough Guide to 134 Clever Algorithms. Springer International Publishing, Switzerland (2014)
11. Madamanchi, D.: Evaluation of a new bio-inspired algorithm: krill herd. Comput. Sci. **11**, 96–115 (2014)
12. Tinós, R., Zhao, L., Chicano, F., Whitley, D.: NK hybrid genetic algorithm for clustering. Evol. Comput. **5**, 1089–778 (2018)
13. Zabihi, F., Nasiri, B.: A novel history-driven artificial bee colony algorithm for data clustering. Appl. Soft Comput. **71**, 226–241 (2018)
14. Das, P., Das, D.K., Dey, S.: A modified bee colony optimization (MBCO) and it's hybridization with k-means for an application to data clustering: Appl. Soft Comput. **70**, 590–603 (2018)
15. Tripathi, A.K., Sharma, K., Bala, M.: A novel clustering method using enhanced grey wolf optimizer and mapreduce. Big Data Res. **14**, 93–100 (2018)
16. Sardar, T.H., Ansari, Z.: An analysis of MapReduce efficiency in document clustering using parallel K-means algorithm. Future Comput. Inform. J. (2018)
17. Patel, V., Tiwari, A., Patel, A.: A comprehensive survey on hybridization of artificial bee colony with particle swarm optimization algorithm and ABC applications to data clustering. In: Proceedings of the International Conference on Informatics and Analytics (2016)

Issues Faced During RPL Protocol Analysis in Contiki-2.7

Sharwari S. Solapure, Harish H. Kenchannavar and Ketki P. Sarode

Abstract In Internet of Things (IoT), sensor nodes sense the data from an environment and process that data for specific applications such as industry, home automation and weather forecasting. These sensor nodes are less in terms of power, storage and bandwidth. Routing Protocol for Low Power and Lossy Network (RPL) is the most commonly used distance vector protocol for IoT applications. To analyse such protocols, some network simulators are designed, i.e. Cooja in Contiki OS, OpenWSN, NS-2 and OMNeT++. Most of the researchers are using Cooja for their research purpose. Some of the issues are faced under Cooja simulator while analysing RPL for dense network simulation. In this paper, different issues faced along with some solutions are highlighted. This analysis is done with Contiki-2.7 operating system and Cooja simulator.

Keywords Internet of things (IoT) · RPL · Cooja

1 Introduction

Internet of Things (IoT) is the most important technological expansion in today's world [1]. IoT is basically a global network of various types of wireless sensor networks [2]. The global network consists of physical devices, home appliances and other items.

S. S. Solapure (✉) · K. P. Sarode
Department of Computer Science and Engineering, Walchand College of Engineering, Sangli, India
e-mail: solapuress21@gmail.com

K. P. Sarode
e-mail: ketki.saroode0@gmail.com

H. H. Kenchannavar
Department of Information Science and Engineering, Gogate Institute of Technology, Belgaum 590008, India
e-mail: harishhk@git.edu

M. Tuba et al. (eds.), *ICT Systems and Sustainability*,
Advances in Intelligent Systems and Computing 1077,
https://doi.org/10.1007/978-981-15-0936-0_51

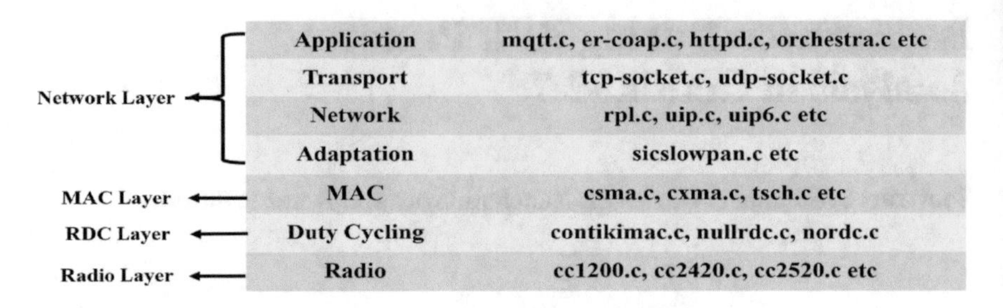

Fig. 1 Contiki OS network stack

Different network simulators are available to analyse this type of network such as Cooja, OpenWSN, NS-2, OMNeT++ and many more [3]. OpenWSN is an open-source project implementing the IEEE 802.15.4e based on the concept of Time Slotted Channel Hopping (TSCH) [4]. OMNeT++ is C++ based simulator used for non-commercial use such as academics and research-oriented institutions [5]. NS-2 and NS-3 are open-source simulators written in C++ and Otcl/tcl, designed specifically for research in communication networks [6].

One of the most popular OSs designed for resource-constrained devices, i.e. Contiki is considered for discussion. Java-based Cooja simulator is used by Contiki OS [7]. There are various versions of Contiki: Instant Contiki, Contiki-2.6, 2.7, 3.0, and NG. Each one is having different features. Most commonly used version is Contiki-2.7.

The network stack of Contiki OS is shown in Fig. 1 [8]. Four layers are present in Contiki protocol stack. Network layer further gets divided into four sub-layers such as application layer, transport layer, network layer and adaptation layer.

Adaptation layer allows IPv6 packets to run on IEEE 802.15.4, which is a standard designed for network of resource-constrained devices. Compiler used for this OS is MSP430. There are various sensor nodes available. Researcher's favourite sky mote is considered for simulation. Memory requirement of sky mote is more than other sensor motes.

There are different protocols supported by four layers of Contiki. RPL is distance vector routing protocol present at network layer, which provides different traffic flows, i.e. point-to-multipoint, multipoint-to-point and point-to-point communication [9]. RPL constructs Destination-Oriented Directed Acyclic Graph (DODAG), where many sender nodes will send data to sink node (server node). The construction of DODAG can be done using control messages such as DODAG Information Object (DIO) and Destination Advertisement Object (DAO) as shown in Fig. 2.

RPL protocol has control message overhead due to which Cooja simulation does not work properly for dense network simulation. The network overhead affects the network lifetime, stability, packet delivery ratio and latency delay.

Section 2 gives the brief literature survey related to issues of Cooja simulator. The detailed methodology is mentioned about different problems faced during RPL simulation in Sect. 3. Finally, conclusion is given along with the references.

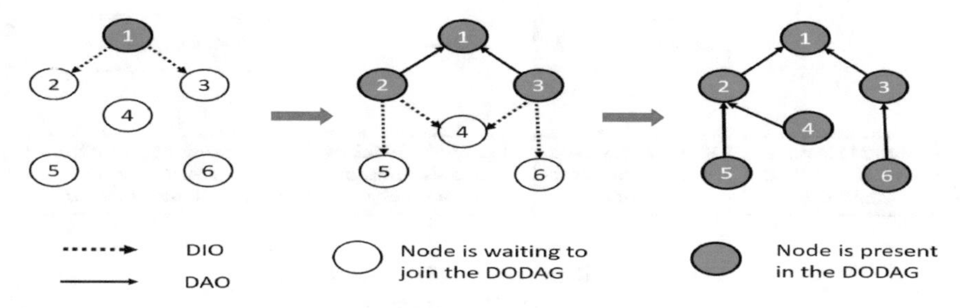

Fig. 2 Experimental setup

2 Literature Survey

Farhana Javed [10] presented an overview, architecture of different operating systems (OS) used in IoT such as Contiki, TinyOS, RIOT (contrast to TinyOS and Contiki), Nano-RK and LiteOS. RIOT and Contiki are mostly used OS for IoT applications as they fulfill the need. Protocols, innovative application and challenges related to these OS such as privacy, security and interoperability for IoT are also focused in this paper.

Accettura [11] evaluated the performance of RPL using Cooja simulator in Contiki OS. Author highlighted overhead problem of RPL-based simulation due to high traffic of control messages of sink to sender node (downward traffic). Improvement in RPL protocol in terms of overhead and performance analysis of RPL compared with existing protocols in WSN is mentioned for future scope.

Kvin Roussel [12] explained about extensibility of Cooja/MSPSim network simulator in Contiki OS to perform experiments on other OS platforms like RTOS. The problem of inaccurate timing of Cooja simulator is also discussed in this paper with comparative study between virtual MSP430-based motes and real hardware. SkyMote/TelosB and Zolertia Z1 are the motes used for analysis purpose.

Harsh Sundani [13] had done a detailed survey of different simulation tools for wireless sensor network. Key features and limitations of each simulator are highlighted. According to author, Cooja simulator has a support for limited number of simulation node types. Time-dependent simulation is difficult is mentioned [13].

3 Methodology

3.1 COOJA Simulator for Contiki-2.7

Cooja simulator is used on top of Contiki-2.7 for simulation of IoT scenarios [7]. The design characteristics of Cooja are shown in Fig. 3. Node type, node memory, observers and interface are the major components of Cooja.

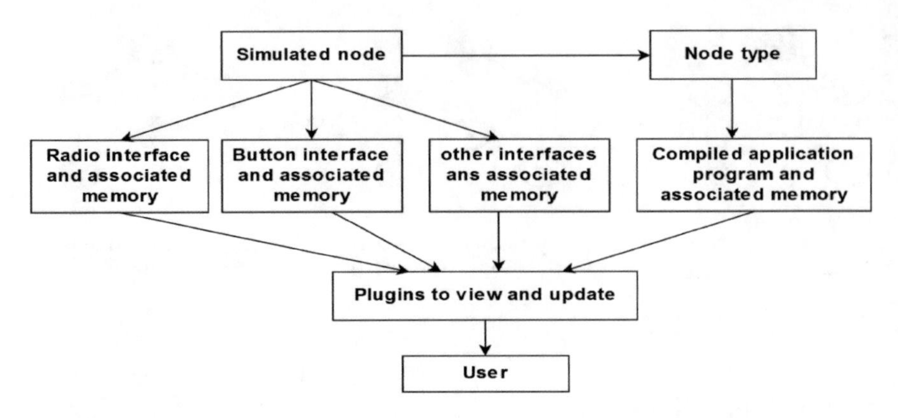

Fig. 3 Simulator design of COOJA

Cooja simulator consists of different 'nodes types'. Even though nodes have same application running, separate memory will be allocated to each node. The interaction between OS nodes and hardware can be done using 'interface'. Interfaces such as button can be used to trigger the events, call OS functions. 'Observers' are the active listeners to the interface, for example, radio medium acts as an observer for radio transmitters in Cooja simulation. 'Plugings' are the graphical user interface for the 'user' to see the expected output.

3.2 RPL for Dense Network in Contiki-2.7

RPL is network layer protocol present in Contiki-2.7 [14]. The location of RPL folder is 'Contiki-2.7/core/net/rpl'. DODAG creation of RPL is already discussed in introduction and shown in Fig. 2. The files present in RPL protocol are given in Table 1.

The most important problem faced by RPL-based simulation is overhead of control messages in network. Increase in control messages sent by sink node to sender node

Table 1 RPL files

File Name	Description
rpl.c	Main RPL function
rpl.h, rpl-conf.h, rpl-private.h	declaration
rpl-dag.c	DODAG creation
rpl-ext-header.c	Extended headers
rpl-icmp6.c	Control messages
rpl-mrhof.c, rpl-ofo	Objective functions

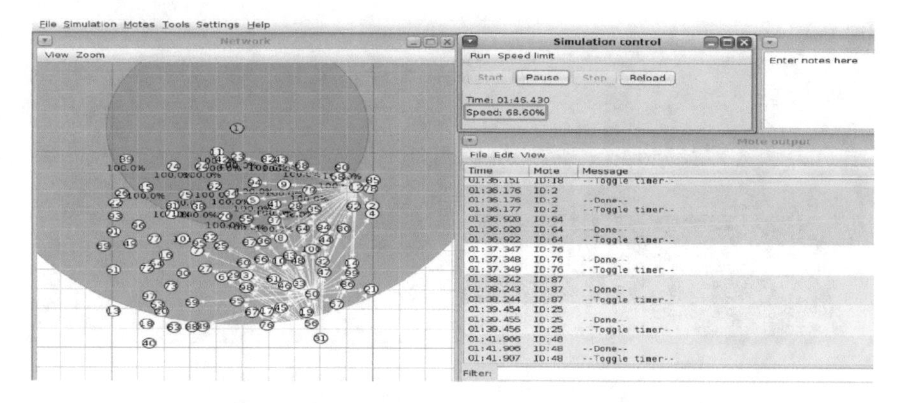

Fig. 4 CoAP-based Cooja simulation for 100 nodes

as discussed in Sect. 1 leads to overload of the network. It decreases the speed of simulation. Applications-based protocols such as CoAP, MQTT will not get affected due to such condition as control packets overhead is not possible there. Simulation based on MQTT protocol is shown in Fig. 4.

If the simulation of RPL is done for less number of nodes for more time, it will work fine as shown in Fig. 5a. The simulation is done for 10 sender nodes and 1 sink node with speed of 360.89%. Figure 5b and c shows the simulation for 50 and 100 nodes with speed of 30.21% and 5.02%, respectively. It shows that speed of simulation decreases with increase in number of nodes. To solve the slow simulation problem of dense RPL network, some solutions are tried.

3.3 *Tried Solutions*

By taking help from Contiki-developer blog [15], numerous good solutions were found. Some are listed with explanation as follows:

- A COOJA.log(Contiki-2.7/tools/Cooja/build/COOJA.log) gets generated automatically at the time of every simulation. All the outputs of print statement get added to that log file. To improve the speed of a simulation for some extent, try to disable the unnecessary print statements from the code so that the size of log file will not increase. Also try to delete that log file every time whenever new simulation starts.
- There are some default settings made by Contiki-2.7 in configuration files of each protocol. Making changes in those configuration files may also help to improve the performance of RPL simulation. Some of the important configuration settings are as follows:

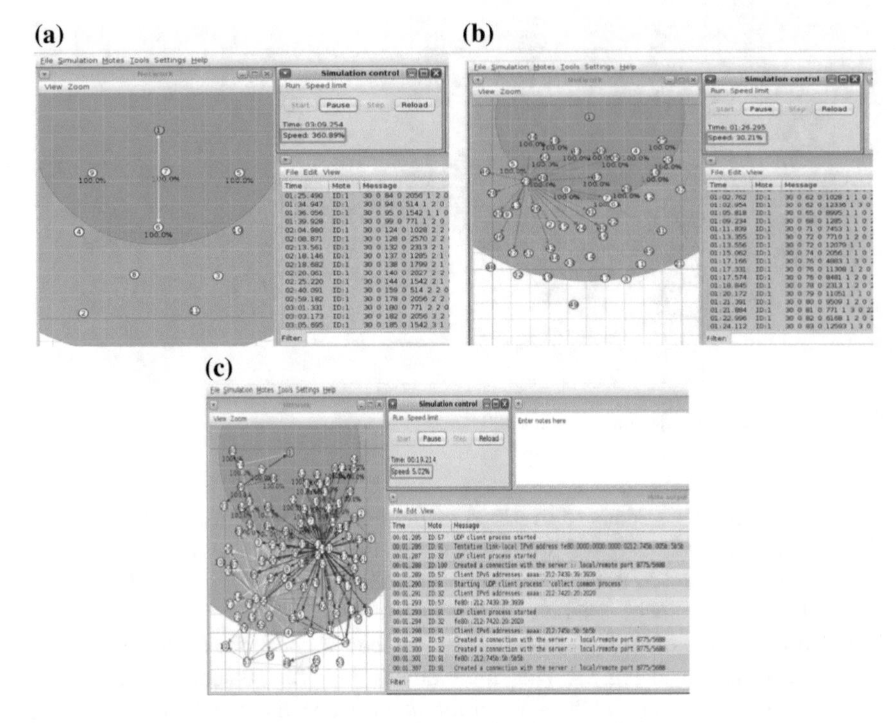

Fig. 5 RPL-based Cooja simulation for **a** 10 nodes, **b** 50 nodes and **c** 100 nodes

- /contiki-2.7/core/contiki-default-conf.h:
  * Disabling TCP by replacing 1 with 0 helps to save the memory.
  ifndef UIP CONF TCP
  define UIP CONF TCP 1
  endif /* UIP CONF TCP */
  * Increasing maximum number of neighbours in neighbour table of each node makes them able to handle more number of nodes. Add maximum neighbour node count greater than 8.
  ifndef NBR TABLE CONF MAX NBRS
  define NBR TABLE CONF MAX NBRS 8
  endif /* NBR TABLE CONF MAX NBRS */

- Changes are made in memory model and related support for 20 bits may help to improve the performance of simulation. To do this, default compiler of Contiki-2.7 is needed to be updated. Default compiler version of MSP430 in Contiki-2.7 is '4.6.3' as shown in Fig. 6. Certain steps to update MSP430 compiler version are stated below:

```
ketki@ketki-HP-Notebook:~$ msp430-gcc --version
msp430-gcc (GCC) 4.6.3 20120301 (mspgcc LTS 20120406 unpatched)
Copyright (C) 2011 Free Software Foundation, Inc.
This is free software; see the source for copying conditions.  There is NO
warranty; not even for MERCHANTABILITY or FITNESS FOR A PARTICULAR PURPOSE.
```

Fig. 6 MSP430-gcc version in Contiki-2.7

- Install the MSP430-GCC version '4.7.3' in bin drive.
- Add the location of extracted binary file path to your system's environment. The updated version of MSP430, i.e. version '4.7.3' successfully gets installed as shown. Some problems arrived for path setting but those are solved.
- Add necessary flags in some files to enable 20-bit support from msp430-gcc 4.7.3 such as cpu/msp430/Makefile.msp430, platform/exp5438 Makefile. exp5438, platform/wismote/Makefile.sky, platform/sky/ Makefile.common.

It has been found that Cooja simulator is not supporting the updated version of compiler. There is an error of Cooja extension due to which this solution does not work well with Contiki-2.7. It may work well if command prompt is used to run the simulation instead of user interface of Cooja. Shell script can be used to retrieve the useful information from the generated log file. To run simulation on command line without GUI, 'ant run_nogui' command can be used.

- go to folder contiki/tools/cooja/ and write command on terminal: ant run_nogui $-Dargs = / < absolute\_path >$/mytest.csc.
- Logs during simulation will appear in contiki/tools/cooja/build/. Inside build log will be present in two files, COOJA.log (Cooja logs) and COOJA.testlog (Simulator's own log).

- Upgradation of the Contiki version is also a good choice for researchers, but many of them are already working on Contiki-2.7 and to make their code compatible with upgraded version is also a very tedious task. Even though you switch your code to new versions of Contiki, for example, Contiki-3.0, Contiki-Ng and Instant Contiki, you may face the same problem for RPL dense simulation.
- Timer(Contiki-2.7/core/net/RPL/rpl-timers.c) plays an important role in RPL. The scheduling of all control messages in RPL is done by this timer file. To reduce the overhead in the network created by RPL, some effective changes must be made. The intensity of the DIO messages must be reduced, when it is not required.

4 Conclusion

IoT applications require efficient routing to distribute the real-time data. Contiki-2.7 along with Cooja is supporting the most commonly used RPL protocol for IoT applications. In the paper, problems related to RPL-based dense network simulation such as control message overhead and low-speed are discussed.

Analysis is done on various types of motes available in Cooja simulator such as sky mote, z1 mote. Memory requirement for sky mote in cooja is high. Due to this, more problems occurred for sky and RPL simulation as compared to other mote types in Cooja.

Many solutions have been tried in this paper to overcome these issues of Cooja simulator. But the result states that there is still no single solution available to solve these issues. No research paper is available to get any perpetual solution to the discussed problem. In future, more efforts are needed to improve the performance of Cooja simulator in terms of memory handling along with Contiki OS and sky motes such as the design of a good memory model.

References

1. Atzori, L., Iera, A., Morabito, G.: The internet of things: a survey. Elsevier **54**(15), 2787–2805 (2010)
2. Pavai, K., Sivagami, A., Sridharan, D.: Study of routing protocols in wireless sensor networks. In: International Conference on Advances in Computing, Control, and Telecommunication Technologies, pp. 522–525. IEEE, Trivandrum, Kerala, India (2009)
3. Chernyshev, M., Baig, Z., Bello, O., Zeadally, S.: Internet of things (IoT): research, simulators, and testbeds. IEEE Internet Things J. **5**(3), 1637–1647 (2017)
4. Watteyne, T., Vilajosana, X., Kerkez, B., Chraim, F., Weekly, K., Wang, Q., Glaser, S., Pister, K.: OpenWSN: a standards-based low-power wireless development environment. Trans. Emerg. Telecommun. Technol. **23**(5), 480–493 (2012)
5. Li, B., Zhang, X.: Research of development in wireless sensor network routing protocols based on NS2. In: International Conference on Electronic and Mechanical Engineering and Information Technology, pp. 1913–1916. IEEE, Harbin, China (2011)
6. Österlind, F., Dunkels, A., Eriksson, J., Finne, N., Voigt, T.: Cross-level sensor network simulation with COOJA. In: 31st IEEE Conference on Local Computer Networks, IEEE, Tampa, FL, USA (2006)
7. Zikria, Y.B., Afzal, M.K., Ishmanov, F., Kim, S.W., Yu, H.: A survey on routing protocols supported by the Contiki internet of things operating system. Futur. Gener. Comput. Syst. **82**, 200–219 (2018)
8. Winter, T., Thubert, P., Brandt, A., Clausen, T., Hui, J., Kelsey, R., Levis, P., Pister, K., Struik, R., Vasseur, J.P.: RPL: IPv6 Routing Protocol for Low Power and Lossy Networks. draft-ietf-roll-rpl-19 (2011)
9. Javed, F., Afzal, M.K., Sharif, M., Kim, B.S.: Internet of things (IoT) operating systems support, networking technologies, applications, and challenges: a comparative review. IEEE Commun. Surv. Tutor. **20**(3), 2062–2100 (2018)
10. Accettura, N., Grieco, L.A., Boggia, G., Camarda, P.: Performance analysis of the RPL routing protocol. In: International Conference on Mechatronics, pp. 767–772. IEEE, Istanbul, Turkey (2011)
11. Roussel, K., Song, Y.Q., Zendra, O.: Using Cooja for WSN simulations: some new uses and limits. In: EWSN 2016-NextMote Workshop pp. 319–324. Junction Publishing (2016)
12. Sundani, H., Li, H., Devabhaktuni, V., Alam, M., Bhattacharya, P.: Wireless sensor network simulators a survey and comparisons. Int. J. Comput. Netw. **2**(5), 249–265 (2011)
13. Zhang, T., Li, X.: Evaluating and Analyzing the Performance of RPL in Contiki. In: First International Workshop on Mobile Sensing, Computing and Communication, pp. 19–24. ACM (2014)

14. Source Forge.: https://sourceforge.net/p/contiki/mailman/contiki-developers. Last Accessed 12 April 2019

Human Identification System Using Finger Vein Image Using Minutia and Local Binary Pattern

Vinayak M. Sadare and Sachin D. Ruikar

Abstract This work explores the Human identification system using the unique vein pattern of the finger image capturing through sensor module. In today's life fingerprint-based biometrics system is commonly used in society but it can be forged, instead vein patterns cannot be forged and it does not affected by any skin damage. Training data is created by capturing finger vein information through infrared sensor. Feature extraction carried using minutia and local binary pattern. Minutia is used to train the model, followed with k means clustering algorithms using Euclidian distance metrics. Local Binary Pattern (LBP) gives the various statistical feature vector. The feature vector model is trained by combing features of the minutia and LBP. Linear discriminant (LD), Ensemble subspace and Support Vector Machine (SVM) are used for classification. It is observed that linear discriminant classifier work better for finger vein classification.

Keywords Gaussian filter · Median filter · Minutia feature · LBP · Machine learning

1 Introduction

In past of years biometric-based personal identification technology can be done using fingerprint, face detection and recognition, iris recognition, etc., but now a days these techniques are common technique because of high use for personal and social security. This finger vein biometric technique empowers more advantageous recognizable proof than other secret phrase or ID cards. A finger vein recognizable proof framework appropriate for acknowledgment process that is sheltered, straightforward and progressively exact [1]. Hence this capturing sensor module can be easily installed at public and private place.

V. M. Sadare (✉) · S. D. Ruikar
Walchand College of Engineering, Sangli, India
e-mail: vinayak.sadare@gmail.com

S. D. Ruikar
e-mail: ruikarsachin@gmail.com

M. Tuba et al. (eds.), *ICT Systems and Sustainability*,
Advances in Intelligent Systems and Computing 1077,
https://doi.org/10.1007/978-981-15-0936-0_52

The principle bit of leeway of this framework is finger vein data cannot be replicated or stolen. Therefore, the finger vein biometrics gives secure and fiducial for an authentication system. The personal identification applications in some special applications, e.g., bank, airport, Aadhaar ID section, we can reduce the hacking attack by using this type of biometry. Thus as of late, finger vein based individual identification innovation become significant research subject. Wang [2] proposes technique for finger vein capture. Its working is based on automated self-adaptive illuminance control. Hsia [3] used different methods of feature extraction from finger vein image such as SURF, SIFT, ORB, BRIEF etc. It represents the feature matching method based on hamming distance. Ajay Kumar [4] proposes a consummate and automatic finger image matching. Author used two techniques for recognition like finger print and finger vein image at the same time. Yang [5] presents various image processing existing method and provide better result assumption. Further it gives brief idea about matching technique based on integration matching. Hu [6] proposes the method of identification using Local Binary Pattern texture feature algorithm.

2 Methodology

Biometric recognition is that the method of creating associate degree identity of a personal supported their physical, chemical or activity characteristics. A biometric system is generally a pattern recognition system that extracts a feature set from the computer file, compares the extracted feature set against feature sets keep within the information, then executes associate degree action supported this result.

The sensor module used to obtain raw biometric data from the user. In the case of vein recognition, it consists of an infrared (IR) camera and an IR illumination source (such as a bank of IRLEDs). Finger vein sensor module consists of two hardware setup first is IR LEDs and IR sensible camera module (see Fig. 1). Infrared light source is used for distinguish the vein pattern inside finger. Hemoglobin inside blood absorbs the infrared radiation hence vein part get visible as compare to surrounding tissue.

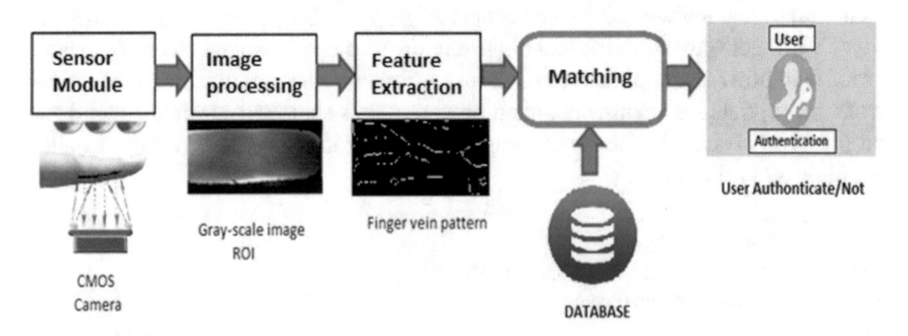

Fig. 1 System architecture/Working flow

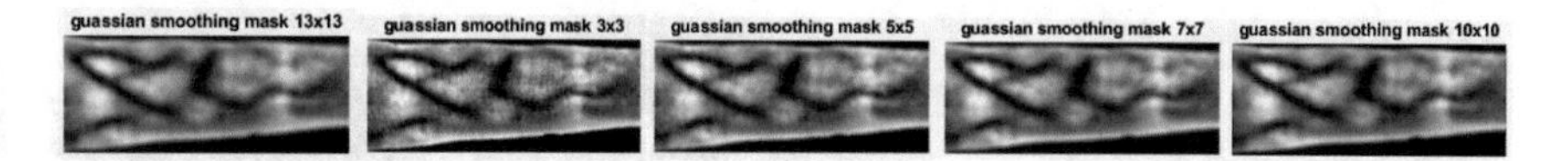

Fig. 2 Gaussian smoothing with 3 × 3, 5 × 5, 7 × 7, 10 × 10 and 13 × 13 mask

This IR spectrum is about 850 nm, which deeply penetrate into tissue, thus allowing the vein pattern, which is suitable to human without hazardous. Feature Extraction module deals with feature extraction from preprocessed image.

Image processing refers to signal processing operations which are performed on images. It is used extensively in the pre-processing and feature extraction stages of the biometric pipeline.

2.1 *Gaussian Smoothing*

The Gaussian smoothing consists an operator that utilizes 2-D and 3-D convolution operator [7]. This type of image consists of random noise and banding noise. It is necessary to remove this type of noise.

Regulation of smoothing is comparable to the mean filter, however it utilizes a unique kernel. The Gaussian operate represents the form of a Gaussian hump. Gaussian smoothing results are explained with mask (see Fig. 2). In 2-D, associate degree identical Gaussian has the form:

$$G(x, y) = \frac{1}{2\pi\sigma^2} e^{\frac{x^2+y^2}{2\sigma^2}} \tag{1}$$

2.2 *Median Filter*

The median filter is basically used to diminish salt and pepper type of noise. A noise present in an image considered as vein or line pattern. Like the Gaussian filter, the median filter considers mask value that define the kernel at which median filtering takes place [7]. The median operation depends on the mask value, the variation of mask value shows different result (see Fig. 3).

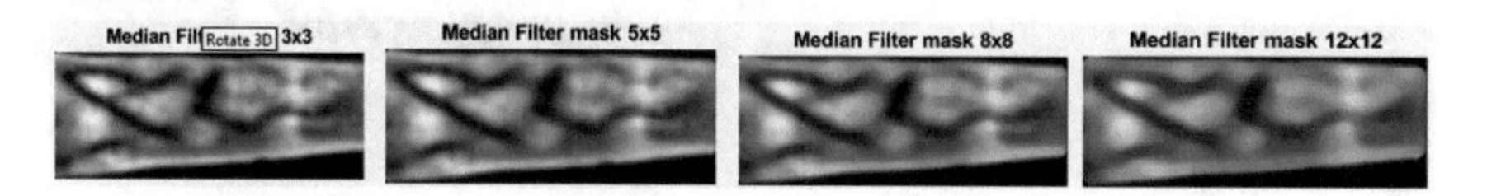

Fig. 3 Median filter with 3 × 3 mask, 5 × 5 mask 8 × 8 mask and 12 × 12 mask

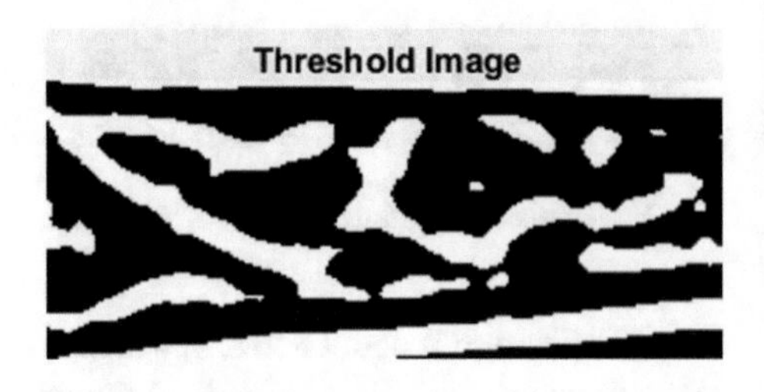

Fig. 4 Thresholded image

2.3 Thresholding of Gray Scale Image

Segmentation involves disuniting a picture into regions equivalent to objects. It involves the distinguishing foreground and background pixels. Thresholding provide result as binary pictures from gray-level ones by setting the threshold with picture element intensity (see Fig. 4). There is another technique of thresholding is Otsu's method.

TH(x, y) is a thresholding for an image with thresholding value T.

$$TH(x, y) = \begin{cases} 1, & f(x, y) \geq T \\ 0, & Otherwise \end{cases} \tag{2}$$

2.4 Dilation

The veins thresholded from the input image contain minor discontinuities. This can lead to bifurcations that extracted incorrectly. To combat this, the image was first dilated (see Fig. 5). It was observed empirically that an appropriate structuring element for the morphological operation was a 3 × 3 matrix of unit value. Dilation has the adverse effect of thickening the veins. However, this was irrelevant as the binary image is thinned immediately afterwards using the Zhang-Suen thinning algorithm.

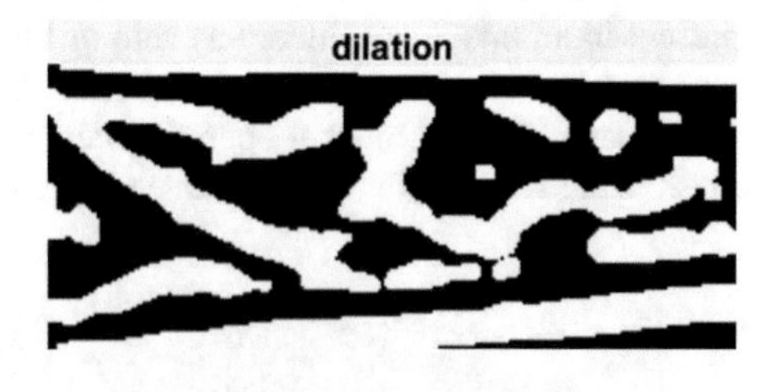

Fig. 5 Dilation operation

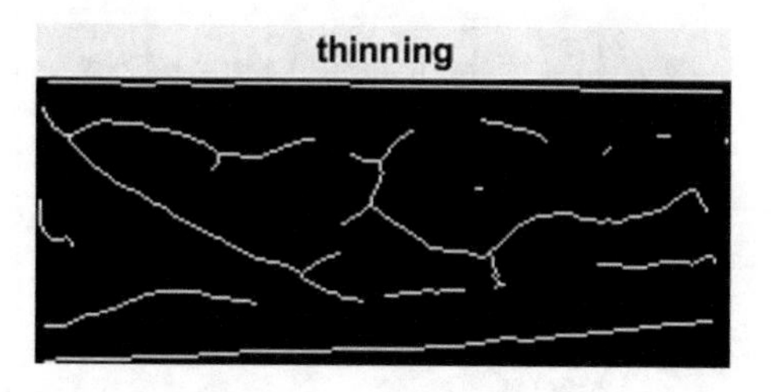

Fig. 6 Thinned image

2.5 Thinning

Morphological thinning method is like erosion or gap operation on image. Thinning of an image deals with removing the foreground pixels from dilated image. Hence this operation gives single line formatted output (see Fig. 6). The thinning procedure given on image x by structuring element y as:

$$\text{thin}(x, y) = x - \text{hit} - \text{and} - \text{miss}(x, y) \tag{3}$$

3 Feature Extraction

The output of the preprocessing stage is a binary image where the white, foreground pixels represent the actual vein. This section deals with extraction of positions of bifurcation points as our first set of features. Bifurcations are the points where one vein splits into two.

3.1 Minutia Points or Bifurcation Points

Minutia points are defined as point where two or more line meet each other (see Fig. 7). Crossing number methodology is generally usually used methodology for detection the bifurcation purpose [5]. This methodology offers higher simplicity and procedure potency. This methodology, works on a dilute binarized image that consists of eight-neighborhood picture element. Estimation of bifurcation points area unit obtained by scanning native neighborhood with three window.

Minutia feature extraction algorithm gives the branch point in the form of (x, y) co-ordinate (see Fig. 8.). This point cannot be trained directly hence k means clustering algorithm is implemented on minutia points. This algorithm extracts centroid of cluster in the form of (x, y) co-ordinate, named as feature vector utilized for training database.

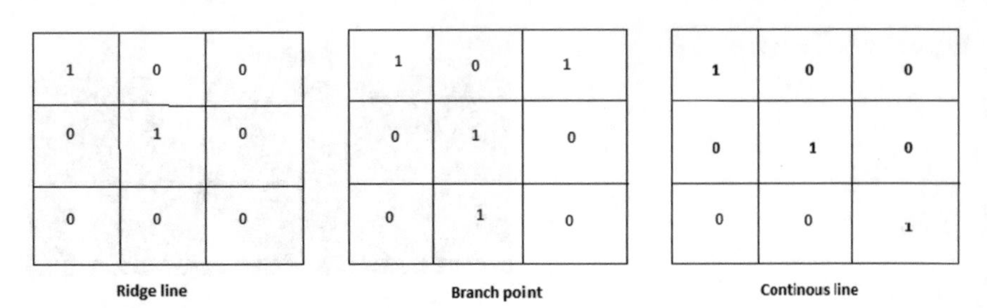

Fig. 7 Minutia points

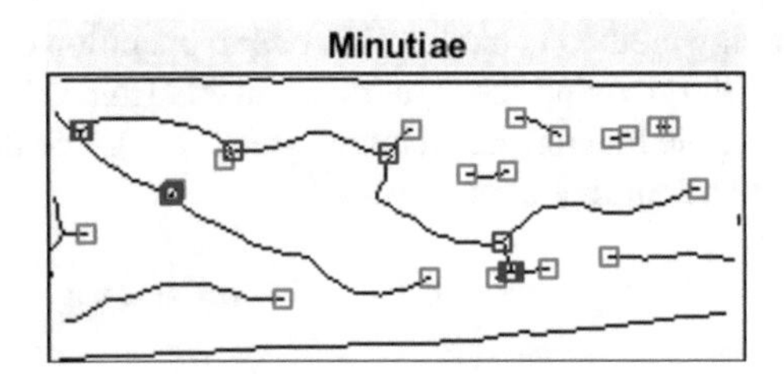

Fig. 8 Minutia feature extraction

3.2 *Local Binary Pattern*

Local Binary Patterns technique commonly utilized as descriptor for the texture analysis. Description in LBP of a pixel in its canonical form is simple and obtained by comparing the pixel's intensity value with eight neighbors to the center pixel intensity (see Fig. 9). These eight values are then obtained by setting threshold with value of the center pixel, and the result is taken in the form of binary pattern.

$$LBP(x, y) = \begin{cases} 1, & x \geq Cp \\ 0, & Otherwise \end{cases} \tag{4}$$

3.2.1 Gray Level Co-occurrence Matrix

The texture analysis approach mostly used in image processing and pattern recognition. This method is categorized into different classes such as statistical, geometrical, model-predicated and signal processing. Statistical feature obtained as

$$\text{Contrast} = \sum_i \sum_j (i - j)^2 P_d(i, j) \tag{5}$$

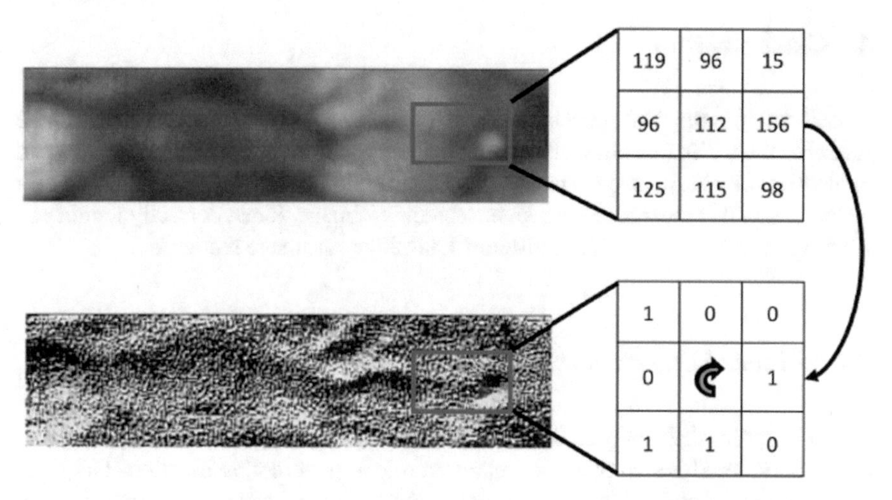

Fig. 9 Local binary pattern

$$\text{Homogeneity} = \sum_{i}\sum_{j} \frac{1}{1+(\mathrm{i}-\mathrm{j})} \mathrm{P_d}\,(\mathrm{i},\mathrm{j}) \tag{6}$$

$$\text{Entropy} = \sum_{i}\sum_{j} P_d(i,\ j) \ln P_d(i,\ j) \tag{7}$$

$$\text{Energy} = \sqrt{ASM} \tag{8}$$

$$\text{ASM} = \sum_{i}\sum_{j} P_d^2(i,j)$$

$$\text{Correlation} = \sum_{i}\sum_{j} P_d(i,j) \frac{(\mathrm{i}-\mu_\mathrm{x})(\mathrm{j}-\mu_\mathrm{y})}{\sigma \mathrm{x}\, \sigma \mathrm{y}} \tag{9}$$

μ and σ are mean and variance

$$\text{Mean} = \mu = \frac{1}{\mathrm{MN}} \sum_{\mathrm{i}}^{\mathrm{M}} \sum_{\mathrm{j}}^{\mathrm{N}} \mathrm{P_d}\,(\mathrm{i},\mathrm{j})^2 \tag{10}$$

$$\text{Standard deviation} = \sqrt{\frac{1}{\mathrm{MN}} \sum\sum (\mathrm{Pd}\,(\mathrm{i},\mathrm{j}) - \mu)\hat{}2} \tag{11}$$

$$\text{Root Mean Square} = \text{RMS} = \frac{\sqrt{\sum |\mu \mathrm{ij}|\hat{}2}}{\mathrm{M}} \tag{12}$$

4 Classification

Classification is the final stage in the biometric system and relies heavily on Machine Learning theory. In this phase, the features obtained from the previous stage are used to identify the query image. The classifier must learn the pattern of the training data so that when it is presented with an unseen query image, it can correctly identify it. Here we describe **Linear Discriminant Classifier** for image recognition.

4.1 Linear Discriminant

Linear Discriminant Analysis (LDA) is most widely utilized as dimensionality reduction technique. This step plays an important role in pattern-classification. The algorithm consists of projection of dataset onto a lower dimensional area with smart class-separability so as vanishing overfitting and additionally cut back machine prices. LDA consists of following steps:

- Calculate the scatter matrices.
- For the scatter matrices, calculate the eigenvectors and corresponding eigenvalues.
- Sort the eigenvectors by decreasing eigenvalues and choose k eigenvectors with the largest eigenvalues to form a $d \times k$ dimensional matrix W.
- Use this $d \times k$ eigenvector matrix to remodel the samples onto the new mathematical space. This will be evaluated by the matrix multiplication:

$$Y = X \times W \tag{13}$$

4.2 Prediction of Class

LDA classifier defines the axes for separates or divides the features (see Fig. 10). Consider example, suppose there are two sets of knowledge points to two completely different categories that have to classify. As shown (see Fig. 11) within the given second graph, once the info point's area unit afore thought on the second plane, there is no line, which will separate classes.

Two criteria area unit employed by LDA to make a replacement axis:

- Maximize the gap between suggests that of the two categories.
- Minimize the variation among every category.

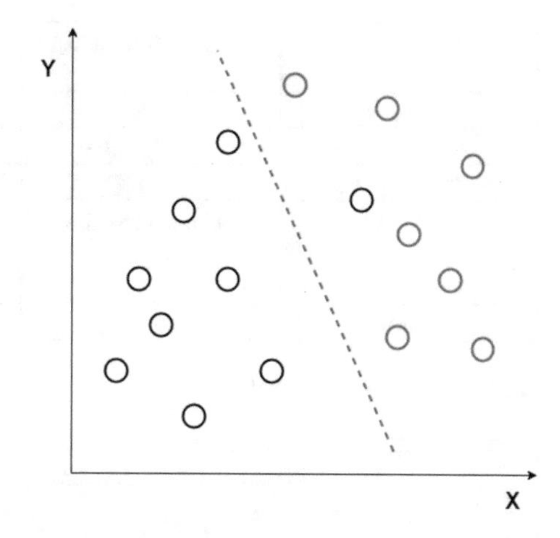

Fig. 10 Prediction of class

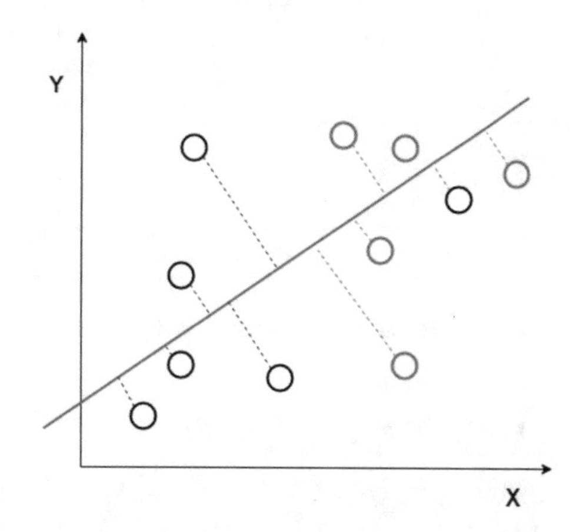

Fig. 11 Prediction of class

5 Result

Feature is converted into feature vector point they are collected in dataset. Dataset consists of 11 features. Feature consists of minutia based feature (two features) and statistical feature (nine features). Minutia based feature consists of number of features in the form n × 2 matrix but this is difficult to store large feature in the dataset. To minimize this matrix we used k-means clustering algorithm on that n × 2 data. Next feature is from Local Binary Pattern algorithm (see Fig. 12). By performing this algorithm we get an output image in the form of Local Binary Pattern. Further this image is converted into GLCMS matrix, which returns the statistical features. Lastly all this features are stored in database. Database consists of twenty class with five sample each of class (see Fig. 13). The GUI of the project shows detail process of execution of the algorithm (see Fig. 14).

Characteristic	Linear Discriminant	Ensemble subspace	Support vector machine
Accuracy	85	84	77
Prediction Speed	1200 obs/sec	4.8641 obs/sec	170 obs/sec.
Training Time	5.9683	4.8641	13.37 sec
Model	Linear discriminant analysis	Ensemble subspace Discriminant	Cubic SVM
PCA	Disabled	Disabled	Disabled

Fig. 12 Classifier performance

Input Samples	Classifiers Result						
	Class	Linear Discriminant		Ensamble Subspace		SVM(Cubic)	
		Detected Class	Accuracy (%)	Detected Class	Accuracy (%)	Detected Class	Accuracy (%)
1	Class 1	1	87	1	83	1	75
2		4	84	1	83	6	71
3	Class 2	2	86	2	88	2	71
4		2	87	2	87	2	74
5	Class 3	3	85	3	86	3	74
6		2	85	2	82	2	73
7	Class 4	4	86	4	84	10	67
8		4	88	4	85	10	67
9	Class 5	5	86	12	85	5	73
10		5	86	5	86	5	73
11	Class 6	6	83	19	84	19	73
12		8	85	6	86	6	72
13	Class 7	7	86	7	87	7	72
14		8	86	8	84	8	72
15	Class 8	8	90	8	85	8	72
16		8	84	8	86	8	72
17	Class 9	9	86	9	84	9	74
18		9	85	9	83	16	73
19	Class 10	10	87	10	86	10	70
20		10	87	10	83	12	70

Input Samples	Classifiers Result						
	Class	Linear Discriminant		Ensamble Subspace		SVM(Cubic)	
		Detected Class	Accuracy (%)	Detected Class	Accuracy (%)	Detected Class	Accuracy (%)
21	Class 11	11	86	11	85	11	70
22		11	86	11	85	11	71
23	Class 12	12	86	12	85	12	73
24		5	85	10	85	10	70
25	Class 13	13	86	13	85	1	74
26		1	90	1	88	1	74
27	Class 14	14	87	1	84	14	72
28		14	85	14	84	14	73
29	Class 15	4	86	15	83	15	73
30		4	88	10	88	10	76
31	Class 16	16	85	16	84	16	69
32		16	84	16	85	16	74
33	Class 17	17	87	17	83	17	73
34		17	83	17	85	17	72
35	Class 18	18	82	18	87	18	73
36		18	84	18	84	18	74
37	Class 19	19	88	19	85	19	74
38		19	87	1	85	9	73
39	Class 20	18	82	1	82	1	72
40		20	89	20	85	20	76

Fig. 13 Different classifiers result analysis

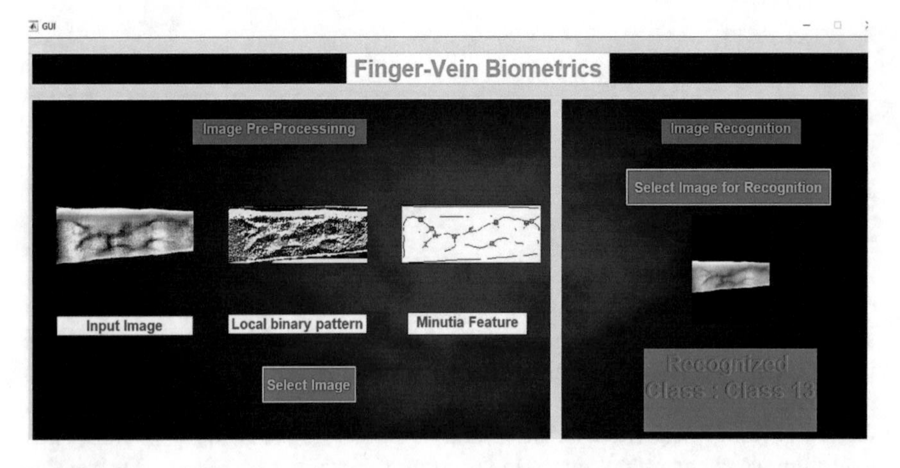

Fig. 14 GUI result

6 Conclusion

The Human identification system using the unique vein pattern of the finger image capturing through sensor module executed in this report. It is observed that the vein patterns are repeated from person to person. The finger vein information taking from infrared sensor is highly secured and not reproducible. Minutia algorithm used to obtain the bifurcation in vein pattern for classification. Local Binary Pattern (LBP) used for training the feature. Linear discriminant (LD), Ensemble subspace and Support Vector Machine (SVM) classifier work better as compare to accuracy near about 85%. The linear discriminant classifier is better as compare to Ensemble subspace and Support Vector Machine (SVM) for finger vein classification.

References

1. Lu, Y., Xie, S.J., Yoon, S., Wang, Z., Park, D.S.: An available database for the research of finger vein recognition. In: 2013 6th International Congress on Image and Signal Processing (CISP), pp. 410–415, Dec 2013
2. Wang, J., Chen, L.: A finger vein image-based personal identification system with self-adaptive illuminance control. **66**(2) (2017)
3. Hsia, C.-H.: New verification strategy for finger-vein recognition system. IEEE Sens. J. **18**(2) (2018)
4. Kumar, A., Zhou, Y.: Human identification using finger images. IEEE Trans. Image Process. **21**(4), 2228–2244 (2012)
5. Yang, L., Yang, G., Yin, Y.: Finger vein recognition with anatomy structure analysis. IEEE (2017). https://doi.org/10.1109/tcsvt.2017.2684833
6. Hu, N., Ma, H., Zhan, T.: A new finger vein recognition method based on LBP and 2DPCA. In: 37th Chinese Control Conference, 25–27 July 2018
7. Sapkale, M.: A finger vein recognition system. In: 2016 Conference on Advances in Signal Processing (CASP), Cummins College of Engineering for Women, Pune, 9–11 June 2016
8. Verma, N.K., et al.: Object matching using speeded up robust features. Intelligent and Evolutionary Systems. Springer International Publishing, pp. 415– 427 (2016)
9. Pug, A., Hartung, D., Busch, C.: Feature extraction from vein images using spatial information and chain codes. Information Security Technical Report, vol. 17, pp. 26–35, Feb 2012
10. Djerouni, A., Hamada, H., Loukil, A., Berrached, N.: Dorsal hand vein image contrast enhancement techniques. Int. J. Comput. Sci. Iss. (IJCSI) **11**(1), 137–142 (2014)
11. Kang, B.J., Park, K.R.: Multimodal biometric method based on vein and geometry of a single finger (2009). https://doi.org/10.1049/iet-cvi.2009.008

Deep Learning Based Vehicle License Plate Recognition System for Complex Conditions

Priya Ingawale and Latika Desai

Abstract Automatic License Plate Recognition (ALPR) system that could be used as a root for several existent world Intelligent Transport System applications. For ALPR difference between jurisdictions on character dimension, spacing and therefore the existence of noise sources like heavy shades, non-uniform brightness, many optical geometries, poor distinction and so on present in license plate images makes it difficult for the recognition accuracy and scalability of ALPR system. The proposed approach offerings a strong and full proof technique with the target of precisely pinpointing vehicle name plates from critical sections in actual world. The approach is deliberate to covenant by blurred vehicle plates, changes happening climate, illumination situations as well as different traffic situations. A unique background removal method is planned to excerpt candidate areas through mainly minimizing the field to be analyzed for license plate selection. To recognize the correct license plate between the candidate areas a gushed license plate identifier based on in lines PNN consuming color saliency structures is presented. For recital assessment, a dataset containing some pictures captured from various sights under different circumstances such as road crossings, roads and express ways, daytime and night-time, numerous climate circumstances and various plate clearness. Various experimentations on the mostly applied vehicle license plate dataset and our recently added dataset proves that the planned tactic significantly overtakes modern strategies in positions of both recognition correctness and run-time efficiency.

Keywords Deep learning · Probabilistic neural network · Automatic license plate recognition

P. Ingawale (✉) · L. Desai
Department of Computer Engineering, Savitribai Phule Pune University, Pune, Maharashtra, India
e-mail: ingawale.priya20@gmail.com

L. Desai
e-mail: latikadesai@gmail.com

M. Tuba et al. (eds.), *ICT Systems and Sustainability*,
Advances in Intelligent Systems and Computing 1077,
https://doi.org/10.1007/978-981-15-0936-0_53

1 Introduction

A vehicle license plate number represents the issue of legal license by the competent authority for the use of public roads. Automatic License Plate Recognition (ALPR) devoid of manual interference or incense is of main importance in the arena of image handling and outline acknowledgment, this is because it plays important role in Intelligent Transport System. Challenges arise when the license plate number and environmental conditions are bad. Some circumstances challenges (Complex Conditions) included in the ALPR process

1. Location: plate number can be anywhere within a frame and allows more than one plate in an image frame.
2. Size and Shape: plate's size depending on the distance of the camera to the street, the plates shape depending on the camera angle to the road.
3. Occlusion: plates could also be coated by dust.
4. Inclination: plates could also be tilted.
5. Illumination: Input vehicle picture could have changed kinds of light, primarily because of ecological lighting or due to vehicle own or other lighting.
6. Color: the color of plate varies according to the type of vehicles and regulations [1].

How the Machine Identify given input image is number or character? One way is to train our machine with lots of images with different types of character and number. Once training is done provided an input image to deep learning algorithm as shown in Fig. 1. It skips the manual steps of extraction of feature, you can directly fed images to deep learning procedure. The algorithm automatically determine the certain feature even if it do not provide, gives certain probabilities and highest probability is answer.

Vehicle license plate is the final identity of vehicle Intelligent Transportation System (ITS) and proper vehicle identification depends on the accuracy of automatic number plate recognition (ANPR) system. Intelligent transport systems plays an vital role in associating smart cities owing to their applications in numerous fields like automatic toll, roads and express ways examinations, urban transportation, car parking in airport, access control, border control, identifying stolen car, electronic toll systems, lane departure warning system and intelligent traffic control system etc.

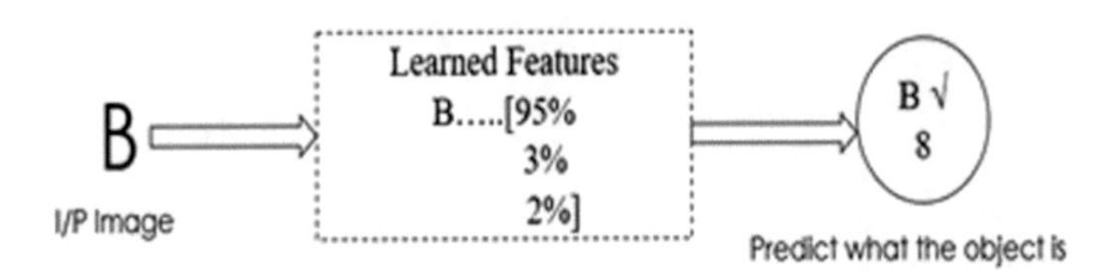

Fig. 1 Deep learning approach for character recognition

2 Literature Survey

This section discussed different methodologies and algorithms of several researchers regarding Automatic License Plate Recognition.

Kakani et al. [2] presented a unified system for recognition of license plate to improve accuracy it is compared with existing technique. Most of the demand is justified by performance evaluation of this system after testing 300 motor vehicle LP images. Though this system takes the large processing time for real-time image processing, the system still incorporates a large scope for additional expansions. Jain and Kundargi [3] planned ANPR system through categorize it according to the features employed in each stage. Future scope involves Multi-plates ANPR system and vehicle producer and model outlining. The feed onward rear broadcast neural network gives an identification amount of 94.12% with treating time (400 ms). Jain et al. [4] proposed an approach of implementing ALPR system using Free Software with Python and Open Computer Vision Library. The future scope involves implementation of system on Open CV library and also does the performance check of the system designed.

Dwivedi et al. [5] demonstrated an approach for identification of license plates of moving vehicles nevertheless further investigation was required on this because of complex backgrounds faced troubles in the recognitions of the license plates of the vehicle throughout the world. Sharma [6] discovered a system for real image capturing of vehicle nameplates and processed using various algorithms through 90 patterns and he has concluded that the normalized cross-correlation technique was extra correct to recognize the number plate than phase-correlation technique besides recognition accuracy of normalized cross correlation was 67.98% and phase correlation was 63.46%. The system was verified through 90 patterns under various circumstances. Qadri and Asif [7] given the automatic vehicle identification method using vehicle license plate. The OCR strategies in their approach were more thoughtful to the mismatch and to dissimilar sizes of name plate outline. Chen et al. [8] discovered a license plate recognition technique for license plate detection of vehicles moving on the streets in different climate conditions through employing a car camera, which was attached on the car itself. Proposed edge outlining and gradient based binarization are used for capturing images at attainable regions in an opening move. Trial results show that the correctness rate of the license plate location and therefore license plate identification are able to do the 91.7% and 88.5% respectively.

Wang and Liui [9] proposed a license plate recognition system using neural network. Test data can be autonomously distinguished over nearly 90% of the license plate components. Sharma et al. [10] researched the technique of mobile cameras utilization for real-world license plate images. Future scope contain to develop a whole system which will show process and recognize vehicle license plate numbers using just a cell phone and application installed on it. Cheang et al. [11] projected a unified ConvNet-RNN model to acknowledge real-world captured license plate images. In future experiments may replace the RNN module with a long short-term memory (LSTM) module. This might improve the performance by keeping the ability to remember the long term dependencies. Panahi and Gholampour [12] discovered a system for license plate recognition through using the data sets of Persian license plates for evaluations, comparison plus expansion of numerous procedures. The accuracy got with this system on dirty plate's portion of is 91.4%. Saleem et al. [13] provided the approach using vertical edge detection algorithmic program. This was with the image normalization technique for removing the unwanted edges where statistical and morphological image processing techniques were used for the LP region is extraction. Template matching algorithms were used for the optical character recognition. This approach provides efficiency 84.8%.

3 Proposed Methodology

3.1 Architecture

Step-1

Image Acquisition: The main task in this system is to detect vehicles and correct localization of number plate. Different conditions like light image, blur image, very high contrast image, multiple images in one frame can be captured during the actual capturing of the images.

Step-2

Image Pre-processing: This stage consist of improving the contrast of the input image, to reducing the noise in the image for enhancing the processing speed, visibility and quality improvement. Different pre-processing algorithm can be applied on the original image. By using different algorithms on image remove noise, blurriness and get a clear view of image.

Step-3

Edge detection: Here the edges of the images and characters will be improved by detecting discontinuities in brightness using canny edge detection.

Step-4

Number Plate Localization: Number plate localization means finding out ROI i.e. Region of Interest from entire image so it will be easier to recognize the number by using only that part of the captured image (Fig. 2).

Step-5

Character Segmentation: Here every character will be segmented without losing the components of character from the captured image. All the alphabets as well as digits of the plate are sectioned then every character or digit is spread in various picture.

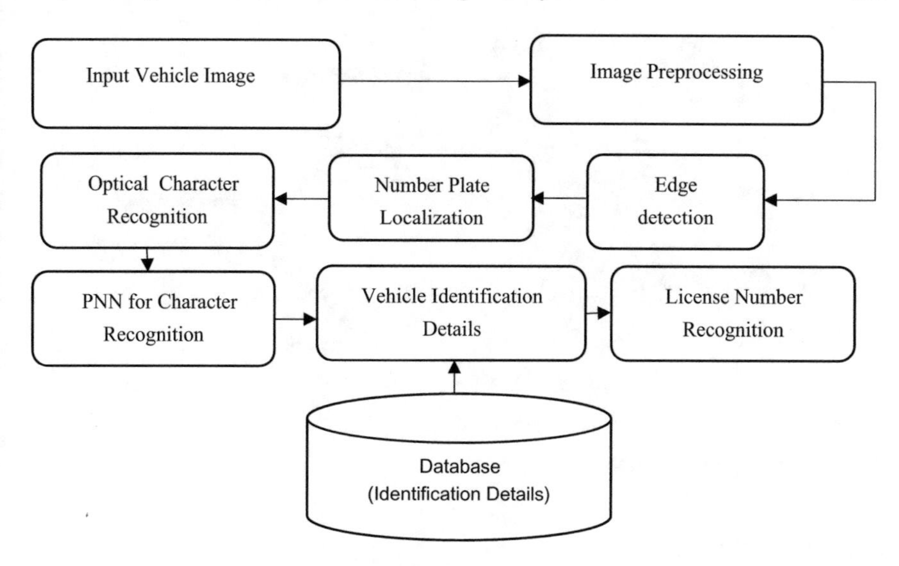

Fig. 2 Block diagram of system architecture

Step-6

Character Recognition: In character recognition stage segmented characters are taken as an input and output of this phase will be displayed as a license plate number. Then at a last character recognition is done by the Probabilistic Neural Networks (PNN). The PNN is used to classify the alphanumeric characters from vehicle license plates built on the data obtained from image processing. PNN often used in classification challenges. The extracted numbers from the character recognition is found out the database and actual vehicle identification details. In PNN, the procedures are structured into a multi-layered feed-forward network with the four layers:

- Input layer
- Pattern layer
- Summation layer
- Output layer

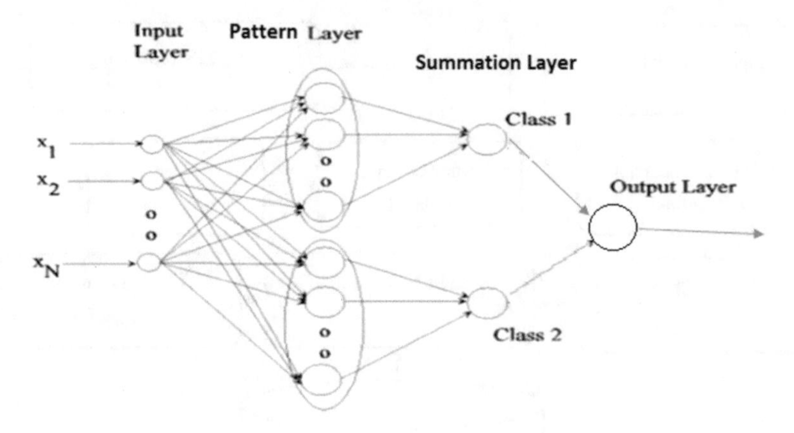

Fig. 3 Block diagram of PNN

3.2 *Algorithms*

Probabilistic Neural Network approach for License Plate Recognition:

1. Once an input is present, the distance from the input vector to the training input vectors are considered in first layer. This produces a vector where its elements identify how nearby the input is to the training input.
2. To sums the involvement for every class of inputs and yields its net outcome as a vector of probabilities second layer is used.
3. Lastly, a compete transfer function on the outcome of the second layer preferences the extreme of these probabilities, and produces a 1 (positive identification) for that class and a 0 (negative identification) for no targeted classes (Figs. 3, 4 and 5).

4 Result and Discussion

Figure 6 demonstrated database of different number of images with various Optical Character Recognition (OCR) algorithm used for simulation compare with proposed system. It furthermore shows successful recognition of number plate which is present in database after applying different optical character recognition algorithms such as Template Matching, Normalized Cross correlation, Phase Correlation, ANN, etc., with recognition rate of number plate from given images and accuracy of system with respect to different OCR algorithm. As shown in figure it is observed that there is still expansion of algorithm require for achieving better accuracy so the Probabilistic neural network approach is used for character identification. Recognition Rate of system depends on database size, change in environmental conditions and quality of images.

Read The new Input Data Xnew

PNN Step-1

Let P feature vector {X*:P=1,...P}Labelled as class 1.
Q Feature Vector {Y':r=1,...R} Labelled as class 2.

Calculate Gauassian Centered of each known Input Vector.

$$g_1(x) = [1/\sqrt{(2\pi\sigma^2)^N}]\exp\{-\|x - x^{(p)}\|^2/(2\sigma^2)\}$$

$$g_2(y) = [1/\sqrt{(2\pi\sigma^2)^N}]\exp\{-\|y - y^{(q)}\|^2/(2\sigma^2)\}$$

PNN Step-2

Sums are defined by,

$$f_1(x) = [1/\sqrt{(2\pi\sigma^2)^N}](1/P)\Sigma_{(p=1,P)} \exp\{-\|x - x^{(p)}\|^2/(2\sigma^2)\}$$

$$f_2(y) = [1/\sqrt{(2\pi\sigma^2)^N}](1/Q)\Sigma_{(q=1,Q)} \exp\{-\|y - y^{(q)}\|^2/(2\sigma^2)\}$$

where **x** is any input feature vector, σ_1 and σ_2 are the spread parameters (standard deviations) for Gaussians in Classes 1 and 2, respectively, N is the dimension of the input vectors, P is the number of center vectors in Class 1 and R is the number of centers in Class 2, $x^{(p)}$ and $y^{(r)}$ are centers in the respective Classes 1 and 2, and $\|x - x^{(p)}\|$ is the Euclidean distance (square root of the sum of squared differences) between **x** and $x^{(p)}$.

PNN Step-3

Select the class with higher conditional probability. Assign the selected class as the class of new input data.

Fig. 4 PNN workflow

Fig. 5 Image preprocessing, localization and recognition of license plate

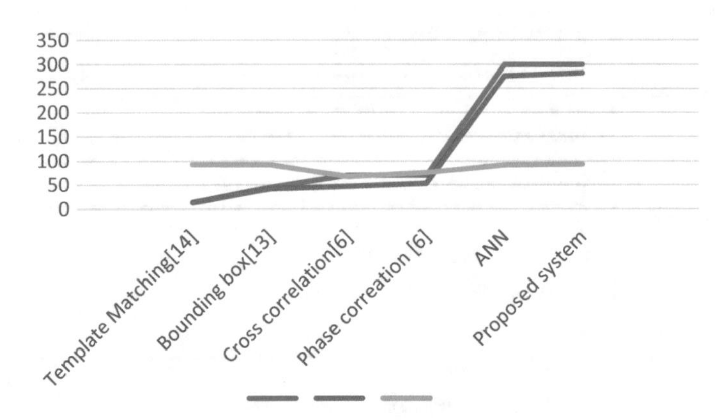

Fig. 6 Analysis and computation of different OCR algorithm

5 Conclusions

For implementation of each sub-part of ALPR system the proposed system declare which algorithms are best suited. In order to reduce computational cost the predefined functions available for pre-processing in Python are to be used for first two steps of implementation. The vehicle image is processed until the vehicle license plate is recognized and the vehicle is identified by taking the image of number plate. This system offers a new and corrective way to license plate finding. The planned method mainly contains three parts: image pre-processing, character extraction means finding out region of interest using edge detection and license plate identification. This technique is able to significantly decline the run-time intricacy of license plate localization without losing discovery preciseness. Efficient image enhancement and deep learning approach enhances the performance of system as it increases accuracy and decreases processing time of image. In future, by adapting deep neural networks for image classification and detection with convolutional layers can get better result.

References

1. Du, S., Ibrahim, M., Shehata, M., Badawy, W.: Automatic license plate recognition (ALPR): a state of the art review. IEEE Trans. Circuits Syst. Video Technol. 1–14 (2011)
2. Kakani, B.V., Gandhi, D., Jani, S.: Improved OCR based automatic vehicle number plate recognition using features trained neural network. In: International Conference on Communication and Network Technology, pp. 1–6. IEEE (2017)
3. Jain, A.S., Kundargi, J.M.: Automatic number plate recognition using artificial neural network. Int. Res. J. Eng. Technol. 1072–1078 (2015)
4. Jai, P., Chopra, N., Gupta, V.: Automatic license plate recognition using openCV. Int. J. Comput. Appl. Technol. Res. 756–761 (2014)
5. Dwivedi, U., Rajput, P., Sharma, M.K.: License plate recognition system for moving vehicles using Laplacian edge detector and feature extraction. Int. Res. J. Eng. Technol. 407–412 (2017)

6. Sharma, G.: Performance analysis of vehicle number plate recognition system using template matching techniques. J. Inf. Technol. Softw. Eng. **8**(2), 1–9 (2018)
7. Qadri, M.T., Asif, M.: Automatic number plate recognition system for vehicle identification using optical character recognition. In: International Conference on Education Technology and Computer, pp. 335–338. IEEE (2009)
8. Chen, C.-H., Chen, T.-Y., Wu, M.-T., Tang, T.-T., Hu, W.-C.: License plate recognition for moving vehicles using a moving camera. In: International Conference on Intelligent Information Hiding and Multimedia Signal Processing, pp. 497–500. IEEE (2013)
9. Wang, C.-M., Liui, J.-H.: License plate recognition system. In: International Conference on Fuzzy Systems and Knowledge Discovery, pp. 1708–1710 (2015)
10. Sharma, A., Dharwadker, A., Kasar, T.: MobLP: a CC-based approach to vehicle license plate number segmentation from images acquired with a mobile phone camera. In: 2010 Annual IEEE India Conference (INDICON), pp. 1–4. IEEE (2010)
11. Cheang, T.K., Chong, Y.S., Tay, H.: Segmentation-free vehicle license plate recognition using ConvNet-RNN. In: International Workshop on Advanced Image Technology, pp. 1–5 (2017)
12. Panahi, R., Gholampour, I.: Accurate detection and recognition of dirty vehicle plate numbers for high-speed applications. IEEE Trans. Intell. Transp. Syst. 767–769 (2016)
13. Saleem, N., Hassam, M., Tahi, H.M., Farooq U.: Automatic license plate recognition using extracted features. In: 2016 4th International Symposium on Computational and Business Intelligence (ISCBI), pp. 221–225 (2016). IEEE
14. Babu, K.M., Raghunadh, M.V.: Vehicle number plate detection and recognition using bounding box method. In: International Conference on Advanced Communication Control and Computing Technologies (ICACCCT), pp. 106–110 (2016)

RAAGANG—A Proposed Model of Tutoring System for Novice Learners of Hindustani Classical Music

Kunjal Gajjar and Mukesh Patel

Abstract The primary objective of work is to build an expert evaluation and feedback system which can be used as tutoring software for novice learners of Hindustani Classical Music (HCM). Computational Musicology from the perspective of HCM deals with Raga recognition and music generation, however to the best of our knowledge there is no work to understand the structure of Raga and evaluate rendered song based on Raga conventions. We propose a model as tutoring system which listens to the vocal input of the singer and evaluate it based on the rules/conventions of Raga. Further, the expert system also provides correct suggestions for the error note. In our model we propose the use of transition matrix for note evaluation and variable length n-gram modeling for error correction. The performance evaluation of error correction module is done using accuracy and aesthetic correctness scale.

Keywords Computational musicology · Tutoring system · Hindustani classical music · Note evaluation · Error correction · Raga structure

1 Introduction

The area of Artificial Intelligence which deals with music and computation is referred as computational musicology. The research in this area mainly focuses on the music generation, genre recognition, pitch-tempo analysis, music information retrieval, voice separation, music grammar and Hierarchical structure, classification patterns etc. [1–3].

The western music is under the gaze of computational musicology since long. However, Indian Classical Music (ICM) is different from Western Music structurally [4]. Also, as ICM allows the singer to improvise the composition at the runtime,

K. Gajjar (✉)
Ahmedabad University, Ahmedabad, Gujarat, India
e-mail: kunjal.gajjar@ahduni.edu

M. Patel
Charotar University of Science and Technology, Changa, Gujarat, India
e-mail: mukeshpatel.mca@charusat.ac.in

M. Tuba et al. (eds.), *ICT Systems and Sustainability*,
Advances in Intelligent Systems and Computing 1077,
https://doi.org/10.1007/978-981-15-0936-0_54

recognition/evaluation of song becomes a difficult task. ICM has two variations: Hindustani Music practiced in northern part of India and Carnatic Music practiced in southern part of India. This paper deals with Hindustani Music where Raga and Taal are most important aspects of composition.

Hindustani Classical Music follows "Guru Shishya Padhati" learning method. The student practices different Ragas in presence of teacher. However, in today's period, it is difficult to practice music constantly in the presence of teacher. In general scenario, after attending the music session, student practices on his own. There are high chances of making mistakes, especially if the student is an early stage learner [4].

In this paper, we propose a model which can work as a tutoring system for the novice learner of Hindustani Music. The student can render a song, which can be evaluated for the correctness following the rules and conventions of Hindustani Classical Music (HCM). Also, using error correction technique, the suggestion for improvisation can be provided to the student.

1.1 Study of Previous Work

The existing systems in context with Indian Classical music mostly target the area of music recognition and composition as explained below:

In a paper, authors have proposed a system named TANSEN for recognition of Raga. The system is implemented for two Ragas: Yaman Kalyan and Bhupali. Authors have used note duration heuristics and hill peak duration for note transcription. Further, they have used Hidden Markov Model for plain Raga identification and Pakkad matching for improvising the results of Raga identification [5].

A model is proposed for Raga identification from notes played on Harmonium. The note identification module performed very well as the frequencies of notes played by the instrument do not fluctuate as would be the case of vocal. The model is designed for identifying eight Ragas. The authors have created database of all the possible combinations of transition for a Raga. The identified notes (sequence) are compared with the raga database with a simple template matching algorithm [6].

Further, in another work, computer-aided generation of Hindustani Classical music is designed considering Aroha-Avaroha feature of the Raga. The system generates note sequence using Finite State Model designed considering the grammar of a Raga. The algorithm gives better results compared to random generation of notes, however the results are aesthetically weak and do not match the compositions generated by musician [7].

In another research work, authors have used n-Gram modeling of Tabla sequences for compositions and improvisation [8]. They have considered note and duration as multiple viewpoints and predicted the next tabla note in the sequence. There is a model on similar concept for North Indian Classical vocal compositions [9]. The research focuses on Multiple Viewpoint models to predict the next note in the sequence.

2 Conventions of HCM Used in Proposed Model

It is important to understand that any composition of Indian Classical Music (ICM) is based on the Octave (Saptak) which is comprised of 12 notes (swar) which are as shown in Table 1. These notes are the basic and most important component of a Raga.

They form the structure of a Raga which is not just the sequence of notes. There are several rules and conventions that form the structure of Raga in HCM. Some of the rules which are used in the proposed model are as follows [4, 10]:

- Aroha-Avaroha—The ascending and descending order of the notes in the Raga. This helps in understanding how a Raga moves in the ascending/descending order. There are scenarios where a note is allowed in the ascending order of the Raga but cannot be sung if composition is moving in the descending order. For example in Raga Bhimpalasi,
  - Aroha—Sa Ga Ma Pa Ni Sa'
  - Avaroha—Sa' Ni Dha Pa Ma Ga Re Sa

Table 1 Indian notes along with the frequency ratio [4]

Sr. no.	Notes	Indian notation	Frequency ratio (natural)
1	Shadaj	Sa	1/1
2	Komal Rishabh	Re	16/15
3	Shuddha Rishabh	Re	9/8
4	Komal Gandhar	Ga	6/5
5	Shuddha Gandhar	Ga	5/4
6	Shuddha Madhyam	Ma	4/3
7	Tivra Madhyam	Ma'	7/5
8	Pancham	P	3/2
9	Komal Dhaivat	Dh	8/5
10	Shuddha Dhaivat	Dh	5/3
11	Komal Nishad	Ni	9/5
12	Shuddha Nishad	Ni	15/8
13	Shadaj (Next Octave)	Sa' (Next octave)	2/1

- Varjit swars—These are the notes which are not allowed in rendition of Raga. For example in Raga Bhupali, notes Ma and Ni are varjit swars. Hence, they are not allowed in the compositions of Raga Bhupali.
- Pakkad—These are the prominent phrases (sequence of notes) which describe a Raga. They bring out the aesthetic beauty of a Raga. They also differentiate it from other Ragas composed from the same set of notes. For example, "Pa Ga Dha Pa Ga Re Ga Re Sa" is a pakkad of Raga Bhupali.
- Vadi swar—It is the king note of a Raga. The note is generally (not compulsory) used with higher frequency compared to other notes.
- Samvadi swar—This is the queen note. It is used with second highest frequency in the Raga.
- Vivadi swars—These are the notes which are not allowed in a Raga composition. However, what differentiates them from varjit swar is that they might be used by experts at times to improve the aesthetic feel of the Raga [11].

3 Architecture of Proposed System

The proposed system is a tutoring system for learners of Hindustani Classical music which accepts the audio input from the user and evaluates it according to the conventions of Raga as explained in Sect. 2. It also suggests the correction of errors identified. The architecture of system is graphical represented in Fig. 1.

There are three sub-modules in the System: Note Identification, Note Evaluation and Error Correction. These sub modules use frequency ratios for note identification, transition matrix for note evaluation and training set which comprises of Raga compositions for error correction.

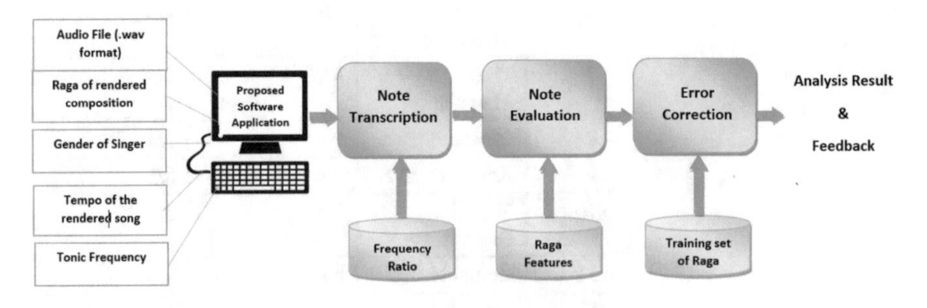

Fig. 1 Overall system architecture

3.1 Input

The proposed system takes properly trimmed (no extra blank space in starting Audio file (in.wav format) as input. Along with it, user also provides name of Raga for which the rendered song is to be evaluated, tempo of song for segmentation, and taal of rendered song for display is also inputted. There are two ways in which tonic frequency i.e. frequency of 'Sa' of middle octave can be entered. The singer can either provide direct input for tonic frequency in Hertz or it can be identified based on the scale of tanpura and gender of the singer. It is very important to select correct tonic frequency as based on this frequency, rest of the note are identified from the audio file.

3.2 Note Identification

This module accepts the audio file as input and generates list of notes using the note duration heuristics. The frequency list is generated based on the autocorrelation algorithm used in PRAAT for pitch listing [12]. For the purpose of implementation, the minimum note duration is fixed to one-fourth of a single beat. Based on the tempo entered by the user, the frequency list is fragmented and notes are identified from the fragments. The approach of note duration heuristics is described in detail in System for automatic Raga identification 5. Figure 2 shows the sub modules of Note Identification.

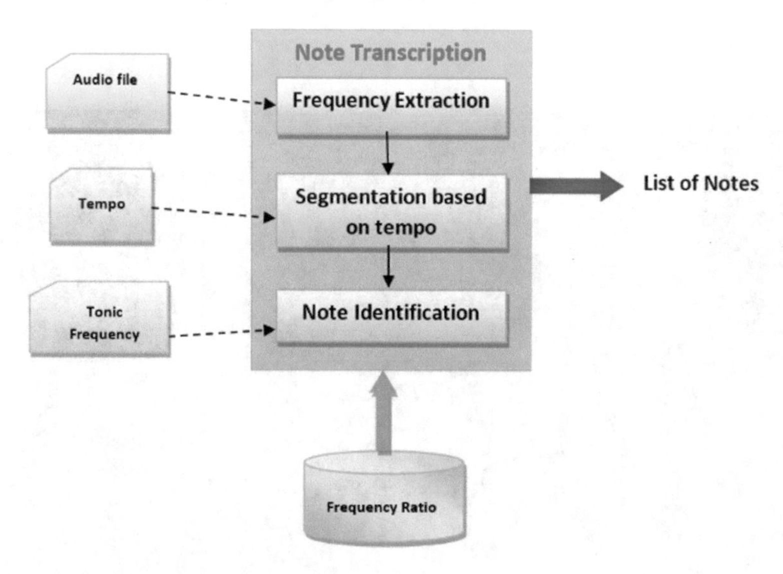

Fig. 2 Note identification module

3.3 Note Evaluation

This module accepts the list of notes generated from note identification module. There are two phases of evaluation as shown in Fig. 3 [13]. In first phase, the notes are evaluated based on aroha-avaroha and varjit swar rules of a Raga. These rules are represented in a form of 36 × 36 transition matrix for each Raga. These 36 rows and columns represent the notes of 3 octaves: Lower, Middle and Upper. The sequence of notes is evaluated for varjit swars, vivadi swars and transition of notes according to aroha-avaroha. Figure 4 shows the transition table generated for Bhupali Raga where 'X' value in the cell represents non-possibility of transition and 'Y' represents possibility of a transition [13].

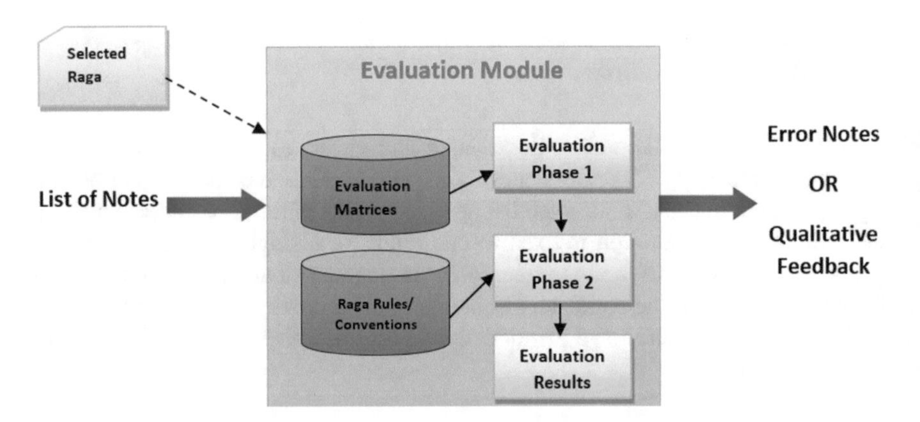

Fig. 3 Note evaluation module

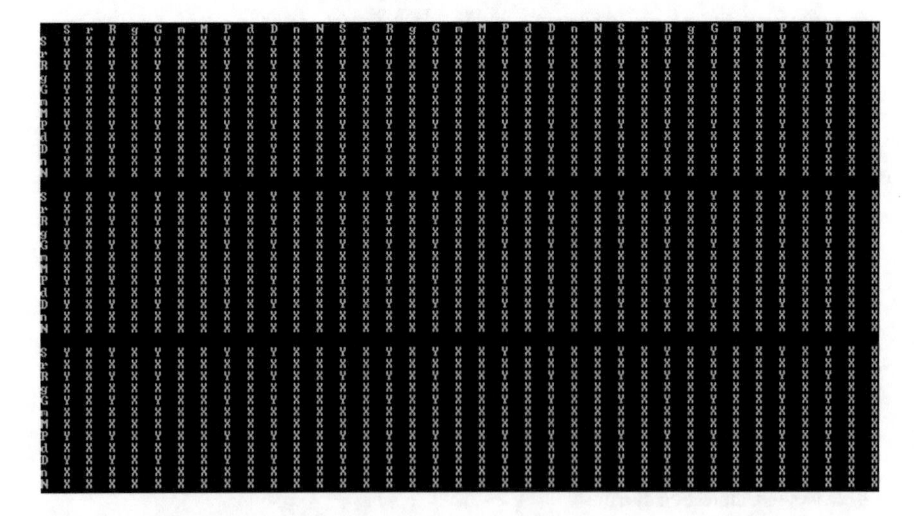

	S	r	R	g	G	m	M	P	d	D	n	N	S	r	R	g	G	m	M	P	d	D	n	N	S	r	R	g	G	m	M	P	d	D	n	N
S	Y	X	Y	X	Y	X	X	Y	X	Y	X	X	Y	X	Y	X	Y	X	X	Y	X	Y	X	X	Y	X	Y	X	Y	X	X	Y	X	Y	X	X
r	X	X	X	X	X	X	X	X	X	X	X	X	X	X	X	X	X	X	X	X	X	X	X	X	X	X	X	X	X	X	X	X	X	X	X	X
R	Y	X	Y	X	Y	X	X	Y	X	Y	X	X	Y	X	Y	X	Y	X	X	Y	X	Y	X	X	Y	X	Y	X	Y	X	X	Y	X	Y	X	X
g	X	X	X	X	X	X	X	X	X	X	X	X	X	X	X	X	X	X	X	X	X	X	X	X	X	X	X	X	X	X	X	X	X	X	X	X
G	Y	X	Y	X	Y	X	X	Y	X	Y	X	X	Y	X	Y	X	Y	X	X	Y	X	Y	X	X	Y	X	Y	X	Y	X	X	Y	X	Y	X	X
m	X	X	X	X	X	X	X	X	X	X	X	X	X	X	X	X	X	X	X	X	X	X	X	X	X	X	X	X	X	X	X	X	X	X	X	X
M	X	X	X	X	X	X	X	X	X	X	X	X	X	X	X	X	X	X	X	X	X	X	X	X	X	X	X	X	X	X	X	X	X	X	X	X
P	Y	X	Y	X	Y	X	X	Y	X	Y	X	X	Y	X	Y	X	Y	X	X	Y	X	Y	X	X	Y	X	Y	X	Y	X	X	Y	X	Y	X	X
d	X	X	X	X	X	X	X	X	X	X	X	X	X	X	X	X	X	X	X	X	X	X	X	X	X	X	X	X	X	X	X	X	X	X	X	X
D	Y	X	Y	X	Y	X	X	Y	X	Y	X	X	Y	X	Y	X	Y	X	X	Y	X	Y	X	X	Y	X	Y	X	Y	X	X	Y	X	Y	X	X
n	X	X	X	X	X	X	X	X	X	X	X	X	X	X	X	X	X	X	X	X	X	X	X	X	X	X	X	X	X	X	X	X	X	X	X	X
N	X	X	X	X	X	X	X	X	X	X	X	X	X	X	X	X	X	X	X	X	X	X	X	X	X	X	X	X	X	X	X	X	X	X	X	X
S	Y	X	Y	X	Y	X	X	Y	X	Y	X	X	Y	X	Y	X	Y	X	X	Y	X	Y	X	X	Y	X	Y	X	Y	X	X	Y	X	Y	X	X
r	X	X	X	X	X	X	X	X	X	X	X	X	X	X	X	X	X	X	X	X	X	X	X	X	X	X	X	X	X	X	X	X	X	X	X	X
R	Y	X	Y	X	Y	X	X	Y	X	Y	X	X	Y	X	Y	X	Y	X	X	Y	X	Y	X	X	Y	X	Y	X	Y	X	X	Y	X	Y	X	X
g	X	X	X	X	X	X	X	X	X	X	X	X	X	X	X	X	X	X	X	X	X	X	X	X	X	X	X	X	X	X	X	X	X	X	X	X
G	Y	X	Y	X	Y	X	X	Y	X	Y	X	X	Y	X	Y	X	Y	X	X	Y	X	Y	X	X	Y	X	Y	X	Y	X	X	Y	X	Y	X	X
m	X	X	X	X	X	X	X	X	X	X	X	X	X	X	X	X	X	X	X	X	X	X	X	X	X	X	X	X	X	X	X	X	X	X	X	X
M	X	X	X	X	X	X	X	X	X	X	X	X	X	X	X	X	X	X	X	X	X	X	X	X	X	X	X	X	X	X	X	X	X	X	X	X
P	Y	X	Y	X	Y	X	X	Y	X	Y	X	X	Y	X	Y	X	Y	X	X	Y	X	Y	X	X	Y	X	Y	X	Y	X	X	Y	X	Y	X	X
d	X	X	X	X	X	X	X	X	X	X	X	X	X	X	X	X	X	X	X	X	X	X	X	X	X	X	X	X	X	X	X	X	X	X	X	X
D	Y	X	Y	X	Y	X	X	Y	X	Y	X	X	Y	X	Y	X	Y	X	X	Y	X	Y	X	X	Y	X	Y	X	Y	X	X	Y	X	Y	X	X
n	X	X	X	X	X	X	X	X	X	X	X	X	X	X	X	X	X	X	X	X	X	X	X	X	X	X	X	X	X	X	X	X	X	X	X	X
N	X	X	X	X	X	X	X	X	X	X	X	X	X	X	X	X	X	X	X	X	X	X	X	X	X	X	X	X	X	X	X	X	X	X	X	X
S	Y	X	Y	X	Y	X	X	Y	X	Y	X	X	Y	X	Y	X	Y	X	X	Y	X	Y	X	X	Y	X	Y	X	Y	X	X	Y	X	Y	X	X
r	X	X	X	X	X	X	X	X	X	X	X	X	X	X	X	X	X	X	X	X	X	X	X	X	X	X	X	X	X	X	X	X	X	X	X	X
R	Y	X	Y	X	Y	X	X	Y	X	Y	X	X	Y	X	Y	X	Y	X	X	Y	X	Y	X	X	Y	X	Y	X	Y	X	X	Y	X	Y	X	X
g	X	X	X	X	X	X	X	X	X	X	X	X	X	X	X	X	X	X	X	X	X	X	X	X	X	X	X	X	X	X	X	X	X	X	X	X
G	Y	X	Y	X	Y	X	X	Y	X	Y	X	X	Y	X	Y	X	Y	X	X	Y	X	Y	X	X	Y	X	Y	X	Y	X	X	Y	X	Y	X	X
m	X	X	X	X	X	X	X	X	X	X	X	X	X	X	X	X	X	X	X	X	X	X	X	X	X	X	X	X	X	X	X	X	X	X	X	X
M	X	X	X	X	X	X	X	X	X	X	X	X	X	X	X	X	X	X	X	X	X	X	X	X	X	X	X	X	X	X	X	X	X	X	X	X
P	Y	X	Y	X	Y	X	X	Y	X	Y	X	X	Y	X	Y	X	Y	X	X	Y	X	Y	X	X	Y	X	Y	X	Y	X	X	Y	X	Y	X	X
d	X	X	X	X	X	X	X	X	X	X	X	X	X	X	X	X	X	X	X	X	X	X	X	X	X	X	X	X	X	X	X	X	X	X	X	X
D	Y	X	Y	X	Y	X	X	Y	X	Y	X	X	Y	X	Y	X	Y	X	X	Y	X	Y	X	X	Y	X	Y	X	Y	X	X	Y	X	Y	X	X
n	X	X	X	X	X	X	X	X	X	X	X	X	X	X	X	X	X	X	X	X	X	X	X	X	X	X	X	X	X	X	X	X	X	X	X	X
N	X	X	X	X	X	X	X	X	X	X	X	X	X	X	X	X	X	X	X	X	X	X	X	X	X	X	X	X	X	X	X	X	X	X	X	X

Fig. 4 Transition matrix of Raga Bhupali

After first phase of evaluation, if no errors are identified, second phase of evaluation is performed. In second phase, vadi, samvadi, and pakkad features of a Raga are considered. Based on these features it is verified if the song belongs to the selected Raga and not to the Raga having similar set of notes and aroha-avaroha sequence.

3.4 Context Sensitive Error Correction

This module accepts the error list generated in the note evaluation module. It is responsible for replacing the error note there by suggesting the correct note. The suggestions for correction are context sensitive as it depends on the position of note in the sequence. Unlike generation algorithms which considers only past sequence of notes to generate n-gram model and predict next note from it, this module uses past sequence of notes as well as future sequence of notes to identify the context of error note and accordingly provide suggestion.

As shown in Fig. 5, short-term model (STM) and long term model (LTM) predictions are generated for correct note. The long term model is generated from the training set of each Raga. The training set for each Raga comprises of 15 compositions with average 1300 notes. The short term model uses the test composition itself for generating n-gram model. The long term model gives general view of the Raga

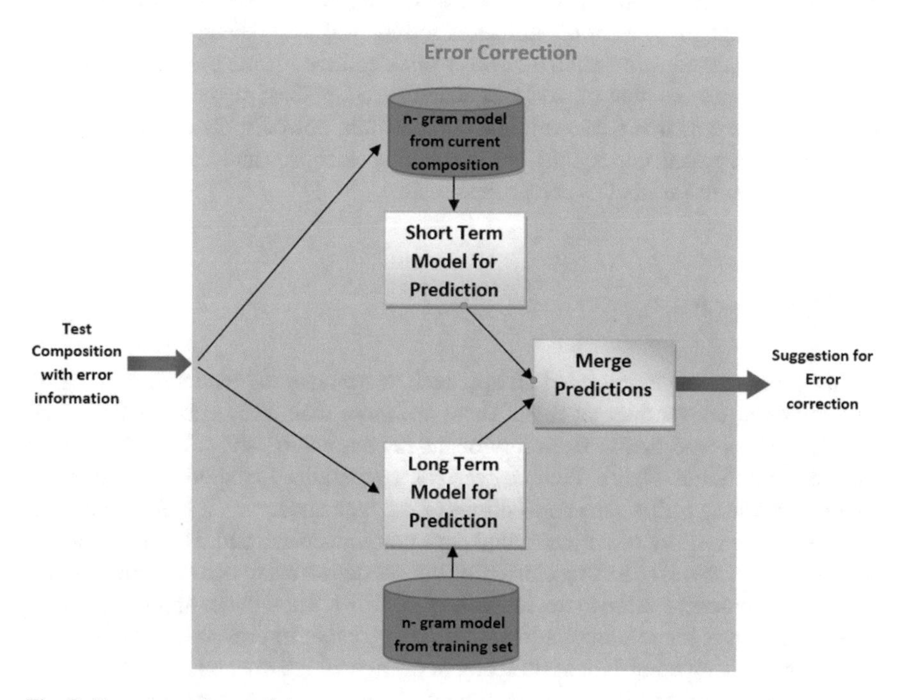

Fig. 5 Error correction module

whereas short term model focuses on single composition there by giving detailed view of it. Using the training set and test composition, a variable length n-gram model is generated based on which the suggestion for correct note is provided to the singer. The predictions made by models are merged considering the probability of notes and Euclidian distance of the suggested note from the identified wrong note. If STM and LTM give different predictions for the wrong note, first the probabilities of the suggestions are considered. If difference between the values of probabilities generated from STM and LTM is ≤ 1, the Euclidian distance of suggestions from the wrong note is calculated and nearest suggestion is considered in such scenario.

4 Results

This section demonstrates the results achieved by implementing the prototype of proposed model. The performance of Note Identification module is tested by providing 40 recordings as input and comparing machine generated list of notes with expert's identified notes by using BLEU Score Evaluator. It is observed that 89.89% correct note transcription is achieved. For note evaluation, 180 compositions in form of text files for 8 different Ragas (Bhupali, Saarang, Kafi, Khamaj, Bhairavi, Desh, Bhimpalasi, Durga) which comprised of errors in terms of aroha-avaroha sequence and varjit swar are generated. They are tested using the transition matrix. The test results show that 100% error identification is achieved with 0% false positives. The error correction module is tested for 180 errors identified in note evaluation module. The results concludes that on scale of accuracy, 56.31% of suggested corrections are exactly same as that in the original composition, however, for aesthetic beauty 70.96% of suggested corrections are excellently adequate and 25.08% are fairly adequate as per the aesthetics correctness scale.

5 Conclusion

In this paper, we have presented an approach to evaluate the rendered song based on features or conventions of raga. There are more than 200 ragas in Hindustani Classical Music and due to flexibility of the raga structure, songs from same raga may sound different. Hence, there is need for a generalized method for checking if the rendered song follows the conventions of the raga under which it is categorized.

Given the enormity of a system that is as comprehensive and dynamic as a live teacher, we necessarily had to constraint our model to what was feasible and yet would be of practical value to novice learner of HCM. Though our approach is very general and gives the required output, there are scope of improvements as there are restriction to the type of input and features of raga which are considered. First, as the audio is segmented based on the tempo of the song, Alap (opening section of the song which is unmetered) [14] and taan (singing of very rapid melodic phrases where duration of note can be less than one-fourth of the beat) [15] are not considered. Second, feature of raga such as meend [16] (slide from one note to other) which might use vivadi and varjit swars for a fraction of beat is ignored. However, although there are constraints, the system would work reasonably well for novice learners of music assuming that they would not be involved in the advance practices such as meend, alap or use of vivadi swars for improvisation of a Raga (generally used by experts).

As future enhancement, considering duration of note along with its position in sequence can give better results for error correction. Lastly, RAAGANG -the proposed system can be used as tutoring system for novice learners and also by music academy to evaluate early stage learners.

References

1. Agarwal, P., Karnick, H., Raj, B.: A comparative study of Indian and Western music forms. In: ISMIR, pp. 29–34 (2013)
2. Scaringella, N., Zoia, G., Mlynek, D.: Automatic genre classification of music content: a survey. IEEE Signal Process. Mag. **23**(2), 133–141 (2006)
3. Meredith, D. (ed.): Computational Music Analysis. Springer (2016)
4. Gajjar, K., Patel, M.: Computational musicology for Raga analysis in Indian classical music: a critical review. Int. J. Comput. Appl. (2017)
5. Pandey, G,, Mishra, C., Ipe, P.: TANSEN: a system for automatic Raga identification. In: IICAI 2003, pp. 1350–1363
6. Pendekar, R., Mahajan, S.P., Mujumdar, R., Ganoo, P.: Harmonium raga recognition. Int. J. Mach. Learn. Comput. **3**(4), 352 (2013)
7. Das, D., Choudhury, M.: Finite state models for generation of Hindustani classical music. In: Proceedings of International Symposium on Frontiers of Research in Speech and Music 2005, pp. 59–64
8. Chordia, P., Sastry, A., Mallikarjuna, T., Albin, A.: Multiple viewpoints modeling of tabla sequences. In: ISMIR 2010, vol. 2010, p. 11
9. Srinivasamurthy, A., Parag, C.: Multiple viewpoint modeling of north Indian classical vocal compositions. In: Proceedings of the International Symposium on Computer Music Modeling and Retrieval (2012)
10. Indian Classical Music and Sikh Kirtan. http://fateh.sikhnet.com/sikhnet/gurbani.nsf/d9c75ce4db27be328725639a0063aecc/085885984cfaafcb872565bc004de79f!OpenDocument. Accessed 11 Feb 2015
11. Vivadiswar. http://www.ragopedia.com/raga/vadi.html. Accessed 10 Nov 2018
12. PRAAT. http://www.fon.hum.uva.nl/praat/. Accessed 10 Oct 2018
13. Gajjar, K., Patel, M.: A matrix based approach for evaluation of vocal renditions in Hindustani classical music. In: Proceedings of the 5th International Conference on Computational Intelligence & Communication Technology (2019)
14. Alpa. https://en.wikipedia.org/wiki/Alap. Accessed 22 Mar 2018
15. Taan. https://en.wikipedia.org/wiki/Taan_(music). Accessed 22 Mar 2018
16. Meend. http://www.itcsra.org/meend. Accessed 10 Nov 2018

Optimizing Root Cause Analysis Time Using Smart Logging Framework for Unix and GNU/Linux Based Operating System

Vyanktesh S. Bharkad and Manik K. Chavan

Abstract The computer activity records are used for statistical purposes, backup, recovery, and root cause analysis of failure on application. These records are referred as a log. The log files are written for recording incoming dialogs, debug, error, status of an application and certain transaction details, by the operating system or other control program. The logs generated by an application that can be referred by user that may be helpful in the event of failure. For example, in a file transfer, FTP program generates a log file consist of date, time, source and destination, etc. In this number of logs generated by the application uses too much disk space. If the logging is tuned down (e.g., by lowering the log level) then the disk space usage is less, but then not enough information is available for debugging issue. To address this problem we proposed a Smart Logging Framework. The Smart Logging Framework provides the feature such as In-memory logging, In-memory packet capturing and Zoom-in log viewer.

Keywords RCA · Root cause analysis · In-memory logging

1 Introduction

In a Unix, Linux-based operating system the logging plays an important role. Logging to the file by an operating system OS is helpful tool for analyzing the behavior of computer system, other OS components and event happened as per time stamp. In order to identify root cause of particular incidents, system/application failure user need to analyze a sequence of log, such that users can see whether the computer system behave properly or not. If we consider the case for web server, file server or application server that works with computer network, logs and network packet captured .pcap files are even more important and the only way to find the root cause and analyze the system problems. In this paper, we are focusing on how to optimize

V. S. Bharkad (✉) · M. K. Chavan
Department of Computer Science and Engineering, Walchand College of Engineering, Sangli 416415, Maharashtra, India
e-mail: vyanky.bharkad@gmail.com

M. Tuba et al. (eds.), *ICT Systems and Sustainability*,
Advances in Intelligent Systems and Computing 1077,
https://doi.org/10.1007/978-981-15-0936-0_55

the RCA (Root Cause Analysis) time by improving earlier logging mechanism. Traditional Unix/Linux applications log messages are directly written to disk whenever an event occurred. There are certain levels of logs that are as follows:

Level 1 Alert: Alert level messages indicate that respective one should be corrected immediately.
Level 2 Critical: Critical level messages, is written in the log file when a critical situation occurs in the normal execution of the system.
Level 3 Error: Occurred error information is shown in this kind of log messages.
Level 4 Warning: Warning messages indicate that an error may occur if action is not taken.
Level 5 Notice: Unusual event is mentioned, but not, an error is shown.
Level 6 Information: Normal operation messages are at this level, no action is required to take.
Level 7 Debug: Messages at debug level contains more details of events, debug level log messages are more useful for developers and for debugging an application.

In of Unix/Linux-based OS having default log level at Level 3, i.e., error level of a log message in order to reduce the log file size while writing a log to disk. If the default log level is increased such as at debug level that causes flooding of the debug level logs in a log file which is expensive to write it on disk as well as writing such huge information on disk takes so much time which is in terms of microseconds. If we compared writing logs on main memory and disk, the writing speed of RAM is more than disk. But RAM capacity of storage is less than disk.

In some network related problem, the debugger required network packet capturing, i.e., TCP dump in order to analyze the problem and to resolve it.

The idea behind a Smart Logging Framework is to be able to have a logging mechanism that helps diagnose and reduces RCA time. The framework will be the pluggable API. It will lay the foundation for dynamic and predictive logging/tracing for the product by providing common interfaces that can be consumed by any module within an application to quickly leverage new features being offered. In traditional logging mechanism such as SYSLOG in Linux-based system produces log written on disk directly, which is expensive and causes high latency, thus it is not capturing debug level logs. For some applications, the user also required network packet capturing for some specific events.

Since applications already have a logging mechanism to persist logs to files, the Framework will intercept the flow of log messages from the originating module to the application endpoints by providing new C interfaces/macros. The Smart Logging Framework will provide the following features:

- In-memory logging.
- In-memory network packet capturing.
- Zoom-in log viewer.

2 Related Work

In this section, we briefly review previous works related to the topics.

This paper proposes a high-performance logging system for embedded UNIX and GNU/Linux applications. Compared to the standard UNIX and GNU/Linux logging method, SYSLOG, our method has two orders of magnitude lower latency and an order of magnitude higher message throughput. This speedup is mainly due to the use of a memory-mapped file as the means of interprocess communication, fewer memory copies and the batching of output messages in the logging daemon. In addition, this logging system also accepts SYSLOG messages, providing compatibility with existing applications. High-performance logging system is in production use in the Cisco UCS Virtual Interface Card [7].

3 Architecture

Figure 1 represents architecture of proposed API. In that the pluggable API takes place in between running application that produces the logs and network packet capturing command as well.

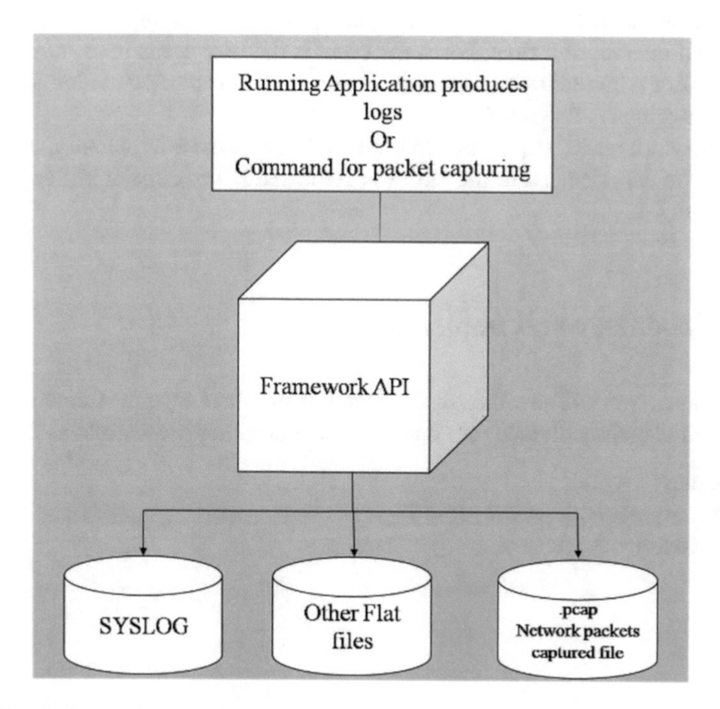

Fig. 1 Block diagram for smart logging framework

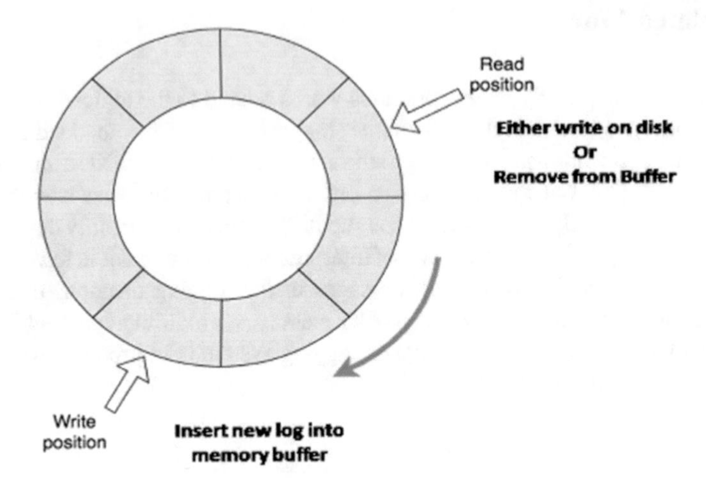

Fig. 2 Block diagram of a circular memory buffer for smart logging framework

The Smart Logging Framework will capture all the upcoming logs from running application and keep it in-memory. Depend on configuration at the time of framework initialization; it will dump the logs that are present in-memory buffer. Here the dump configuration refers to dump on demand, dump on buffer full or dump on a timer of some fixed interval of a time. For some event if the user needs to capture network packets. User required to initialize the network packet capturing module integrated in smart logging framework API.

The network packet capturing module also keeps packets in memory, dump it in .pcap file on user demand or the buffer is getting full. The circular buffer works as shown in Fig. 2.

4 Methodology and Implementation

In this paper, we will see the implementation of smart logging framework. The framework is mainly divided into three main features they are as follows:

- In-memory logging.
- In-memory network packet capturing.
- Zoom-in log viewer.

4.1 In-Memory Logging

Writing log messages on disk is expensive and causes high latency. Due to which the default log level is INFO in order to avoid flooding of DEBUG/TRACE logs. Every time a user does not require the DEBUG level logs. At the time of failure or RCA DEBUG level logs, i.e., details around time are required. The writing speed of RAM is 5–10 times faster than writing the log to disk. So instead of writing all the DEBUG level logs or all level logs directly on disk SMART LOGGING FRAMEWORK will keep these logs in-memory depend on the configuration at the time of initializing it will dump on disk accordingly. Figure 3 shows the flow of in-memory logging of Smart Logging Framework. The following steps describe the flow of Fig. 3 as follows:

Step 1: User will be required to include the Smart Logging Framework API in an application. The Smart Logging Framework unit will specify the in-memory configuration, such as buffer size in bytes, Log file path, the timer for dumping logs on a fixed interval of time, etc.

Step 2: While an application is producing a logs, check the configuration of in-memory is ENABLE/DISABLE. If it is enabled, then add the log into the memory

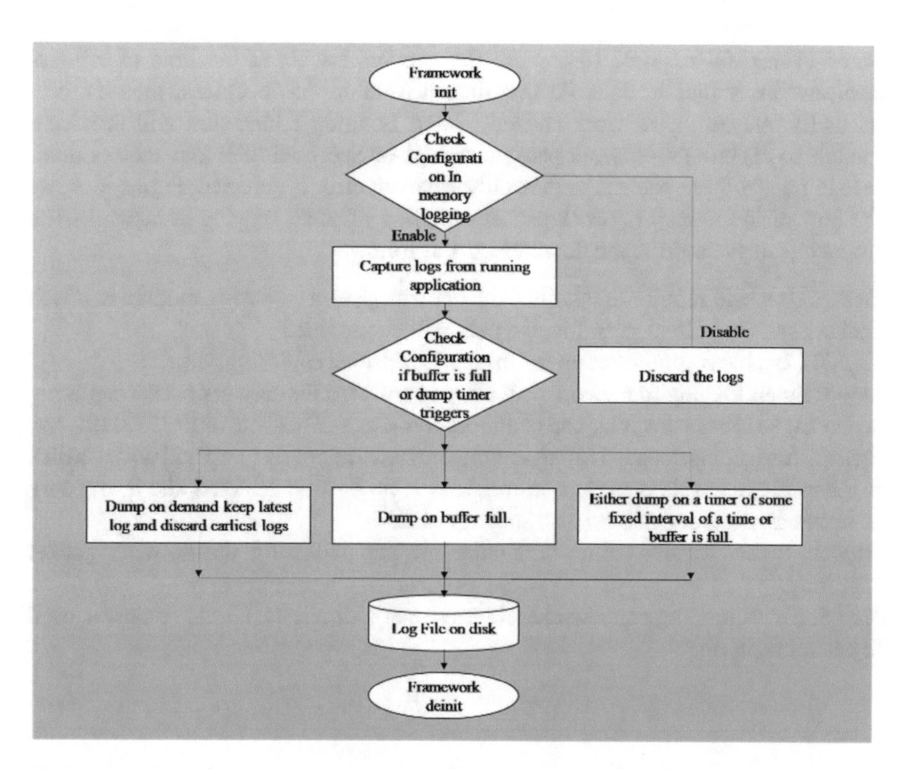

Fig. 3 Flow chart of feature in-memory logging feature of smart logging framework

buffer. If in-memory logging is disabled, then discard the logs incoming from a running application.
Step 3: Capture the log and save it into memory buffer with its original timestamp.
Step 4: If memory buffer is getting full or dump timer gets triggered. The specified, configured action will take place. The action will be are as follows:

- Dump the logs present in memory on demand here the user is required to specify explicitly dump the logs.
- Dump on log buffer full, in which the logs will dump if memory buffer is getting full.
- Dump either on the timer of some fixed interval of a time or buffer is full.

Step 5: After a successful dump of log buffer will get ready to capture next upcoming logs.
Step 6: Smart logging framework deinit will destroy all active modules and release resources as well.

4.2 In-memory Network Packet Capturing

Some events are required to capture the network packet at the time of transaction/transfer of files or data. At that time if a failure has occurred, then the user needs to analyze the network packets. Smart Logging Framework will provide a module to capture the network packets with given specified filter and another interface to persist these packets to disk only when an error is detected. Figure 4 shows the flow of in-memory network packet capturing of smart logging framework. The following steps describe the flow of Fig. 4 as follows.

Step 1: User will require to initialize the network packet capturing module in which packet capturing filter, .pcap file, the path will be provided.
Step 2: Check the configuration for in-memory packet capturing.
Step 3: Smart logging framework start nw_cap will start the network packet capturing according to species nw_pkt_cap configuration. e.g. TCP packet, or IP 127.0.0.1, etc.
Step 4: If an error is detected Smart logging framework nw_pkt_cap will write on disk in .pcap file the packets present in-memory. If no error is detected, the in-memory packets are discarded and no disk space is used.
Step 5: Smart logging framework stop_nw_cap will stop the network packet capturing.
Step 6: Smart logging framework deinit nw_cap will release all the resources used in packet capturing.

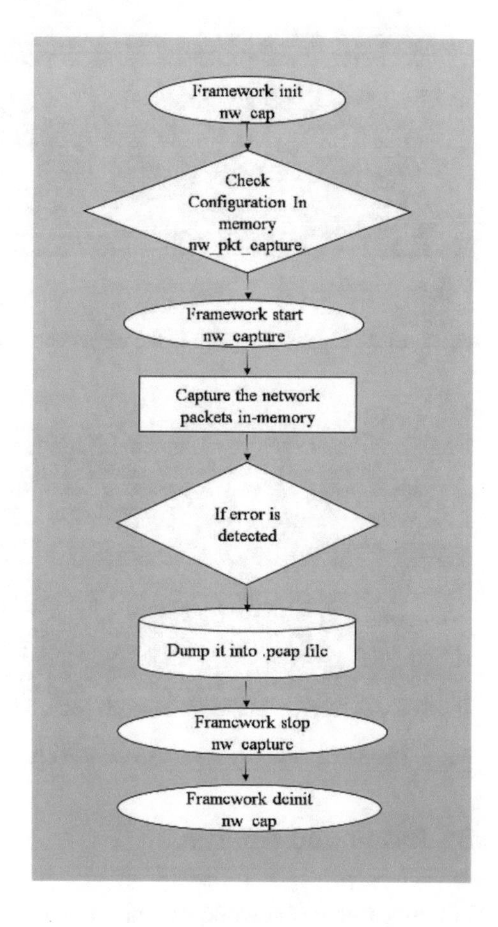

Fig. 4 Flow chart smart logging framework in-memory network packet capturing feature

4.3 Zoom-in Log Viewer

The Smart Logging Framework provides feature to sort a merge specified component's log file into a single log file with respect to timestamp. That makes easy reading logs for a particular time stamp and debugger can see what are the events occurred at a particular time.

Zoom-in log view is a third feature of Smart Logging Framework, which can be used using one script. The script reads all logs from specified log files of each component that produces logs and written from in-memory logging, sort it according to time stamp. The sorted log is written into excel sheet with one reference log file as an abstract view of the event. The detail log of the rest of the component is placed in subgroup with respect to same time stamp. The abstract view of Zoom-in log viewer as shown in Fig. 5 and the actual view is shown in Fig. 6.

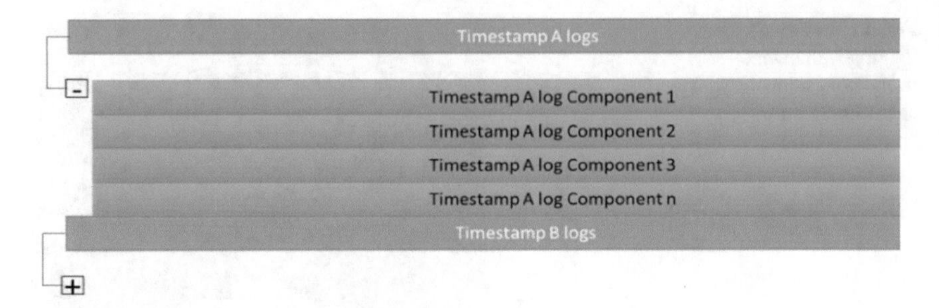

Fig. 5 Block diagram for zoom-in log viewer concept

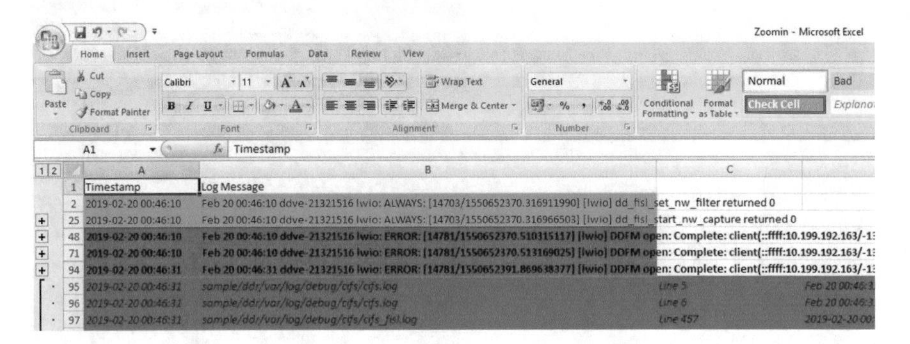

Fig. 6 Result for zoom-in log view where logs are dumped through In-memory logging

5 Result and Analysis

Proposed framework works on Unix-based operating system. The in-memory logging feature writes a log into log file according to the given configuration of the framework at the time of initialization. From Fig. 6 at the last end of the framework, the result of in-memory logging is shown in Zoom-in feature of Smart Logging Framework.

The framework also provides the feature of network packet capturing. Figure 7 the captured network packet in Wireshark tool. The shown in Fig. 7 is output of network communication using port number 445, i.e., SMB protocol.

6 Conclusion

We have presented one of the useful API for logging mechanism, which will improve the logging mechanism of application with respect to time, log levels that help to reduce the time of RCA. In this API internally we have used data structure to store log and network packet in-memory. That makes helping to an in-memory logging operation. Packet capturing feature removes overhead to separately capture network

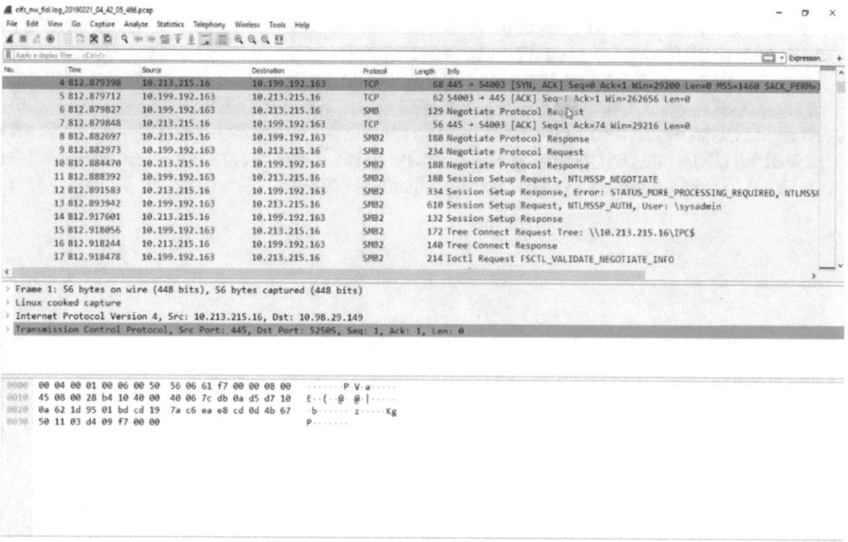

Fig. 7 Result for in-memory packet capturing in Wireshark

packets for the specific operation. Zoom-in feature reduces the effort of checking the logs of other component at the same time stamp. The aim of the Smart Logging Framework API is to improve the previous logging technique.

Acknowledgements I express my deepest thanks to Mrs. Abhidnya Joshi and Prof. M. K. Chavan for taking part in useful decision giving necessary advices and guidance. My deepest sense of gratitude to Prof. A. R. Surve and Mrs. H. V. Gandhifor their careful and precious guidance which was extremely valuable for my study both theoretically and practically.

References

1. Sundaravadivel, P., Kesavan, K., Kesavan, L., Mohanty, S.P., Kougianos, E.: Smart-log: a deep-learning based automated nutrition monitoring system in the IoT. IEEE Trans. Consum. Electron. **64**(3) (2018)
2. Spolaor, R., Dal Santo, E., Conti, M.: DELTA-data extraction and logging tool for android. IEEE Trans. Mob. Comput. (2017)
3. Wang, K., Fung, C., Ding, C., Pei, P., Huang, S., Luan, Z., Qian, D.: A methodology for root-cause analysis in component based systems. In: IEEE 23rd International Symposium on Quality of Service (IWQoS) (2015)
4. Wan, H., Lu, Y., Xu, Y., Shu, J.: Empirical study of redo and undo logging in persistent memory. In: IEEE Conference 2016 5th Non-Volatile Memory Systems and Applications Symposium (NVMSA)

5. Ryu, S., Lee, K., Han, H.: In-memory writeahead logging for mobile smart devices with NVRAM. IEEE Trans. Consum. Electron. **61**(1) (2015)
6. Anamika, Bisht, G., Kanhaiya, Poonam: Log aggregator for better root-cause-analysis. Anamika et al., (IJCSIT) Int. J. Comput. Sci. Inf. Technol. **6**(2), 1100–1102 (2015)
7. Jeong, J.: Cisco Systems San Jose, California 95134, USA. High performance logging system for embedded UNIX and GNU/Linux applications. In: IEEE International Conference on Embedded and Real-Time Computing Systems and Applications (2013)

MongoDB Indexing for Performance Improvement

Rupali Chopade and Vinod Pachghare

Abstract For any digital application, database positions at the heart of that application. Today with the big data requirement, databases are roaming from traditional relational databases towards NoSQL databases. The diverse numbers of database options are available under the NoSQL category. As per the database engine survey, MongoDB is the preferred NoSQL database among other databases. Due to numerous features available in MongoDB, this database is widely used in different applications. This database is fulfilling the needed requirements for upcoming applications. This paper is a study of indexing, which is one of the important artifacts of the database. Indexing is one of the special forms of the data structure. It plays an important role in performance improvement by saving execution time in terms of document scan. Use of indexing and its effect on query result is highlighted here. Another reason for selecting this artifact is indexing study is also important from a database forensics perspective. Database forensics is a detailed analysis of the database to find the origin of the problem.

Keywords Index · NoSQL · MongoDB · Database forensics

1 Introduction

In today's digital era vast data is produced every day. Various sources of these data are social media, healthcare, educational, the stock market, electronic media, business transactions, etc. This big data requirement is attracting NoSQL databases [1, 2]. NoSQL database consists of four types of databases namely Document based, Column based, Key-Value pair and Graph-based as shown in Fig. 1 [3, 4]. Example of each type is also shown in the corresponding rectangle. Each category of NoSQL database has its own format for data storage. MongoDB is a NoSQL database having

R. Chopade (✉) · V. Pachghare
College of Engineering Pune, Pune, India
e-mail: rmc18.comp@coep.ac.in

V. Pachghare
e-mail: vkp.comp@coep.ac.in

M. Tuba et al. (eds.), *ICT Systems and Sustainability*,
Advances in Intelligent Systems and Computing 1077,
https://doi.org/10.1007/978-981-15-0936-0_56

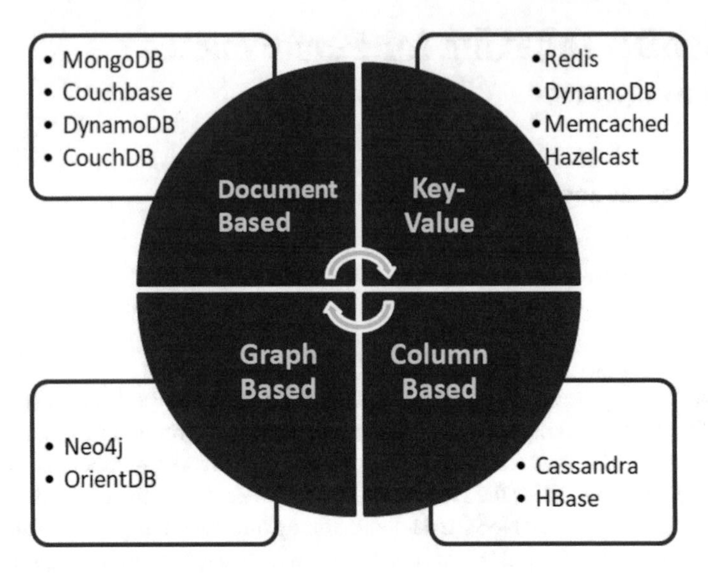

Fig. 1 NoSQL database types

document-based category, in which data is represented as documents and documents are managed by collection inside the database. This relationship between database, collection, document, and index is as shown in Fig. 2. As per database engine ranking, MongoDB is found at 5th position among all categories of databases and it stands first in the NoSQL category [5]. MongoDB is one of the popular database under NoSQL [6, 7].

Outline of Paper

The paper is outlined as related work in Sects. 2, 3 gives a brief idea about MongoDB Indexing with index types, execution statistics, and index management and finally paper ends with the conclusion.

2 Related Work

Extensible Storage Engine called as ESE database is mostly useful for storing web browsers or windows data. Database header and pages are the internal parts of this database structure. Pages are represented as B-tree structure format. Kim et al. have given technique and tool [8] to recover the deleted records and tables. To recover deleted pages and deleted tables MsysObject table is useful. The entire table information is managed by MsysObject catalog table. Accuracy of this tool is proved by comparing it with an existing tool. If record header is damaged then this tool cannot recover the data.

A B^+ trees is a commonly used data structure in database systems. Here authors defined this data structure with a signature [9] and they developed an algorithm for

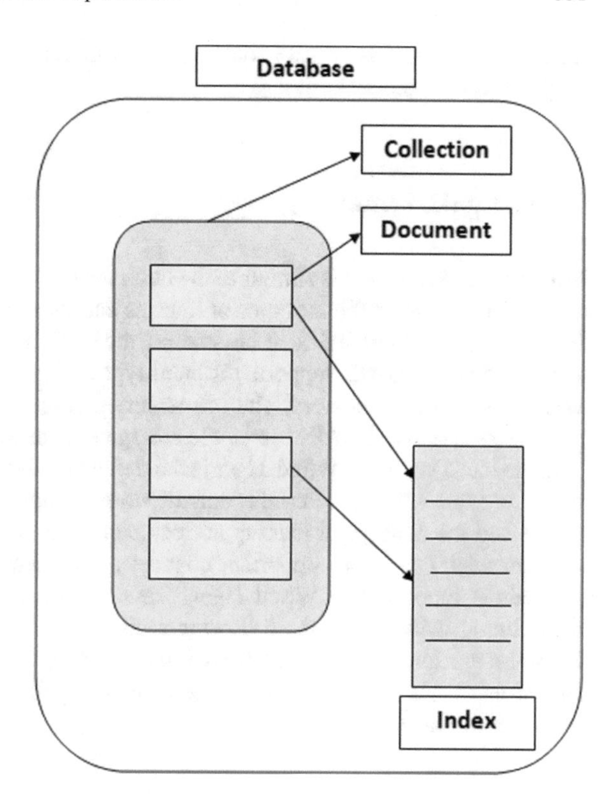

Fig. 2 MongoDB structure

the same. The signature logging mechanism is used for the forensic purpose and this signature concept supports it. Changes made to B^+ tree will affect the B^+ tree signature also. To observe these changes the signature log can be verified.

B^+ trees are most popular data structure for indexing purpose. This is presented by Peter Kieseberg et al. to show the importance of it in the database whenever forensics process comes in picture [10]. Indexing concept is used to speed up I/O operations in databases. In database storage engines, most of the indexes are built using B^+ or B trees. In this paper, the authors presented a B^+ tree structure for Insert operation only and limitation of this approach is that insert operation is not bijective.

To find out the changes in structure when insert operations are used authors proved that B^+ trees can be used [11]. The limitation is that to find the previous deleted data using B^+ trees is not possible. Through this paper authors showed their focus on leaf nodes of trees. The entire structure is not considered.

MySQL database contains MyISAM and InnoDB storage engines. Normally used storage engine is InnoDB. This storage engine use the B-tree indexing for pointing the pages. Indexing is an important concept to search data faster. Primary and secondary are the two types of indexing. Primary key table uses primary index. Most of the times, the secondary index is used for searching purpose. Using indexing internally data is represented in a singly linked list format. In this list indexing is a pointer to the specific record. Sensitive or secret data [12] can be hidden using data hiding.

Secondary indexing is useful to achieve this data hiding. It will not remove it but it only unlinks the index entries.

3 MongoDB Indexing

Database index concept is similar to the book index, which we use for faster searching. In this section different types of indexes and its execution statistics are shown. In MongoDB default index is created on the _id attribute, which is represented as ObjectId. MongoDB supports different types of storage engines like MMAPv1, Wired Tiger and in-memory. The in-memory storage engine is available in enterprise version only. MMAPv1 is an old storage engine which is not used now. Now with the latest versions, Wired Tiger is the default engine of MongoDB. From MongoDB version 3.0 it became the default storage engine. This storage engine has introduced the concept of locking at document level as well as CPU scaling in an efficient way. One more important concept of compression with snappy and zlib methods is supported by Wired Tiger. The overall performance of Wired Tiger is better than MMAPv1. In the following section, queries are executed on Windows 10, 64-bit machine, MongoDB 4.0 version supporting WiredTiger storage engine [13, 14]. MongoDB's _mdb_catalog.wt file maintains collection and index information as shown in Fig. 3.

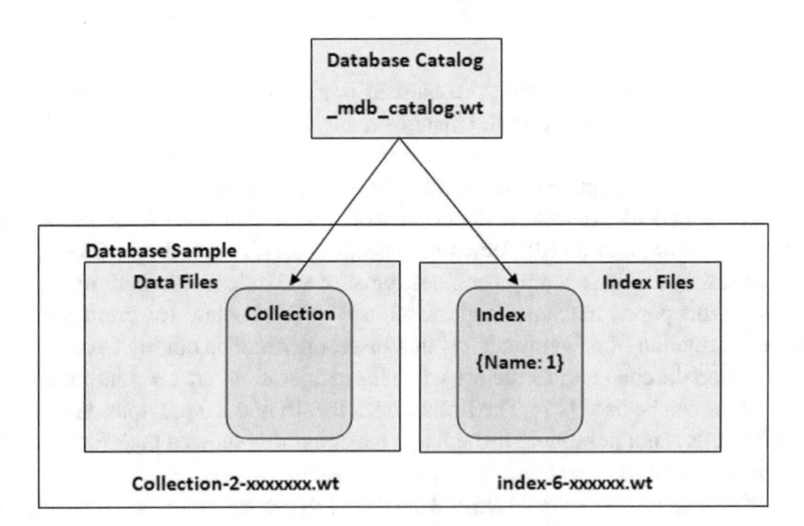

Fig. 3 WiredTiger file maintaining collection and indexes

3.1 Index Types

Different types of indexes offered by MongoDB are discussed here [15, 16].

- **Single Field Index**

In a single field index, the index is created on one field of document. Along with the _id field, MongoDB allows users to create an index in ascending or in descending order. For ascending order index is created with value 1 and −1 is used for descending order. Figure 4 gives an example of the creation of a single field index.

- **Compound Index**

Index creation on multiple fields is possible using a compound index. Following is the example for the same, shown in Fig. 5.

- **Multikey Index**

Few MongoDB collections may consist of documents in which the array of data is stored. When we have to create the index on an array field, then the multikey index concept is useful. Creating a multikey index is shown in Fig. 6.

The multikey index is designed to use on only one array field. Creating Index on multiple array fields will generate an error message as shown in Fig. 7.

Covered queries are also not supported by multi key indexing. Covered queries are the queries in which all attributes are part of the index.

- **Hashed Index**

Creating a hash value on a specific field will store the hash value of that field. MongoDB supports hashed index on single field only. This indexing can be used

```
MongoDB Enterprise > db.products.createIndex({item:1})
{
        "createdCollectionAutomatically" : false,
        "numIndexesBefore" : 1,
        "numIndexesAfter" : 2,
        "ok" : 1
}
```

Fig. 4 Single field index creation

```
MongoDB Enterprise > db.employee.createIndex({First_Name:1,Last_Name:-1})
{
        "createdCollectionAutomatically" : false,
        "numIndexesBefore" : 1,
        "numIndexesAfter" : 2,
        "ok" : 1
}
```

Fig. 5 Compound field index creation

```
MongoDB Enterprise > db.employee.createIndex({"Address.zip":1})
{
        "createdCollectionAutomatically" : false,
        "numIndexesBefore" : 2,
        "numIndexesAfter" : 3,
        "ok" : 1
}
```

Fig. 6 Multikey index creation

```
MongoDB Enterprise > db.employee.createIndex({"Address.zip":1,"project_loc":1})
{
        "ok" : 0,
        "errmsg" : "cannot index parallel arrays [project_loc] [Address]",
        "code" : 171,
        "codeName" : "CannotIndexParallelArrays"
}
```

Fig. 7 Multikey index creation on two array fields

with the shard key. MongoDB shards are used in a distributed environment. Creation of hashed indexing is shown in Fig. 8.

- **Text Indexing**

Text indexing supports an index on text string values. The document which consists of text value, on that field text indexing is useful. The syntax of text index creation is as shown in Fig. 9.

Suppose we want to create an index on all text field documents then the following query can be issued as shown in Fig. 10.

$** indicates the wildcard characters, to specify text documents.

```
MongoDB Enterprise > db.example.createIndex({_id:"hashed"})
{
        "createdCollectionAutomatically" : false,
        "numIndexesBefore" : 1,
        "numIndexesAfter" : 2,
        "ok" : 1
}
```

Fig. 8 Hashed indexing

```
MongoDB Enterprise > db.example.createIndex({"name":"text"})
{
        "createdCollectionAutomatically" : false,
        "numIndexesBefore" : 1,
        "numIndexesAfter" : 2,
        "ok" : 1
}
```

Fig. 9 Text indexing

```
MongoDB Enterprise > db.example.createIndex({"$**":"text"})
{
        "createdCollectionAutomatically" : false,
        "numIndexesBefore" : 1,
        "numIndexesAfter" : 2,
        "ok" : 1
}
```

Fig. 10 Text indexing on all text fields

There are other types of indexes are also available like 2d, 2dsphere, and geoheystack indexes. These indexes are not discussed here.

3.2 *Execution Statistics*

The queries which are executed in the absence of an index and its effect in presence of index can be observed in this section.

It can be verified from Fig. 11 that to search any value, complete collection scan is required, as highlighted with a red box. This stage is COLLSCAN, which is a collection scan. There are a total of six documents available in the collection, so total documents examined are 6, which is also highlighted with a red box. Figure 12 is a snapshot of after index creation. Now collection scan has changed to index scan highlighted as IXSCAN. Also, the total documents examined are only 1.

In execution statistics, details can be checked from **"winningPlan**" and **"executionStats**" section.

Total indexes and index size can be checked from a query from Fig. 13.

3.3 *Index Management*

MongoDB indexes can be managed by dropping the index. Following are the few examples of the same. To drop the all indexes from the collection, use the query of Fig. 14.

To drop the specific index, query as shown in Fig. 15 can be used.

To check the indexes available on the collection, getIndexes query can be used. Whenever any collection is created one file corresponding to that collection is created and similarly index file is also created. So it is important to understand how these indexes are managed [17].

```
MongoDB Enterprise > db.products.find({item:"pen"}).explain("executionStats")
{
        "queryPlanner" : {
                "plannerVersion" : 1,
                "namespace" : "mydb.products",
                "indexFilterSet" : false,
                "parsedQuery" : {
                        "item" : {
                                "$eq" : "pen"
                        }
                },
                "winningPlan" : {
                        "stage" : "COLLSCAN",
                        "filter" : {
                                "item" : {
                                        "$eq" : "pen"
                                }
                        },
                        "direction" : "forward"
                },
                "rejectedPlans" : [ ]
        },
        "executionStats" : {
                "executionSuccess" : true,
                "nReturned" : 1,
                "executionTimeMillis" : 0,
                "totalKeysExamined" : 0,
                "totalDocsExamined" : 6,
                "executionStages" : {
                        "stage" : "COLLSCAN",
                        "filter" : {
                                "item" : {
                                        "$eq" : "pen"
                                }
```

Fig. 11 Execution statistics before index creation

4 Conclusion

The efficient execution of MongoDB queries can be performed by using the index. Basically, indexes are available for performance enhancement, so understanding index use is supreme important. If indexes are not available, then during specific search option, the entire collection scan is performed by MongoDB. I.e. every document is compared with the search field, to find the search result. Instead of that, using an index on the field limits the number of documents to be examined. By using index ordering, results can be obtained in a sorted manner. MongoDB indexing uses a B-tree data structure. The attribute selection for index creation is also very much important. If we compare birth_date and gender attributes, the index should be created on birth_date, rather than gender. MongoDB is a commonly used document based NoSQL database so performance of this database is most important. Indexing plays an important role, in performance improvement of this database. Understanding the index cost associated with a specific query is a future work of this study.

```
MongoDB Enterprise > db.products.find({item:"pen"}).explain("executionStats")
{
        "queryPlanner" : {
                "plannerVersion" : 1,
                "namespace" : "mydb.products",
                "indexFilterSet" : false,
                "parsedQuery" : {
                        "item" : {
                                "$eq" : "pen"
                        }
                },
                "winningPlan" : {
                        "stage" : "FETCH",
                        "inputStage" : {
                                "stage" : "IXSCAN",
                                "keyPattern" : {
                                        "item" : 1
                                },
                                "indexName" : "item_1",
                                "isMultiKey" : false,
                                "multiKeyPaths" : {
                                        "item" : [ ]
                                },
                                "isUnique" : false,
                                "isSparse" : false,
                                "isPartial" : false,
                                "indexVersion" : 2,
                                "direction" : "forward",
                                "indexBounds" : {
                                        "item" : [
                                                "[\"pen\", \"pen\"]"
                                        ]
"executionStats" : {
        "executionSuccess" : true,
        "nReturned" : 1,
        "executionTimeMillis" : 1,
        "totalKeysExamined" : 1,
        "totalDocsExamined" : 1,
        "executionStages" : {
                "stage" : "FETCH",
                "nReturned" : 1,
```

Fig. 12 Execution statistics after index creation

```
MongoDB Enterprise > db.stats()
{
        "db" : "sample",
        "collections" : 7,
        "views" : 0,
        "objects" : 101,
        "avgObjSize" : 528.1485148514852,
        "dataSize" : 53343,
        "storageSize" : 196608,
        "numExtents" : 0,
        "indexes" : 9,
        "indexSize" : 200704,
        "fsUsedSize" : 83602808832,
        "fsTotalSize" : 267911163904,
        "ok" : 1
}
```

Fig. 13 Query to get index size

```
MongoDB Enterprise > db.employee.dropIndexes()
{
        "nIndexesWas" : 2,
        "msg" : "non-_id indexes dropped for collection",
        "ok" : 1
}
```

Fig. 14 Dropping all indexes

```
MongoDB Enterprise > db.employee.dropIndex({"Address.zip":1})
{ "nIndexesWas" : 3, "ok" : 1 }
```

Fig. 15 Drop specific index

References

1. Sablatura, J., Zhou, B.: Forensic database reconstruction. In: 2017 IEEE International Conference on Big Data (Big Data), pp. 3700–3704 (2017)
2. Qi, M.: Digital forensics and NoSQL databases. In: 2014 11th International Conference on Fuzzy Systems and Knowledge Discovery (FSKD), pp. 734–739 (2014)
3. Hauger, W.K., Olivier, M.S.: NoSQL databases: forensic attribution implications. SAIEE Africa Res. J. **109**, 119–132 (2018)
4. Hauger, W.K., Olivier, M.S.: Forensic attribution in NoSQL databases. In: Inf. Secur. S. Afr. (ISSA), 74–82 (2017)
5. DB-Engines Ranking—popularity ranking of database management systems. https://db-engines.com/en/ranking. Accessed 20 June 2019
6. The MongoDB 4.0 Manual—MongoDB Manual. https://docs.mongodb.com/manual/. Accessed 27 Feb 2019
7. Mango DB Top 5 considerations when evaluating NoSQL Databases. White Pap

8. Kim, J., Park, A., Lee, S.: Recovery method of deleted records and tables from ESE database. Digit. Investig. **18**, S118–S124 (2016)
9. Kieseberg, P., Schrittwieser, S., Morgan, L., et al.: Using the structure of b^+-trees for enhancing logging mechanisms of databases. Int. J. Web Inf. Syst. **9**, 53–68 (2013)
10. Kieseberg, P., Schrittwieser, S., Mulazzani, M., et al.: Trees cannot lie: using data structures for forensics purposes. In: 2011 European Intelligence and Security Informatics Conference (EISIC), pp. 282–285 (2011)
11. Kieseberg, P., Schrittwieser, S., Weippl, E.: Structural limitations of B^+-tree forensics. In: Proceedings of the Central European Cybersecurity Conference 2018, p. 9 (2018)
12. Fruhwirt, P., Kieseberg, P., Weippl, E.: Using internal MySQL/InnoDB B-tree index navigation for data hiding. In: IFIP International Conference on Digital Forensics, pp. 179–194 (2015)
13. Yoon, J., Jeong, D., Kang, C., Lee, S.: Forensic investigation framework for the document store NoSQL DBMS: MongoDB as a case study. Digit. Investig. **17**, 53–65 (2016)
14. Yoon, J., Lee, S.: A method and tool to recover data deleted from a MongoDB. Digit. Investig. **24**, 106–120 (2018)
15. Effective MongoDB Indexing (Part 1)—DZone Database. https://dzone.com/articles/effective-mongodb-indexing-part-1. Accessed 2 Mar 2019
16. Effective MongoDB Indexing (Part 2)—DZone Database. https://dzone.com/articles/effective-mongodb-indexing-part-2. Accessed 24 June 2019
17. Chopade, R., Pachghare, V.K.: Ten years of critical review on database forensics research. Digit. Investig. (2019)

Author Index

M. Tuba et al. (eds.), *ICT Systems and Sustainability*,
Advances in Intelligent Systems and Computing 1077,
https://doi.org/10.1007/978-981-15-0936-0

Printed in the United States
By Bookmasters